Nonlinear Ordinary Differential Equations

An introduction for Scientists and Engineers
FOURTH EDITION

D. W. Jordan and P. Smith
Keele University

OXFORD
UNIVERSITY PRESS

OXFORD
UNIVERSITY PRESS

Great Clarendon Street, Oxford OX2 6DP

Oxford University Press is a department of the University of Oxford.
It furthers the University's objective of excellence in research, scholarship,
and education by publishing worldwide in

Oxford New York

Auckland Cape Town Dar es Salaam Hong Kong Karachi
Kuala Lumpur Madrid Melbourne Mexico City Nairobi
New Delhi Shanghai Taipei Toronto

With offices in

Argentina Austria Brazil Chile Czech Republic France Greece
Guatemala Hungary Italy Japan Poland Portugal Singapore
South Korea Switzerland Thailand Turkey Ukraine Vietnam

Oxford is a registered trade mark of Oxford University Press
in the UK and in certain other countries

Published in the United States
by Oxford University Press Inc., New York

First published 1977
Second edition 1987
Third edition 1999
Fourth edition 2007

British Library Cataloguing in Publication Data

Data available

Library of Congress Cataloging in Publication Data

Jordan, D.W. (Dominic William)
Nonlinear ordinary differential equations / D.W. Jordan and
P. Smith. — 3rd ed.
(Oxford applied and engineering mathematics)
1. Differential equations, Nonlinear. I. Smith, Peter, 1935–
II. Title, III. Series.
QA372.J58   1999   515'.352—dc21   99-17648.

Typeset by Newgen Imaging Systems (P) Ltd., Chennai, India
Printed in Great Britain
on acid-free paper by
Biddles Ltd., King's Lynn, Norfolk

ISBN 978–0–19–920824–1 (Hbk)
ISBN 978–0–19–920825–8 (Pbk)

10 9 8 7 6 5 4 3 2 1

Contents

Preface to the fourth edition

This book is a revised and reset edition of *Nonlinear ordinary differential equations*, published in previous editions in 1977, 1987, and 1999. Additional material reflecting the growth in the literature on nonlinear systems has been included, whilst retaining the basic style and structure of the textbook. The wide applicability of the subject to the physical, engineering, and biological sciences continues to generate a supply of new problems of practical and theoretical interest.

The book developed from courses on nonlinear differential equations given over many years in the Mathematics Department of Keele University. It presents an introduction to dynamical systems in the context of ordinary differential equations, and is intended for students of mathematics, engineering and the sciences, and workers in these areas who are mainly interested in the more direct applications of the subject. The level is about that of final-year undergraduate, or master's degree courses in the UK. It has been found that selected material from Chapters 1 to 5, and 8, 10, and 11 can be covered in a one-semester course by students having a background of techniques in differential equations and linear algebra. The book is designed to accommodate courses of varying emphasis, the chapters forming fairly self-contained groups from which a coherent selection can be made without using significant parts of the argument.

From the large number of citations in research papers it appears that although it is mainly intended to be a textbook it is often used as a source of methods for a wide spectrum of applications in the scientific and engineering literature. We hope that research workers in many disciplines will find the new edition equally helpful.

General solutions of nonlinear differential equations are rarely obtainable, though particular solutions can be calculated one at a time by standard numerical techniques. However, this book deals with *qualitative methods* that reveal the novel phenomena arising from nonlinear equations, and produce good numerical estimates of parameters connected with such general features as stability, periodicity and chaotic behaviour without the need to solve the equations. We illustrate the reliability of such methods by graphical or numerical comparison with numerical solutions. For this purpose the Mathematica™software was used to calculate particular exact solutions; this was also of great assistance in the construction of perturbation series, trigonometric identities, and for other algebraic manipulation. However, experience with such software is not necessary for the reader.

Chapters 1 to 4 mainly treat plane autonomous systems. The treatment is kept at an intuitive level, but we try to encourage the reader to feel that, almost immediately, useful new investigative techniques are readily available. The main features of the phase plane—equilibrium points, linearization, limit cycles, geometrical aspects—are investigated informally. Quantitative estimates for solutions are obtained by energy considerations, harmonic balance, and averaging methods.

Various perturbation techniques for differential equations which contain a small parameter are described in Chapter 5, and singular perturbations for non-uniform expansions are treated extensively in Chapter 6. Chapter 7 investigates harmonic and subharmonic responses, and entrainment, using mainly the van der Pol plane method. Chapters 8, 9, and 10 deal more formally with stability. In Chapter 9 its is shown that solution perturbation to test stability can lead to linear equations with periodic coefficients including Mathieu's equation, and Floquet theory is included Chapter 10 presents. Liapunov methods for stability for presented. Chapter 11 includes criteria for the existence of periodic solutions. Chapter 12 contains an introduction to bifurcation methods and manifolds. Poincaré sequences, homoclinic bifurcation; Melnikov's method and Liapunov exponents are explained, mainly through examples, in Chapter 13.

The text has been subjected to a thorough revision to improve, we hope, the understanding of nonlinear systems for a wide readership. The main new features of the subject matter include an extended explanation of Mathieu's equation with particular reference to damped systems, more on the exponential matrix and a detailed account of Liapunov exponents for both difference and differential equations.

Many of the end-of-chapter problems, of which there are over 500, contain significant applications and developments of the theory in the chapter. They provide a way of indicating developments for which there is no room in the text, and of presenting more specialized material. We have had many requests since the first edition for a solutions manual, and simultaneously with the publication of the fourth edition, there is now available a companion book, *Nonlinear Ordinary Differential Equations: Problems and Solutions* also published by Oxford University Press, which presents, in detail, solutions of all end-of-chapter problems. This opportunity has resulted in a re-working and revision of these problems. In addition there are 124 fully worked examples in the text. We felt that we should include some routine problems in the text with selected answers but no full solutions. There are 88 of these new "Exercises", which can be found at the end of most sections. In all there are now over 750 examples and problems in the book.

On the whole we have we have tried to keep the text free from scientific technicality and to present equations in a simple reduced from where possible, believing that students have enough to do to follow the underlying arguments.

We are grateful to many correspondents for kind words, for their queries, observations and suggestions for improvements. We wish to express our appreciation to Oxford University Press for giving us this opportunity to revise the book, and to supplement it with the new solutions handbook.

Dominic Jordan
Peter Smith

Keele
June 2007

Second-order differential equations in the phase plane

Very few ordinary differential equations have explicit solutions expressible in finite terms. This is not simply because ingenuity fails, but because the repertory of standard functions (polynomials, exp, sin, and so on) in terms of which solutions may be expressed is too limited to accommodate the variety of differential equations encountered in practice. Even if a solution can be found, the 'formula' is often too complicated to display clearly the principal features of the solution; this is particularly true of implicit solutions and of solutions which are in the form of integrals or infinite series.

The qualitative study of differential equations is concerned with how to deduce important characteristics of the solutions of differential equations without actually solving them. In this chapter we introduce a geometrical device, the phase plane, which is used extensively for obtaining directly from the differential equation such properties as equilibrium, periodicity, unlimited growth, stability, and so on. The classical pendulum problem shows how the phase plane may be used to reveal all the main features of the solutions of a particular differential equation.

1.1 Phase diagram for the pendulum equation

The simple pendulum (see Fig. 1.1) consists of a particle P of mass m suspended from a fixed point O by a light string or rod of length a, which is allowed to swing in a vertical plane. If there is no friction the equation of motion is

$$\ddot{x} + \omega^2 \sin x = 0, \qquad (1.1)$$

where x is the inclination of the string to the downward vertical, g is the gravitational constant, and $\omega^2 = g/a$.

We convert eqn (1.1) into an equation connecting $\dot{x}$ and x by writing

$$\ddot{x} = \frac{\mathrm{d}\dot{x}}{\mathrm{d}t} = \frac{\mathrm{d}\dot{x}}{\mathrm{d}x}\frac{\mathrm{d}x}{\mathrm{d}t}$$

$$= \frac{\mathrm{d}}{\mathrm{d}x}\left(\frac{1}{2}\dot{x}^2\right). \qquad (1.2)$$

This representaion of $\ddot{x}$ is called the **energy transformation**. Equation (1.1) then becomes

$$\frac{\mathrm{d}}{\mathrm{d}x}\left(\frac{1}{2}\dot{x}^2\right) + \omega^2 \sin x = 0.$$

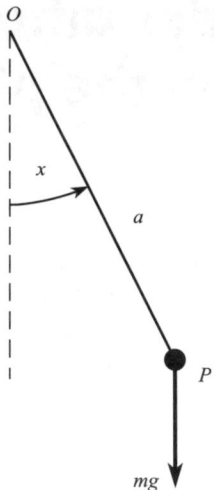

Figure 1.1   The simple pendulum, with angular displacement x.

By integrating this equation with respect to x we obtain

$$\tfrac{1}{2}\dot{x}^2 - \omega^2 \cos x = C, \tag{1.3}$$

where C is an arbitrary constant. Notice that this equation expresses **conservation of energy** during any particular motion, since if we multiply through eqn (1.3) by a constant ma^2, we obtain

$$\tfrac{1}{2}ma^2\dot{x}^2 - mga\cos x = E,$$

where E is another arbitrary constant. This equation has the form

$$E = \text{kinetic energy of } P + \text{potential energy of } P,$$

and a particular value of E corresponds to a particular free motion.

Now write $\dot{x}$ in terms of x from eqn (1.3):

$$\dot{x} = \pm\surd 2 (C + \omega^2 \cos x)^{1/2}. \tag{1.4}$$

This is a *first-order* differential equation for $x(t)$. It cannot be solved in terms of elementary functions (see McLachlan 1956), but we shall show that it is possible to reveal the main features of the solution by working directly from eqn (1.4) without actually solving it.

Introduce a new variable, y, defined by

$$\dot{x} = y. \tag{1.5a}$$

Then eqn (1.4) becomes

$$\dot{y} = \pm\surd 2 (C + \omega^2 \cos x)^{1/2}. \tag{1.5b}$$

Set up a frame of Cartesian axes x, y, called the **phase plane**, and plot the one-parameter family of curves obtained from (1.5b) by using different values of C. We obtain Fig. 1.2. This is called

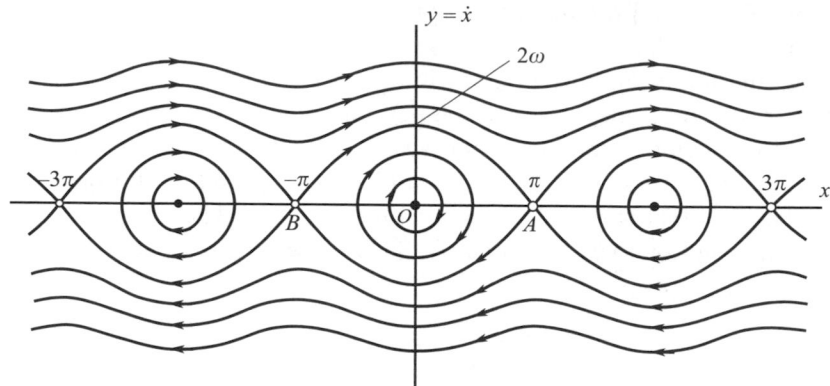

Figure 1.2   Phase diagram for the simple pendulum equation (1.1).

phase diagram for the problem, and the curves are called the **phase paths**. Various types of phase path can be identified in terms of C. On the paths joining $(-\pi, 0)$ and $(\pi, 0)$, $C = \omega^2$; for paths within these curves $\omega^2 > C > -\omega^2$; and for paths outside $C > \omega^2$. Equation (1.56) implies the 2π-periodicity in x shown in Fig. 1.2. The meaning of the arrowheads will be explained shortly.

A given pair of values (x, y), or $(x, \dot{x})$, represented by a point P on the diagram is called a **state of the system**. A state gives the angular velocity $\dot{x} = y$ at a particular inclination x, and these variables are what we sense when we look at a swinging pendulum at any particular moment. A given state $(x, \dot{x})$ serves also as a pair of **initial conditions** for the original differential equation (1.1); therefore a given state determines all subsequent states, which are obtained by following the phase path that passes through the point P: (x, y), where (x, y) is the initial state.

The **directions** in which we must proceed along the phase paths for increasing time are indicated by the arrowheads in Fig. 1.2. This is determined from (1.5a): when $y > 0$, then $\dot{x} > 0$, so that x must increase as t increases. Therefore the required direction is always *from left to right in the upper half-plane*. Similarly, the direction is always *from right to left in the lower half-plane*. The complete picture, Fig. 1.2, is the phase diagram for this problem.

Despite the non-appearance of the time variable in the phase plane display, we can deduce several physical features of the pendulum's possible motions from Fig. 1.2. Consider first the possible states of the physical equilibrium of the pendulum. The obvious one is when the pendulum hangs without swinging; then $x = 0$, $\dot{x} = 0$, which corresponds to the origin in Fig. 1.2. The corresponding *time*-function $x(t) = 0$ is a perfectly legitimate **constant solution** of eqn (1.1); the phase path degenerates to a single point.

If the suspension consists of a light rod there is a second position of equilibrium, where it is balanced vertically on end. This is the state $x = \pi$, $\dot{x} = 0$, another constant solution, represented by point A on the phase diagram. The same physical condition is described by $x = -\pi$, $x = 0$, represented by the point B, and indeed the state $x = n\pi$, $\dot{x} = 0$, where n is any integer, corresponds physically to one of these two equilibrium states. In fact we have displayed in Fig. 1.2 only part of the phase diagram, whose pattern repeats periodically; there is not in this case a one-to-one relationship between the *physical* state of the pendulum and points on its phase diagram.

Since the points O, A, B represent states of physical equilibrium, they are called **equilibrium points** on the phase diagram.

Now consider the family of closed curves immediately surrounding the origin in Fig. 1.2. These indicate **periodic motions**, in which the pendulum swings to and fro about the vertical. The **amplitude** of the swing is the maximum value of x encountered on the curve. For small enough amplitudes, the curves represent the usual 'small amplitude' solutions of the pendulum equation in which eqn (1.1) is simplified by writing $\sin x \approx x$. Then (1.1) is approximated by $\ddot{x} + \omega^2 x = 0$, having solutions $x(t) = A \cos \omega t + B \sin \omega t$, with corresponding phase paths

$$x^2 + \frac{y^2}{\omega^2} = \text{constant}$$

The phase paths are nearly ellipses in the small amplitude region.

The wavy lines at the top and bottom of Fig. 1.2, on which $\dot{x}$ is of constant sign and x continuously increases or decreases, correspond to whirling motions on the pendulum. The fluctuations in $\dot{x}$ are due to the gravitational influence, and for phase paths on which $\dot{x}$ is very large these fluctuations become imperceptible: the phase paths become nearly straight lines parallel to the x axis.

We can discuss also the **stability** of the two typical equilibrium points O and A. If the initial state is displaced slightly from O, it goes on to one of the nearby closed curves and the pendulum oscillates with small amplitude about O. We describe the equilibrium point at O as being **stable**. If the initial state is slightly displaced from A (the vertically upward equilibrium position) however, it will normally fall on the phase path which carries the state far from the equilibrium state A into a large oscillation or a whirling condition (see Fig. 1.3). This equilibrium point is therefore described as **unstable**.

An exhaustive account of the pendulum can be found in the book by Baker and Blackburn (2005).

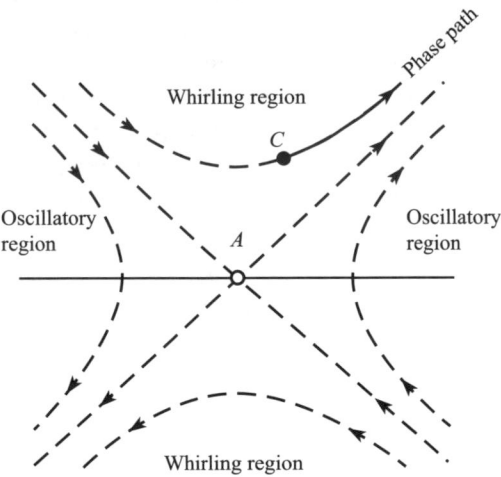

Figure 1.3   Unstable equilibrium point for the pendulum: typical displaced initial state C.

1.2 Autonomous equations in the phase plane

The second-order differential equation of general type

$$\ddot{x} = f(x, \dot{x}, t)$$

with **initial conditions**, say $x(t_0)$ and $\dot{x}(t_0)$, is an example of a **dynamical system**. The evolution or future states of the system are then given by $x(t)$ and $\dot{x}(t)$. Generally, dynamical systems are **initial-value problems** governed by ordinary or partial differential equations, or by difference equations. In this book we consider mainly those nonlinear systems which arise from ordinary differential equations.

The equation above can be interpreted as an equation of motion for a mechanical system, in which x represents displacement of a particle of unit mass, $\dot{x}$ its velocity, $\ddot{x}$ its acceleration, and f the applied force, so that this general equation expresses Newton's law of motion for the particle:

$$\text{acceleration} = \text{force per unit mass}$$

A mechanical system is in equilibrium if its state does not change with time. This implies that *an equilibrium state corresponds to a constant solution* of the differential equation, and conversely. A constant solution implies in particular that $\dot{x}$ and $\ddot{x}$ must be simultaneously zero. Note that $\dot{x} = 0$ is not alone sufficient for equilibrium: a swinging pendulum is instantaneously at rest at its maximum angular displacement, but this is obviously not a state of equilibrium. Such constant solutions are therefore the constant solutions (if any) of the equation

$$f(x, 0, t) = 0.$$

We distinguish between two types of differential equation:

(i) the **autonomous type** in which f does not depend explicitly on t;

(ii) the **non-autonomous** or **forced equation** where t appears explicitly in the function f.

A typical non-autonomous equation models the damped linear oscillator with a harmonic forcing term

$$\ddot{x} + k\dot{x} + \omega_0^2 x = F \cos \omega t,$$

in which $f(x, \dot{x}, t) = -k\dot{x} - \omega_0^2 x + F \cos \omega t$. There are no equilibrium states. Equilibrium states are not usually associated with non-autonomous equations although they can occur as, for example, in the equation (Mathieu's equation, Chapter 9)

$$\ddot{x} + (\alpha + \beta \cos t)x = 0.$$

which has an equilibrium state at $x = 0$, $\dot{x} = 0$.

In the present chapter we shall consider only *autonomous systems*, given by the differential equation

$$\ddot{x} = f(x, \dot{x}), \tag{1.6}$$

in which t is absent on the right-hand side. To obtain the representation on the phase plane, put

$$\dot{x} = y, \tag{1.7a}$$

so that

$$\dot{y} = f(x, y). \tag{1.7b}$$

This is a pair of simultaneous first-order equations, equivalent to (1.6).

The **state** of the system at a time t_0 consists of the pair of numbers $(x(t_0), \dot{x}(t_0))$, which can be regarded as a pair of initial conditions for the original differential equation (1.6). The initial state therefore determines all the subsequent (and preceding) states in a particular free motion.

In the **phase plane** with axes x and y, the state at time t_0 consists of the pair of values $(x(t_0), y(t_0))$. These values of x and y, represented by a point P in the phase plane, serve as initial conditions for the simultaneous first-order differential equations (1.7a), (1.7b), and therefore determine all the states through which the system passes in a particular motion. The succession of states given parametrically by

$$x = x(t), \qquad y = y(t), \tag{1.8}$$

traces out a curve through the initial point P: $(x(t_0), y(t_0))$, called a **phase path**, a **trajectory** or an **orbit**.

The **direction** to be assigned to a phase path is obtained from the relation $\dot{x} = y$ (eqn 1.7a). When $y > 0$, then $\dot{x} > 0$, so that x is increasing with time, and when $y < 0$, x is decreasing with time. Therefore the directions are from *left to right in the upper half-plane, and from right to left in the lower half-plane.*

To obtain a relation between x and y that defines the phase paths, eliminate the parameter t between (1.7a) and (1.7b) by using the identity

$$\frac{\dot{y}}{\dot{x}} = \frac{dy}{dx}.$$

Then the **differential equation for the phase paths** becomes

$$\frac{dy}{dx} = \frac{f(x, y)}{y}. \tag{1.9}$$

A particular phase path is singled out by requiring it to pass through a particular point P: (x, y), which corresponds to an initial state (x_0, y_0), where

$$y(x_0) = y_0. \tag{1.10}$$

The complete pattern of phase paths including the directional arrows constitutes the **phase diagram**. The time variable t does not figure on this diagram.

The **equilibrium points** in the phase diagram correspond to **constant solutions** of eqn (1.6), and likewise of the equivalent pair (1.7a) and (1.7b). These occur when $\dot{x}$ *and* $\dot{y}$ *are simultaneously zero*; that is to say, at points on the x axis where

$$y = 0 \quad \text{and} \quad f(x, 0) = 0. \tag{1.11}$$

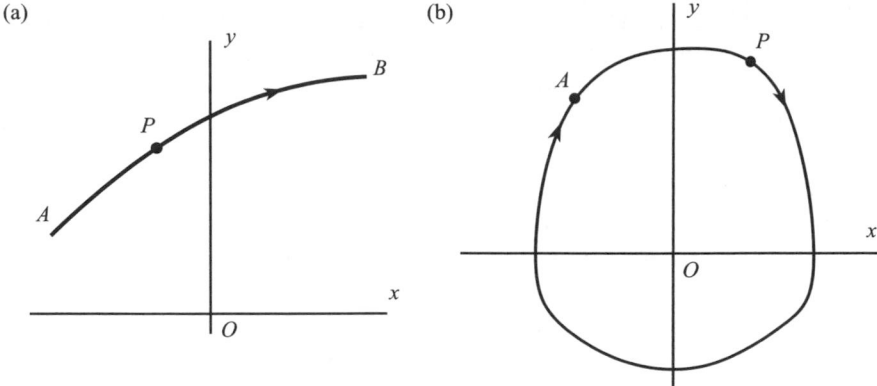

Figure 1.4   (a) The representative point P on a segment of a phase path. (b) A closed path: P leaves A and returns to A an infinite number of times.

Equilibrium points can be regarded as degenerate phase paths. At equilibrium points we obtain, from eqn (1.9),

$$\frac{\mathrm{d}y}{\mathrm{d}x} = \frac{0}{0},$$

so they are singular points of eqn (1.9), although they are not singular points of the time-dependent equations (1.7) (see Appendix A for a description of singular points).

In the representation on the phase plane the time t is not involved quantitatively, but can be featured by the following considerations. Figure 1.4(a) shows a segment $\widehat{AB}$ of a phase path. Suppose that the system is in a state A at time $t = t_A$. The moving point P represents the states at times $t \geq t_A$; it moves steadily along $\widehat{AB}$ (from left to right in $y > 0$) as t increases, and is called a **representative point** on $\widehat{AB}$.

The velocity of P along the curve $\widehat{AB}$ is given in component form by

$$(\dot{x}(t), \dot{y}(t)) = (y, f(x, y))$$

(from (1.7)): this depends *only* on its position $P: (x, y)$, and not at all on t and t_A (this is true only for *autonomous equations*). If t_B is the time P reaches B, the time T_{AB} taken for P to move from A to B,

$$T_{AB} = t_B - t_A, \tag{1.12}$$

is *independent of the initial time t_A*. The quantity T_{AB} is called the **elapsed time** or **transit time** from A to B along the phase path.

We deduce from this observation that if $x(t)$ represents any particular solution of $\ddot{x} = f(x, \dot{x})$, then the *family of solutions $x(t - t_1)$*, where t_1 may take any value, is represented by the *same phase path and the same representative point*. The graphs of the functions $x(t)$ and $x(t - t_1)$, and therefore of $y(t) = \dot{x}(t)$ and $y(t - t_1)$, are identical in shape, but are displaced along the time axis by an interval t_1, as if the system they represent had been switched on at two different times of day.

Consider the case when a phase path is a **closed curve**, as in Fig. 1.4(b). Let A be any point on the path, and let the representative point P be at A at time t_A. After a certain interval of time T, P returns to A, having gone once round the path. Its second circuit starts at A at time $t_A + T$, but since its subsequent positions depend only on the time elapsed from its starting point, and not on its starting time, the second circuit will take the same time as the first circuit, and so on. A *closed phase path therefore represents a motion which is periodic in time.*

The converse is not true—a path that is not closed may also describe a periodic motion. For example, the time-solutions corresponding to the whirling motion of a pendulum (Fig. 1.2) are periodic.

The transit time $T_{AB} = t_B - t_A$ of the representative point P from state A to state B along the phase path can be expressed in several ways. For example,

$$T_{AB} = \int_{t_A}^{t_B} \mathrm{d}t = \int_{t_A}^{t_B} \left(\frac{\mathrm{d}x}{\mathrm{d}t}\right)^{-1} \frac{\mathrm{d}x}{\mathrm{d}t}\,\mathrm{d}t$$

$$= \int_{\widehat{AB}} \frac{\mathrm{d}x}{\dot{x}} = \int_{\widehat{AB}} \frac{\mathrm{d}x}{y(x)}. \tag{1.13}$$

This is, in principle, calculable, given y as a function of x on the phase path. Notice that the final integral depends only on the path $\widehat{AB}$ and not on the initial time t_A, which confirms the earlier conclusion. The integral is a line integral, having the usual meaning in terms of infinitesimal contributions:

$$\int_{\widehat{AB}} \frac{\mathrm{d}x}{y} = \lim_{N \to \infty} \sum_{i=0}^{N-1} \frac{\delta x_i}{y(x_i)},$$

in which we follow values of x in the direction of the path by increments δx_i, appropriately signed. Therefore the δx_i are positive in the upper half-plane and negative in the lower half-plane. It may therefore be necessary to split up the integral as in the following example.

Example 1.1   *The phase paths of a system are given by the family $x + y^2 = C$, where C is an arbitrary constant. On the path with $C = 1$ the representative point moves from $A : (0, 1)$ to $B : (-1, -\sqrt{2})$. Obtain the transit time T_{AB}.*

The path specified is shown in Fig. 1.5. It crosses the x axis at the point $C : (1, 0)$, and at this point δx changes sign. On $\widehat{AC}$, $y = (1 - x)^{1/2}$, and on $\widehat{CB}$, $y = -(1 - x)^{1/2}$. Then

$$T_{AB} = \int_{\widehat{AC}} \frac{\mathrm{d}x}{y} + \int_{\widehat{CB}} \frac{\mathrm{d}x}{y} = \int_0^1 \frac{\mathrm{d}x}{(1-x)^{1/2}} + \int_1^{-1} \frac{\mathrm{d}x}{[-(1-x)^{1/2}]}$$

$$= [-2(1-x)^{1/2}]_0^1 + [2(1-x)^{1/2}]_1^{-1} = 2 + 2\sqrt{2}.$$

For an expression alternative to eqn (1.13), see Problem 1.8.   ●

Here we summarize the main properties of autonomous differential equations $\ddot{x} = f(x, \dot{x})$, as represented in the phase plane by the equations

$$\dot{x} = y, \qquad \dot{y} = f(x, y). \tag{1.14}$$

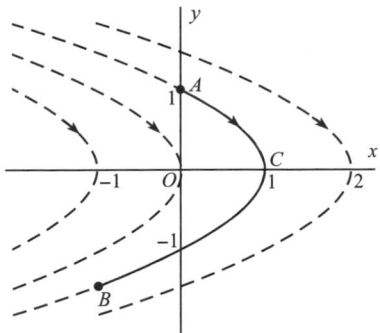

Figure 1.5   Path AB along which the transit time is calculated.

(i) *Equation for the phase paths:*

$$\frac{dy}{dx} = \frac{f(x, y)}{y}.$$ (1.15)

(ii) *Directions of the phase paths:* from left to right in the upper half-plane; from right to left in the lower half-plane.

(iii) *Equilibrium points:* situated at points $(x, 0)$ where $f(x, 0) = 0$; representing constant solutions.

(iv) *Intersection with the x axis:* the phase paths cut the x axis at right angles, except possibly at equilibrium points (see (ii)).

(v) *Transit times:* the transit time for the representative point from a point A to a point B along a phase path is given by the line integral

$$T_{AB} = \int_{\widehat{AB}} \frac{dx}{y}.$$ (1.16)

(vi) *Closed paths:* closed phase paths represent periodic time-solutions $(x(t), y(t))$.

(vii) *Families of time-solutions:* let $x_1(t)$ be any particular solution of $\ddot{x} = f(x, \dot{x})$. Then the solutions $x_1(t - t_1)$, for any t_1, give the same phase path and representative point.

The examples which follow introduce further ideas.

Example 1.2   *Construct the phase diagram for the simple harmonic oscillator equation $\ddot{x} + \omega^2 x = 0$.*

This approximates to the pendulum equation for small-amplitude swings. Corresponding to equations (1.14) we have

$$\dot{x} = y, \qquad \dot{y} = -\omega^2 x.$$

There is a single equilibrium point, at $x = 0$, $y = 0$. The phase paths are the solutions of (1.15):

$$\frac{dy}{dx} = -\omega^2 \frac{x}{y}.$$

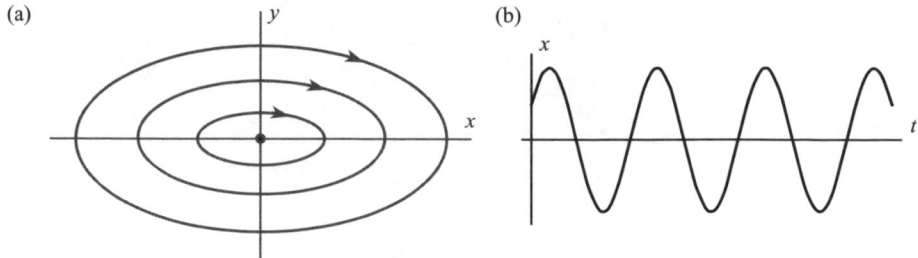

Figure 1.6   (a) centre for the simple harmonic oscillator. (b) Typical solution.

This is a separable equation, leading to

$$y^2 + \omega^2 x^2 = C,$$

where C is arbitrary, subject to $C \geq 0$ for real solutions. The phase diagram therefore consists of a family of ellipses concentric with the origin (Fig. 1.6(a)). All solutions are therefore periodic. Intuitively we expect the equilibrium point to be stable since phase paths near to the origin remain so. Figure 1.6(b) shows one of the periodic time-solutions associated with a closed path.   ●

An equilibrium point surrounded in its immediate neighbourhood (not necessarily over the whole plane) by closed paths is called a **centre**. A centre is stable equilibrium point.

Example 1.3   *Construct the phase diagram for the equation $\ddot{x} - \omega^2 x = 0$.*
The equivalent first-order pair (1.14) is

$$\dot{x} = y, \qquad \dot{y} = \omega^2 x.$$

There is a single equilibrium point $(0, 0)$. The phase paths are solutions of (1.15):

$$\frac{dy}{dx} = \omega^2 \frac{x}{y}.$$

Therefore their equations are

$$y^2 - \omega^2 x^2 = C, \tag{1.17}$$

where the parameter C is arbitrary. These paths are hyperbolas, together with their asymptotes $y = \pm\omega x$, as shown in Fig. 1.7.   ●

Any equilibrium point with paths of this type in its neighbourhood is called a **saddle point**. Such a point is unstable, since a small displacement from the equilibrium state will generally take the system on to a phase path which leads it far away from the equilibrium state.

The question of stability is discussed precisely in Chapter 8. In the figures, stable equilibrium points are usually indicated by a full dot ●, and unstable ones by an 'open' dot ○.

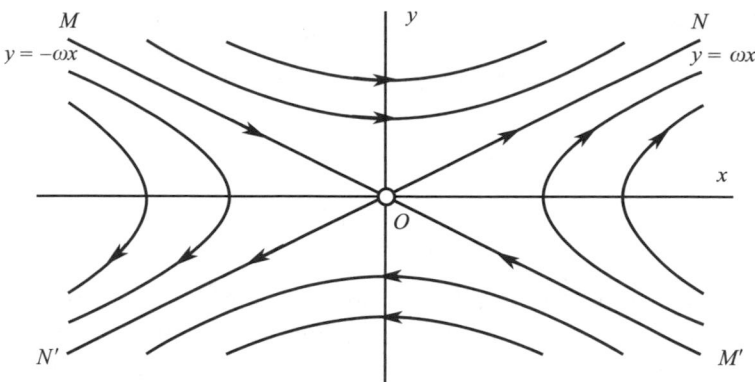

Figure 1.7 Saddle point: only the paths MO and $M'O$ approach the origin.

The differential equations in Examples 1.2 and 1.3 can be solved explicitly for x in terms of t. For Example 1.2, the general solution of $\ddot{x} + \omega^2 x = 0$ is

$$x(t) = A \cos \omega t + B \sin \omega t, \tag{1.18}$$

where A and B are arbitrary constants. This can be written in another form by using the ordinary trigonometric identities. Put

$$\kappa = (A^2 + B^2)^{1/2}$$

and let ϕ satisfy the equations

$$\frac{A}{\kappa} = \cos \phi, \qquad \frac{B}{\kappa} = \sin \phi.$$

Then (1.18) becomes

$$x(t) = \kappa \cos(\omega t - \phi), \tag{1.19}$$

where the amplitude κ and the phase angle ϕ are arbitrary. Figure 1.6(b) shows an example of this time-solution: all values of ϕ produce the same phase path (see (vii) in the summary above). The period of every oscillation is $2\pi/\omega$, which is independent of initial conditions (known as an **isochronous oscillation**).

For Example 1.3, the time-solutions of $\ddot{x} - \omega^2 x = 0$ are given by

$$x(t) = Ae^{\omega t} + Be^{-\omega t}, \tag{1.20}$$

where A and B are arbitrary. To make a correspondence with Fig. 1.7, we require also

$$y = \dot{x}(t) = A\omega e^{\omega t} - B\omega e^{-\omega t}. \tag{1.21}$$

Assume that $\omega > 0$. Then from eqns (1.20) and (1.21), all the solutions approach infinity as $t \to \infty$, except those for which $A = 0$ in (1.20) and (1.21). The case $A = 0$ is described in the phase plane parametrically by

$$x = Be^{-\omega t}, \qquad y = -B\omega e^{-\omega t}. \tag{1.22}$$

For these paths we have

$$\frac{y}{x} = -\omega;$$

these are the paths MO and $M'O$ in Fig. 1.7, and they approach the origin as $t \to \infty$. Note that these paths represent not just one time-solution, but a whole family of time-solutions $x(t) = Be^{-\omega t}$, and this is the case for every phase path (see (vii) in the summary above: for this case put $B = \pm e^{-\omega t_1}$, for any value of t_1).
    Similarly, if $B = 0$, then we obtain the solutions

$$x = Ae^{\omega t}, \qquad y = A\omega e^{\omega t},$$

for which $y = \omega x$: this is the line NN'. As $t \to -\infty$, $x \to 0$ and $y \to 0$. The origin is an example of a **saddle point**, characterised by a pair of incoming phase paths MO, $M'O$ and outgoing paths ON, ON', known as **separatrices**.

Example 1.4   *Find the equilibrium points and the general equation for the phase paths of $\ddot{x} + \sin x = 0$. Obtain the particular phase paths which satisfy the initial conditions (a) $x(t_0) = 0$, $y(t_0) = \dot{x}(t_0) = 1$; (b) $x(t_0) = 0$, $y(t_0) = 2$.*

    This is a special case of the pendulum equation (see Section 1.1 and Fig. 1.2). The differential equations in the phase plane are, in terms of t,

$$\dot{x} = y, \qquad \dot{y} = -\sin x.$$

Equilibrium points lie on the x axis at points where $\sin x = 0$; that is at $x = n\pi \, (n = 0, \pm 1, \pm 2, \ldots)$. When n is even they are centres; when n is odd they are saddle points.
    The differential equation for the phase paths is

$$\frac{dy}{dx} = -\frac{\sin x}{y}.$$

This equation is separable, leading to

$$\int y \, dy = -\int \sin x \, dx,$$

or

$$\tfrac{1}{2}y^2 = \cos x + C, \tag{i}$$

where C is the parameter of the phase paths. Therefore the equation of the phase paths is

$$y = \pm\sqrt{2}(\cos x + C)^{1/2}. \tag{ii}$$

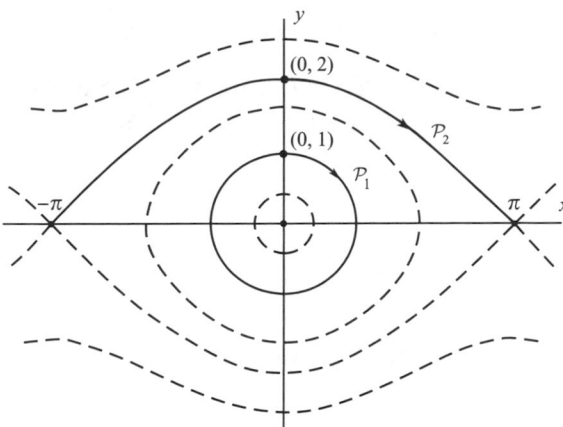

Figure 1.8   Phase paths for $\ddot{x} + \sin x = 0$.

Since y must be real, C may be chosen arbitrarily, but in the range $C \geq 1$. Referring to Fig. 1.8, or the extended version Fig. 1.2, the permitted range of C breaks up as follows:

Values of C	Type of motion
$C = -1$	Equilibrium points at $(n\pi, 0)$ (centres for n even; saddle points for n odd)
$-1 < C < 1$	Closed paths (periodic motions)
$C = 1$	Paths which connect equilibrium points (separatrices)
$C > 1$	Whirling motions

(a) $x(t_0) = 0, y(t_0) = 1$. From (i) we have $\frac{1}{2} = 1 + C$, so that $C = -\frac{1}{2}$. The associated phase path is (from (ii))

$$y = \sqrt{2} \left(\cos x - \tfrac{1}{2} \right)^{1/2},$$

shown as $\mathcal{P}_1$ in Fig. 1.8. The path is closed, indicating a periodic motion.

(b) $x(t_0) = 0, y(t_0) = 2$. From (i) we have $2 = 1 + C$, or $C = 1$. The corresponding phase path is

$$y = \sqrt{2}(\cos x + 1)^{1/2}.$$

On this path $y = 0$ at $x = \pm n\pi$, so that the path connects two equilibrium points (note that it does not continue beyond them). As $t \to \infty$, the path approaches $(\pi, 0)$ and emanates from $(-\pi, 0)$ at $t = -\infty$. This path, shown as $\mathcal{P}_2$ in Fig. 1.8, is called a **separatrix**, since it separates two modes of motion; oscillatory and whirling. It also connects two saddle points. ●

Exercise 1.1
Find the equilibrium points and the general equation for the phase paths of $\ddot{x} + \cos x = 0$. Obtain the equation of the phase path joining two adjacent saddles. Sketch the phase diagram.

Exercise 1.2
Find the equilibrium points of the system $\ddot{x} + x - x^2 = 0$, and the general equation of the phase paths. Find the elapsed time between the points $(-\frac{1}{2}, 0)$ and $(0, \frac{1}{\sqrt{3}})$ on a phase path.

1.3 Mechanical analogy for the conservative system $\ddot{x} = f(x)$

Consider the family of autonomous equations having the more restricted form

$$\ddot{x} = f(x). \tag{1.23}$$

Replace $\dot{x}$ by the new variable y to obtain the equivalent pair of first-order equations

$$\dot{x} = y, \qquad \dot{y} = f(x). \tag{1.24}$$

In the (x, y) phase plane, the states and paths are defined exactly as in Section 1.2, since eqn (1.23) is a special case of the system (1.6).

When $f(x)$ is nonlinear the analysis of the solutions of (1.23) is sometimes helped by considering a mechanical model whose equation of motion is the same as eqn (1.23). In Fig. 1.9, a particle P having *unit* mass is free to move along the axis Ox. It is acted on by a force $f(x)$ which depends only on the displacement coordinate x, and is counted as positive if it acts in the positive direction of the x axis. The equation of motion of P then takes the form (1.23). Note that frictional forces are excluded since they are usually functions of the velocity $\dot{x}$, and their direction of action depends on the sign of $\dot{x}$; but the force $f(x)$ depends *only* on position.

Sometimes physical intuition enables us to predict the likely behaviour of the particle for specific force functions. For example, suppose that

$$\ddot{x} = f(x) = 1 + x^2.$$

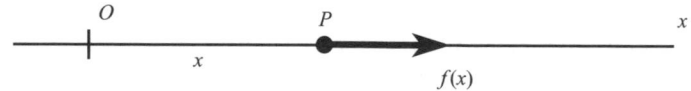

Figure 1.9   Unit particle P under the force $f(x)$.

Then $f(x) > 0$ always, so f always acts from left to right in Fig. 1.9. There are no equilibrium points of the system, so we expect that, whatever the initial conditions, P will be carried away to infinity, and there will be no oscillatory behaviour.

Next, suppose that

$$f(x) = -\lambda x, \quad \lambda > 0,$$

where λ is a constant. The equation of motion is

$$\ddot{x} = -\lambda x.$$

This is the force on the unit particle exerted by a linear spring of stiffness λ, when the origin is taken at the point where the spring has its natural length l (Fig. 1.10). We know from experience that such a spring causes oscillations, and this is confirmed by the explicit solution (1.19), in which $\lambda = \omega^2$. The cause of the oscillations is that $f(x)$ is a **restoring force**, meaning that its direction is always such as to try to drive P towards the origin.

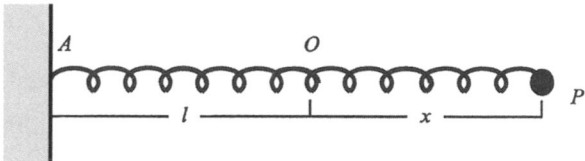

Figure 1.10   Unit particle P attached to a spring of natural length $l = AO$. The displacement of P from O is x.

Now consider a spring having a nonlinear relation between tension and extension:

$$\text{tension} = -f(x),$$

where $f(x)$ has the restoring property; that is

$$f(x) > 0 \quad \text{for} \ x < 0,$$

$$f(0) = 0, \tag{1.25}$$

$$f(x) < 0 \quad \text{for} \ x > 0.$$

We should expect oscillatory behaviour in this case also. The equation

$$\ddot{x} = -x^3 \tag{1.26}$$

is of this type, and the phase paths are shown in the lower diagram in Fig. 1.11 (the details are given in Example 1.5). However, the figure tells us a good deal more; that the oscillations do not consist merely of to-and-fro motions, but are strictly regular and periodic. The result is obtained from the more detailed analysis which follows.

Returning to the general case, let $x(t)$ represent a particular solution of eqn (1.23). When the particle P in Fig. 1.9 moves from a position x to a nearby position $x + \delta x$ the work δW done on P by $f(x)$ is given by

$$\delta W = f(x)\delta x.$$

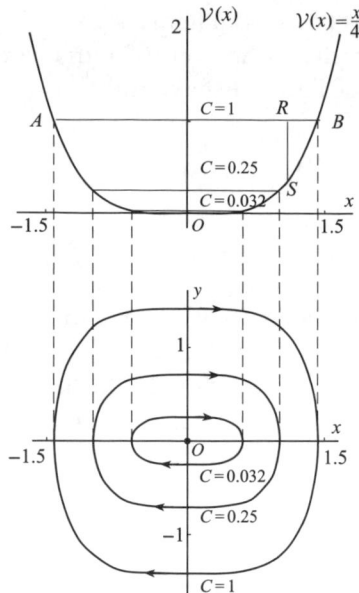

Figure 1.11

This work goes to increment the kinetic energy $\mathcal{T}$ of the (unit) particle, where $\mathcal{T} = \frac{1}{2}\dot{x}^2$:

$$\delta\mathcal{T} = \delta\mathcal{W} = f(x)\delta x.$$

Divide by δx and let $\delta x \to 0$; we obtain

$$\frac{d\mathcal{T}}{dx} = f(x). \tag{1.27}$$

Now define a function $\mathcal{V}(x)$ by the relation

$$\frac{d\mathcal{V}}{dx} = -f(x), \tag{1.28}$$

where $\mathcal{V}(x)$ is called a **potential function** for $f(x)$. Specifically,

$$\mathcal{V}(x) = -\int f(x)\, dx, \tag{1.29}$$

where $\int f(x)\, dx$ stands for any *particular* indefinite integral of $f(x)$. (Indefinite integrals involve an arbitrary constant: any constant may be chosen here for convenience, but it is necessary to stick with it throughout the problem so that $\mathcal{V}(x)$ has a single, definite, meaning.) If we specify a self-contained device that will generate the force $f(x)$, such as a spring in Fig. 1.10, or the earth's gravitation field, then $\mathcal{V}(x)$ is equal to the physical potential energy stored in the device, at any rate up to an additive constant. From (1.27) and (1.28) we obtain

$$\frac{d}{dt}(\mathcal{T} + \mathcal{V}) = 0,$$

so that during any particular motion

$$T + V = \text{constant}, \tag{1.30}$$

or, explicitly,

$$\frac{1}{2}\dot{x}^2 - \int f(x)\mathrm{d}x = C, \tag{1.31}$$

where C is a parameter that depends upon the particular motion and the particular potential function which has been chosen. As we range through all possible values of C consistent with real values of $\dot{x}$ we cover all possible motions. (Note that eqn (1.31) can also be obtained by using the energy transformation (1.2); or the phase-plane equation (1.9) with $\dot{x} = y$.)

In view of eqn (1.30), systems governed by the equation $\ddot{x} = f(x)$ are called **conservative systems**. From (1.31) we obtain

$$\dot{x} = \pm\sqrt{2}(C - V(x))^{1/2}. \tag{1.32}$$

The equivalent equations in the phase plane are

$$\dot{x} = y, \qquad \dot{y} = f(x),$$

and (1.32) becomes

$$y = \pm\sqrt{2}(C - V(x))^{1/2}, \tag{1.33}$$

which is the equation of the phase paths.

Example 1.5   *Show that all solutions of the equation*

$$\ddot{x} + x^3 = 0$$

are periodic.

Here $f(x) = -x^3$, which is a restoring force in terms of the earlier discussion, so we expect oscillations. Let

$$V(x) = -\int f(x)\mathrm{d}x = \frac{1}{4}x^4,$$

in which we have set the usual arbitrary constant to zero for simplicity. From (1.33) the phase paths are given by

$$y = \pm\sqrt{2}(C - V(x))^{1/2} = \pm\sqrt{2}(C - \tfrac{1}{4}x^4)^{1/2}. \tag{1.33a}$$

Figure 1.11 illustrates how the structure of the phase diagram is constructed from eqn (1.33a). In order to obtain any real values for y, we must have $C \geq 0$. In the top frame the graph of $V(x) = \frac{1}{4}x^4$ is shown, together with three horizontal lines for representative values of $C > 0$. The distance RS is equal to $C - \frac{1}{4}x^4$ for $C = 1$ and a typical value of x. The relevant part of the graph, for which y in eqn (1.33a) takes real values, is the part below the line AB. Then, at the typical point on the segment, $y = \pm\sqrt{2}(RS)^{1/2}$. These two values are placed in the lower frame on Fig. 1.11.

The complete process for $C = 1$ produces a closed curve in the phase diagram, representing a periodic motion. For larger or smaller C, larger or smaller ovals are produced. When $C = 0$ there is only one point—the equilibrium point at the origin, which is a centre.   ●

Equilibrium points of the system $\dot{x} = y$, $\dot{y} = f(x)$ occur at points where $y = 0$ and $f(x) = 0$, or alternatively where

$$y = 0, \qquad \frac{dV}{dx} = 0,$$

from (1.28). The values of x obtained are therefore those where $V(x)$ has a minimum, maximum or point of inflection, and the type of equilibrium point is different in these three cases. Figure 1.12 shows how their nature can be established by the method used in Example 1.5:

$$\left.\begin{array}{l} \text{a } \mathbf{minimum \ of} \ V(x) \text{ generates a } \mathbf{centre} \text{ (stable);} \\ \text{a } \mathbf{maximum} \text{ of } V(x) \text{ generates a } \mathbf{saddle} \text{ (unstable);} \\ \text{a } \mathbf{point \ of \ inflection} \text{ leads to a } \mathbf{cusp}, \text{ as shown in Fig. 1.12(c),} \end{array}\right\} \qquad (1.34)$$

Consider these results in terms of the force function $f(x)$ in the mechanical model. Suppose that $f(x_e) = 0$, so that $x = x_e$, $y = 0$ is an equilibrium point. If x changes sign from positive to negative as x increases through the value x_e, then it is a restoring force (eqn (1.25)). Since $dV/dx = -f(x)$, this is also the condition for $V(x_e)$ to be minimum. Therefore a restoring force always generates a centre.

If $f(x)$ changes from negative to positive through x_e the particle is repelled from the equilibrium point, so we expect an unstable equilibrium point on general grounds. Since $V(x)$ has a maximum at x_e in this case, the point $(x_e, 0)$ is a saddle point, so the expectation is confirmed.

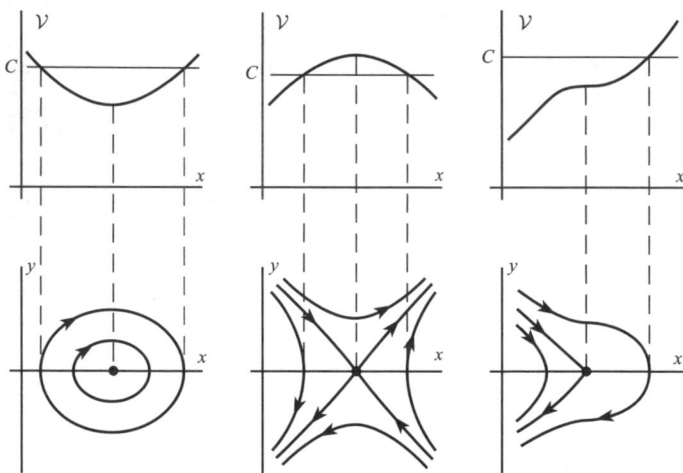

Figure 1.12   Typical phase diagrams arising from the stationary points of the potential energy.

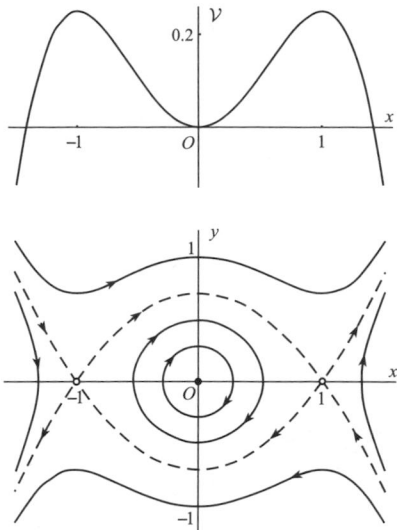

Figure 1.13   The dashed phase paths are separatrices associated with the equilibrium points at $(-1, 0)$ and $(1, 0)$.

Example 1.6   *Sketch the phase diagram for the equation* $\ddot{x} = x^3 - x$.

This represents a conservative system (the pendulum equation (1.1) reduces to a similar one after writing $\sin x \approx x - \frac{1}{6}x^3$ for moderate amplitudes). We have $f(x) = x^3 - x$, so by eqn (1.29)

$$\mathcal{V}(x) = \tfrac{1}{2}x^2 - \tfrac{1}{4}x^4.$$

Figure 1.13 shows the construction of the phase diagram.

There are three equilibrium points: a centre at $(0, 0)$ since $\mathcal{V}(0)$ is a minimum; and two saddle points, at $(-1, 1)$ and $(1, 0)$ since $\mathcal{V}(-1)$ and $\mathcal{V}(1)$ are maxima. The reconciliation between the types of phase path originating around these points is achieved across special paths called separatrices, shown as broken lines (see Example 1.4 for an earlier occurrence). They correspond to values of C in the equation

$$y = \pm\sqrt{2}(C - \mathcal{V}(x))^{1/2}$$

of $C = \frac{1}{4}$ and $C = 0$, equal to the ordinates of the maxima and minimum of $\mathcal{V}(x)$. They start or end on equilibrium points, and must not be mistaken for closed paths. ●

Example 1.7   *A unit particle P is attached to a long spring having the stress–strain property*

$$\text{tension} = xe^{-x},$$

where x is the extension from its natural length. Show that the point $(0, 0)$ *on the phase diagram is a centre, but that for large disturbances P will escape to infinity.*

The equation of motion is $\ddot{x} = f(x)$, where $f(x) = xe^{-x}$, so this is a restoring force (eqn (1.25)). Therefore we expect oscillations over a certain range of amplitude. However, the spring becomes very weak as x increases, so the question arises as to whether it has the strength to reverse the direction of motion if P is moving rapidly towards the right.

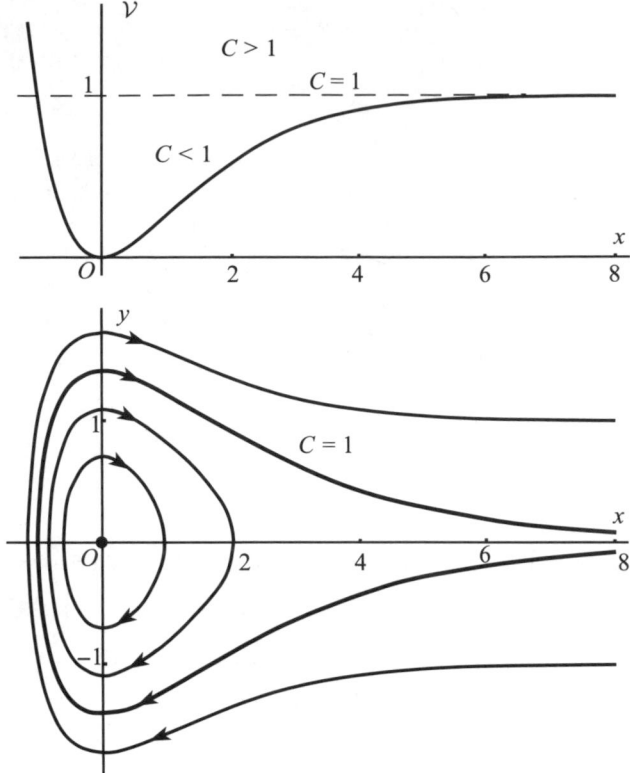

Figure 1.14

We have

$$\mathcal{V}(x) = -\int f(x)\mathrm{d}x = \int x\mathrm{e}^{-x}\mathrm{d}x = 1 - \mathrm{e}^{-x}(1+x),$$

having chosen, for convenience, a value of the arbitrary constant that causes $\mathcal{V}(0)$ to be zero. The upper frame in Fig. 1.14 shows the graph of $\mathcal{V}(x)$.

The function $\mathcal{V}(x)$ has a minimum at $x = 0$, so the origin is a centre, indicating periodic motion. As $x \to \infty$, $\mathcal{V}(x) \to 1$. The phase diagram is made up of the curves

$$y = \pm\sqrt{2}(C - \mathcal{V}(x))^{1/2}.$$

The curves are constructed as before: any phase path occupies the range in which $\mathcal{V}(x) \le C$.

The value $C = 1$ is a critical value: it leads to the path separating the oscillatory region from the region in which P goes to infinity, so this path is a separatrix. For values of C approaching $C = 1$ from below, the ovals become increasingly extended towards the right. For $C \ge 1$ the spring stretches further and further and goes off to infinity.

The transition takes place across the path given by

$$\tfrac{1}{2}y^2 + \mathcal{V}(x) = \tfrac{1}{2}y^2 + \{1 - \mathrm{e}^{-x}(1+x)\} = C = 1.$$

The physical interpretation of $\frac{1}{2}y^2$ is the kinetic energy of P, and $\mathcal{V}(x)$ is the potential energy stored in the spring due to its displacement from equilibrium. Therefore for any motion in which the total energy is greater than 1, P will go to infinity. There is a parallel between this case and the escape velocity of a space vehicle.

•

Exercise 1.3
Find the potential function $\mathcal{V}(x)$ for the conservative system $\ddot{x} - x + x^2 = 0$. Sketch $\mathcal{V}(x)$ against x, and the main features of the phase diagram.

1.4 The damped linear oscillator

Generally speaking, equations of the form

$$\ddot{x} = f(x, \dot{x}) \tag{1.35}$$

do not arise from conservative systems, and so can be expected to show new phenomena. The simplest such system is the linear oscillator with linear damping, having the equation

$$\ddot{x} + k\dot{x} + cx = 0, \tag{1.36}$$

where $c > 0$, $k > 0$. An equation of this form describes a spring–mass system with a damper in parallel (Fig. 1.15(a)); or the charge on the capacitor in a circuit containing resistance, capacitance, and inductance (Fig. 1.15(b)). In Fig. 1.15(a), the spring force is proportional to the extension x of the spring, and the damping, or frictional force, is proportional to the velocity $\dot{x}$. Therefore

$$m\ddot{x} = -mcx - mk\dot{x}$$

by Newton's law, where c and k are certain constants relating to the stiffness of the spring and the degree of friction in the damper respectively. Since the friction generates heat, which

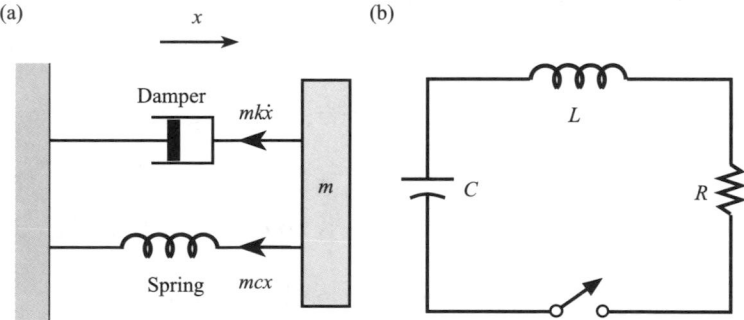

Figure 1.15   (a) Block controlled by a spring and damper. (b) Discharge of a capacitor through an (L, R, C) circuit.

is irrecoverable energy, the system is not conservative. These devices serve as models for many other oscillating systems. We shall show how the familiar features of damped oscillations show up on the phase plane.

Equation (1.36) is a standard type of differential equation, and the procedure for solving it goes as follows. Look for solutions of the form

$$x(t) = e^{pt}, \tag{1.37}$$

where p is a constant, by substituting (1.37) into (1.36). We obtain the **characteristic equation**

$$p^2 + kp + c = 0. \tag{1.38}$$

This has the solutions

$$\left.\begin{array}{l} p_1 \\ p_2 \end{array}\right\} = \frac{1}{2}\{-k \pm \sqrt{(k^2 - 4c)}\}. \tag{1.39}$$

where p_1 and p_2 may be both real, or complex conjugates depending on the sign of $k^2 - 4c$.

Unless $k^2 - 4c = 0$, we have found two solutions of (1.36); $e^{p_1 t}$ and $e^{p_2 t}$, and the general solution is

$$x(t) = Ae^{p_1 t} + Be^{p_2 t}, \tag{1.40}$$

where A and B are arbitrary constants which are real if $k^2 - 4c > 0$, and complex conjugates if $k^2 - 4c < 0$. If $k^2 - 4c = 0$ we have only one solution, of the form $e^{-\frac{1}{2}kt}$; we need a second one, and it can be checked that this takes the form $te^{-\frac{1}{2}kt}$. Therefore, in the case of coincident solutions of the characteristic equation, the general solution is

$$x(t) = (A + Bt)e^{-\frac{1}{2}kt}, \tag{1.41}$$

where A and B are arbitrary real constants.

Put

$$k^2 - 4c = \Delta, \tag{1.42}$$

where Δ is called the discriminant of the characteristic equation (1.38). The physical character of the motion depends upon the nature of the parameter Δ, as follows:

Strong damping ($\Delta > 0$)

In this case p_1 and p_2 are real, distinct and negative; and the general solution is

$$x(t) = Ae^{p_1 t} + Be^{p_2 t}; \quad p_1 < 0, \ p_2 < 0. \tag{1.43}$$

Figure 1.16(a) shows two typical solutions. There is no oscillation and the t axis is cut at most once. Such a system is said to be deadbeat.

To obtain the differential equation of the phase paths, write as usual

$$\dot{x} = y, \quad \dot{y} = -cx - ky; \tag{1.44}$$

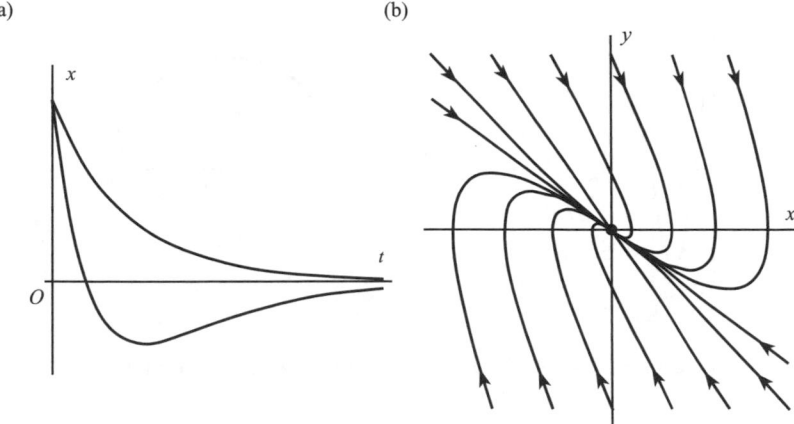

(a)

(b)

Figure 1.16   (a) Typical damped time solutions for strong damping. (b) Phase diagram for a stable node.

then

$$\frac{\mathrm{d}y}{\mathrm{d}x} = -\frac{cx + ky}{y}. \tag{1.45}$$

There is a single equilibrium point, at $x = 0$, $y = 0$. A general approach to linear systems such as (1.44) is developed in Chapter 2: for the present the solutions of (1.45) are too complicated for simple interpretation. We therefore proceed in the following way. From (1.43), putting $y = \dot{x}$,

$$x = Ae^{p_1 t} + Be^{p_2 t}, \qquad y = Ap_1 e^{p_1 t} + Bp_2 e^{p_2 t} \tag{1.46}$$

for fixed A and B, there can be treated as a parametric representation of a phase path, with parameter t. The phase paths in Fig. 1.16(b) are plotted in this way for certain values of $k > 0$ and $c > 0$. This shows a new type of equilibrium point, called a **stable node**. All paths start at infinity and terminate at the origin, as can be seen by putting $t = \pm\infty$ into (1.43). More details on the structure of nodes is given in Chapter 2.

Weak damping ($\Delta < 0$)

The exponents p_1 and p_2 are complex with negative real part, given by

$$p_1, p_2 = -\tfrac{1}{2}k \pm \tfrac{1}{2}\mathrm{i}\sqrt{(-\Delta)},$$

where $\mathrm{i} = \sqrt{-1}$. The expression (1.40) for the general solution is then, in general, complex. To extract the cases for which (1.40) delivers **real solutions**, allow A and B to be arbitrary and complex, and put

$$A = \tfrac{1}{2}Ce^{\mathrm{i}\alpha},$$

where α is real and $C = 2|A|$; and

$$B = \bar{A} = \tfrac{1}{2}Ce^{-\mathrm{i}\alpha},$$

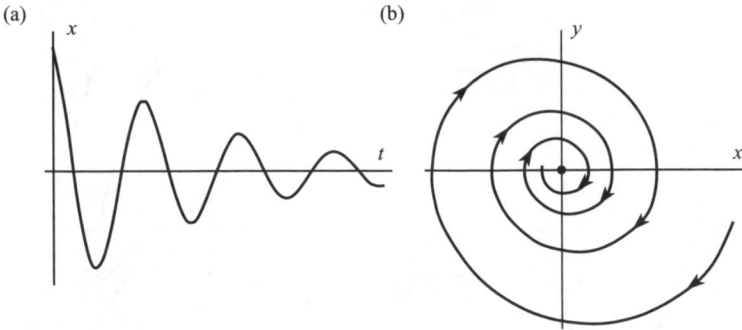

Figure 1.17   (a) Typical time solution for weak damping. (b) Phase diagram for a stable spiral showing just one phase path.

where $\bar{A}$ is the complex conjugate of A. Then (1.40) reduces to

$$x(t) = Ce^{-\frac{1}{2}kt} \cos\{\tfrac{1}{2}\sqrt{(-\Delta)}t + \alpha\};$$

C and α are real and arbitrary, and $C > 0$.

A typical solution is shown in Fig. 1.17(a); it represents an oscillation of frequency $(-\Delta)^{\frac{1}{2}}/(4\pi)$ and exponentially decreasing amplitude $Ce^{-\frac{1}{2}kt}$. Its image on the phase plane is shown in Fig. 1.17(b). The whole phase diagram would consist of a *family* of such spirals corresponding to different time solutions.

The equilibrium point at the origin is called a **stable spiral** or a **stable focus**.

Critical damping ($\Delta = 0$)

In this case $p_1 = p_2 = -\frac{1}{2}k$, and the solutions are given by (1.41). The solutions resemble those for strong damping, and the phase diagram shows a stable node.

We may also consider cases where the signs of k and c are negative:

Negative damping ($k < 0$, $c > 0$)

Instead of energy being lost to the system due to friction or resistance, energy is generated within the system. The node or spiral is then unstable, the directions being outward (see Fig. 1.18). A slight disturbance from equilibrium leads to the system being carried far from the equilibrium state (see Fig. 1.18).

Spring with negative stiffness ($c < 0$, k takes any value)

The phase diagram shows a saddle point, since p_1, p_2 are real but of opposite signs.

Exercise 1.4
For the linear system $\ddot{x} - 2\dot{x} + 2x = 0$, classify its equilibrium point and sketch the phase diagram.

(a)   (b)

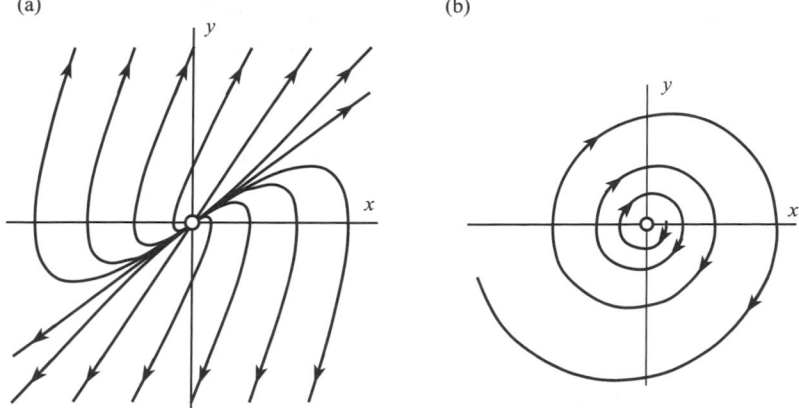

Figure 1.18 Phase diagrams for (a) the unstable node ($k < 0$, $\Delta > 0$); (b) unstable spiral ($k < 0$, $\Delta < 0$) showing just one phase path.

Exercise 1.5
Show that every phase path of

$$\ddot{x} + \varepsilon |x| \operatorname{sgn} \dot{x} + x = 0, \quad 0 < \varepsilon < 1,$$

is an isochronous spiral (that is, every circuit of the origin on every path occur in the same time).

1.5 Nonlinear damping: limit cycles

Consider the autonomous system

$$\ddot{x} = f(x, \dot{x}),$$

where f is a nonlinear function; and f takes the form

$$f(x, \dot{x}) = -h(x, \dot{x}) - g(x),$$

so that the differential equation becomes

$$\ddot{x} + h(x, \dot{x}) + g(x) = 0. \tag{1.47}$$

The equivalent pair of first-order equations for the phase paths is

$$\dot{x} = y, \qquad \dot{y} = -h(x, y) - g(x). \tag{1.48}$$

For the purposes of interpretation we shall assume that there is a single equilibrium point, and that it is at the origin (having been moved to the origin, if necessary, by a change of axes). Thus

$$h(0,0) + g(0) = 0$$

and the only solution of $h(x,0) + g(x) = 0$ is $x = 0$. We further assume that

$$g(0) = 0, \tag{1.49}$$

so that

$$h(0,0) = 0. \tag{1.50}$$

Under these circumstances, by writing eqn (1.47) in the form

$$\ddot{x} + g(x) = -h(x, \dot{x}), \tag{1.51}$$

we can regard the system as being modelled by a unit particle on a spring whose free motion is governed by the equation $\ddot{x} + g(x) = 0$ (a conservative system), but is also acted upon by an external force $-h(x, \dot{x})$ which supplies or absorbs energy. If $g(x)$ is a restoring force (eqn (1.25) with $-g(x)$ in place of $f(x)$), then we should expect a tendency to oscillate, modified by the influence of the external force $-h(x, \dot{x})$. In both the free and forced cases, equilibrium occurs when $x = \dot{x} = 0$.

Define a potential energy function for the spring system by

$$\mathcal{V}(x) = \int g(x)\,dx, \tag{1.52a}$$

and the kinetic energy of the particle by

$$\mathcal{T} = \tfrac{1}{2}\dot{x}^2. \tag{1.52b}$$

The total energy $\mathcal{E}$ for the particle and spring alone is

$$\mathcal{E} = \mathcal{T} + \mathcal{V} = \tfrac{1}{2}\dot{x}^2 + \int g(x)\,dx, \tag{1.53}$$

so that the rule of change of energy

$$\frac{d\mathcal{E}}{dt} = \dot{x}\ddot{x} + g(x)\dot{x}.$$

Therefore, by (1.49)

$$\frac{d\mathcal{E}}{dt} = \dot{x}(-g(x) - h(x, \dot{x}) + g(x)) = -\dot{x}h(x, \dot{x}) \tag{1.54}$$

$$= -yh(x, y)$$

in the phase plane. This expression represents external the rate of supply of energy generated by the term $-h(x, \dot{x})$ representing the external force.

Suppose that, in some connected region $\mathcal{R}$ of the phase plane which contains the equilibrium point $(0, 0)$, $d\mathcal{E}/dt$ is negative:

$$\frac{d\mathcal{E}}{dt} = -yh(x, y) < 0 \tag{1.55}$$

(except on $y = 0$, where it is obviously zero). Consider any phase path which, after a certain point, lies in $\mathcal{R}$ for all time. Then $\mathcal{E}$ continuously decreases along the path. The effect of h resembles damping or resistance; energy is continuously withdrawn from the system, and this produces a general decrease in amplitude until the initial energy runs out. We should expect the path to approach the equilibrium point.

If

$$\frac{d\mathcal{E}}{dt} = -yh(x, y) > 0 \tag{1.56}$$

in $\mathcal{R}$ (for $y \neq 0$), the energy increases along every such path, and the amplitude of the phase paths increases so long as the paths continue to remain in $\mathcal{R}$. Here h has the effect of negative damping, injecting energy into the system for states lying in $\mathcal{R}$. ●

Example 1.8 *Examine the equation*
$$\ddot{x} + |\dot{x}|\dot{x} + x = 0$$

for damping effects.
The free oscillation is governed by $\ddot{x} + x = 0$, and the external force is given by

$$-h(x, \dot{x}) = -|\dot{x}|\dot{x}.$$

Therefore, from (1.55), the rate of change of energy

$$\frac{d\mathcal{E}}{dt} = -|\dot{x}|\dot{x}^2 = -|y|y^2 < 0$$

everywhere (except for $y = 0$). There is loss of energy along every phase path no matter where it goes in the phase plane. We therefore expect that from any initial state the corresponding phase path enters the equilibrium state as the system runs down. ●

A system may possess both characteristics; energy being injected in one region and extracted in another region of the phase plane. On any common boundary to these two regions, $\dot{x}h(x, y) = 0$ (assuming that $h(x, y)$ is continuous). The common boundary *may* constitute a phase path, and if so the energy $\mathcal{E}$ is constant along it. This is illustrated in the following example.

Example 1.9 *Examine the equation*
$$\ddot{x} + (x^2 + \dot{x}^2 - 1)\dot{x} + x = 0$$

for energy input and damping effects.
Put $\dot{x} = y$; then

$$h(x, y) = (x^2 + \dot{x}^2 - 1)\dot{x},$$

and, from (1.55),

$$\frac{d\mathcal{E}}{dt} = -yh(x, y) = -(x^2 + y^2 - 1)y^2.$$

Therefore the energy in the particle–spring system is governed by:

$$\frac{d\mathcal{E}}{dt} > 0 \quad \text{along the paths in the region } x^2 + y^2 < 1;$$

$$\frac{d\mathcal{E}}{dt} < 0 \quad \text{along the paths in the region } x^2 + y^2 > 1;$$

The regions concerned are shown in Fig. 1.19. It can be verified that $x = \cos t$ satisfies the differential equation given above. Therefore $\dot{x} = y = -\sin t$; so that the boundary between the two regions, the circle

$$x^2 + y^2 = 1,$$

is a phase path, as shown. Along it

$$\mathcal{T} + \mathcal{V} = \tfrac{1}{2}\dot{x}^2 + \tfrac{1}{2}x^2 = \tfrac{1}{2}(x^2 + y^2) = \tfrac{1}{2}$$

is constant, so it is a curve of constant energy $\mathcal{E}$, called an energy level.

The phase diagram consists of this circle together with paths spiralling towards it from the interior and exterior, and the (unstable) equilibrium point at the origin. All paths approach the circle. Therefore the system moves towards a state of steady oscillation, whatever the (nonzero) initial conditions.   ●

The circle in Fig. 1.19 is an **isolated closed path**: 'isolated' in the sense that there is no other closed path in its immediate neighbourhood. An isolated closed path is called a **limit cycle**, and when it exists it is always one of the most important features of a physical system. Limit cycles can only occur in nonlinear systems. The limit cycle in Fig. 1.19 is a **stable limit cycle**, since if the system is disturbed from its regular oscillatory state, the resulting new path, on

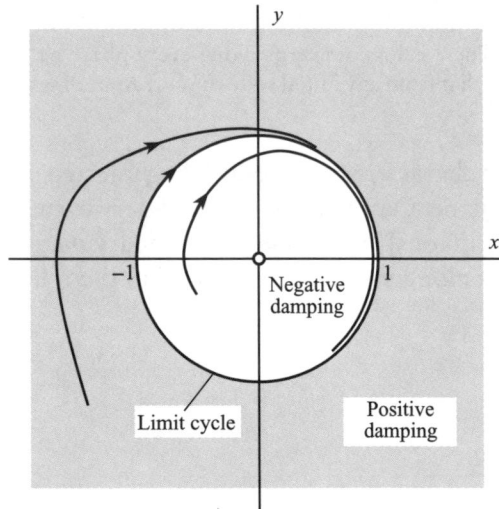

Figure 1.19   Approach of two phase paths to the stable limit cycle $x^2 + y^2 = 1$ generated by $\ddot{x} + (x^2 + \dot{x}^2 - 1)\dot{x} + x = 0$.

either side, will be attracted back to the limit cycle. There also exist *unstable* limit cycles, where neighbouring phase paths, on one side or the other are repelled from the limit cycle.

An important example of the significance of a stable limit cycle is the pendulum clock (for details see Section 1.6(iii)). Energy stored in a hanging weight is gradually supplied to the system by means of the escapement; this keeps the pendulum swinging. A balance is *automatically* set up between the rate of supply of energy and the loss due to friction in the form of a stable limit cycle which ensures strict periodicity, and recovery from any sudden disturbances.

An equation of the form

$$\ddot{x} = f(x),$$

which is the 'conservative' type treated in Section 1.3, cannot lead to a limit cycle. From the argument in that section (see Fig. 1.12), there is no shape for $\mathcal{V}(x)$ that could produce an *isolated* closed phase path.

We conclude this section by illustrating several approaches to equations having the form

$$\ddot{x} + h(x, \dot{x}) + g(x) = 0, \tag{1.57}$$

which do not involve any necessary reference to mechanical models or energy.

(i) Polar coordinates

We shall repeat Example 1.9 using polar coordinates. The structure of the phase diagram is made clearer, and other equations of similar type respond to this technique. Let r, θ be polar coordinates, where $x = r \cos \theta, y = r \sin \theta$, so that

$$r^2 = x^2 + y^2, \qquad \tan \theta = \frac{y}{x}.$$

Then, differentiating these equations with respect to time,

$$2r\dot{r} = 2x\dot{x} + 2y\dot{y}, \qquad \dot{\theta} \sec^2 \theta = \frac{x\dot{y} - \dot{x}y}{x^2}$$

so that

$$\dot{r} = \frac{x\dot{x} + y\dot{y}}{r}, \qquad \dot{\theta} = \frac{x\dot{y} - \dot{x}y}{r^2}. \tag{1.58}$$

We then substitute

$$x = r \cos \theta, \qquad \dot{x} = y = r \sin \theta$$

into these expressions, using the form for $\dot{y}$ obtained from the given differential equation.

Example 1.10 *Express the equation (see Example 1.9)*

$$\ddot{x} + (x^2 + \dot{x}^2 - 1)\dot{x} + x = 0$$

on the phase plane, in terms of polar coordinates r, θ.

We have $x = r \cos \theta$ and $\dot{x} = y = r \sin \theta$, and

$$\dot{y} = -(x^2 + \dot{x}^2 - 1)\dot{x} - x = -(r^2 - 1)r \sin \theta - r \cos \theta.$$

By substituting these functions into (1.58) we obtain

$$\dot{r} = -r(r^2 - 1) \sin^2 \theta,$$

$$\dot{\theta} = -1 - (r^2 - 1) \sin \theta \cos \theta.$$

One particular solution is

$$r = 1, \qquad \theta = -t,$$

corresponding to the limit cycle, $x = \cos t$, $y = -\sin t$, observed in Example 1.9. Also (except when $\sin \theta = 0$; that is, except on the x axis)

$$\dot{r} > 0 \quad \text{when } 0 < r < 1$$

$$\dot{r} < 0 \quad \text{when } r > 1,$$

showing that the paths approach the limit cycle $r = 1$ from both sides. The equation for $\dot{\theta}$ also shows a steady clockwise spiral motion for the representative points, around the limit cycle.   ●

(ii) Topographic curves

We shall introduce topographic curves through an example.

Example 1.11   *Investigate the trend of the phase paths for the differential equation*

$$\ddot{x} + |\dot{x}|\dot{x} + x^3 = 0.$$

The system has only one equilibrium point, at $(0,0)$. Write the equation in the form

$$\ddot{x} + x^3 = -|\dot{x}|\dot{x},$$

and multiply through by $\dot{x}$:

$$\dot{x}\ddot{x} + x^3\dot{x} = -|\dot{x}|\dot{x}^2.$$

In terms of the phase plane variables x, y this becomes

$$y\frac{dy}{dt} + x^3\frac{dx}{dt} = -|y|y^2.$$

Consider a phase path that passes through an arbitrary point A at time t_A and arrives at a point B at time $t_B > t_A$. By integrating this last equation from t_A to t_B we obtain

$$\left[\tfrac{1}{2}y^2 + \tfrac{1}{4}x^4\right]_{t=t_A}^{t_B} = -\int_{t_A}^{t_B} |y|y^2 \, dt.$$

The right-hand side is negative everywhere, so

$$\left(\tfrac{1}{2}y^2 + \tfrac{1}{4}x^4\right)_{t=t_B} > \left(\tfrac{1}{2}y^2 + \tfrac{1}{4}x^4\right)_{t=t_A}$$

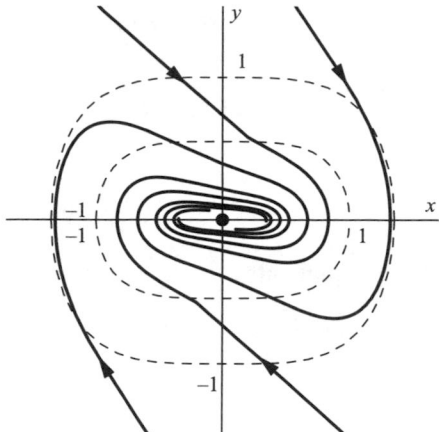

Figure 1.20   Phase diagram for $\dot{x} = y,\ \dot{y} = -|y|y - x^3$: the broken lines represent constant level curves.

along the phase path. Therefore the values of the bracketed expression, $\frac{1}{2}y^2 + \frac{1}{4}x^4$, constantly diminishes along every phase path. But the family of curves given by

$$\tfrac{1}{2}y^2 + \tfrac{1}{4}x^4 = \text{constant}$$

is a family of ovals closing in on the origin as the constant diminishes. The paths cross these ovals successively in an inward direction, so that the phase paths all move towards the origin as shown in Fig. 1.20. In mechanical terms, the ovals are curves of constant energy.   ●

Such familes of closed curves, which can be used to track the paths to a certain extent, are called **topographic curves**, and are employed extensively in Chapter 10 to determine stability. The 'constant energy' curves, or **energy levels**, in the example constitute a special case.

(iii) Equations of motion in generalized coordinates

Suppose we have a conservative mechanical system, which may be in one, two, or three dimensions, and may contain solid elements as well as particles, but such that its configuration is completely specified by the value of a certain single variable x. The variable need not represent a displacement; it might, for example, be an angle, or even the reading on a dial forming part of the system. It is called a **generalized coordinate**.

Generally, the kinetic and potential energies $\mathcal{T}$ and $\mathcal{V}$ will take the forms

$$\mathcal{T} = p(x)\dot{x}^2 + q(x), \qquad \mathcal{V} = \mathcal{V}(x),$$

where $p(x) > 0$. The equation of motion can be derived using Lagrange's equation

$$\frac{\mathrm{d}}{\mathrm{d}t}\left(\frac{\partial \mathcal{T}}{\partial \dot{x}}\right) - \frac{\partial \mathcal{T}}{\partial x} = -\frac{\mathrm{d}\mathcal{V}}{\mathrm{d}x}.$$

Upon substituting for $\mathcal{T}$ and $\mathcal{V}$ we obtain the equation of motion in terms of x:

$$2p(x)\ddot{x} + p'(x)\dot{x}^2 + (\mathcal{V}'(x) - q'(x)) = 0. \tag{1.59}$$

This equation is not of the form $\ddot{x} = f(x)$ discussed in Section 1.3. To reduce to this form substitute u for x from the definition

$$u = \int p^{1/2}(x)\, \mathrm{d}x. \tag{1.60}$$

Then

$$\dot{u} = p^{1/2}(x)\dot{x} \quad \text{and} \quad \ddot{u} = \tfrac{1}{2}p^{-1/2}(x)p'(x)\dot{x}^2 + p^{1/2}(x)\ddot{x}.$$

After obtaining $\dot{x}$ and $\ddot{x}$ from these equations and substituting in eqn (1.59) we have

$$\ddot{u} + g(u) = 0,$$

where

$$g(u) = \tfrac{1}{2}p^{-1/2}(x)(\mathcal{V}'(x) - q'(x)).$$

This is of the conservative type discussed in Section 1.3.

Exercise 1.6
Find the equation of the limit cycle of

$$\ddot{x} + (4x^2 + \dot{x}^2 - 4)\dot{x} + 4x = 0.$$

What is its period?

1.6   Some applications

(i) Dry friction

Dry (or Coulomb) friction occurs when the surfaces of two solids are in contact and in relative motion without lubrication. The model shown in Fig. 1.21 illustrates dry friction. A continuous belt is driven by rollers at a constant speed v_0. A block of mass m connected to a fixed support by a spring of stiffness c rests on the belt. If F is the frictional force between the block and the belt and x is the extension of the spring, then the equation of motion is

$$m\ddot{x} + cx = F.$$

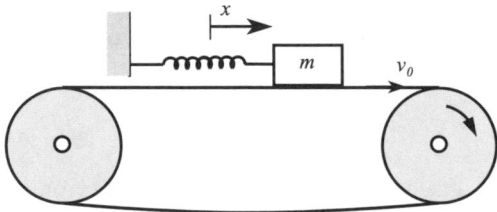

Figure 1.21   A device illustrating dry friction.

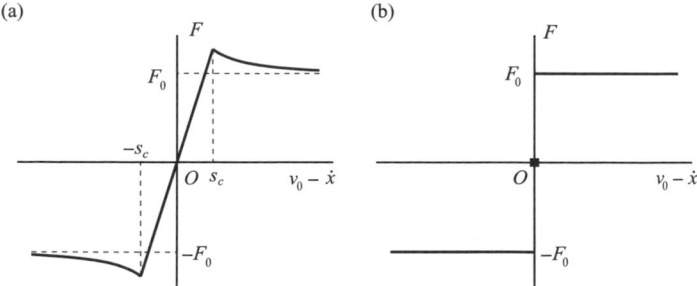

Figure 1.22 (a) Typical dry friction/slip velocity profile. (b) Idealized approximation with discontinuity.

Assume that F depends on the **slip velocity**, $v_0 - \dot{x}$; a typical relation is shown in Fig. 1.22(a). For small slip velocities the frictional force is proportional to the slip velocity. At a fixed small value of the slip speed s_c the magnitude of the frictional force peaks and then gradually approaches a constant F_0 or $-F_0$ for large slip speeds. We will replace this function by a simpler one having a discontinuity at the origin:

$$F = F_0 \, \mathrm{sgn}(v_0 - \dot{x})$$

where F_0 is a positive constant (see Fig. 1.22(b)) and the sgn (signum) function is defined by

$$\mathrm{sgn}(u) = \begin{cases} 1, & u > 0, \\ 0, & u = 0, \\ -1, & u < 0. \end{cases}$$

The equation of motion becomes

$$m\ddot{x} + cx = F_0 \, \mathrm{sgn}(v_0 - \dot{x}).$$

The term on the right is equal to F_0 when $v_0 > \dot{x}$, and $-F_0$ when $v_0 < \dot{x}$, and we obtain the following solutions for the phase paths in these regions:

$$y = \dot{x} > v_0: \quad my^2 + c\left(x + \frac{F_0}{c}\right)^2 = \text{constant},$$

$$y = \dot{x} < v_0: \quad my^2 + c\left(x - \frac{F_0}{c}\right)^2 = \text{constant}.$$

These are families of displaced ellipses, the first having its centre at $(-F_0/c, 0)$ and the second at $(F_0/c, 0)$. Figure 1.23 shows the corresponding phase diagram, in non-dimensional form with $x' = x\sqrt{c}$ and $y' = y\sqrt{m}$. In terms of these variables the paths are given by

$$y' > v_0\sqrt{m}: \quad y'^2 + \left(x' + \frac{F_0}{\sqrt{c}}\right)^2 = \text{constant},$$

$$y' < v_0\sqrt{m}: \quad y'^2 + \left(x' - \frac{F_0}{\sqrt{c}}\right)^2 = \text{constant},$$

which are arcs of displaced circles.

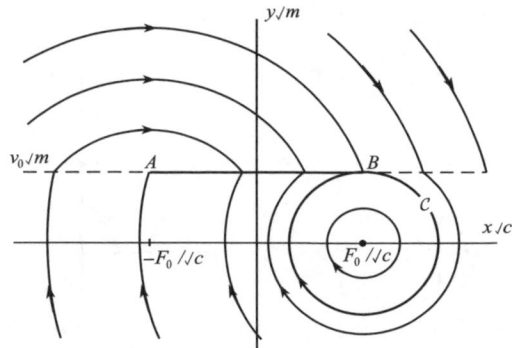

Figure 1.23   Phase diagram for the stick–slip dry friction oscillator. (Note that the axes are scaled as $x\sqrt{c}, y\sqrt{m}$.)

There is a single equilibrium point, at $(F_0/c, 0)$, which is a centre. Points on y (or $\dot{x}$) $= v_0$ are not covered by the differential equation since this is where F is discontinuous, so the behaviour must be deduced from other physical arguments. On encountering the state $\dot{x} = v_0$ for $|x| < F_0/c$, the block will move with the belt along AB until the maximum available friction, F_0, is insufficient to resist the increasing spring tension. This is at B when $x = F_0/c$; the block then goes into an oscillation represented by the closed path $\mathcal{C}$ through $(F_0/c, v_0)$. In fact, for any initial conditions lying outside this ellipse, the system ultimately settles into this oscillation. A computed phase diagram corresponding to a more realistic frictional force as in Fig. 1.21(a) is displayed in Problem 3.50, Fig 3.32. This kind of motion is often described as a **stick–slip oscillation**.

(ii) The brake

Consider a simple brake shoe applied to the hub of a wheel as shown in Fig. 1.24. The friction force will depend on the pressure and the angular velocity of the wheel, $\dot{\theta}$. We assume again a simplified dry-friction relation corresponding to constant pressure

$$F = -F_0 \, \text{sgn}(\dot{\theta})$$

so if the wheel is otherwise freely spinning its equation of motion is

$$I\ddot{\theta} = -F_0 a \, \text{sgn}(\dot{\theta})$$

where I is the moment of inertia of the wheel and a the radius of the brake drum. The phase paths are found by rewriting the differential equation, using the transformation (1.2):

$$I\dot{\theta}\frac{\mathrm{d}\dot{\theta}}{\mathrm{d}\theta} = -F_0 a \, \text{sgn}(\dot{\theta}),$$

whence for $\dot{\theta} > 0$

$$\tfrac{1}{2}I\dot{\theta}^2 = -F_0 a\theta + C$$

and for $\dot{\theta} < 0$

$$\tfrac{1}{2}I\dot{\theta}^2 = F_0 a\theta + C.$$

These represent two families of parabolas as shown in Fig. 1.25. $(\theta, 0)$ is an equilibrium point for every θ.

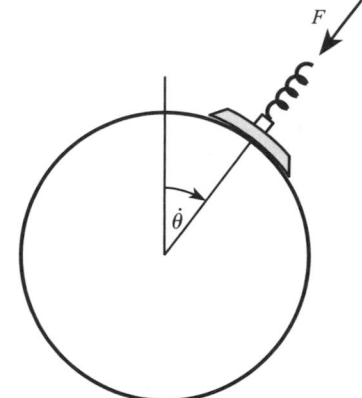

Figure 1.24   A brake model.

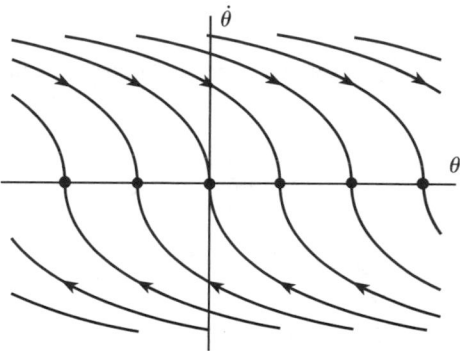

Figure 1.25   Phase diagram for the brake model in Fig. 1.24.

(iii) The pendulum clock: a limit cycle

Figure 1.26 shows the main features of the pendulum clock. The 'escape wheel' is a toothed wheel, which drives the hands of the clock through a succession of gears. It has a spindle around which is wound a wire with a weight at its free end. The escape wheel is intermittently arrested by the 'anchor', which has two teeth. The anchor is attached to the shaft of the pendulum and rocks with it, controlling the rotation of the escape wheel. The anchor and teeth on the escape wheel are so designed that as one end of the anchor loses contact with a tooth, the other end engages a tooth but allows the escape wheel to turn through a small angle, which turns the hands of the clock. Every time this happens the anchor receives small impulses, which is heard as the 'tick' of the clock. These impulses maintain the oscillation of the pendulum, which would otherwise die away. The loss of potential energy due to the weight's descent is therefore fed periodically into the pendulum via the anchor mechanism. Although the impulses push the pendulum in the same direction at each release of the anchor, the anchor shape ensures that they are slightly different in magnitude.

It can be shown that the system will settle into steady oscillations of fixed amplitude independently of sporadic disturbance and of initial conditions. If the pendulum is swinging with

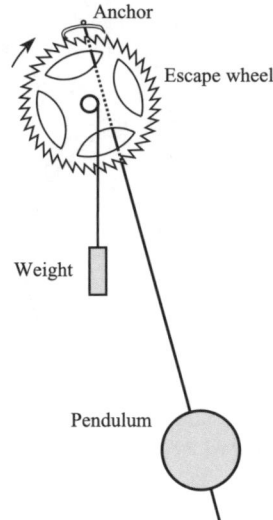

Figure 1.26 The driving mechanism (escapement) of weight-driven pendulum clock.

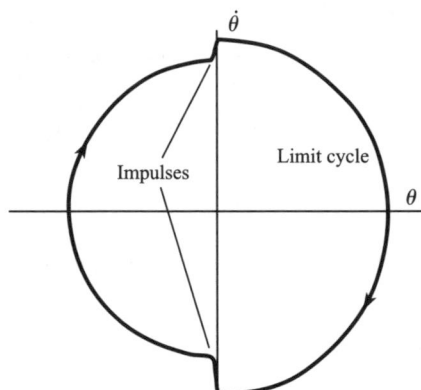

Figure 1.27 Phase diagram for a damped impulse-driven oscillator model for a pendulum clock.

too great an amplitude, its loss of energy per cycle due to friction is large, and the impulse supplied by the escapement is insufficient to offset this. The amplitude consequently decreases. If the amplitude is too small, the frictional loss is small; the impulses will over-compensate and the amplitude will build up. A balanced state is therefore approached, which appears in the $\theta, \dot{\theta}$ plane (Fig. 1.27) as an isolated closed curve $\mathcal{C}$. Such an isolated periodic oscillation, or **limit cycle** (see Section 1.5) can occur only in systems described by nonlinear equations, and the following simple model shows where the nonlinearity is located.

The motion can be approximated by the equation

$$I\ddot{\theta} + k\dot{\theta} + c\theta = f(\theta, \dot{\theta}), \tag{1.61}$$

where I is the moment of inertia of the pendulum, k is a small damping constant, c is another constant determined by gravity, θ is the angular displacement, and $f(\theta, \dot{\theta})$ is the moment, supplied twice per cycle by the escapement mechanism. The moment $f(\theta, \dot{\theta})$ will be a nonlinear function in θ and $\dot{\theta}$. The model function

$$f(\theta, \dot{\theta}) = \tfrac{1}{2}[(k_1 + k_2) + (k_1 - k_2)\,\mathrm{sgn}(\dot{\theta})]\delta(\theta),$$

where $\delta(\theta)$ is the Dirac delta, or impulse, function, and k_1, k_2 are positive constraints, delivers impulses to the pendulum when $\theta = 0$. If $\dot{\theta} > 0$, then $f(\theta, \dot{\theta}) = k_1\delta(\theta)$, and if $\dot{\theta} < 0$, then $f(\theta, \dot{\theta}) = k_2\delta(\theta)$. The pendulum will be driven to overcome damping if $k_2 > k_1$. The pendulum clock was introduced by Huyghens in 1656. The accurate timekeeping of the pendulum represented a great advance in clock design over the earlier weight-driven clocks (see Baker and Blackburn (2005)).

Such an oscillation, generated by an energy source whose input is not regulated externally, but which *automatically* synchronizes with the existing oscillation, is called a **self-excited oscillation**. Here the build-up is limited by the friction.

Exercise 1.7
A smooth wire has the shape of a cycloid given parametrically by $x = a(\phi + \sin\phi)$, $y = a(1 - \cos\phi)$, $(-\pi < \phi < \pi)$. A bead is released from rest where $\phi = \phi_0$. Using conservation of energy confirm that

$$a\dot{\phi}^2 \cos^2 \tfrac{1}{2}\phi = g(\sin^2 \tfrac{1}{2}\phi_0 - \sin^2 \tfrac{1}{2}\phi).$$

Hence show that the period of oscillation of the bead is $4\pi\sqrt{(a/g)}$ (i.e., independent of ϕ_0). This is known as the **tautochrone**.

1.7 Parameter-dependent conservative systems

Suppose $x(t)$ satisfies

$$\ddot{x} = f(x, \lambda) \tag{1.62}$$

where λ is a parameter. The equilibrium points of the system are given by $f(x, \lambda) = 0$, and in general their location will depend on the parameter λ. In mechanical terms, for a particle of unit mass with displacement x, $f(x, \lambda)$ represents the force experienced by the particle. Define a function $\mathcal{V}(x, \lambda)$ such that $f(x, \lambda) = -\partial\mathcal{V}/\partial x$ for each value of λ; then $\mathcal{V}(x, \lambda)$ is the potential energy per unit mass of the equivalent mechanical system and equilibrium points correspond to stationary values of the potential energy. As indicated in Section 1.3, we expect a minimum of potential energy to correspond to a stable equilibrium point, and other stationary values (the maximum and point of inflexion) to be unstable. In fact, $\mathcal{V}$ is a minimum at $x = x_1$ if $\partial\mathcal{V}/\partial x$ changes from negative to positive on passing through x_1; this implies that $f(x, \lambda)$ changes sign from positive to negative as x increases through $x = x_1$. It acts as a restoring force.

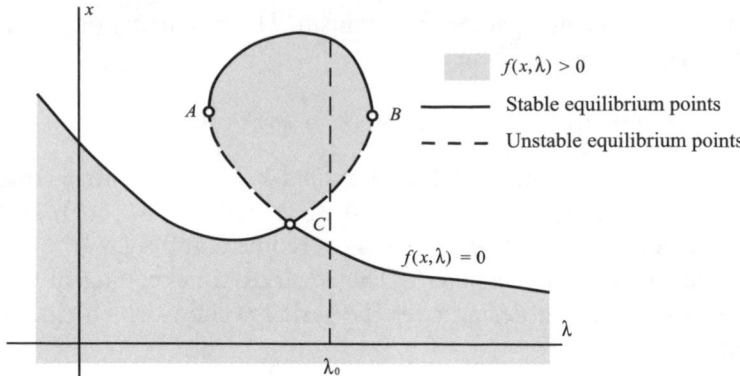

Figure 1.28   Representative stability diagram showing stability curves for the equilibrium points of $\ddot{x} = f(x, \lambda)$.

There exists a simple graphical method of displaying the stability of equilibrium points for a parameter-dependent system, in which both the number and stability of equilibrium points may vary with λ. We assume $f(x, \lambda)$ to be continuous in both x and λ. Plot the curve $f(x, \lambda) = 0$ in the λ, x plane; this curve represents the equilibrium points. Shade the domains in which $f(x, \lambda) > 0$ as shown in Fig. 1.28: If a segment of the curve has shading below it, the corresponding equilibrium points are stable, since for fixed λ, f changes from positive to negative as x increases.

The solid line between A and B corresponds to stable equilibrium points. A and B are unstable: C is also unstable since f is positive on both sides of C. The nature of the equilibrium points for a given value of λ can easily be read from the figure; for example when $\lambda = \lambda_0$ as shown, the system has three equilibrium points, two of which are stable. A, B and C are known as **bifurcation points**. As λ varies through such points the equilibrium point may split into two or more, or several equilibrium points may merge into a single one. More information on bifurcation can be found in Chapter 12.

Example 1.12   *A bead slides on a smooth circular wire of radius a which is constrained to rotate about a vertical diameter with constant angular velocity ω. Analyse the stability of the bead.*

The bead has a velocity component $a\dot{\theta}$ tangential to the wire and a component $a\omega \sin\theta$ perpendicular to the wire, where θ is the inclination of the radius to the bead to the downward vertical as shown in Fig. 1.29. The kinetic energy $\mathcal{T}$ and potential energy $\mathcal{V}$ are given by

$$\mathcal{T} = \tfrac{1}{2}ma^2(\dot{\theta}^2 + \omega^2 \sin^2\theta), \quad \mathcal{V} = -mga\cos\theta.$$

Since the system is subject to a moving constraint (that is, the angular velocity of the wire is imposed), the usual energy equation does not hold. Lagrange's equation for the system is

$$\frac{\mathrm{d}}{\mathrm{d}t}\left(\frac{\partial \mathcal{T}}{\partial \dot{\theta}}\right) - \frac{\partial \mathcal{T}}{\partial \theta} = -\frac{\partial \mathcal{V}}{\partial \theta},$$

which gives

$$a\ddot{\theta} = a\omega^2 \sin\theta \cos\theta - g\sin\theta.$$

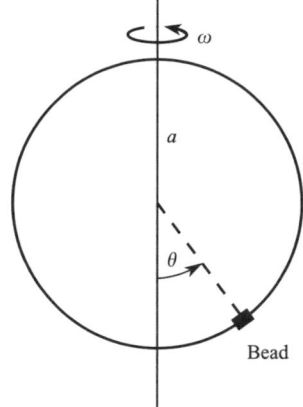

Figure 1.29   Bead on a rotating wire.

Set $a\omega^2/g = \lambda$. Then

$$a\ddot{\theta} = a\dot{\theta}\frac{\mathrm{d}\dot{\theta}}{\mathrm{d}\theta} = g\sin\theta(\lambda\cos\theta - 1)$$

which, after integration, becomes

$$\tfrac{1}{2}a\dot{\theta}^2 = g(1 - \tfrac{1}{2}\lambda\cos\theta)\cos\theta + C,$$

the equation of the phase paths.

In the notation of eqn (1.62), we have from (i):

$$f(\theta,\lambda) = \frac{g\sin\theta(\lambda\cos\theta - 1)}{a}.$$

The equilibrium points are given by $f(\theta,\lambda) = 0$, which is satisfied when $\sin\theta = 0$ or $\cos\theta = \lambda^{-1}$. From the periodicity of the problem, $\theta = \pi$ and $\theta = -\pi$ correspond to the same state of the system.

The regions where $f < 0$ and $f > 0$ are separated by curves where $f = 0$, and can be located, therefore, by checking the sign at particular points; for example, $f(\tfrac{1}{2}\pi, 1) = -g/a < 0$. Figure 1.30 shows the stable and unstable equilibrium positions of the bead. The point A is a **bifurcation point**, and the equilibrium there is stable. It is known as a **pitchfork bifurcation** because of its shape.

Phase diagrams for the system may be constructed as in Section 1.3 for fixed values of λ. Two possibilities are shown in Fig. 1.31. Note that they confirm the stability predictions of Fig. 1.30.   ●

Exercise 1.8
Sketch the stability diagram for the parameter-dependent equation

$$\ddot{x} = \lambda^3 + \lambda^2 - x^2,$$

and discuss the stability of the equilibrium points.

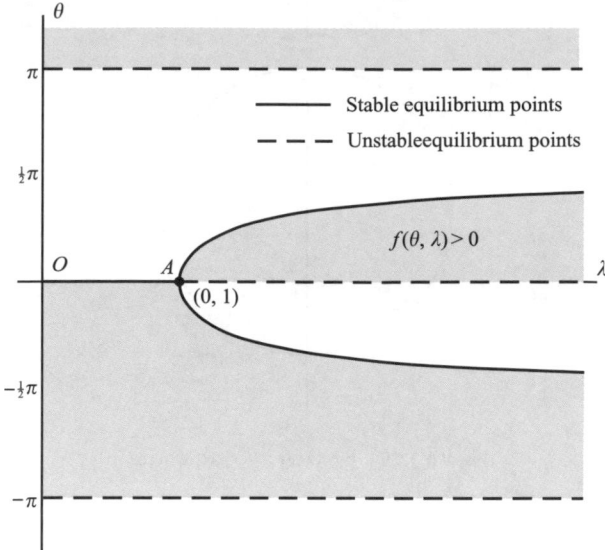

Figure 1.30   Stability diagram for a bead on a rotating wire.

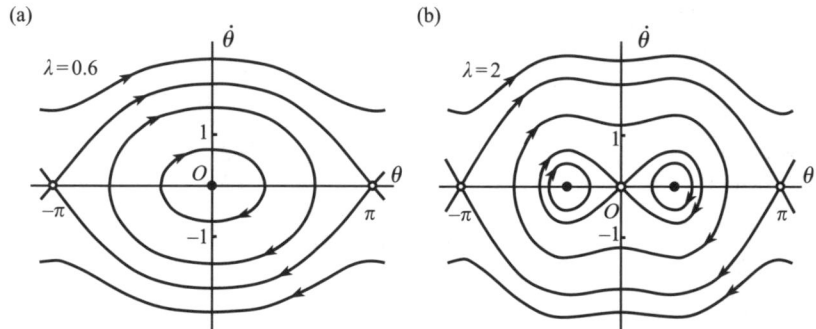

Figure 1.31   Typical phase diagrams for the rotating bead equation $\ddot{\theta} = (g/a) \sin\theta (\lambda \cos\theta - 1)$ for the cases (a) $\lambda < 0$; (b) $\lambda > 0$, with $a = g$ in both cases.

1.8   Graphical representation of solutions

Solutions and phase paths of the system

$$\dot{x} = y, \qquad \dot{y} = f(x, y)$$

can be represented graphically in a number of ways. As we have seen, the solutions of $dy/dx = f(x, y)/y$ can be displayed as paths in the phase plane (x, y). Different ways of viewing paths and solutions of the pendulum equation $\ddot{x} = -\sin x$ are shown in Fig. 1.32. Figure 1.32(a) shows typical phase paths including separatrices.

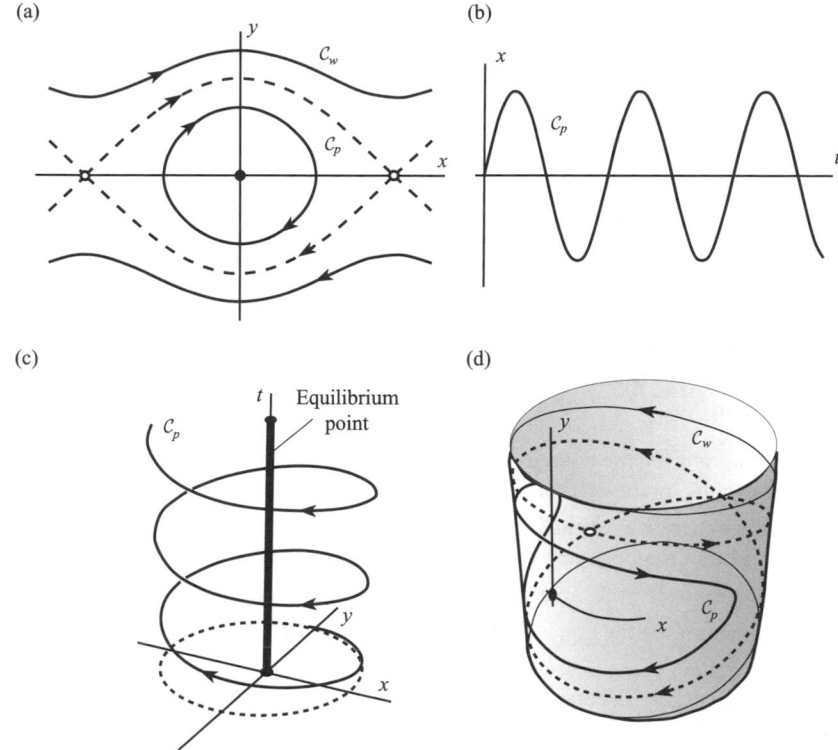

Figure 1.32   Different views of solutions of the pendulum equation $\ddot{x} + \sin x = 0$. In particular, the periodic solution $\mathcal{L}_p$ and a whirling solution $\mathcal{L}_w$ are shown (a) in the phase plane; (b) as an (x, t) solution; (c) in the solution space (x, y, t); (d) on a cylindrical phase surface which can be used for differential equations periodic in x.

If the solutions of $\ddot{x} = f(x, \dot{x})$ are known, either exactly or numerically, then the behaviour of x in terms of t can be shown in an (x, t) graph as in Fig. 1.32(b) for a periodic solution of the pendulum. Alternatively time (t) can be added as a third axis to the phase plane, so that solutions can be plotted parametrically as $(x(t), y(t), t)$ in three dimensions: solutions of the pendulum equation are shown in Fig. 1.32(c). This representation is particularly appropriate for the general phase plane (Chapter 2) and for forced systems.

If $f(x, \dot{x})$ is periodic in x, that is if there exists a number C such that $f(x + C, \dot{x}) = f(x, \dot{x})$ for all x, then phase paths on any x-interval of length C are repeated on any prior or succeeding intervals of length C. Hence solutions can be wrapped round a cylinder of circumference C. Figure 1.32(d) shows the phase paths in Fig. 1.32(a) plotted on to the cylinder, the x-axis now being wrapped round the cylinder.

Exercise 1.9
Sketch phase paths and solution of the damped oscillations
(i) $\ddot{x} + 2\dot{x} + 2x = 0$,   (ii) $\ddot{x} - 3\dot{x} + 2x = 0$,
as in Fig. 1.32(a)–(c).

Problems

1.1 Locate the equilibrium points and sketch the phase diagrams in their neighbourhood for the following equations:

(i) $\ddot{x} - k\dot{x} = 0$.

(ii) $\ddot{x} - 8x\dot{x} = 0$.

(iii) $\ddot{x} = k(|x| > 1), \quad \ddot{x} = 0(|x| < 1)$.

(iv) $\ddot{x} + 3\dot{x} + 2x = 0$.

(v) $\ddot{x} - 4\dot{x} + 40x = 0$.

(vi) $\ddot{x} + 3|\dot{x}| + 2x = 0$.

(vii) $\ddot{x} + k\,\mathrm{sgn}(\dot{x}) + c\,\mathrm{sgn}(x) = 0, c > k$. Show that the path starting at $(x_0, 0)$ reaches $((c-k)^2 x_0/(c+k)^2, 0)$ after one circuit of the origin. Deduce that the origin is a spiral point.

(viii) $\dot{x} + x\,\mathrm{sgn}(x) = 0$.

1.2 Sketch the phase diagram for the equation $\ddot{x} = -x - \alpha x^3$, considering all values of α. Check the stability of the equilibrium points by the method of Section 1.7.

1.3 A certain dynamical system is governed by the equation $\ddot{x} + \dot{x}^2 + x = 0$. Show that the origin is a centre in the phase plane, and that the open and closed paths are separated by the path $2y^2 = 1 - 2x$.

1.4 Sketch the phase diagrams for the equation $\ddot{x} + e^x = a$, for $a < 0$, $a = 0$, and $a > 0$.

1.5 Sketch the phase diagram for the equation $\ddot{x} - e^x = a$, for $a < 0$, $a = 0$, and $a > 0$.

1.6 The potential energy $V(x)$ of a conservative system is continuous, and is strictly increasing for $x < -1$, zero for $|x| \le 1$, and strictly decreasing for $x > 1$. Locate the equilibrium points and sketch the phase diagram for the system.

1.7 Figure 1.33 shows a pendulum striking an inclined wall. Sketch the phase diagram of this 'impact oscillator', for α positive and α negative, when (i) there is no loss of energy at impact, (ii) the magnitude of the velocity is halved on impact.

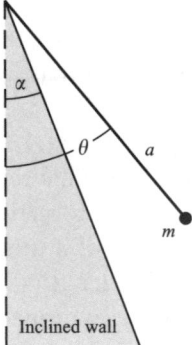

Figure 1.33 Impact pendulum.

1.8 Show that the time elapsed, T, along a phase path C of the system $\dot{x} = y, \dot{y} = f(x, y)$ is given, in a form alternative to (1.13), by

$$T = \int_C (y^2 + f^2)^{1/2}\,\mathrm{d}s,$$

where $\mathrm{d}s$ is an element of distance along C.

By writing $\delta s \simeq (y^2 + f^2)^{1/2} \delta t$, indicate, very roughly, equal time intervals along the phase paths of the system $\dot{x} = y$, $\dot{y} = 2x$.

1.9 On the phase diagram for the equation $\ddot{x} + x = 0$, the phase paths are circles. Use (1.13) in the form $\delta t \simeq \delta x / y$ to indicate, roughly, equal time steps along several phase paths.

1.10 Repeat Problem 1.9 for the equation $\ddot{x} + 9x = 0$, in which the phase paths are ellipses.

1.11 The pendulum equation, $\ddot{x} + \omega^2 \sin x = 0$, can be approximated for moderate amplitudes by the equation $\ddot{x} + \omega^2 (x - \frac{1}{6}x^3) = 0$. Sketch the phase diagram for the latter equation, and explain the differences between it and Fig. 1.2.

1.12 The displacement, x, of a spring-mounted mass under the action of Coulomb dry friction is assumed to satisfy $m\ddot{x} + cx = -F_0 \operatorname{sgn}(\dot{x})$, where m, c, and F_0 are positive constants (Section 1.6). The motion starts at $t = 0$, with $x = x_0 > 3F_0/c$ and $\ddot{x} = 0$. Subsequently, whenever $x = -\alpha$, where $2F_0/c - x_0 < -\alpha < 0$ and $\dot{x} > 0$, a trigger operates, to increase suddenly the forward velocity so that the kinetic energy increases by a constant amount E. Show that if $E > 8F_0^2/c$, a periodic motion exists, and show that the largest value of x in the periodic motion is equal to $F_0/c + E/4F_0$.

1.13 In Problem 1.12, suppose that the energy is increased by E at $x = -\alpha$ for both $\ddot{x} < 0$ and $\ddot{x} > 0$; that is, there are two injections of energy per cycle. Show that periodic motion is possible if $E > 6F_0^2/c$, and find the amplitude of the oscillation.

1.14 The 'friction pendulum' consists of a pendulum attached to a sleeve, which embraces a close-fitting cylinder (Fig. 1.34). The cylinder is turned at a constant rate Ω. The sleeve is subject to Coulomb dry friction through the couple $G = -F_0 \operatorname{sgn}(\dot{\theta} - \Omega)$. Write down the equation of motion, find the equilibrium states, and sketch the phase diagram.

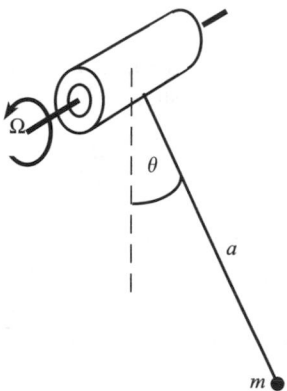

Figure 1.34   Friction-driven pendulum.

1.15 By plotting the 'potential energy' of the nonlinear conservative system $\ddot{x} = x^4 - x^2$, construct the phase diagram of the system. A particular path has the initial conditions $x = \frac{1}{2}, \dot{x} = 0$ at $t = 0$. Is the subsequent motion periodic?

1.16 The system $\ddot{x} + x = -F_0 \operatorname{sgn}(\dot{x})$, $F_0 > 0$, has the initial conditions $x = x_0 > 0, \dot{x} = 0$. Show that the phase path will spiral exactly n times before entering equilibrium (Section 1.6) if $(4n - 1)F_0 < x_0 < (4n + 1)F_0$.

1.17 A pendulum of length a has a bob of mass m which is subject to a horizontal force $m\omega^2 a \sin\theta$, where θ is the inclination to the downward vertical. Show that the equation of motion is $\ddot{\theta} = \omega^2 (\cos\theta - \lambda) \sin\theta$, where $\lambda = g/\omega^2 a$. Investigate the stability of the equilibrium states by the method of Section 1.7 for parameter-dependent systems. Sketch the phase diagrams for various λ.

1.18 Investigate the stability of the equilibrium points of the parameter-dependent system $\ddot{x} = (x - \lambda)(x^2 - \lambda)$.

1.19 If a bead slides on a smooth parabolic wire rotating with constant angular velocity ω about a vertical axis, then the distance x of the particle from the axis of rotation satisfies $(1 + x^2)\ddot{x} + (g - \omega^2 + \dot{x}^2)x = 0$. Analyse the motion of the bead in the phase plane.

1.20 A particle is attached to a fixed point O on a smooth horizontal plane by an elastic string. When unstretched, the length of the string is $2a$. The equation of motion of the particle, which is constrained to move on a straight line through 0, is

$$\ddot{x} = -x + a \operatorname{sgn}(x), \quad |x| > a \text{ (when the string is stretched)},$$

$$\ddot{x} = 0 \quad |x| \le a \text{ (when the string is slack)},$$

x being the displacement from 0. Find the equilibrium points and the equations of the phase paths, and sketch the phase diagram.

1.21 The equation of motion of a conservative system is $\ddot{x} + g(x) = 0$, where $g(0) = 0$, $g(x)$ is strictly increasing for all x, and

$$\int_0^x g(u)\, du \to \infty \quad \text{as } x \to \pm\infty. \tag{a}$$

Show that the motion is always periodic.

By considering $g(x) = xe^{-x^2}$, show that if (a) does not hold, the motions are not all necessarily periodic.

1.22 The wave function $u(x, t)$ satisfies the partial differential equation

$$\frac{\partial^2 u}{\partial x^2} + \alpha \frac{\partial u}{\partial x} + \beta u^3 + \gamma \frac{\partial u}{\partial t} = 0,$$

where α, β, and γ are positive constants. Show that there exist travelling wave solutions of the form $u(x, t) = U(x - ct)$ for any c, where $U(\zeta)$ satisfies

$$\frac{d^2 U}{d\zeta^2} + (\alpha - \gamma c)\frac{dU}{d\zeta} + \beta U^3 = 0.$$

Using Problem 1.21, show that when $c = \alpha/\gamma$, all such waves are periodic.

1.23 The linear oscillator $\ddot{x} + \dot{x} + x = 0$ is set in motion with initial conditions $x = 0$, $\dot{x} = v$, at $t = 0$. After the first and each subsequent cycle the kinetic energy is instantaneously increased by a constant, E, in such a manner as to increase $\dot{x}$. Show that if $E = \frac{1}{2}v^2(1 - e^{4\pi/\sqrt{3}})$, a periodic motion occurs. Find the maximum value of x in a cycle.

1.24 Show how phase paths of Problem 1.23 having arbitrary initial conditions spiral on to a limit cycle. Sketch the phase diagram.

1.25 The kinetic energy, $\mathcal{T}$, and the potential energy, $\mathcal{V}$, of a system with one degree of freedom are given by

$$\mathcal{T} = T_0(x) + \dot{x}T_1(x) + \dot{x}^2 T_2(x), \quad \mathcal{V} = \mathcal{V}(x).$$

Use Lagrange's equation

$$\frac{d}{dt}\left(\frac{\partial \mathcal{T}}{\partial \dot{x}}\right) - \frac{\partial \mathcal{T}}{\partial x} = -\frac{\partial \mathcal{V}}{\partial x}$$

to obtain the equation of motion of the system. Show that the equilibrium points are stationary points of $T_0(x) - \mathcal{V}(x)$, and that the phase paths are given by the energy equation

$$T_2(x)\dot{x}^2 - T_0(x) + \mathcal{V}(x) = \text{constant}.$$

1.26 Sketch the phase diagram for the equation $\ddot{x} = -f(x + \dot{x})$, where

$$f(u) = \begin{cases} f_0, & u \geq c, \\ f_0 u/c, & |u| \leq c, \\ -f_0, & u \leq -c, \end{cases}$$

where f_0, c are constants, $f_0 > 0$, and $c > 0$. How does the system behave as $c \to 0$?

1.27 Sketch the phase diagram for the equation $\ddot{x} = u$, where

$$u = -\text{sgn}\left(\sqrt{2}|x|^{1/2}\text{sgn}(x) + \dot{x}\right).$$

(u is an elementary control variable which can switch between $+1$ and -1. The curve $\sqrt{2}|x|^{1/2}\text{sgn}(x) + y = 0$ is called the switching curve.)

1.28 The relativistic equation for an oscillator is

$$\frac{\mathrm{d}}{\mathrm{d}t}\left\{\frac{m_0\dot{x}}{\sqrt{[1 - (\dot{x}/c)^2]}}\right\} + kx = 0, \quad |\dot{x}| < c$$

where m_0, c, and k are positive constants. Show that the phase paths are given by

$$\frac{m_0 c^2}{\sqrt{[1 - (y/c)^2]}} + \tfrac{1}{2}kx^2 = \text{constant}.$$

If $y = 0$ when $x = a$, show that the period, T, of an oscillation is given by

$$T = \frac{4}{c\sqrt{\varepsilon}}\int_0^a \frac{[1 + \varepsilon(a^2 - x^2)]\mathrm{d}x}{\sqrt{(a^2 - x^2)}\sqrt{[2 + \varepsilon(a^2 - x^2)]}}, \quad \varepsilon = \frac{k}{2m_0 c^2}.$$

The constant ε is small; by expanding the integrand in powers of ε show that

$$T \approx \frac{\pi\sqrt{2}}{c}\left(\frac{1}{\varepsilon^{1/2}} + \frac{3}{8}\varepsilon^{1/2}a^2\right).$$

1.29 A mass m is attached to the mid-point of an elastic string of length $2a$ and stiffness λ (Fig. 1.35). There is no gravity acting, and the tension is zero in the equilibrium position. Obtain the equation of motion for transverse oscillations and sketch the phase paths.

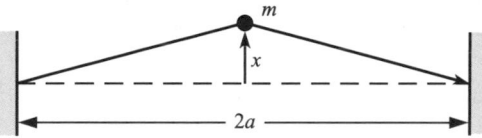

Figure 1.35   Transverse oscillations.

1.30 The system $\ddot{x} + x = F(v_0 - \dot{x})$ is subject to the friction law

$$F(u) = \begin{cases} 1, & u > \varepsilon, \\ u/\varepsilon, & -\varepsilon \leq u \leq \varepsilon, \\ -1, & u < -\varepsilon, \end{cases}$$

where $u = v_0 - \dot{x}$ is the slip velocity and $v_0 > \varepsilon > 0$. Find explicit equations for the phase paths in the $(x, y = \dot{x})$ plane. Compute a phase diagram for $\varepsilon = 0.2, v_0 = 1$ (say). Explain using the phase diagram that the equilibrium point at $(1,0)$ is a centre, and that all paths which start outside the circle $(x-1)^2 + y^2 = (v_0 - \varepsilon)^2$ eventually approach this circle.

1.31 The system $\ddot{x} + x = F(\dot{x})$ where

$$F(\dot{x}) = \begin{cases} k\dot{x} + 1, & \dot{x} < v_0, \\ 0, & \dot{x} = v_0, \\ -k\dot{x} - 1, & \dot{x} > v_0, \end{cases}$$

and $k > 0$, is a possible model for Coulomb dry friction with damping. If $k < 2$, show that the equilibrium point is an unstable spiral. Compute the phase paths for, say, $k = 0.5$, $v_0 = 1$. Using the phase diagram discuss the motion of the system, and describe the limit cycle.

1.32 A pendulum with a magnetic bob oscillates in a vertical plane over a magnet, which repels the bob according to the inverse square law, so that the equation of motion is (Fig. 1.36)

$$ma^2\ddot{\theta} = -mga \sin\theta + Fh \sin\phi,$$

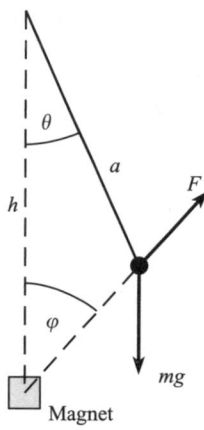

Figure 1.36 Magnetically repelled pendulum.

where $h > a$ and $F = c/(a^2 + h^2 - 2ah \cos\theta)$ and c is a constant. Find the equilibrium positions of the bob, and classify them as centres and saddle points according to the parameters of the problem. Describe the motion of the pendulum.

1.33 A pendulum with equation $\ddot{x} + \sin x = 0$ oscillates with amplitude a. Show that its period, T, is equal to $4K(\beta)$, where $\beta = \sin^2 \frac{1}{2}a$ and

$$K(\beta) = \int_0^{\pi/2} \frac{d\phi}{\sqrt{(1 - \beta \sin^2 \phi)}}.$$

The function $K(\beta)$ has the power series representation

$$K(\beta) = \frac{1}{2}\pi \left[1 + \left(\frac{1}{2}\right)^2 \beta + \left(\frac{1.3}{2.4}\right)^2 \beta^2 + \cdots \right], \quad |\beta| < 1.$$

Deduce that, for small amplitudes,

$$T = 2\pi \left(1 + \frac{1}{16}a^2 + \frac{11}{3072}a^4 \right) + O(a^6).$$

1.34 Repeat Problem 1.33 with the equation $\ddot{x} + x - \varepsilon x^3 = 0$ ($\varepsilon > 0$), and show that

$$T = \frac{4\sqrt{2}}{\sqrt{(2 - \varepsilon a^2)}} K(\beta), \quad \beta = \frac{\varepsilon a^2}{2 - \varepsilon a^2},$$

and that

$$T = 2\pi \left(1 + \tfrac{3}{8}\varepsilon a^2 + \tfrac{57}{256}\varepsilon^2 a^4 \right) + O(\varepsilon^3 a^6),$$

as $\varepsilon a^2 \to 0$.

1.35 Show the equation of the form $\ddot{x} + g(x)\dot{x}^2 + h(x) = 0$ are effectively conservative. (Find a transformation of x which puts the equations into the usual conservative form. Compare with eqn (1.59).)

1.36 Sketch the phase diagrams of the following: (i) $\dot{x} = y$, $\dot{y} = 0$, (ii) $\dot{x} = y$, $\dot{y} = 1$, (iii) $\dot{x} = y$, $\dot{y} = y$.

1.37 Show that the phase plane for the equation $\ddot{x} - \varepsilon x \dot{x} + x = 0$ has a centre at the origin, by finding the equation of the phase paths.

1.38 Show that the equation $\ddot{x} + x + \varepsilon x^3 = 0$ ($\varepsilon > 0$) with $x(0) = a, \dot{x}(0) = 0$ has phase paths given by

$$\dot{x}^2 + x^2 + \tfrac{1}{2}\varepsilon x^4 = \left(1 + \tfrac{1}{2}\varepsilon a^2 \right) a^2.$$

Show that the origin is a centre. Are all phase paths closed, and hence all solutions periodic?

1.39 Locate the equilibrium points of the equation $\ddot{x} + \lambda + x^3 - x = 0$, in the x, λ plane. Show that the phase paths are given by

$$\tfrac{1}{2}\dot{x}^2 + \lambda x + \tfrac{1}{4}\lambda x^4 - \tfrac{1}{2}x^2 = \text{constant}.$$

Investigate the stability of the equilibrium points.

1.40 Burgers' equation

$$\frac{\partial \phi}{\partial t} + \phi \frac{\partial \phi}{\partial x} = c \frac{\partial^2 \phi}{\partial x^2}$$

shows diffusion and nonlinear effects in fluid mechanics (see Logan (1994)). Find the equation for permanent waves by putting $\phi(x,t) = U(x - ct)$, where c is the constant wave speed. Find the equilibrium points and the phase paths for the resulting equation and interpret the phase diagram.

1.41 A uniform rod of mass m and length L is smoothly pivoted at one end and held in a vertical position of equilibrium by two unstretched horizontal springs, each of stiffness k, attached to the other end as shown in Fig. 1.37. The rod is free to oscillate in a vertical plane through the springs and the rod. Find the potential energy $V(\theta)$ of the system when the rod is inclined at an angle θ to the upward vertical. For

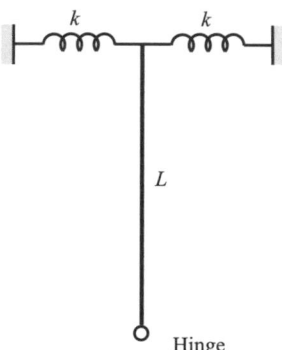

Figure 1.37 Spring restrained inverted pendulum.

small θ confirm that

$$\mathcal{V}(\theta) \approx (kL - \tfrac{1}{4}mg)L\theta^2.$$

Sketch the phase diagram for small $|\theta|$, and discuss the stability of this inverted pendulum.

1.42 Two stars, each with gravitational mass μ, are orbiting each other under their mutual gravitational forces in such a way that their orbits are circles of radius a. A satellite of relatively negligible mass is moving on a straight line through the mass centre G such that the line is perpendicular to the plane of the mutual orbits of this binary system. Explain why the satellite will continue to move on this line. If z is the displacement of the satellite from G, show that

$$\ddot{z} = -\frac{2\mu z}{(a^2 + z^2)^{3/2}}.$$

Obtain the equations of the phase paths. What type of equilibrium point is $z = 0$?

1.43 A long wire is bent into the shape of a smooth curve with equation $z = f(x)$ in a fixed vertical (x, z) plane (assume that $f'(x)$ and $f''(x)$ are continuous). A bead of mass m can slide on the wire: assume friction is negligible. Find the kinetic and potential energies of the bead, and write down the equation of the phase paths. Explain why the method of Section 1.3 concerning the phase diagrams for stationary values of the potential energy still holds.

1.44 In the previous problem suppose that friction between the bead and the wire is included. Assume linear damping in which motion is opposed by a frictional force proportional (factor k) to the velocity. Show that the equation of motion of the bead is given by

$$m(1 + f'(x)^2)\ddot{x} + mf''(x)\dot{x}^2 + k\dot{x}(1 + f'(x)^2) + mgf'(x) = 0,$$

where m is its mass.

   Suppose that the wire has the parabolic shape given by $z = x^2$ and that dimensions are chosen so that $k = m$ and $g = 1$. Compute the phase diagram in the neighbourhood of the origin, and explain general features of the diagram near and further away from the origin. (Further theory and experimental work on motion on tracks can be found in the book by Virgin (2000).)

2 Plane autonomous systems and linearization

Chapter 1 describes the application of phase-plane methods to the equation $\ddot{x} = f(x, \dot{x})$ through the equivalent first-order system $\dot{x} = y$, $\dot{y} = f(x, y)$. This approach permits a useful line of argument based on a mechanical interpretation of the original equation. Frequently, however, the appropriate formulation of mechanical, biological, and geometrical problems is not through a second-order equation at all, but directly as a more general type of first-order system of the form $\dot{x} = X(x, y)$, $\dot{y} = Y(x, y)$. The appearance of these equations is an invitation to construct a phase plane with x, y coordinates in which solutions are represented by curves $(x(t), y(t))$ where $x(t)$, $y(t)$ are the solutions. The constant solutions are represented by equilibrium points obtained by solving the equations $X(x, y) = 0$, $Y(x, y) = 0$, and these may now occur anywhere in the plane. Near the equilibrium points we may make a linear approximation to $X(x, y)$, $Y(x, y)$, solve the simpler equations obtained, and so determine the local character of the paths. This enables the stability of the equilibrium states to be settled and is a starting point for global investigations of the solutions.

2.1 The general phase plane

Consider the general autonomous first-order system

$$\dot{x} = X(x, y), \qquad \dot{y} = Y(x, y) \tag{2.1}$$

of which the type considered in Chapter 1,

$$\dot{x} = y, \qquad \dot{y} = f(x, y),$$

is a special case. Assume that the functions $X(x, y)$ and $Y(x, y)$ are smooth enough to make the system **regular** (see Appendix A) in the region of interest. As in Section 1.2, the system is called **autonomous** because the time variable t does not appear in the right-hand side of (2.1). We shall give examples later of how such systems arise.

The solutions $x(t)$, $y(t)$ of (2.1) may be represented on a plane with Cartesian axes x, y. Then as t increases $(x(t), y(t))$ traces out a directed curve in the plane called a **phase path**.

The appropriate form for the initial conditions of (2.1) is

$$x = x_0, \quad y = y_0 \qquad \text{at } t = t_0 \tag{2.2}$$

where x_0 and y_0 are the **initial values** at time t_0; by the existence and uniqueness theorem (Appendix A) there is one and only one solution satisfying this condition when (x_0, y_0) is an 'ordinary point'. This does not at once mean that there is one and only one phase path through the point (x_0, y_0) on the phase diagram, because this same point could serve as the initial conditions for other starting times. Therefore it might seem that other phase paths through the same point could result: the phase diagram would then be a tangle of criss-crossed curves. We may see that this is not so by forming the differential equation for the phase paths. Since $\dot{y}/\dot{x} = dy/dx$ on a path the required equation is

$$\frac{dy}{dx} = \frac{Y(x, y)}{X(x, y)}. \tag{2.3}$$

Equation (2.3) does not give any indication of the **direction** to be associated with a phase path for increasing t. This must be settled by reference to eqns (2.1). The signs of X and Y at any particular point determine the direction through the point, and generally the directions at all other points can be settled by the requirement of *continuity of direction* of adjacent paths.

The diagram depicting the phase paths is called the **phase diagram**. A typical point (x, y) is called a **state** of the system, as before. The phase diagram shows the evolution of the states of the system, starting from arbitrary initial states.

Points where the right-hand side of (2.3) satisfy the conditions for regularity (Appendix A) are called the **ordinary points** of (2.3). There is one and only one phase path through any ordinary point (x_0, y_0), no matter at what time t_0 the point (x_0, y_0) is encountered. Therefore infinitely many solutions of (2.1), differing only by time displacements, produce the same phase path.

However, eqn (2.3) may have singular points where the conditions for regularity do not hold, even though the time equations (2.1) have no peculiarity there. Such singular points occur where $X(x, y) = 0$. Points where $X(x, y)$ and $Y(x, y)$ are both zero,

$$X(x, y) = 0, \qquad Y(x, y) = 0 \tag{2.4}$$

are called **equilibrium points**. If x_1, y_1 is a solution of (2.4), then

$$x(t) = x_1, \qquad y(t) = y_1$$

are **constant solutions** of (2.1), and are degenerate phase paths. The term **fixed point** is also used.

Since $dy/dx = Y(x, y)/X(x, y)$ is the differential equation of the phase paths, phase paths which cut the curve defined by the equation $Y(x, y) = cX(x, y)$ will do so with the same slope c: such curves are known as **isoclines**. The two particular isoclines $Y(x, y) = 0$, which paths cut with zero slope, and $X(x, y) = 0$, which paths cut with infinite slope, are helpful in phase diagram sketching. The points where these isoclines intersect define the equilibrium points. Between the isoclines, $X(x, y)$ and $Y(x, y)$ must each be of one sign. For example, in a region in the (x, y) plane in which $X(x, y) > 0$ and $Y(x, y) > 0$, the phase paths must have positive slopes. This will also be the case if $X(x, y) < 0$ and $Y(x, y) < 0$. Similarly, if $X(x, y)$ and $Y(x, y)$ have opposite signs in a region, then the phase paths must have negative slopes.

Example 2.1   *Locate the equilibrium points, and sketch the phase paths of*

$$\dot{x} = y(1 - x^2), \qquad \dot{y} = -x(1 - y^2).$$

The equilibrium points occur at the simultaneous solutions of

$$y(1 - x^2) = 0, \qquad x(1 - y^2) = 0.$$

The solutions of these equations are, respectively, $x = \pm 1$, $y = 0$ and $x = 0$, $y = \pm 1$, so that there are five solution pairs, $(0, 0), (1, 1), (1, -1), (-1, 1)$ and $(-1, -1)$ which are equilibrium points.

The phase paths satisfy the differential equation

$$\frac{dy}{dx} = -\frac{x(1 - y^2)}{y(1 - x^2)},$$

which is a first-order separable equation. Hence

$$-\int \frac{x\,dx}{1 - x^2} = \int \frac{y\,dy}{1 - y^2},$$

so that

$$\tfrac{1}{2} \ln |1 - x^2| = -\tfrac{1}{2} \ln |1 - y^2| + C,$$

which is equivalent to

$$(1 - x^2)(1 - y^2) = A, \text{a constant,}$$

(the modulus signs are no longer necessary). Notice that there are special solutions along the lines $x = \pm 1$ and $y = \pm 1$ where $A = 0$. These solutions and the locations of the equilibrium points help us to plot the phase diagram, which is shown in Fig. 2.1. Paths cross the axis $x = 0$ with zero slope, and paths cross $y = 0$ with

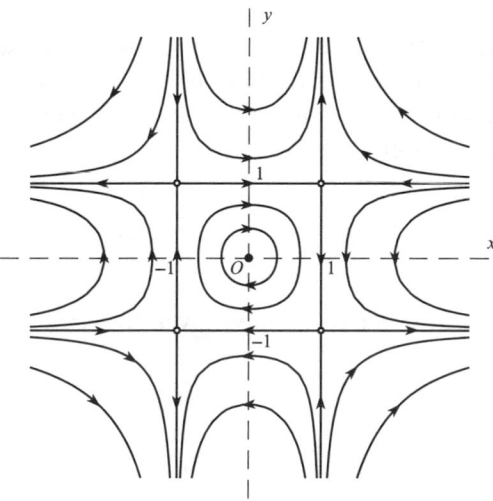

Figure 2.1   Phase diagram for $\dot{x} = y(1 - x^2)$; $\dot{y} = -x(1 - y^2)$; the dashed lines are isoclines of zero slope and infinite slope.

infinite slope. The **directions** of the paths may be found by continuity, starting at the point $(0, 1)$, say, where $\dot{x} > 0$, and the phase path therefore runs from left to right. ●

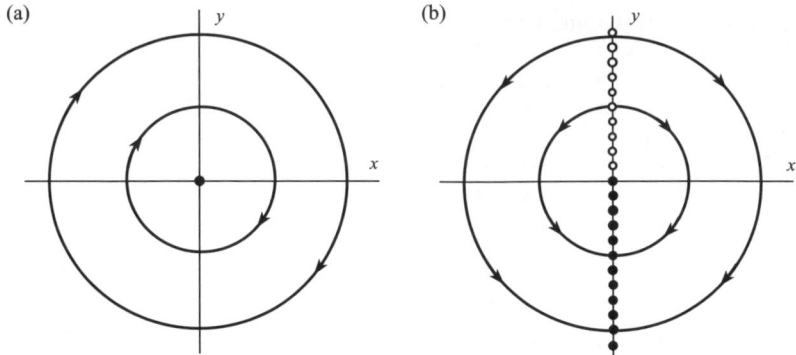

Figure 2.2   Phase diagrams for (a) $\dot{x} = y, \dot{y} = -x$; (b) $\dot{x} = xy, \dot{y} = -x^2$.

Example 2.2   *Compare the phase diagrams of the systems*
(a) $\dot{x} = y, \dot{y} = -x$; (b) $\dot{x} = xy, \dot{y} = -x^2$.
The equation for the paths is the same for both, namely

$$\frac{dy}{dx} = -\frac{x}{y}$$

(strictly, for $y \neq 0$, and for $x \neq 0$ in the second case), giving a family of circles in both cases (Fig. 2.2). However, in case (a) there is an equilibrium point only at the origin, but in case (b) every point on the y axis is an equilibrium point. The directions, too, are different. By considering the signs of $\dot{x}, \dot{y}$ in the various quadrants the phase diagram of Fig. 2.2(b) is produced. ●

A second-order differential equation can be reduced to the general form (2.1) in an arbitrary number of ways, and this occasionally has advantages akin to changing the variable to simplify a differential equation. For example the straightforward reduction $\dot{x} = y$ applied to $x\ddot{x} - \dot{x}^2 - x^3 = 0$ leads to the system

$$\dot{x} = y, \qquad \dot{y} = \frac{y^2}{x} + x^2. \tag{2.5}$$

Suppose, instead of y, we use another variable y_1 given by $y_1(t) = \dot{x}(t)/x(t)$, so that

$$\dot{x} = xy_1. \tag{2.6}$$

Then from (2.6) $\ddot{x} = x\dot{y}_1 + \dot{x}y_1 = x\dot{y}_1 + xy_1^2$ (using (2.6) again). But from the differential equation, $\ddot{x} = (\dot{x}^2/x) + x^2 = xy_1^2 + x^2$. Therefore,

$$\dot{y}_1 = x. \tag{2.7}$$

The pair of eqns (2.6) and (2.7) provide a representation alternative to (2.5). The phase diagram using x, y_1 will, of course, be different in appearance from the x, y diagram.

Returning to the general equation (2.1), the time T elapsing along a segment C of a phase path connecting two states (compare Fig. 1.4(a) and eqn (1.13)) is given by

$$T = \int_C dt = \int_C \left(\frac{dx}{dt}\right)^{-1} \left(\frac{dx}{dt}\right) dt = \int_C \frac{dx}{X(x,y)}. \tag{2.8}$$

Alternatively, let ds be a length element of C. Then $ds^2 = dx^2 + dy^2$ on the path, and

$$T = \int_C \left(\frac{ds}{dt}\right)^{-1} \left(\frac{ds}{dt}\right) dt = \int_C \frac{ds}{(X^2 + Y^2)^{1/2}}. \tag{2.9}$$

The integrals above depend only on X and Y and the geometry of the phase path; therefore the time scale is implicit in the phase diagram. **Closed paths** represent periodic solutions.

Exercise 2.1
Locate all equilibrium points of $\dot{x} = \cos y$, $\dot{y} = \sin x$. Find the equation of all phase paths.

2.2 Some population models

In the following examples systems of the type (2.1) arise naturally. Further examples from biology can be found in Pielou (1969), Rosen (1973) and Strogatz (1994).

Example 2.3   *A predator–prey problem* (Volterra's model)
In a lake there are two species of fish: A, which lives on plants of which there is a plentiful supply, and B (the predator) which subsists by eating A (the prey). We shall construct a crude model for the interaction of A and B.

Let $x(t)$ be the population of A and $y(t)$ that of B. We assume that A is relatively long-lived and rapidly breeding if left alone. Then in time δt there is a population increase given by

$$ax\delta t, \quad a > 0$$

due to births and 'natural' deaths, and 'negative increase'

$$-cxy\delta t, \quad c > 0$$

owing to A's being eaten by B (the number being eaten in this time being assumed proportional to the number of encounters between A and B). The net population increase of A, δx, is given by

$$\delta x = ax\delta t - cxy\delta t,$$

so that in the limit $\delta t \to 0$

$$\dot{x} = ax - cxy. \tag{2.10}$$

Assume that, in the absence of prey, the starvation rate of B predominates over the birth rate, but that the compensating growth of B is again proportional to the number of encounters with A. This gives

$$\dot{y} = -by + xyd \tag{2.11}$$

with $b > 0, d > 0$. Equations (2.10) and (2.11) are a pair of simultaneous nonlinear equations of the form (2.1).

We now plot the phase diagram in the x, y plane. Only the quadrant

$$x \geq 0, \qquad y \geq 0$$

is of interest. The equilibrium points are where

$$X(x, y) \equiv ax - cxy = 0, \qquad Y(x, y) \equiv -by + xyd = 0;$$

that is at $(0, 0)$ and $(b/d, a/c)$. The phase paths are given by $\mathrm{d}y/\mathrm{d}x = Y/X$, or

$$\frac{\mathrm{d}y}{\mathrm{d}x} = \frac{(-b + xd)y}{(a - cy)x},$$

which is a separable equation leading to

$$\int \frac{(a - cy)}{y} \, \mathrm{d}y = \int \frac{(-b + xd)}{x} \, \mathrm{d}x.$$

or

$$a \log_e y + b \log_e x - cy - xd = C, \qquad (2.12)$$

where C is an arbitrary constant, the parameter of the family. Writing (2.12) in the form $(a \log_e y - cy) + (b \log_e x - xd) = C$, the result of Problem 2.25 shows that this is a system of closed curves centred on the equilibrium point $(b/d, a/c)$.

Figure 2.3 shows the phase paths for a particular case. The direction on the paths can be obtained from the sign of $\dot{x}$ at a single point, even on $y = 0$. This determines the directions at all points by continuity. From (2.11) and (2.10) the isoclines of zero slope occur on $\dot{y} = 0$, that is, on the lines $y = 0$ and $y = xd/b$, and those of infinite slope on $\dot{x} = 0$, that is, on the lines $x = 0$ and $y = ax/c$.

Since the paths are closed, the fluctuations of $x(t)$ and $y(t)$, starting from any initial population, are periodic, the maximum population of A being about a quarter of a period behind the maximum population of B. As A gets eaten, causing B to thrive, the population x of A is reduced, causing eventually a drop in that of B. The shortage of predators then leads to a resurgence of A and the cycle starts again. A sudden change in state

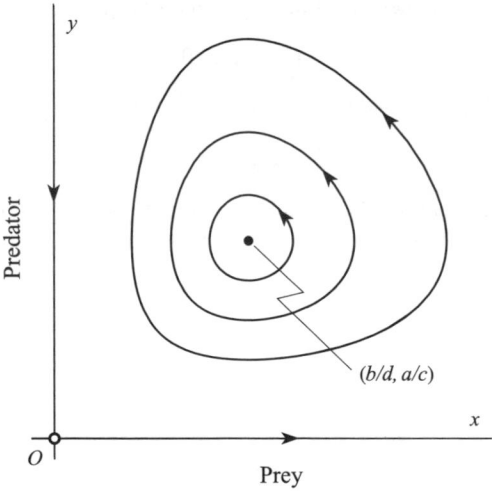

Figure 2.3   Typical phase diagram for the predator–prey model.

due to external causes, such as a bad season for the plants, puts the state on to another closed curve, but no tendency to an equilibrium population, nor for the population to disappear, is predicted. If we expect such a tendency, then we must construct a different model (see Problems 2.12 and 2.13).   ●

Example 2.4 *A general epidemic model*
Consider the spread of a non-fatal disease in a population which is assumed to have constant size over the period of the epidemic. At time t suppose the population consists of

$x(t)$ *susceptibles*: those so far uninfected and therefore liable to infection;

$y(t)$ *infectives*: those who have the disease and are still at large;

$z(t)$ who are isolated, or who have recovered and are therefore immune.

Assume there is a steady contact rate between susceptibles and infectives and that a constant proportion of these contacts result in transmission. Then in time δt, δx of the susceptibles become infective, where

$$\delta x = -\beta x y \delta t,$$

and β is a positive constant.

If $\gamma > 0$ is the rate at which current infectives become isolated, then

$$\delta y = \beta x y \delta t - \gamma y \delta t.$$

The number of new isolates δz is given by

$$\delta z = \gamma y \delta t.$$

Now let $\delta t \to 0$. Then the system

$$\dot{x} = -\beta x y, \qquad \dot{y} = \beta x y - \gamma y, \qquad \dot{z} = \gamma y, \qquad (2.13)$$

with suitable initial conditions, determines the progress of the disease. Note that the result of adding the equations is

$$\frac{\mathrm{d}}{\mathrm{d}t}(x + y + z) = 0;$$

that is to say, the assumption of a constant population is built into the model. x and y are defined by the first two equations in (2.13). With the restriction $x \geq 0$, $y \geq 0$, equilibrium occurs for $y = 0$ (all $x \geq 0$). The analysis of this problem in the phase plane is left as an exercise (Problem 2.29).   ●

We shall instead look in detail at a more complicated epidemic:

Example 2.5 *Recurrent epidemic*
Suppose that the problem is as before, except that the stock of susceptibles $x(t)$ is being added to at a constant rate μ per unit time. This condition could be the result of fresh births in the presence of a childhood disease such as measles in the absence of vaccination. In order to balance the population in the simplest way we shall assume that deaths occur naturally and only among the immune, that is, among the $z(t)$ older people most of whom have had the disease. For a constant population the equations become

$$\dot{x} = -\beta x y + \mu, \qquad (2.14)$$
$$\dot{y} = \beta x y - \gamma y, \qquad (2.15)$$
$$\dot{z} = \gamma y - \mu \qquad (2.16)$$

(note that $(\mathrm{d}/\mathrm{d}t)(x + y + z) = 0$: the population size is steady).

Consider the variation of x and y, the active participants, represented on the x, y phase plane. We need only (2.14) and (2.15), which show an equilibrium point $(\gamma/\beta, \mu/\gamma)$.

Instead of trying to solve the equation for the phase paths we shall try to get an idea of what the phase diagram is like by forming linear approximations to the right-hand sides of (2.14), (2.15) in the neighbourhood of the equilibrium point. Near the equilibrium point we write

$$x = \gamma/\beta + \xi, \qquad y = \mu/\gamma + \eta$$

(ξ, η small) so that $\dot{x} = \dot{\xi}$ and $\dot{y} = \dot{\eta}$. Retaining only the *linear terms* in the expansion of the right sides of (2.14), (2.15), we obtain

$$\dot{\xi} = -\frac{\beta\mu}{\gamma}\xi - \gamma\eta, \qquad (2.17)$$

$$\dot{\eta} = \frac{\beta\mu}{\gamma}\xi. \qquad (2.18)$$

We are said to have **linearized** (2.14) and (2.15) near the equilibrium point. Elimination of ξ gives

$$\gamma\ddot{\eta} + (\beta\mu)\dot{\eta} + (\beta\mu\gamma)\eta = 0. \qquad (2.19)$$

This is the equation for the damped linear oscillator (Section 1.4), and we may compare (2.19) with eqn (1.36) of Chapter 1, but it is necessary to remember that eqn (2.19) only holds as an approximation close to the equilibrium point of (2.14) and (2.15). When the 'damping' is light ($\beta\mu/\gamma^2 < 4$) the phase path is a spiral. Figure 2.4 shows some phase paths for a particular case. All starting conditions lead to the stable equilibrium point E: this point is called the **endemic state** for the disease.   ●

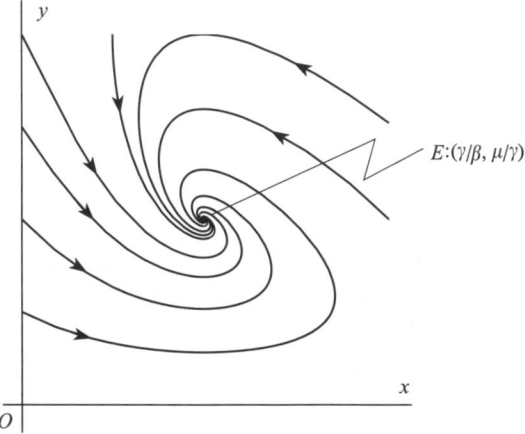

Figure 2.4   Typical spiral phase paths for the recurrent epidemic.

Exercise 2.2
The populations x and y of two species satisfy the equations

$$\dot{x} = x(3 - x - 2y), \quad \dot{y} = y(3 - 2x - y), \quad (x, y \geq 0),$$

(after scaling). Find the equilibrium points of the system. Confirm that $y = x$ is a phase path. Sketch the phase diagram. What happens to the species in the cases with initial populations (a) $x = 10, y = 9$, (b) $x = 9, y = 10$?

2.3   Linear approximation at equilibrium points

Approximation to a nonlinear system by linearizing it at near equilibrium point, as in the last example, is a most important and generally useful technique. If the geometrical nature of the equilibrium points can be settled in this way the broad character of the phase diagram often becomes clear. Consider the system

$$\dot{x} = X(x, y), \qquad \dot{y} = Y(x, y). \tag{2.20}$$

Suppose that the *equilibrium point to be studied has been moved to the origin* by a translation of axes, if necessary, so that

$$X(0, 0) = Y(0, 0) = 0.$$

We can therefore write, by a Taylor expansion,

$$X(x, y) = ax + by + P(x, y), \qquad Y(x, y) = cx + dy + Q(x, y),$$

where

$$a = \frac{\partial X}{\partial x}(0, 0), \quad b = \frac{\partial X}{\partial y}(0, 0), \quad c = \frac{\partial Y}{\partial x}(0, 0), \quad d = \frac{\partial Y}{\partial y}(0, 0) \tag{2.21}$$

and $P(x, y), Q(x, y)$ are of lower order of magnitude than the linear terms as (x, y) approaches the origin $(0, 0)$. The **linear approximation** to (2.21) in the neighbourhood at the origin is defined as the system

$$\dot{x} = ax + by, \qquad \dot{y} = cx + dy. \tag{2.22}$$

We expect that the solutions of (2.22) will be geometrically similar to those of (2.20) near the origin, an expectation fulfilled in most cases (but see Problem 2.7: a centre may be an exception).

Over the next two sections we shall show how simple relations between the coefficients a, b, c, d enable the equilibrium point of the system (2.22) to be classified.

Exercise 2.3
Find the linear approximations of

$$\dot{x} = \sin x + 2y, \qquad \dot{y} = xy + 3ye^x + x$$

near the origin.

2.4   The general solution of linear autonomous plane systems

The following two examples illustrate techniques for solving the system of linear differential equations with constant coefficients:

$$\dot{x} = ax + by, \qquad \dot{y} = cx + dy \tag{2.23}$$

for $x(t), y(t)$. They are followed by a more general treatment.

Example 2.6 *Solve the differential equations*

$$\dot{x} = x - 2y, \qquad \dot{y} = -3x + 2y.$$

Look for two **linearly independent solutions** (that is, solutions which are not simply constant multiples of each other) each of which takes the form of the pair

$$x = re^{\lambda t}, \qquad y = se^{\lambda t} \tag{i}$$

where r, s, λ are certain constants. Suppose that the two solutions are $(x_1(t), y_1(t))$ and $(x_2(t), y_2(t))$; then the general solution will be given by

$$x(t) = C_1 x_1(t) + C_2 x_2(t), \qquad y(t) = C_1 y_1(t) + C_2 y_2(t). \tag{ii}$$

To obtain the basic solutions, substitute (i) into the given differential equations. After cancelling the common factor $e^{\lambda t}$ and rearranging the terms, we obtain the pair of algebraic equations for the three unknowns λ, r, and s:

$$(1 - \lambda)r - 2s = 0, \qquad -3r + (2 - \lambda)s = 0. \tag{iii}$$

Regarding these as linear simultaneous equations for r and s, it is known that the determinant of the coefficients must be zero, or else the only solution is $r = 0, s = 0$. Therefore we require

$$\det \begin{bmatrix} 1 - \lambda & -2 \\ -3 & 2 - \lambda \end{bmatrix} = \lambda^2 - 3\lambda - 4 = (\lambda - 4)(\lambda + 1) = 0.$$

There are two permissible values of λ:

$$\lambda = \lambda_1 = 4 \quad \text{and} \quad \lambda = \lambda_2 = -1. \tag{iv}$$

For each of these values in turn we have to solve (iii) for r and s:
    The case $\lambda = \lambda_1 = 4$. Equations (iii) become

$$-3r - 2s = 0, \qquad -3r - 2s = 0. \tag{iv}$$

The equations are identical in this case; and in *every* case it turns out that one is simply a multiple of the other, so that essentially the two equations become a single equation. Choose *any* nonzero solution of (iv), say

$$r = r_1 = -2, \qquad s = s_1 = 3,$$

and we have found a solution having the form (i):

$$x(t) = x_1(t) = -2e^{4t}, \qquad y(t) = y_1(t) = 3e^{4t}. \tag{v}$$

The case $\lambda = \lambda_2 = -1$. Equations (iii) become

$$2r - 2s = 0, \qquad -3r + 3s = 0. \tag{vi}$$

These are both equivalent to $r - s = 0$. We take the simplest solution

$$r = r_2 = 1, \qquad s = s_2 = 1.$$

We now have as the second, independent solution of the differential equations:

$$x(t) = x_2(t) = e^{-t}, \qquad y(t) = y_2(t) = e^{-1}. \tag{vii}$$

Finally, from (ii), (v), and (vii), the general solution is given by

$$x(t) = -2C_1 e^{4t} + C_2 e^{-t}, \qquad y(t) = 3C_1 e^{4t} + C_2 e^{-t},$$

where C_1 and C_2 are arbitrary constants.   ●

In the following example the exponents λ_1 and λ_2 are complex numbers.

Example 2.7 *Obtain the general solution of the system*

$$\dot{x} = x + y, \qquad \dot{y} = -5x - 3y.$$

Proceed exactly as in Example 2.6. Substitute

$$x = re^{\lambda t}, \qquad y = se^{\lambda t}$$

into the differential equations specified, obtaining

$$(1 - \lambda)r + s = 0, \qquad -5r - (3 + \lambda)s = 0. \tag{i}$$

There exist nonzero solutions (r, s) only if

$$\det \begin{bmatrix} 1 - \lambda & 1 \\ -5 & -3 - \lambda \end{bmatrix} = \lambda^2 + 2\lambda + 2 = 0. \tag{ii}$$

The permitted values of λ are therefore

$$\lambda_1 = -1 + i, \qquad \lambda_2 = -1 - i. \tag{iii}$$

These are complex numbers, and since (ii) is a quadratic equation, they are complex conjugate:

$$\lambda_2 = \bar{\lambda}_1. \tag{iv}$$

The case $\lambda = \lambda_1 = -1 + i$. Equations (i) become

$$(2 - i)r + s = 0, \qquad -5r - (2 + i)s = 0;$$

(as always, these equations are actually multiples of each other). A particular solution of the differential equations associated with these is

$$x_1(t) = e^{(-1+i)t}, \qquad y_1(t) = (-2 + i)e^{(-1+i)t}. \tag{v}$$

The case $\lambda = \lambda_2 = \bar{\lambda}_1$. There is no need to rework the equations for r and s: eqn (i) shows that since $\lambda_2 = \bar{\lambda}$, we may take

$$r_2 = \bar{r}_1 = 1, \qquad s_2 = \bar{s}_1 = -2 - i.$$

The corresponding solution of the differential equations is

$$x_2(t) = e^{(-1-i)t}, \qquad y_2(t) = (-2 - i)e^{(-1-i)t},$$

(which are the complex conjugates of $x_1(t), y_1(t)$). The general solution of the system is therefore

$$x(t) = C_1 e^{(-1+i)t} + C_2 e^{(-1-i)t},$$

$$y(t) = C_1(-2+i)e^{(-1+i)t} + C_2(-2-i)e^{(-1-i)t}.$$ (iv)

If we allow C_1 and C_2 to be arbitrary *complex* numbers, then (vi) gives us all the real and complex solutions of the equations. We are interested only in real solutions, but we must be sure that we extract all of them from (vi). This is done by allowing C_1 to be arbitrary and complex, and requiring that

$$C_2 = \bar{C}_1.$$

The terms on the right of eqn (vi) are then complex conjugate, so their sums are real; we obtain

$$x(t) = 2\text{Re}\{C_1 e^{(-1+i)t}\},$$

$$y(t) = 2\text{Re}\{C_1(-2+i)e^{(-1+i)t}\},$$

By putting

$$2C_1 = c_1 + ic_2,$$

where c_1 and c_2 are *real* arbitrary constants, we obtain the general real solution in the form

$$x(t) = e^{-t}(c_1 \cos t - c_2 \sin t),$$

$$y(t) = -e^{-t}\{(2c_1 + c_2)\cos t + (c_1 - 2c_2)\sin t\}. \qquad \bullet$$

The general linear autonomous case is more manageable (particularly for higher order systems) when the algebra is expressed in matrix form. Define the column vectors

$$\boldsymbol{x}(t) = \begin{bmatrix} x(t) \\ y(t) \end{bmatrix}, \qquad \dot{\boldsymbol{x}}(t) = \begin{bmatrix} \dot{x}(t) \\ \dot{y}(t) \end{bmatrix}.$$

The system to be solved is

$$\dot{x} = ax + by, \qquad \dot{y} = cx + dy,$$

which may be written as

$$\dot{\boldsymbol{x}} = \boldsymbol{A}\boldsymbol{x} \quad \text{with } \boldsymbol{A} = \begin{bmatrix} a & b \\ c & d \end{bmatrix}. \qquad (2.24)$$

We shall only consider cases where there is a single equilibrium point, at the origin, the condition for this being

$$\det \boldsymbol{A} = ad - bc \neq 0. \qquad (2.25)$$

(If $\det \boldsymbol{A} = 0$, then one of its rows is a multiple of the other, so that $ax + by = 0$ (or $cx + dy = 0$) consists of a line of equilibrium points.)

We seek a **fundamental solution** consisting of two linearly independent solutions of (2.24), having the form

$$x_1(t) = v_1 e^{\lambda_1 t}, \qquad x_2(t) = v_2 e^{\lambda_2 t}, \tag{2.26}$$

where λ_1, λ_2 are constants, and v_1, v_2 are constant vectors. It is known (see Chapter 8) that the general solution is given by

$$x(t) = C_1 x_1(t) + C_2 x_2(t), \tag{2.27}$$

where C_1 and C_2 are arbitrary constants.

To determine $\lambda_1, v_1, \lambda_2, v_2$ in (2.26) substitute

$$x(t) = v e^{\lambda t} \tag{2.28}$$

into the system equations (2.24). After cancelling the common factor $e^{\lambda t}$, we obtain

$$\lambda v = A v,$$

or

$$(A - \lambda I) v = 0, \tag{2.29}$$

where I is the identity matrix. If we put

$$v = \begin{bmatrix} r \\ s \end{bmatrix}, \tag{2.30}$$

eqn (2.29) represents the pair of scalar equations

$$(a - \lambda) r + b s = 0, \qquad c r + (d - \lambda) s = 0, \tag{2.31}$$

for λ, r, s.

It is known from algebraic theory that eqn (2.29) has nonzero solutions for v only if the determinant of the matrix of the coefficients in eqns (2.31) is zero. Therefore

$$\det \begin{bmatrix} a - \lambda & b \\ c & d - \lambda \end{bmatrix} = 0, \tag{2.32}$$

or

$$\lambda^2 - (a + d)\lambda + (ad - bc) = 0. \tag{2.33}$$

This is called the **characteristic equation**, and its solutions, λ_1 and λ_2, the **eigenvalues** of the matrix A, or the **characteristic exponents** for the problem. For the purpose of classifying the solutions of the characteristic equation (2.33) it is convenient to use the following notations:

$$\lambda^2 - p\lambda + q = 0,$$

where

$$p = a + d, \qquad q = ad - bc. \tag{2.34}$$

Also put

$$\Delta = p^2 - 4q. \tag{2.35}$$

The eigenvalues $\lambda = \lambda_1$ and $\lambda = \lambda_2$ are given by

$$\lambda_1, \lambda_2 = \tfrac{1}{2}(p \pm \Delta^{1/2}). \tag{2.36}$$

These are to be substituted successively into (2.31) to obtain corresponding values for the constants r and s.

There are two main classes to be considered; when the eigenvalues are real and different, and when they are complex. These cases are distinguished by the sign of the **discriminant** Δ (we shall not consider the special case when $\Delta = 0$). If $\Delta > 0$ the eigenvalues are real, and if $\Delta < 0$ they are complex. We assume also that $q \neq 0$ (see (2.25)).

Time solutions when $\Delta > 0, q \neq 0$

In this case λ_1 and λ_2 are real and distinct. When $\lambda = \lambda_1$ eqns (2.31) for r and s become

$$(a - \lambda_1)r + bs = 0, \qquad cr + (d - \lambda_1)s = 0. \tag{2.37}$$

Since the determinant (2.32) is zero, its rows are linearly dependent. Therefore one of these eqns (2.37) is simply a multiple of the other; effectively we have only one equation connecting r and s. Let $r = r_1$, $s = s_1$ be any (nonzero) solution of (2.37), and put (in line with (2.30))

$$v_1 = \begin{bmatrix} r_1 \\ s_1 \end{bmatrix} \neq 0. \tag{2.38}$$

This is called an **eigenvector** of A corresponding to the eigenvalue λ_1. We have then obtained one of the two basic time solutions having form (2.26).

This process is repeated starting with $\lambda = \lambda_2$, giving rise to an eigenvector

$$v_2 = \begin{bmatrix} r_2 \\ s_2 \end{bmatrix}.$$

The general solution is then given by (2.27):

$$x(t) = C_1 v_1 e^{\lambda_1 t} + C_2 v_2 e^{\lambda_2 t} \tag{2.39}$$

in vector form, where C_1 and C_2 are arbitrary constants.

Time solutions when $\Delta < 0, q \neq 0$

In this case λ_1 and λ_2, obtained from (2.36), are complex, given by

$$\begin{aligned} \lambda_1 &= \tfrac{1}{2}\{p + i(-\Delta)^{1/2}\} = \alpha + i\beta, \\ \lambda_2 &= \tfrac{1}{2}\{p - i(-\Delta)^{1/2}\} = \alpha - i\beta, \end{aligned} \tag{2.40}$$

where $\alpha = \tfrac{1}{2}p$ and $\beta = \tfrac{1}{2}(-\Delta)^{1/2}$ are real numbers. Therefore λ_1 and λ_2 are complex conjugates.

Obtain an eigenvector corresponding to λ_1 from (2.31),

$$v = v_1 = \begin{bmatrix} r_1 \\ s_1 \end{bmatrix}, \tag{2.41}$$

exactly as before, where r_1 and s_1 are now complex. Since a, b, c, d are all real numbers, a suitable eigenvector corresponding to $\lambda_2 (= \bar{\lambda}_1)$ is given by taking $r_2 = \bar{r}_1, s_2 = \bar{s}_1$ as solutions of (2.31):

$$v_2 = \bar{v}_1 = \begin{bmatrix} \bar{r}_1 \\ \bar{s}_1 \end{bmatrix}.$$

Therefore, two basic **complex** time solutions taking the form (2.26) are

$$v e^{(\alpha+i\beta)t}, \qquad \bar{v} e^{(\alpha-i\beta)t},$$

where v is given by (2.41). The general (complex) solution of (2.24) is therefore

$$x(t) = C_1 v e^{(\alpha+i\beta)t} + C_2 \bar{v} e^{(\alpha-i\beta)t}, \tag{2.42}$$

where C_1 and C_2 are arbitrary constants which are in general complex.

We are interested only in *real solutions*. These are included among those in (2.42); the expression is real if and only if

$$C_2 = \bar{C}_1,$$

in which case the second term is the conjugate of the first, and we obtain

$$x(t) = 2\mathrm{Re}\{C_1 v e^{(\alpha+i\beta)t}\},$$

or

$$x(t) = \mathrm{Re}\{C v e^{(\alpha+i\beta)t}\}, \tag{2.43}$$

where $C(=2C_1)$ is an arbitrary complex constant.

Exercise 2.4
Find the eigenvalues, eigenvectors and general solution of

$$\dot{x} = -4x + y, \qquad \dot{y} = -2x - y$$

2.5   The phase paths of linear autonomous plane systems

For the system

$$\dot{x} = ax + by, \qquad \dot{y} = cx + dy, \tag{2.44}$$

the general character of the phase paths can be obtained from the time solutions (2.39) and (2.43). It might be thought that an easier approach would be to obtain the phase paths directly by solving their differential equation

$$\frac{dy}{dx} = \frac{cx + dy}{ax + by},$$

but the standard methods for solving these equations lead to implicit relations between x and y which are difficult to interpret geometrically. If eqn (2.44) is a linear approximation near the origin then the following phase diagrams will generally approximate to the phase diagram of the nonlinear system. Linearization is an important tool in phase plane analysis.

The phase diagram patterns fall into three main classes depending on the eigenvalues, which are the solutions λ_1, λ_2 of the characteristic equation (2.34):

$$\lambda^2 - p\lambda + q = 0, \tag{2.45}$$

with $p = a + d$ and $q = ad - bc \neq 0$. The three classes are

(A) λ_1, λ_2 real, distinct and having the same sign;
(B) λ_1, λ_2 real, distinct and having opposite signs;
(C) λ_1, λ_2 are complex conjugates.

These cases are now treated separately.

(A) The eigenvalues real, distinct, and having the same sign

Call the greater of the two eigenvalues λ_1, so that

$$\lambda_2 < \lambda_1. \tag{2.46}$$

In component form the general solution (2.39) for this case becomes

$$x(t) = C_1 r_1 e^{\lambda_1 t} + C_2 r_2 e^{\lambda_2 t}, \quad y(t) = C_1 s_1 e^{\lambda_1 t} + C_2 s_2 e^{\lambda_2 t}, \tag{2.47}$$

where C_1, C_2 are arbitrary constants and r_1, s_1 and r_2, s_2 are constants obtained by solving (2.37) with $\lambda = \lambda_1$ and $\lambda = \lambda_2$ respectively. From (2.47) we obtain also

$$\frac{dy}{dx} = \frac{\dot{y}}{\dot{x}} = \frac{C_1 s_1 \lambda_1 e^{\lambda_1 t} + C_2 s_2 \lambda_2 e^{\lambda_2 t}}{C_1 r_1 \lambda_1 e^{\lambda_1 t} + C_2 r_2 \lambda_2 e^{\lambda_2 t}} \tag{2.48}$$

Suppose firstly that λ_1 and λ_2 are *negative*, so that

$$\lambda_2 < \lambda_1 < 0. \tag{2.49}$$

From (2.49) and (2.47), along any phase path

$$\left. \begin{array}{l} x \text{ and } y \text{ approach the origin as } t \to \infty, \\ x \text{ and } y \text{ approach infinity as } t \to -\infty. \end{array} \right\} \tag{2.50}$$

Also there are four radial phase paths, which lie along a pair of straight lines as follows:

$$\left.\begin{array}{l}\text{if } C_2 = 0, \quad \dfrac{y}{x} = \dfrac{s_1}{r_1}; \\[2mm] \text{if } C_1 = 0, \quad \dfrac{y}{x} = \dfrac{s_2}{r_2}.\end{array}\right\} \tag{2.51}$$

From (2.48), the dominant terms being those involving $e^{\lambda_1 t}$ for large positive t, and $e^{\lambda_2 t}$ for large negative t, we obtain

$$\left.\begin{array}{ll}\dfrac{dy}{dx} \to \dfrac{s_1}{r_1} & \text{as } t \to \infty, \\[2mm] \dfrac{dy}{dx} \to \dfrac{s_2}{r_2} & \text{as } t \to -\infty.\end{array}\right\} \tag{2.52}$$

Along with (2.50) and (2.51), this shows that every phase path is tangential to $y = (s_1/r_1)x$ at the origin, and approaches the direction of $y = (s_2/r_2)x$ at infinity. The radial solutions (2.51) are called **asymptotes** of the family of phase paths. These features can be seen in Fig. 2.5(a).

If the eigenvalues λ_1, λ_2 are both positive, with $\lambda_1 > \lambda_2 > 0$, the phase diagram has similar characteristics (Fig. 2.5(b)), but all the phase paths are directed outward, running from the origin to infinity.

These patterns show a new feature called a **node**. Figure 2.5(a) shows a **stable node** and Fig. 2.5(b) an **unstable node**. The conditions on the coefficients which correspond to these cases are:

$$\left.\begin{array}{l}\textbf{stable node: } \Delta = p^2 - 4q > 0, \quad q > 0, \, p < 0; \\[2mm] \textbf{unstable node: } \Delta = p^2 - 4q > 0, \quad q > 0, \, p > 0.\end{array}\right\} \tag{2.53}$$

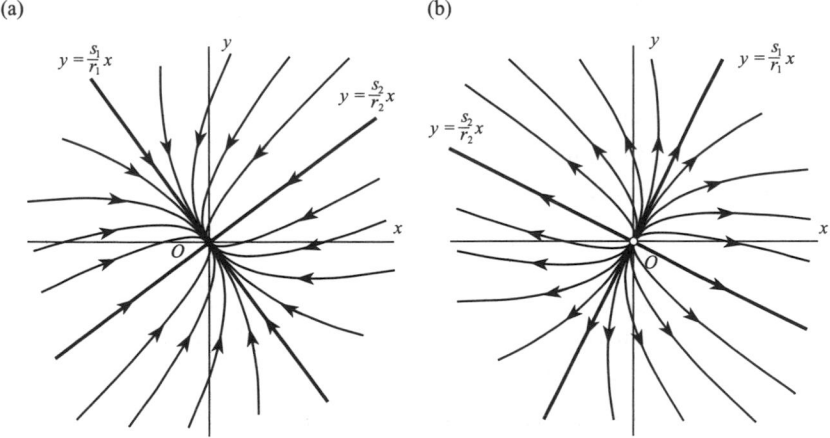

Figure 2.5   (a) Stable node; (b) unstable node.

(B) The eigenvalues are real, distinct, and of opposite signs

In this case

$$\lambda_2 < 0 < \lambda_1,$$

and the solution (2.47) and the formulae (2.48) still apply. In the same way as before, we can deduce that four of the paths are straight lines radiating from the origin, two of them lying along each of the lines

$$\frac{y}{x} = \frac{s_1}{r_1} \quad \text{and} \quad \frac{y}{x} = \frac{s_2}{r_2} \tag{2.54}$$

which are broken by the equilibrium point at the origin.

In this case however there are only *two paths* which *approach the origin*. From (2.47) it can be seen that these are the straight-line paths which lie along $y/x = s_2/r_2$, obtained by putting $C_1 = 0$. The other pair of straight-line paths go to infinity as $t \to \infty$, as do all the other paths. Also, every path (except for the two which lie along $y/x = s_2/r_2$) starts at infinity as $t \to -\infty$. The pattern is like a family of hyperbolas together with its asymptotes, as illustrated in Fig. 2.6.

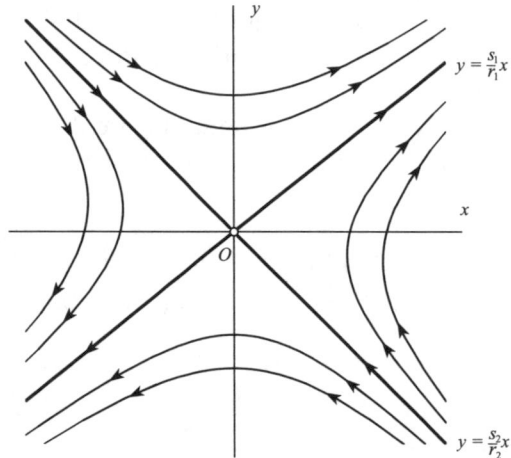

Figure 2.6   Saddle point.

The equilibrium point at the origin is a **saddle**. From (2.45), the conditions on the coefficients of the characteristic equation are

$$\text{saddle point } \Delta = p^2 - 4q > 0, \quad q < 0. \tag{2.55}$$

A saddle is always *unstable*.

The following example illustrates certain short-cuts which are useful for cases when the system coefficients a, b, c, d are given numerically.

Example 2.8   *Sketch the phase diagram and obtain the time solutions of the system*

$$\dot{x} = 3x - 2y, \qquad \dot{y} = 5x - 4y \tag{i}$$

The characteristic equation is $\lambda^2 - p\lambda + q = 0$, where $p = a + d = -1$ and $q = ad - bc = -2$. Therefore

$$\lambda^2 + \lambda - 2 = 0 = (\lambda - 1)(\lambda + 2),$$

so that

$$\lambda_1 = 1, \qquad \lambda_2 = -2. \tag{ii}$$

Since these are of opposite sign the origin is a saddle. (If all we had needed was the phase diagram, we could have checked (2.55) instead:

$$q = -2 < 0 \quad \text{and} \quad p^2 - 4q = 9 > 0;$$

but we need λ_1 and λ_2 for the time solution.)

We know that the asymptotes are straight lines, and therefore have the form $y = mx$. We can find m by substituting $y = mx$ into the equation for the paths:

$$\frac{dy}{dx} = \frac{cx + dy}{ax + by} = \frac{5x - 4y}{3x - 2y},$$

which implies that $m = (5 - 4m)/(3 - 2m)$. Therefore

$$2m^2 - 7m + 5 = 0,$$

from which $m = \frac{5}{2}$ and 1, so the asymptotes are

$$y = \frac{5}{2}x \quad \text{and} \quad y = x. \tag{iii}$$

The pattern of the phase diagram is therefore as sketched in Fig. 2.7. The directions may be found by continuity, starting at any point. For example, at C: $(1, 0)$, eqn (i) gives $\dot{y} = 5 > 0$, so the path through C follows the direction of increasing y. This settles the directions of all other paths.

The general time solution of (i) is

$$\left. \begin{array}{l} x(t) = C_1 r_1 e^{\lambda_1 t} + C_2 r_2 e^{\lambda_2 t} = C_1 r_1 e^t + C_2 r_2 e^{-2t} \\ y(t) = C_1 s_1 e^{\lambda_1 t} + C_2 s_2 e^{\lambda_2 t} = C_1 s_1 e^t + C_2 s_2 e^{-2t}. \end{array} \right\} \tag{iv}$$

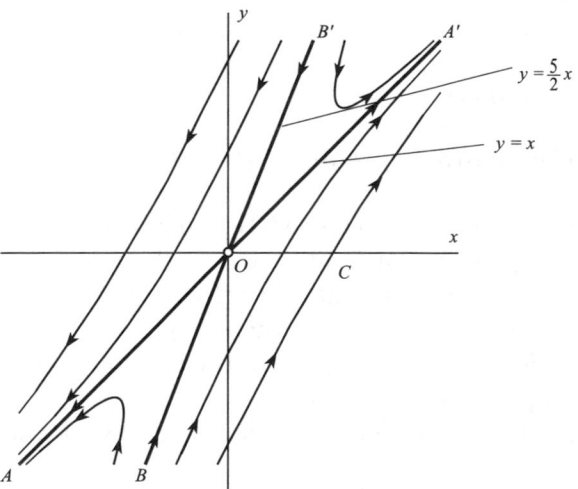

Figure 2.7   Saddle point of $\dot{x} = 3x - 2y$, $\dot{y} = 5x - 4y$.

The paths $B'O, BO$ correspond to solutions with $C_1 = 0$, since they enter the origin as $t \to \infty$. On these paths

$$\frac{y}{x} = \frac{C_2 s_2 e^{-2t}}{C_2 r_2 e^{-2t}} = \frac{s_2}{r_2} = \frac{5}{2}$$

from eqn (iii). Similarly, on OA and OA'

$$\frac{y}{x} = \frac{s_1}{r_1} = 1.$$

We may therefore choose $s_1 = r_1 = 1$, and $s_2 = 5$, $r_2 = 2$. Putting these values into (iv) we obtain the general solution

$$x(t) = C_1 e^t + 2C_2 e^{-2t},$$
$$y(t) = C_1 e^t + 5C_2 e^{-2t},$$

where C_1 and C_2 are arbitrary constants. ●

(C) The eigenvalues are complex

Complex eigenvalues of real matrices always occur as complex conjugate pairs, so put

$$\lambda_1 = \alpha + i\beta, \quad \lambda_2 = \alpha - i\beta \quad (\alpha, \beta \text{ real}). \tag{2.56}$$

By separating the components of (2.43) we obtain for the general solution

$$x(t) = e^{\alpha t} \text{Re}\{Cr_1 e^{i\beta t}\}, \quad y(t) = e^{\alpha t} \text{Re}\{Cs_1 e^{i\beta t}\}, \tag{2.57}$$

where C, r_1, and s_1 are all complex in general.

Suppose firstly that $\alpha = 0$. Put C, r_1, s_1 into polar form:

$$C = |C|e^{i\gamma}, \quad r_1 = |r_1|e^{i\rho}, \quad s_1 = |s_1|e^{i\sigma}.$$

Then (2.57), with $\alpha = 0$, becomes

$$x(t) = |C||r_1| \cos(\beta t + \gamma + \rho), \quad y(t) = |C||s_1| \cos(\beta t + \gamma + \sigma). \tag{2.58}$$

The motion of the representative point $(x(t), y(t))$ in the phase plane consists of two simple harmonic components of equal circular frequency β, in the x and y directions, but they have different phase and amplitude. The phase paths therefore form a family of geometrically similar ellipses which, in general, is inclined at a constant angle to the axes. (The construction is similar to the case of elliptically polarized light; the proof involves eliminating $\cos(\beta t + \gamma)$ and $\sin(\beta t + \gamma)$ between the equations in (2.58).)

This case is illustrated in Fig. 2.8. The algebraic conditions corresponding to the centre at the origin are

$$\textbf{centre } p = 0, \quad q > 0. \tag{2.59}$$

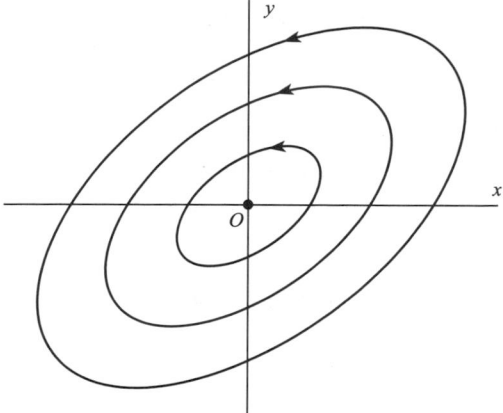

Figure 2.8 Typical centre: rotation may be in either sense.

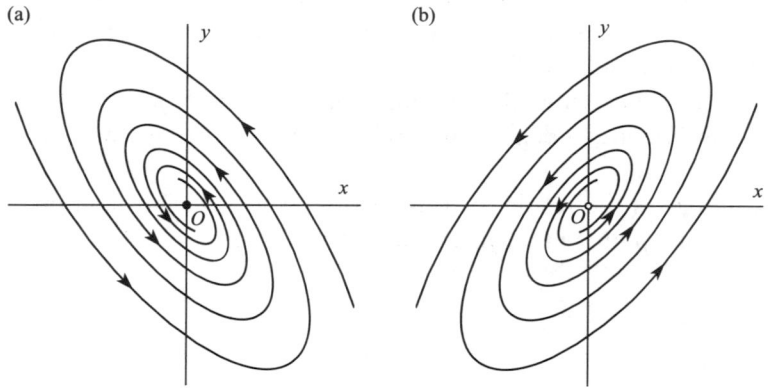

Figure 2.9 (a) Stable spiral; (b) unstable spiral.

Now suppose that $\alpha \neq 0$. As t increases in eqns (2.57), the elliptical paths above are modified by the factor $e^{\alpha t}$. This prevents them from closing, and each ellipse turns into a spiral; a contracting spiral if $\alpha < 0$, and an expanding spiral if $\alpha > 0$ (see Fig. 2.9). The equilibrium point is then called a **spiral** or **focus**, stable if $\alpha < 0$, unstable if $\alpha > 0$. The directions may be *clockwise* or *counterclockwise*.

The algebraic conditions are

$$\left.\begin{array}{ll} \textbf{stable spiral:} & \Delta = p^2 - 4q < 0, \quad q > 0, \quad p < 0; \\ \textbf{unstable spiral:} & \Delta = p^2 - 4q < 0, \quad q > 0, \quad p > 0. \end{array}\right\} \qquad (2.60)$$

Example 2.9 *Determine the nature of the equilibrium point of the system* $\dot{x} = -x - 5y, \dot{y} = x + 3y$.
We have $a = -1, b = -5, c = 1, d = 3$. Therefore

$$p = a + d = 2 > 0, \qquad q = ad - bc = 2 > 0,$$

so that $\Delta = p^2 - 4q = -4 < 0$. These are the conditions (2.60) for an unstable spiral. By putting, say $x > 0, y = 0$ into the equation for $\dot{y}$, we obtain $\dot{y} > 0$ for phase paths as they cross the positive x axis. The spiral paths therefore unwind in the counterclockwise direction.   ●

In addition to the cases discussed there are several **degenerate cases**. These occur when there is a repeated eigenvalue, or when an eigenvalue is zero.

If $q = \det A = 0$, then the eigenvalues are $\lambda_1 = p$, $\lambda_2 = 0$. If $p \neq 0$, then as in the case (2.39), with $\boldsymbol{v}_1$ and $\boldsymbol{v}_2$ the eigenvectors,

$$\boldsymbol{x}(t) = C_1 \boldsymbol{v}_1 e^{pt} + C_2 \boldsymbol{v}_2.$$

There is a line of equilibrium points given by

$$ax + by = 0$$

(which is effectively the same equation as $cx + dy = 0$; the two expressions are linearly dependent). The phase paths form a family of parallel straight lines as shown in Fig. 2.10. A further special case arises if $q = 0$ and $p = 0$.

If $\Delta = 0$, then eigenvalues are real and equal with $\lambda = \frac{1}{2}p$. If $p \neq 0$, it can be shown that the equilibrium point becomes a degenerate node (see Fig. 2.10), in which the two asymptotes have converged.

Figure 2.10 summarizes the results of this section as a pictorial diagram, whilst the table classifies the equilibrium points in terms of the parameters p, q and Δ.

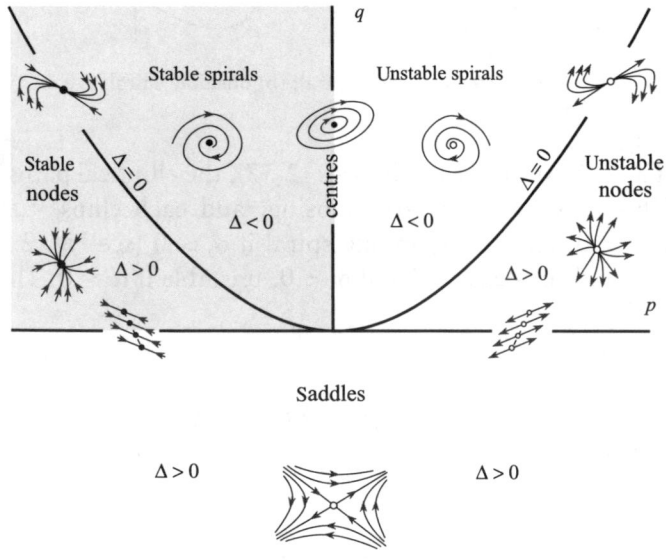

Figure 2.10   Classification for the linear system in the (p, q) plane: $\dot{x} = ax + by$, $\dot{y} = cx + dy$ with $p = a + d$, $q = ad - bc$, $\Delta = p^2 - 4q$.

Classification of equilibrium points of $\dot{x} = ax + by, \dot{y} = cx + dy$

	$p = a + d$	$q = ad - bc$	$\Delta = p^2 - 4q$
Saddle	—	$q < 0$	$\Delta > 0$
Stable node	$p < 0$	$q > 0$	$\Delta > 0$
Stable spiral	$p < 0$	$q > 0$	$\Delta < 0$
Unstable node	$p > 0$	$q > 0$	$\Delta > 0$
Unstable spiral	$p > 0$	$q > 0$	$\Delta < 0$
centre	$p = 0$	$q > 0$	$\Delta < 0$
Degenerate stable node	$p < 0$	$q > 0$	$\Delta = 0$
Degenerate unstable node	$p > 0$	$q > 0$	$\Delta = 0$

$$(2.61)$$

A **centre** may be regarded as a degenerate case, forming a transition between stable and unstable spirals. The existence of a centre depends on there being a particular *exact* relation, namely $a + d = 0$, between coefficients of the system, so a centre is rather a fragile feature. Consequently, if the linear approximation to a nonlinear system predicts a centre it cannot be reliably concluded that the original system has a centre: it might have a stable, or worse, an unstable spiral (see, e.g. Problem 2.7). The same applies to all the *degenerate* cases indicated: if they are used as linear approximations then, taken alone, they are unreliable indicators.

If there exists a neighbourhood of an equilibrium point such that every phase path starting in the neighbourhood ultimately approaches the equilibrium point, the point is known as an **attractor**. (The term is used both for linear and nonlinear systems.) The stable node and stable spiral are attractors. An attractor with all path directions reversed is a **repellor**. Unstable nodes and spirals are repellors, but a saddle point is not. The terms attractor and repellor can also be applied to limit cycles, and to less well defined attracting sets, such as the strange attractor of Chapter 13, from which paths cannot escape.

If the eigenvalues of the linearized equation have nonzero real parts then the equilibrium point is said to be **hyperbolic**. It is shown in Chapter 10 that at hyperbolic points the phase diagrams of the nonlinear equations and the linearized equations are, locally, qualitatively the same. Spirals, nodes, and saddles are hyperbolic but the centre is not.

Exercise 2.5
Using Fig. 2.10 classify the equilibrium points of:

(a) $\dot{x} = -4x + 2y, \ \dot{y} = 4x + 3y$;
(b) $\dot{x} = -6x + 5y, \ \dot{y} = -5x + 2y$;
(b) $\dot{x} = 11x + 6y, \ \dot{y} = -6x - 2y$.

2.6   Scaling in the phase diagram for a linear autonomous system

Consider the system

$$\dot{x} = ax + by, \qquad \dot{y} = cx + dy. \tag{2.62}$$

In Fig. 2.11, $\mathcal{P}$ represents a segment of a phase path, passing through $A : (x_A, y_A)$ at (say) $t = 0$, and $P : (x_P(t), y_P(t))$ is the representative point on $\mathcal{P}$ at time t.

The segment $\mathcal{C}_k$ is a scaled copy of $\mathcal{P}$, constructed in the following way. Choose any constant k; and the two points $B : (x_B, y_B)$ and $Q : (x_Q, y_Q)$ such that

$$(x_B, y_B) = (kx_A, ky_A), \qquad (x_Q, y_Q) = (kx_P, ky_P). \tag{2.63}$$

Then B lies on the radius OA and Q on the radius OP, extended as necessary. Points B and Q are on the same side or opposite sides of the origin according to whether $k > 0$ or $k < 0$, respectively (in Fig. 2.11, $k > 1$). As the representative point P traces the phase path $\mathcal{P}, Q$ traces the curve $\mathcal{C}_k$.

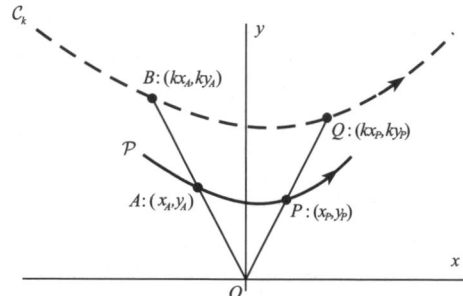

Figure 2.11 $\mathcal{P}$ is a phase path segment. $\mathcal{C}_k$ is the image of $\mathcal{P}$, expanded by a factor k. It is also a phase path.

Since P is the representative point on the phase path $\mathcal{P}$, the system equations (2.62) give

$$k\dot{x}_P = a(kx_P) + b(ky_P), \qquad k\dot{y}_P = c(kx_P) + d(ky_P).$$

Therefore, from (2.63)

$$\dot{x}_Q = ax_Q + by_Q, \qquad \dot{y}_Q = cx_Q + dy_Q. \tag{2.64}$$

The functions $x_Q(t), y_Q(t)$ therefore satisfy the system equations, with Q passing through B at time $t = 0$. Hence *given any value of k, $\mathcal{C}_k$ is another phase path, with Q its representative point.*

Various facts follow easily from this result.

(i) Any phase path segment spanning a sector centred on the origin determines all the rest within the sector, and in the opposite sector. A region consisting of a circle of radius r, centred on the origin, contains the same geometrical pattern of phase paths no matter what the value of r may be.

Benson, Joseph

(ii) All the phase paths spanning a two-sided sector are geometrically similar. They are similarly positioned and directed if $k > 0$, and are reflected in the origin if $k < 0$.

(iii) Any half cycle of a spiral (that is any segment of angular width π) generates the complete spiral structure of the phase diagram.

(iv) All path segments spanning a two-sided radial sector are traversed by the representative points in the same time. In particular, all closed paths have the same period. All complete loops of any spiral path (that is, in a sectorial angle 2π) have the same transit time.

(v) A linear system has no limit cycles (i.e., no *isolated* closed paths).

2.7 Constructing a phase diagram

Suppose that the given system

$$\dot{x} = X(x, y), \qquad \dot{y} = Y(x, y) \tag{2.65}$$

has an equilibrium point at (x_0, y_0):

$$X(x_0, y_0) = 0, \qquad Y(x_0, y_0) = 0. \tag{2.66}$$

The pattern of phase paths close to (x_0, y_0) may be investigated by linearizing the equations at this point, retaining only linear terms of the Taylor series for X and Y there. It is simplest to use the method leading up to eqn (2.22) to obtain the coefficients. If local coordinates are defined by

$$\xi = x - x_0, \qquad \eta = y - y_0,$$

then, approximately,

$$\begin{bmatrix} \dot{\xi} \\ \dot{\eta} \end{bmatrix} = \begin{bmatrix} a & b \\ c & d \end{bmatrix} \begin{bmatrix} \xi \\ \eta \end{bmatrix}, \tag{2.67}$$

where the coefficients are given by

$$\begin{bmatrix} a & b \\ c & d \end{bmatrix} = \begin{bmatrix} \dfrac{\partial X}{\partial x}(x_0, y_0) & \dfrac{\partial X}{\partial y}(x_0, y_0) \\ \dfrac{\partial Y}{\partial x}(x_0, y_0) & \dfrac{\partial Y}{\partial y}(x_0, y_0) \end{bmatrix}. \tag{2.68}$$

The equilibrium point is then classified using the methods of Section 2.4. This is done for each equilibrium point in turn, and it is then possible to make a fair guess at the complete pattern of the phase paths, as in the following example.

Example 2.10   *Sketch the phase diagram for the nonlinear system*

$$\dot{x} = x - y, \qquad \dot{y} = 1 - xy. \tag{i}$$

The equilibrium points are at $(-1, -1)$ and $(1, 1)$. The matrix for linearization, to be evaluated at each equilibrium point in turn, is

$$\begin{bmatrix} \dfrac{\partial X}{\partial x} & \dfrac{\partial X}{\partial y} \\[2mm] \dfrac{\partial Y}{\partial x} & \dfrac{\partial Y}{\partial y} \end{bmatrix} = \begin{bmatrix} 1 & -1 \\ -y & -x \end{bmatrix}. \tag{ii}$$

At $(-1, -1)$ eqns (2.67) becomes

$$\begin{bmatrix} \dot{\xi} \\ \dot{\eta} \end{bmatrix} = \begin{bmatrix} 1 & -1 \\ 1 & 1 \end{bmatrix} \begin{bmatrix} \xi \\ \eta \end{bmatrix}, \tag{iii}$$

where $\xi = x + 1$, $\eta = y + 1$. The eigenvalues of the coefficient matrix are found to be $\lambda_1, \lambda_2 = 1 \pm i$, implying an unstable spiral. To obtain the direction of rotation, it is sufficient to use the linear equations (iii) (or the original equations may be used): putting $\eta = 0$, $\xi > 0$ we find $\dot{\eta} = \xi > 0$, indicating that the rotation is *counterclockwise* as before.

At $(1, 1)$, we find that

$$\begin{bmatrix} \dot{\xi} \\ \dot{\eta} \end{bmatrix} = \begin{bmatrix} 1 & -1 \\ -1 & -1 \end{bmatrix} \begin{bmatrix} \xi \\ \eta \end{bmatrix}, \tag{iv}$$

where $\xi = x - 1$, $\eta = y - 1$. The eigenvalues are given by $\lambda_1, \lambda_2 = \pm\sqrt{2}$, which implies a saddle. The directions of the 'straight-line' paths from the saddle (which become curved separatrices when away from the equilibrium point), are resolved by the technique of Example 2.8: from (iv)

$$\frac{\dot{\eta}}{\dot{\xi}} = \frac{d\eta}{d\xi} = \frac{-\xi - \eta}{\xi - \eta}. \tag{v}$$

We know that two solutions of this equation have the form $\eta = m\xi$ for some values of m. By substituting in (v) we obtain $m^2 - 2m - 1 = 0$, so that $m = 1 \pm \sqrt{2}$.

Finally the phase diagram is put together as in Fig. 2.12, where the phase paths in the neighbourhoods of the equilibrium points are now known. The process can be assisted by sketching in the direction fields on the lines $x = 0$, $x = 1$, etc., also on the curve $1 - xy = 0$ on which the phase paths have zero slopes, and the line $y = x$ on which the paths have infinite slopes.   ●

Exercise 2.6

Find and classify the equilibrium points of

$$\dot{x} = \tfrac{1}{8}(x + y)^3 - y, \qquad \dot{y} = \tfrac{1}{8}(x + y)^3 - x.$$

Verify that lines $y = x$, $y = 2 - x$, $y = -2 - x$, are phase paths. Finally sketch the phase diagram of the system.

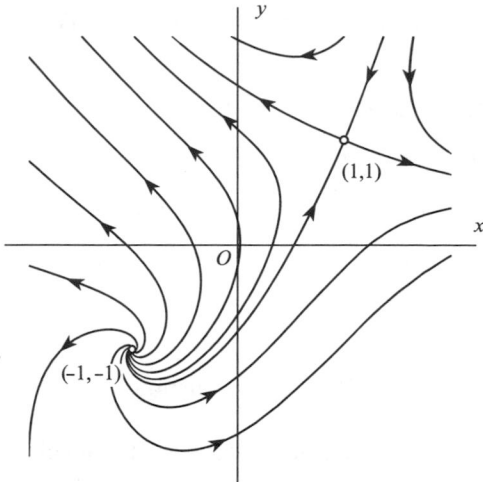

Figure 2.12   Phase diagram for $\dot{x} = x - y$, $\dot{y} = 1 - xy$.

Exercise 2.7
Classify the equilibrium points of the following systems:
(a) $\dot{x} = x^2 + y^2 - 2$, $\dot{y} = y - x^2$;
(b) $\dot{x} = x^2 - y^2 + 1$, $\dot{y} = y - x^2 + 5$;
Using isoclines draw a rough sketch of their phase diagrams. Compare your diagrams with computed phase diagrams.

2.8 Hamiltonian systems

By analogy with the form of Hamilton's canonical equations in mechanics, a system

$$\dot{x} = X(x, y), \qquad \dot{y} = Y(x, y) \tag{2.69}$$

is called a **Hamiltonian system** if there exists a function $H(x, y)$ such that

$$X = \frac{\partial H}{\partial y} \quad \text{and} \quad Y = \frac{\partial H}{\partial x}. \tag{2.70}$$

Then H is called the **Hamiltonian function** for the system. A necessary and sufficient condition for (2.69) to be Hamiltonian is that

$$\frac{\partial X}{\partial x} + \frac{\partial Y}{\partial y} = 0. \tag{2.71}$$

Let $x(t), y(t)$ represent a particular time solution. Then along the corresponding phase path,

$$\frac{\mathrm{d}H}{\mathrm{d}t} = \frac{\partial H}{\partial x}\frac{\mathrm{d}x}{\mathrm{d}t} + \frac{\partial H}{\partial y}\frac{\mathrm{d}y}{\mathrm{d}t}$$

$$= -YX + XY \quad \text{(from (2.69), (2.70))}$$

$$= 0.$$

Therefore,

$$H(x, y) = \text{constant} \tag{2.72}$$

along any phase path. From (2.72), the phase paths are the **level curves**, or **contours**,

$$H(x, y) = C \tag{2.73}$$

of the surface

$$z = H(x, y)$$

in three dimensions.

Suppose that the system has an equilibrium point at (x_0, y_0) so that

$$\frac{\partial H}{\partial x} = \frac{\partial H}{\partial y} = 0 \quad \text{at } (x_0, y_0). \tag{2.74}$$

Then $H(x, y)$ has a stationary point at (x_0, y_0). Sufficient conditions for the three main types of stationary point are given by standard theory; we condense the standard criteria as follows. Put

$$q_0 = \frac{\partial^2 H}{\partial x^2}\frac{\partial^2 H}{\partial y^2} - \left(\frac{\partial^2 H}{\partial x \partial y}\right) \tag{2.75}$$

evaluated at (x_0, y_0). Then

(a) $H(x, y)$ has a maximum or minimum at (x_0, y_0) if

$$q_0 > 0; \tag{2.76}$$

(b) $H(x, y)$ has a saddle at (x_0, y_0) if

$$q_0 < 0. \tag{2.77}$$

(We shall not consider cases where $q_0 = 0$, although similar features may still be present.)

Since the phase paths are the contours of $z = H(x, y)$, we expect that in the case (2.76), the equilibrium point at (x_0, y_0) will be a centre, and that in case (2.77) it will be a saddle point.

There is no case corresponding to a node or spiral: a Hamiltonian system contains only centres and various types of saddle point.

The same prediction is obtained by linearizing the equations at the equilibrium point. From (2.46) the linear approximation at (x_0, y_0) is

$$\dot{x} = a(x - x_0) + b(y - y_0), \quad \dot{y} = c(x - x_0) + d(y - y_0), \tag{2.78}$$

where, in the Hamiltonian case, the coefficients become

$$a = \frac{\partial^2 H}{\partial x \partial y}, \quad b = \frac{\partial^2 H}{\partial y^2}, \quad c = -\frac{\partial^2 H}{\partial x^2}, \quad d = -\frac{\partial^2 H}{\partial x \partial y}, \tag{2.79}$$

all evaluated at (x_0, y_0).

The classification of the equilibrium point is determined by the values of p and q defined in eqn (2.34), and it will be seen that the parameter q is exactly the same as the parameter q_0 defined in (2.75). We have, from (2.45) and (2.79),

$$p = a + d = 0, \quad q = ad - bc = -\left(\frac{\partial^2 H}{\partial x \partial y}\right)^2 + \frac{\partial^2 H}{\partial x^2} \frac{\partial^2 H}{\partial y^2}, \tag{2.80}$$

at (x_0, y_0) (therefore $q = q_0$, as defined in (2.75)). The conditions (2.60) for a centre are that $p = 0$ (automatically satisfied in (2.80)) and that $q > 0$. This is the same as the requirement (2.76), that H should have a maximum or minimum at (x_0, y_0).

Note that the criterion (2.80) for a **centre**, based on the complete geometrical character of $H(x, y)$, is conclusive whereas, as we have pointed out several times, the linearization criterion is not always conclusive.

If $q = 0$ maxima and minima of H still correspondent to centres, but more complicated types of saddle are possible.

Example 2.11 *For the equations*

$$\dot{x} = y(13 - x^2 - y^2), \quad \dot{y} = 12 - x(13 - x^2 - y^2):$$

(a) show that the system is Hamiltonian and obtain the Hamiltonian function $H(x, y)$; (b) obtain the equilibrium points and classify them; (c) sketch the phase diagram.

(a) We have

$$\frac{\partial X}{\partial x} + \frac{\partial Y}{\partial y} = \frac{\partial}{\partial x}\{y(13 - x^2 - y^2)\} + \frac{\partial}{\partial y}\{12 - x(13 - x^2 - y^2)\}$$

$$= -2xy + 2xy = 0.$$

Therefore, by (2.71), this is a Hamiltonian system. From (2.70)

$$\frac{\partial H}{\partial x} = -Y = -12 + x(13 - x^2 - y^2), \tag{i}$$

$$\frac{\partial H}{\partial y} = X = y(13 - x^2 - y^2). \tag{ii}$$

Integrate (i) with respect to x keeping y constant, and (ii) with respect to y keeping x constant: we obtain

$$H = -12x + \tfrac{13}{2}x^2 - \tfrac{1}{4}x^4 - \tfrac{1}{2}x^2y^2 + u(y), \tag{iii}$$

$$H = \tfrac{13}{2}y^2 - \tfrac{1}{2}x^2y^2 - \tfrac{1}{4}y^4 + v(x), \tag{iv}$$

respectively, where $u(y)$ and $v(x)$ are *arbitrary* functions of y and x only, but subject to the consistency of eqns (iii) and (iv). The two equations will match only if we choose

$$u(y) = \tfrac{13}{2}y^2 - \tfrac{1}{4}y^4 - C,$$

$$v(x) = -12x + \tfrac{13}{2}x^2 - \tfrac{1}{4}x^4 - C,$$

where C is any constant (to see why, subtract (iv) from (iii): the resulting expression must be identically zero). Therefore the phase paths are given by

$$H(x, y) = -12x + \tfrac{13}{2}(x^2 + y^2) - \tfrac{1}{4}(x^4 + y^4) - \tfrac{1}{4}(x^4 + y^4) - \tfrac{1}{4} - \tfrac{1}{2}x^2y^2 = C, \tag{v}$$

where C is a parameter.
(b) the equilibrium points occur where

$$y(13 - x^2 - y^2) = 0, \quad 12 - x(13 - x^2 - y^2) = 0.$$

Solutions exist only at points where $y = 0$, and

$$x^3 - 13x + 12 = (x - 1)(x - 3)(x + 4) = 0.$$

Therefore, the coordinates of the equilibrium points are

$$(1, 0), \quad (3, 0), \quad (-4, 0). \tag{vi}$$

The second derivatives of $H(x, y)$ are

$$\frac{\partial^2 H}{\partial x^2} = 13 - 3x^2 - y^2, \quad \frac{\partial^2 H}{\partial y^2} = 13 - x^2 - 3y^2, \quad \frac{\partial^2 H}{\partial x \partial y} = -2xy.$$

We need only compute q (see (2.75)) at the equilibrium points. The results are

Equilibrium points	(1, 0)	(3, 0)	(−4, 0)
Value of q	$120 > 0$	$-56 < 0$	$105 > 0$
Classification	centre	Saddle	centre

(c) A computed phase diagram is shown in Fig. 2.13.   ●

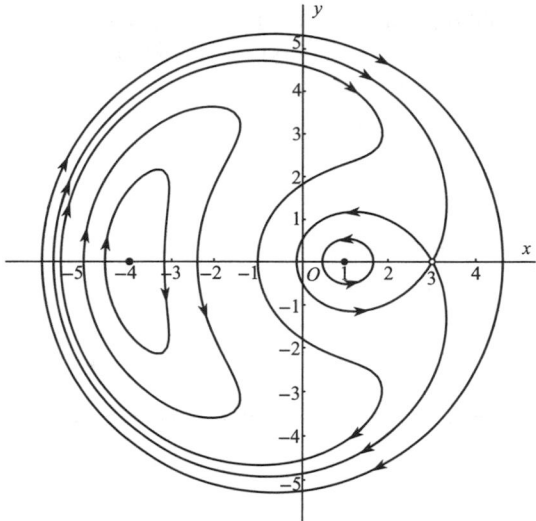

Figure 2.13   Phase diagram for the Hamiltonian system $\dot{x} = y(13 - x^2 - y^2)$, $\dot{y} = 12 - x(13 - x^2 - y^2)$.

Exercise 2.8
Shows that the system

$$\dot{x} = (x^2 - 1)(3y^2 - 1), \qquad \dot{y} = -2xy(y^2 - 1)$$

is Hamiltonian. Find the coordinates of the 8 equilibrium points. Using the obvious exact solutions and the Hamiltonian property draw a rough sketch of the phase diagram.

Problems

2.1 Sketch phase diagrams for the following linear systems and classify the equilibrium point:
  (i) $\dot{x} = x - 5y$, $\dot{y} = x - y$;
  (ii) $\dot{x} = x + y$, $\dot{y} = x - 2y$;
  (iii) $\dot{x} = -4x + 2y$, $\dot{y} = 3x - 2y$;
  (iv) $\dot{x} = x + 2y$, $\dot{y} = 2x + 2y$;
  (v) $\dot{x} = 4x - 2y$, $\dot{y} = 3x - y$;
  (vi) $\dot{x} = 2x + y$, $\dot{y} = -x + y$.

2.2 Some of the following systems either generate a single eigenvalue, or a zero eigenvalue, or in other ways vary from the types illustrated in Section 2.5. Sketch their phase diagrams.

  (i) $\dot{x} = 3x - y$, $\dot{y} = x + y$;
  (ii) $\dot{x} = x - y$, $\dot{y} = 2x - 2y$;
  (iii) $\dot{x} = x$, $\dot{y} = 2x - 3y$;
  (iv) $\dot{x} = x$, $\dot{y} = x + 3y$;
  (v) $\dot{x} = -y$, $\dot{y} = 2x - 4y$;
  (vi) $\dot{x} = x$, $\dot{y} = y$;
  (vii) $\dot{x} = 0$, $\dot{y} = x$.

2.3 Locate and classify the equilibrium points of the following systems. Sketch the phase diagrams: it will often be helpful to obtain isoclines and path directions at other points in the plane.

$$
\begin{aligned}
&\text{(i)}\ \dot{x}=x-y, & &\dot{y}=x+y-2xy;\\
&\text{(ii)}\ \dot{x}=ye^{y}, & &\dot{y}=1-x^{2};\\
&\text{(iii)}\ \dot{x}=1-xy, & &\dot{y}=(x-1)y;\\
&\text{(iv)}\ \dot{x}=(1+x-2y)x, & &\dot{y}=(x-1)y;\\
&\text{(v)}\ \dot{x}=x-y, & &\dot{y}=x^{2}-1;\\
&\text{(vi)}\ \dot{x}=-6y+2xy-8, & &\dot{y}=y^{2}-x^{2};\\
&\text{(vii)}\ \dot{x}=4-4x^{2}-y^{2}, & &\dot{y}=3xy;\\
&\text{(viii)}\ \dot{x}=-y\sqrt{(1-x^{2})}, & &\dot{y}=x\sqrt{(1-x^{2})}\ \text{for}\ |x|\le 1;\\
&\text{(ix)}\ \dot{x}=\sin y, & &\dot{y}=-\sin x;\\
&\text{(x)}\ \dot{x}=\sin x\cos y, & &\dot{y}=\sin y\cos x.
\end{aligned}
$$

2.4 Construct phase diagrams for the following differential equations, using the phase plane in which $y=\dot{x}$.

(i) $\ddot{x}+x-x^{3}=0$;

(ii) $\ddot{x}+x+x^{3}=0$;

(iii) $\ddot{x}+\dot{x}+x-x^{3}=0$;

(iv) $\ddot{x}+\dot{x}+x+x^{3}=0$;

(v) $\ddot{x}=(2\cos x-1)\sin x.$

2.5 Confirm that system $\dot{x}=x-5y$, $\dot{y}=x-y$ consists of a centre. By substitution into the equation for the paths or otherwise show that the family of ellipses given by

$$
x^{2}-2xy+5y^{2}=\text{constant}
$$

describes the paths. Show that the axes are inclined at about $13.3°$ (the major axis) and $-76.7°$ (the minor axis) to the x direction, and that the ratio of major to minor axis length is about 2.62.

2.6 The family of curves which are orthogonal to the family described by the equation $(\mathrm{d}y/\mathrm{d}x)=f(x,y)$ is given by the solution of $(\mathrm{d}y/\mathrm{d}x)=-[1/f(x,y)]$. (These are called orthogonal trajectories of the first family.) Prove that the family which is orthogonal to a centre that is associated with a linear system is a node.

2.7 Show that the origin is a spiral point of the system $\dot{x}=-y-x\sqrt{(x^{2}+y^{2})}$, $\dot{y}=x-y\sqrt{(x^{2}+y^{2})}$ but a centre for its linear approximation.

2.8 Show that the systems $\dot{x}=y$, $\dot{y}=-x-y^{2}$, and $\dot{x}=x+y_{1}$, $\dot{y}_{1}=-2x-y_{1}-(x+y_{1})^{2}$, both represent the equation $\ddot{x}+\dot{x}^{2}+x=0$ in different (x,y) and (x,y_{1}) phase planes. Obtain the equation of the phase paths in each case.

2.9 Use eqn (2.9) in the form $\delta s\simeq\delta t\sqrt{(X^{2}+Y^{2})}$ to mark off approximately equal time steps on some of the phase paths of $\dot{x}=xy$, $\dot{y}=xy-y^{2}$.

2.10 Obtain approximations to the phase paths described by eqn (2.12) in the neighbourhood of the equilibrium point $x=b/d, y=a/c$ for the predator–prey problem $\dot{x}=ax-cxy$, $\dot{y}=-by+dxy$ (see Example 2.3). (Write $x=b/d+\xi, y=a/c+\eta$, and expand the logarithms to second-order terms in ξ and η.)

2.11 For the system $\dot{x}=ax+by, \dot{y}=cx+dy$, where $ad-bc=0$, show that all points on the line $cx+dy=0$ are equilibrium points. Sketch the phase diagram for the system $\dot{x}=x-2y, \dot{y}=2x-4y$.

2.12 The interaction between two species is governed by the deterministic model $\dot{H}=(a_{1}-b_{1}H-c_{1}P)H$, $\dot{P}=(-a_{2}+c_{2}H)P$, where H is the population of the host (or prey), and P is that of the parasite (or predator), all constants being positive. (Compare Example 2.3: the term $-b_{1}H^{2}$ represents interference with the host population when it gets too large.) Assuming that $a_{1}c_{2}-b_{1}a_{2}>0$, find the equilibrium states for the populations, and find how they vary with time from various initial populations.

2.13 With the same terminology as in Problem 2.12, analyse the system $\dot{H} = (a_1 - b_1 H - c_1 P)H, \dot{P} = (a_2 - b_2 P + c_2 H)P$, all the constants being positive. (In this model the parasite can survive on alternative food supplies, although the prevalence of the host encourages growth in population.) Find the equilibrium states. Confirm that the parasite population can persist even if the host dies out.

2.14 Consider the host–parasite population model $\dot{H} = (a_1 - c_1 P)H, \dot{P} = (a_2 - c_2(P/H))P$, where the constants are positive. Analyse the system in the H, P plane.

2.15 In the population model $\dot{F} = -\alpha F + \beta \mu(M)F, \dot{M} = -\alpha M + \gamma \mu(M)F$, where $\alpha > 0, \beta > 0, \gamma > 0, F$ and M are the female and male populations. In both cases the death rates are α. The birth rate is governed by the coefficient $\mu(M) = 1 - e^{-kM}, k > 0$, so that for large M the birth rate of females is βF and that for males is γF, the rates being unequal in general. Show that if $\beta > \alpha$ then there are two equilibrium states, at $(0, 0)$ and at $([-\beta/(\gamma k)]\log[(\beta - \alpha)/\beta], [-1/k]\log[(\beta - \alpha)/\beta])$.

Show that the origin is stable and that the other equilibrium point is a saddle point, according to their linear approximations. Verify that $M = \gamma F/\beta$ is a particular solution. Sketch the phase diagram and discuss the stability of the populations.

2.16 A rumour spreads through a closed population of constant size $N + 1$. At time t the total population can be classified into three categories:

x persons who are ignorant of the rumour;

y persons who are actively spreading the rumour;

z persons who have heard the rumour but have stopped spreading it: if two persons who are spreading the rumour meet then they stop spreading it.

The contact rate between any two categories is a constant, μ.

Show that the equations

$$\dot{x} = -\mu xy, \qquad \dot{y} = \mu[xy - y(y-1) - yz]$$

give a deterministic model of the problem. Find the equations of the phase paths and sketch the phase diagram.

Show that, when initially $y = 1$ and $x = N$, the number of people who ultimately never hear the rumour is x_1, where

$$2N + 1 - 2x_1 + N\log(x_1/N) = 0.$$

2.17 The one-dimensional steady flow of a gas with viscosity and heat conduction satisfies the equations

$$\frac{\mu_0}{\rho c_1}\frac{dv}{dx} = \sqrt{(2v)}[2v - \sqrt{(2v)} + \theta],$$

$$\frac{k}{gR\rho c_1}\frac{d\theta}{dx} = \sqrt{(2v)}\left[\frac{\theta}{\gamma - 1} - v + \sqrt{(2v)} - c\right],$$

where $v = u^2/(2c_1^2), c = c_2^2/c_1^2$ and $\theta = gRT/c_1^2 = p/(\rho c_1^2)$. In this notation, x is measured in the direction of flow, u is the velocity, T is the temperature, ρ is the density, p the pressure, R the gas constant, k the coefficient of thermal conductivity, μ_0 the coefficient of viscosity, γ the ratio of the specific heats, and c_1, c_2 are arbitrary constants. Find the equilibrium states of the system.

2.18 A particle moves under a central attractive force γ/r^α per unit mass, where r, θ are the polar coordinates of the particle in its plane of motion. Show that

$$\frac{d^2 u}{d\theta^2} + u = \frac{\gamma}{h^2}u^{\alpha-2},$$

where $u = r^{-1}$, h is the angular momentum about the origin per unit mass of the particle, and γ is a constant. Find the non-trival equilibrium point in the $u, du/d\theta$ plane and classify it according to its linear approximation. What can you say about the stability of the circular orbit under this central force?

2.19 The relativistic equation for the central orbit of a planet is

$$\frac{d^2u}{d\theta^2} + u = k + \varepsilon u^2,$$

where $u = 1/r$, and r, θ are the polar coordinates of the planet in the plane of its motion. The term εu^2 is the 'Einstein correction', and k and ε are positive constants, with ε very small. Find the equilibrium point which corresponds to a perturbation of the Newtonian orbit. Show that the equilibrium point is a centre in the $u, du/d\theta$ plane according to the linear approximation. Confirm this by using the potential energy method of Section 1.3.

2.20 A top is set spinning at an axial rate n radians/sec about its pivotal point, which is fixed in space. The equations for its motion, in terms of the angles θ and μ are (see Fig. 2.14)

$$A\ddot{\theta} - A(\Omega + \dot{\mu})^2 \sin\theta \cos\theta + Cn(\Omega + \dot{\mu})\sin\theta - Mgh\sin\theta = 0,$$

$$A\dot{\theta}^2 + A(\Omega + \dot{\mu})^2 \sin^2\theta + 2Mgh\cos\theta = E;$$

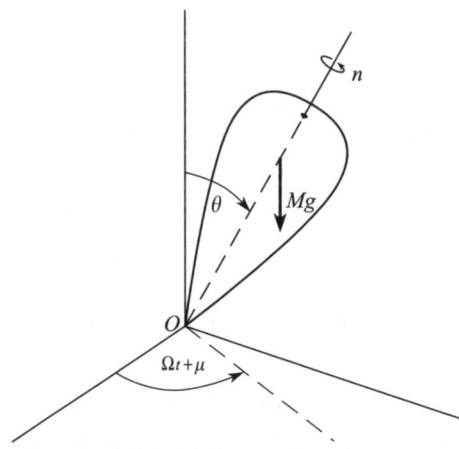

Figure 2.14   Spinning top.

where (A, A, C) are the principal moments of inertia about O, M is the mass of the top, h is the distance between the mass centre and the pivot, and E is a constant. Show that an equilibrium state is given by $\theta = \alpha$, after elimination of Ω between

$$A\Omega^2 \cos\alpha - Cn\Omega + Mgh = 0, \quad A\Omega^2 \sin^2\alpha + 2Mgh\cos\alpha = E.$$

Suppose that $E = 2Mgh$, so that $\theta = 0$ is an equilibrium state. Show that, close to this state, θ satisfies

$$A\ddot{\theta} + [(C - A)\Omega^2 - Mgh]\theta = 0.$$

For what condition on Ω is the motion stable?

2.21 Three freely gravitating particles with gravitational masses μ_1, μ_2, μ_3, move in a plane so that they always remain at the vertices of an equilateral triangle P_1, P_2, P_3 with varying side-length $a(t)$ as shown in Figure 2.15. The triangle rotates in the plane with spin $\Omega(t)$ about the combined mass-centre G. If the position vectors of the particles $\mathbf{r}_1$, $\mathbf{r}_2$, $\mathbf{r}_3$, relative to G, show that the equations of motion are

$$\ddot{\mathbf{r}}_i = -\frac{\mu_1 + \mu_2 + \mu_3}{a^3}\mathbf{r}_i \quad (i = 1, 2, 3).$$

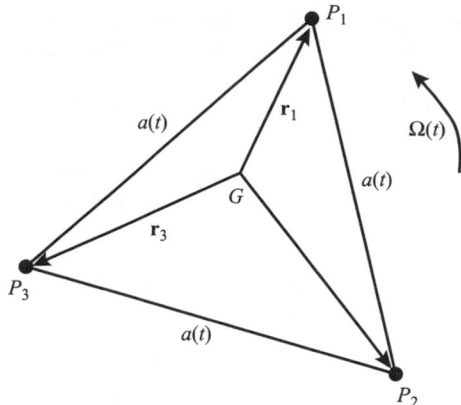

Figure 2.15 Lagrange equilateral configuration for a three-body problem with $P_1 P_2 = P_2 P_3 = P_3 P_1 = a(t)$.

If $|\mathbf{r}_i| = r_i$, deduce the polar equations

$$\ddot{r}_i - r_i \Omega^2 = -\frac{\mu_1 + \mu_2 + \mu_3}{a^3} r_i, \quad r_i^2 \Omega = \text{constant} \qquad (i = 1, 2, 3).$$

Explain why a satisfies

$$\ddot{a} - a\Omega^2 = -\frac{\mu_1 + \mu_2 + \mu_3}{a^2}, \quad a^2 \Omega = \text{constant} = K,$$

say, and that solutions of these equations completely determine the position vectors. Express the equation in non-dimensionless form by the substitutions $a = K^2/(\mu_1 + \mu_2 + \mu_3), t = K^3 \tau/(\mu_1 + \mu_2 + \mu_3)^2$, sketch the phase diagram for the equation in μ obtained by eliminating Ω, and discuss possible motions of this Lagrange configuration.

2.22 A disc of radius a is freely pivoted at its centre A so that it can turn in a vertical plane. A spring, of natural length $2a$ and stiffness λ connects a point B on the circumference of the disc to a fixed point O, distance $2a$ above A. Show that θ satisfies

$$I\ddot{\theta} = -Ta\sin\phi, \qquad T = \lambda a[(5 - 4\cos\theta)^{1/2} - 2],$$

where T is the tension in the spring, I is the moment of inertia of the disc about A, $\widehat{OAB} = \theta$ and $\widehat{ABO} = \phi$. Find the equilibrium states of the disc and their stability.

2.23 A man rows a boat across a river of width a occupying the strip $0 \leq x \leq a$ in the x, y plane, always rowing towards a fixed point on one bank, say $(0, 0)$. He rows at a constant speed u relative to the water, and the river flows at a constant speed v. Show that

$$\dot{x} = -ux/\sqrt{(x^2 + y^2)}, \quad \dot{y} = v - uy/\sqrt{(x^2 + y^2)},$$

where (x, y) are the coordinates of the boat. Show that the phase paths are given by $y + \sqrt{(x^2 + y^2)} = Ax^{1-\alpha}$, where $\alpha = v/u$. Sketch the phase diagram for $\alpha < 1$ and interpret it. What kind of point is the origin? What happens to the boat if $\alpha > 1$?

2.24 In a simple model of a national economy, $\dot{I} = I - \alpha C$, $\dot{C} = \beta(I - C - G)$, where I is the national income, C is the rate of consumer spending, and G the rate of government expenditure; the constants α and β satisfy $1 < \alpha < \infty$, $1 \leq \beta < \infty$. Show that if the rate of government expenditure G is constant G_0 there is an equilibrium state. Classify the equilibrium state and show that the economy oscillates when $\beta = 1$.

Consider the situation when government expenditure is related to the national income by the rule $G = G_0 + kI$, where $k > 0$. Show that there is no equilibrium state if $k \geq (\alpha - 1)/\alpha$. How does the economy then behave?

Discuss an economy in which $G = G_0 + kI^2$, and show that there are two equilibrium states if $G_0 < (\alpha - 1)^2/(4k\alpha^2)$.

2.25 Let $f(x)$ and $g(y)$ have local minima at $x = a$ and $y = b$ respectively. Show that $f(x) + g(y)$ has a minimum at (a, b). Deduce that there exists a neighbourhood of (a, b) in which all solutions of the family of equations

$$f(x) + g(y) = \text{constant}$$

represent closed curves surrounding (a, b).

Show that $(0, 0)$ is a centre for the system $\dot{x} = y^5, \dot{y} = -x^3$, and that all paths are closed curves.

2.26 For the predator–prey problem in Section 2.2, show by using Problem 2.25 that all solutions in $y > 0$, $x > 0$ are periodic.

2.27 Show that the phase paths of the Hamiltonian system $\dot{x} = -\partial H/\partial y$, $\dot{y} = \partial H/\partial x$ are given by $H(x, y) = constant$. Equilibrium points occur at the stationary points of $H(x, y)$. If (x_0, y_0) is an equilibrium point, show that (x_0, y_0) is stable according to the linear approximation if $H(x, y)$ has a maximum or a minimum at the point. (Assume that all the second derivatives of H are nonzero at x_0, y_0.)

2.28 The equilibrium points of the nonlinear parameter-dependent system $\dot{x} = y$, $\dot{y} = f(x, y, \lambda)$ lie on the curve $f(x, 0, \lambda) = 0$ in the x, λ plane. Show that an equilibrium point (x_1, λ_1) is stable and that all neighbouring solutions tend to this point (according to the linear approximation) if $f_x(x_1, 0, \lambda_1) < 0$ and $f_y(x_1, 0, \lambda_1) < 0$.

Investigate the stability of $\dot{x} = y$, $\dot{y} = -y + x^2 - \lambda x$.

2.29 Find the equations for the phase paths for the general epidemic described (Section 2.2) by the system

$$\dot{x} = -\beta xy, \qquad \dot{y} = \beta xy - \gamma y, \qquad \dot{z} = \gamma y.$$

Sketch the phase diagram in the x, y plane. Confirm that the number of infectives reaches its maximum when $x = \gamma/\beta$.

2.30 Two species x and y are competing for a common food supply. Their growth equations are

$$\dot{x} = x(1 - x - y), \quad \dot{y} = y(3 - x - \tfrac{3}{2}y), \qquad (x, y > 0).$$

Classify the equilibrium points using linear approximations. Draw a sketch indicating the slopes of the phase paths in $x \geq 0$, $y \geq 0$. If $x = x_0 > 0$, $y = y_0 > 0$ initially, what do you expect the long term outcome of the species to be? Confirm your conclusions numerically by computing phase paths.

2.31 Sketch the phase diagram for the competing species x and y for which

$$\dot{x} = (1 - x^2 - y^2)x, \qquad \dot{y} = (\tfrac{5}{4} - x - y)y.$$

2.32 A space satellite is in free flight on the line joining, and between, a planet (mass m_1) and its moon (mass m_2), which are at a fixed distance a apart. Show that

$$-\frac{\gamma m_1}{x^2} + \frac{\gamma m_2}{(a - x)^2} = \ddot{x},$$

where x is the distance of the satellite from the planet and γ is the gravitational constant. Show that the equilibrium point is unstable according to the linear approximation.

2.33 The system

$$\dot{V}_1 = -\sigma V_1 + f(E - V_2), \quad \dot{V}_2 = -\sigma V_2 + f(E - V_1), \qquad \sigma > 0, E > 0$$

represents (Andronov and Chaikin 1949) a model of a triggered sweeping circuit for an oscilloscope. The conditions on $f(u)$ are: $f(u)$ continuous on $-\infty < u < \infty$, $f(-u) = -f(u)$, $f(u)$ tends to a limit as $u \to \infty$, and $f'(u)$ is monotonic decreasing (see Fig. 3.20).

Show by a geometrical argument that there is always at least one equilibrium point, (v_0, v_0) say, and that when $f'(E - v_0) < \sigma$ it is the only one; and deduce by taking the linear approximation that it is a stable node. (Note that $f'(E - v) = - \, \mathrm{d}f(E - v)/\mathrm{d}v$.)

Show that when $f'(E - v_0) > \sigma$ there are two others, at $(V', (1/\sigma)f(E - V'))$ and $((1/\sigma)f(E - V'), V')$ respectively for some V'. Show that these are stable nodes, and that the one at (v_0, v_0) is a saddle point.

2.34 Investigate the equilibrium points of $\dot{x} = a - x^2$, $\dot{y} = x - y$. Show that the system has a saddle and a stable node for $a > 0$, but no equilibrium points if $a < 0$. The system is said to undergo a **bifurcation** as a increases through $a = 0$. This bifurcation is an example of a **saddle-node bifurcation**. This will be discussed in more detail in Section 12.4. Draw phase diagrams for $a = 1$ and $a = -1$.

2.35 Figure 2.16 represents a circuit for activating an electric arc A which has the voltage–current characteristic shown. Show that $L\dot{I} = V - V_a(I)$, $RC\dot{V} = - RI - V + E$ where $V_a(I)$ has the general shape shown in Fig. 2.16. By forming the linear approximating equations near the equilibrium points find the conditions on E, L, C, R, and V_a' for stable working assuming that $V = E - RI$ meets the curve $V = V_a(I)$ in three points of intersection.

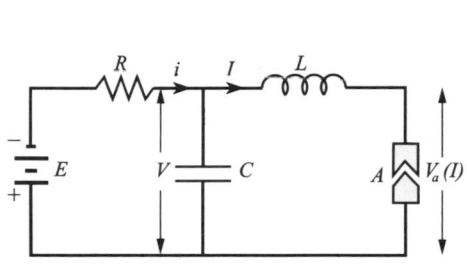

Figure 2.16

2.36 The equation for the current x in the circuit of Fig. 2.17(a) is

$$LC\ddot{x} + RC\dot{x} + x = I.$$

Neglect the grid current, and assume that I depends only on the relative grid potential e_g : $I = I_s$ (saturation current) for $e_g > 0$ and $I = 0$ for $e_g < 0$ (see Fig. 2.17(b)). Assume also that the mutual inductance $M > 0$, so that $e_g \gtrless 0$ according as $\dot{x} \gtrless 0$. Find the nature of the phase paths. By considering their successive intersections with the x axis show that a limit cycle is approached from all initial conditions (Assume $R^2 C < 4L$).

2.37 For the circuit in Fig. 2.17(a) assume that the relation between I and e_g is as in Fig. 2.18; that is $I = f(e_g + ke_p)$, where e_g and e_p are the relative grid and plate potentials, $k > 0$ is a constant, and in the neighbourhood of the point of inflection, $f(u) = I_0 + au - bu^3$, where $a > 0$, $b > 0$. Deduce the equation for x when the D.C. source E is set so that the operating point is the point of inflection. Find when the origin is a stable or an unstable point of equilibrium. (A form of Rayleigh's equation, Example 4.6, is obtained, implying an unstable or a stable limit cycle respectively.)

2.38 Figure 2.19(a) represents two identical D.C. generators connected in parallel, with inductance and resistance L, r. Here R is the resistance of the load. Show that the equations for the currents are

$$L\frac{\mathrm{d}i_1}{\mathrm{d}t} = -(r + R)i_1 - Ri_2 + E(i_1), L\frac{\mathrm{d}i_2}{\mathrm{d}t} = -Ri_1 - (r + R)i_2 + E(i_2).$$

(a)

(b)

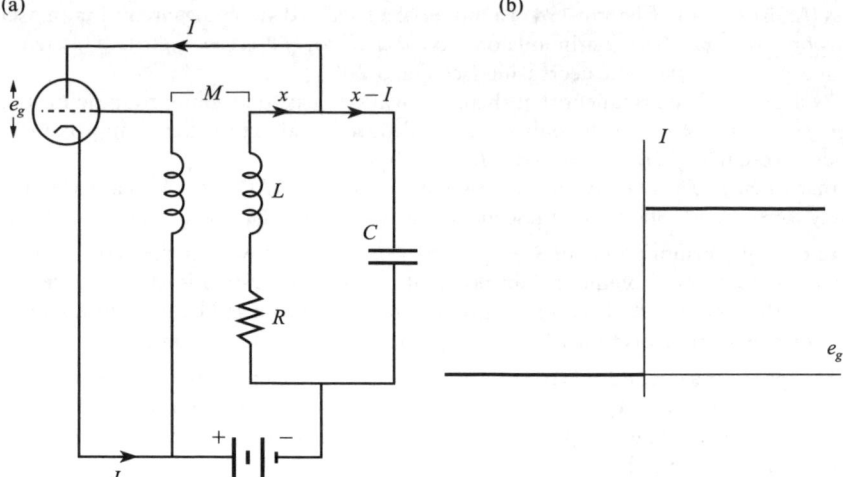

Figure 2.17

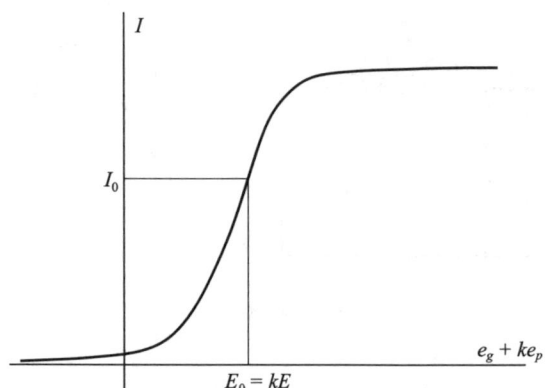

Figure 2.18

(a)

(b)

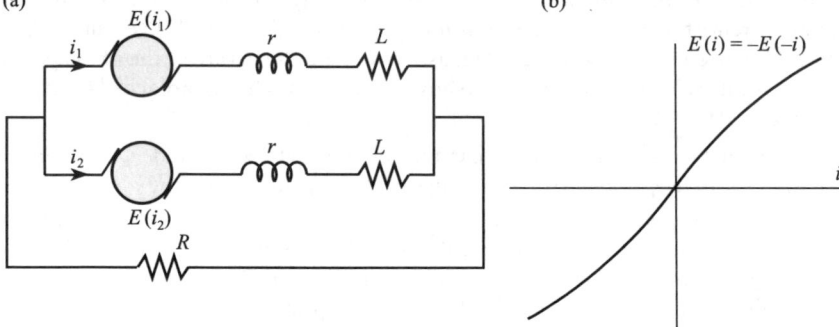

Figure 2.19

Assuming that $E(i)$ has the characteristics indicated by Fig. 2.19(b) show that
  (i) when $E'(0) < r$ the state $i_1 = i_2 = 0$ is stable and is otherwise unstable;
 (ii) when $E'(0) < r$ there is a stable state $i_1 = -i_2$ (no current flows to R);
(iii) when $E'(0) > r + 2R$ there is a state with $i_1 = i_2$, which is unstable.

2.39 Show that the Emden–Fowler equation of astrophysics

$$(\xi^2 \eta')' + \xi^\lambda \eta^n = 0$$

is equivalent to the predator–prey model

$$\dot{x} = -x(1 + x + y), \qquad \dot{y} = y(\lambda + 1 + nx + y)$$

after the change of variable $x = \xi \eta'/\eta$, $y = \xi^{\lambda-1}\eta^n/\eta'$, $t = \log|\xi|$.

2.40 Show that Blasius' equation $\eta''' + \eta\eta'' = 0$ is transformed by $x = \eta\eta'/\eta''$, $y = \eta'^2/\eta\eta''$, $t = \log|\eta'|$ into

$$\dot{x} = x(1 + x + y), \qquad \dot{y} = y(2 + x - y).$$

2.41 Consider the family of linear systems

$$\dot{x} = X\cos\alpha - Y\sin\alpha, \qquad \dot{y} = X\sin\alpha + Y\cos\alpha$$

where $X = ax + by$, $Y = cx + dy$, and a,b,c,d are constants and α is a parameter. Show that the parameters (Table (2.62)) are

$$p = (a + d)\cos\alpha + (b - c)\sin\alpha, \quad q = ad - bc.$$

Deduce that the origin is a saddle point for all α if $ad < bc$.

If $a = 2$, $b = c = d = 1$, show that the origin passes through the sequence stable node, stable spiral, centre, unstable spiral, unstable node, as α varies over range π.

2.42 Show that, given $X(x, y)$, the system equivalent to the equation $\ddot{x} + h(x, \dot{x}) = 0$ is

$$\dot{x} = X(x, y), \qquad \dot{y} = -\left\{h(x, X) + x\frac{\partial X}{\partial x}\right\} \bigg/ \frac{\partial X}{\partial y}.$$

2.43 The following system models two species with populations N_1 and N_2 competing for a common food supply:

$$\dot{N_1} = \{a_1 - d_1(bN_1 + cN_2)\}N_1, \qquad \dot{N_2} = \{a_2 - d_2(bN_1 + cN_2)\}N_2.$$

Classify the equilibrium points of the system assuming that all coefficients are positive. Show that if $a_1 d_2 > a_2 d_1$ then the species N_2 dies out and the species N_1 approaches a limiting size (Volterra's Exclusion Principle).

2.44 Show that the system

$$\dot{x} = X(x, y) = -x + y, \qquad \dot{y} = Y(x, y) = \frac{4x^2}{1 + 3x^2} - y$$

has three equilibrium points at $(0,0)$, $(\frac{1}{3}, \frac{1}{3})$ and $(1, 1)$. Classify each equilibrium point. Sketch the isoclines $X(x, y) = 0$ and $Y(x, y) = 0$, and indicate the regions where dy/dx is positive, and where dy/dx is negative. Sketch the phase diagram of the system.

2.45 Show that the systems (A) $\dot{x} = P(x, y)$, $\dot{y} = Q(x, y)$ and (B) $\dot{x} = Q(x, y)$, $\dot{y} = P(x, y)$ have the same equilibrium points. Suppose that system (A) has three equilibrium points which, according to their linear approximations are, (a) a stable spiral, (b) an unstable node, (c) a saddle point. To what extent can the equilibrium points in (B) be classified from this information?

2.46 The system defined by the equations

$$\dot{x} = a + x^2 y - (1 + b)x, \quad \dot{y} = bx - yx^2 \qquad (a \neq 0, b \neq 0)$$

is known as the **Brusselator** and arises in a mathematical model of a chemical reaction (see Jackson (1990)). Show that the system has one equilibrium point at $(a, b/a)$. Classify the equilibrium point in each of the following cases:

(a) $a = 1, b = 2$;

(b) $a = \frac{1}{2}, b = \frac{1}{4}$.

  In case (b) draw the isoclines of zero and infinite slope in the phase diagram. Hence sketch the phase diagram.

2.47 A Volterra model for the population size $p(t)$ of a species is, in reduced form,

$$\kappa \frac{dp}{dt} = p - p^2 - p \int_0^t p(s) \, ds, \quad p(0) = p_0,$$

where the integral term represents a toxicity accumulation term (see Small (1989)). Let $x = \log p$, and show that x satisfies

$$\kappa \ddot{x} + e^x \dot{x} + e^x = 0.$$

Put $y = \dot{x}$, and show that the system is also equivalent to

$$\dot{y} = -(y + 1)p/\kappa, \quad \dot{p} = yp.$$

Sketch the phase diagram in the (y, p) plane. Also find the exact equation of the phase paths.

3 Geometrical aspects of plane autonomous systems

In this chapter, we discuss several topics which are useful for establishing the structure of the phase diagram of autonomous systems. The 'index' of an equilibrium point provides supporting information on its nature and complexity, particularly in strongly nonlinear cases, where the linear approximation is zero. Secondly, the phase diagram does not give a complete picture of the solutions; it is not sufficiently specific about the behaviour of paths at infinity beyond the boundaries of any diagram, and we show various projections which include 'points at infinity' and give the required overall view. Thirdly, a difficult question is to determine whether there are any limit cycles and roughly where they are; this question is treated again in Chapter 11, but here we give some elementary conditions for their existence or non-existence. Finally, having obtained all the information our methods allow about the geometrical layout of the phase diagram we may want to compute a number of typical paths, and some suggestions for carrying this out are made in Section 3.5.

3.1 The index of a point

Given the system

$$\dot{x} = X(x, y), \qquad \dot{y} = Y(x, y), \tag{3.1}$$

let Γ be any smooth *closed* curve, traversed counterclockwise, consisting only of ordinary points (in particular, it does not pass through any equilibrium points). Let Q be a point on Γ (Fig. 3.1); then there is one and only one phase path passing through Q. The phase paths belong to the family described by the equation

$$\frac{dy}{dx} = \frac{Y(x, y)}{X(x, y)}, \tag{3.2}$$

(see (2.3)). In time $\delta t > 0$ the coordinates of a representative point Q, (x_Q, y_Q) will increase by $\delta x, \delta y$ respectively, where

$$\delta x \approx X(x_Q, y_Q)\delta t, \qquad \delta y \approx Y(x_Q, y_Q)\delta t.$$

Therefore, the vector $S = (X, Y)$ is tangential to the phase path through the point, and points in the direction of increasing t. Its inclination can be measured by the angle ϕ measured counterclockwise from the positive direction of the x axis to the direction of S, so that

$$\tan \phi = Y/X. \tag{3.3}$$

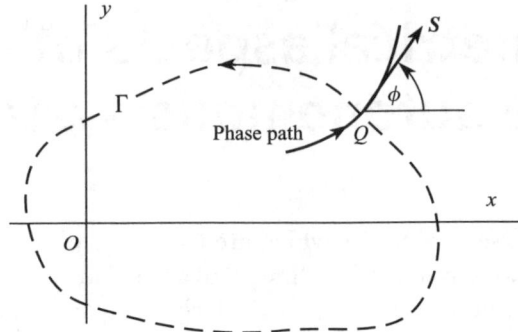

Figure 3.1   The angle of inclination ϕ of a phase path at its intersection with an arbitrarily directed closed curve Γ. The vector $S = (X, Y)$.

The curve Γ is traversed in the *counterclockwise* direction and the variation in ϕ is followed along it. When the value of ϕ at one point has been decided, the value for other points is settled by requiring ϕ to be a continuous function of position, except that ϕ will not in general have its original value on returning to its starting point after a full cycle: the value may differ by an integer multiple of 2π.

Example 3.1   *Trace the variation of the vector S and the angle ϕ when $X(x, y) = 2x^2 - 1$, $Y(x, y) = 2xy$, and Γ is the unit circle centred at the origin.*

The system $\dot{x} = X(x, y)$, $\dot{y} = Y(x, y)$ has equilibrium points at $(1/\sqrt{2}, 0)$ and $(-1/\sqrt{2}, 0)$, which are inside the unit circle Γ. Let θ be the polar angle of the representative point, and put $x = \cos\theta$, $y = \sin\theta$ on Γ. Then $(X, Y) = (\cos 2\theta, \sin 2\theta), 0 \le \theta \le 2\pi$. This vector is displayed in Fig. 3.2. The angle ϕ takes, for example, the values $0, \frac{1}{2}\pi, 2\pi, 4\pi$ at A, B, C, and A' as we track counterclockwise round the circle.

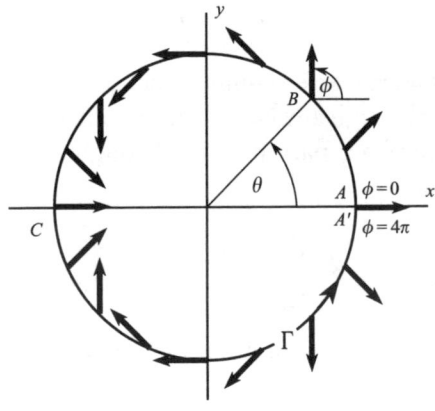

Figure 3.2

In any complete counterclockwise revolution, ϕ increases by 4π.

In every case the change in ϕ, denoted by $[\phi]_\Gamma$, must be a multiple of 2π;

$$[\phi]_\Gamma = 2\pi I_\Gamma, \tag{3.4}$$

say, where I_Γ is an integer, positive, negative, or zero. I_Γ is called the **index of** Γ with respect to the vector field (X, Y), Γ being described counterclockwise. In the previous example $I_\Gamma = 2$.

An expression for I_Γ is obtained as follows. Suppose that, as in Fig. 3.3(a), the curve Γ is described counterclockwise once by the position vector $\boldsymbol{r}$, given parametrically by

$$r(s) = (x(s), y(s)), \quad s_0 \leq s \leq s_1, \tag{3.5}$$

where $x(s_0) = x(s_1)$, $y(s_0) = y(s_1)$, and s is the parameter. From (3.3),

$$\frac{\mathrm{d}}{\mathrm{d}s}(\tan \phi) = \frac{\mathrm{d}}{\mathrm{d}s}\left(\frac{Y}{X}\right),$$

or, after some reduction,

$$\frac{\mathrm{d}\phi}{\mathrm{d}s} = \frac{XY' - YX'}{X^2 + Y^2}. \tag{3.6}$$

(The dash denotes differentiation with respect to s.) Then from (3.4)

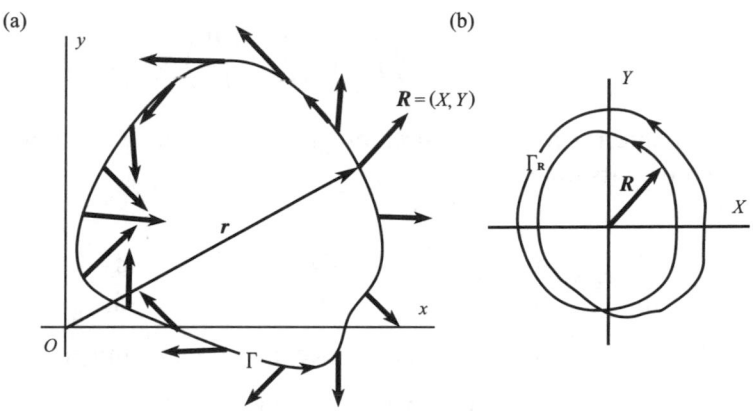

Figure 3.3

$$I_\Gamma = \frac{1}{2\pi} \int_{s_0}^{s_1} \frac{\mathrm{d}\phi}{\mathrm{d}s}\mathrm{d}s = \frac{1}{2\pi} \int_{s_0}^{s_1} \frac{XY' - YX'}{X^2 + Y^2}\mathrm{d}s. \tag{3.7}$$

As $r(s) = (x(s), y(s))$ describes Γ, $R(s) = (X, Y)$, regarded as a position vector on a plane with axes X, Y, describes another curve, Γ_R say. Γ_R is closed, since $\boldsymbol{R}$ returns to its original value after a complete cycle. From eqn (3.4) Γ_R encircles the origin I_Γ times, counterclockwise if I_Γ is positive and clockwise if it is negative. This is illustrated in Fig. 3.3 for $I_\Gamma = 2$. We can replace (3.7) by a line integral around the curve Γ_R so that, with X and Y as variables

$$I_\Gamma = \frac{1}{2\pi} \oint_{\Gamma_R} \frac{X\mathrm{d}Y - Y\mathrm{d}X}{X^2 + Y^2}. \tag{3.8}$$

Example 3.2   *In Example 3.1, $X(x, y) = 2x^2 - 1$, $Y(x, y) = 2xy$, and Γ is the unit circle centred at the origin. Confirm that $I_\Gamma = 2$ using (3.7).*

With Γ defined by $x = \cos\theta$, $y = \sin\theta$ $(0 \le \theta \le 2\pi)$ and $(X, Y) = (\cos 2\theta, \sin 2\theta)$,

$$
\begin{aligned}
I_\Gamma &= \frac{1}{2\pi} \int_0^{2\pi} \frac{XY' - YX'}{X^2 + Y^2} \, \mathrm{d}\theta \\
&= \frac{1}{2\pi} \int_0^{2\pi} \frac{2\cos 2\theta \cos 2\theta - \sin 2\theta(-2\sin 2\theta)}{\cos^2 2\theta + \sin^2 2\theta} \, \mathrm{d}\theta \\
&= \frac{1}{2\pi} \int_0^{2\pi} 2 \, \mathrm{d}\theta = 2.
\end{aligned}
$$

●

Theorem 3.1   *Suppose that Γ lies in a simply connected region on which X, Y and their first derivatives are continuous and X and Y are not simultaneously zero. (In other words there is no equilibrium point there.) Then I_Γ is zero.*

Proof   **Green's theorem** in the plane states that if Γ is a closed, non-self-intersecting curve, lying in a simply connected region on which the functions $P(x, y)$ and $Q(x, y)$ have continuous first partial derivatives, then

$$
\oint_\Gamma (P \, \mathrm{d}x + Q \, \mathrm{d}y) = \iint_{D_\Gamma} \left(\frac{\partial Q}{\partial x} - \frac{\partial P}{\partial y} \right) \mathrm{d}x \mathrm{d}y,
$$

where D_Γ is the region interior to Γ. (The first integral is a line integral round Γ, the second a double integral taken over its interior.)

In (3.7), write

$$
\frac{\mathrm{d}X}{\mathrm{d}s} = X_x \frac{\mathrm{d}x}{\mathrm{d}s} + X_y \frac{\mathrm{d}y}{\mathrm{d}s}, \qquad \frac{\mathrm{d}Y}{\mathrm{d}s} = Y_x \frac{\mathrm{d}x}{\mathrm{d}s} + Y_y \frac{\mathrm{d}y}{\mathrm{d}s},
$$

(where X_x denotes $\partial X / \partial x$, and so on). Then (3.7) becomes the line integral

$$
I_\Gamma = \frac{1}{2\pi} \oint_\Gamma \left(\frac{XY_x - YX_x}{X^2 + Y^2} \mathrm{d}x + \frac{XY_y - YX_y}{X^2 + Y^2} \mathrm{d}y \right).
$$

The functions $P = (XY_x - YX_x)/(X^2 + Y^2)$ and $Q = (XY_y - YX_y)/(X^2 + Y^2)$ satisfy the conditions for Green's theorem, since $X^2 + Y^2 \ne 0$ on Γ and its interior. Therefore

$$
I_\Gamma = \frac{1}{2\pi} \iint_{D_\Gamma} \left[\frac{\partial}{\partial x} \left(\frac{XY_y - YX_y}{X^2 + Y^2} \right) - \frac{\partial}{\partial y} \left(\frac{XY_x - YX_x}{X^2 + Y^2} \right) \right] \mathrm{d}x \, \mathrm{d}y.
$$

By evaluating the partial derivatives it can be verified that the integrand is identically zero on D_Γ. Therefore $I_\Gamma = 0$.   ∎

Corollary to Theorem 3.1   *Let Γ be a simple closed curve, and Γ' a simple closed curve inside Γ. Then if, on Γ, Γ' and the region between them there is no equilibrium point, and if X, Y and their first derivatives are continuous there, then $I_\Gamma = I_{\Gamma'}$.*

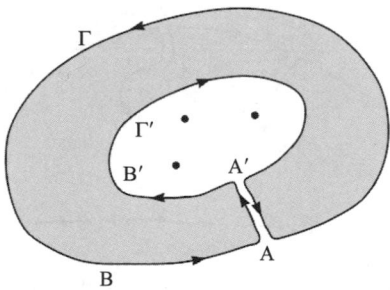

Figure 3.4 The composite simply connected closed curve C bridging Γ and Γ', and traversed in the counterclockwise direction (i.e., with its interior on the left). The dots represent equilibrium points which are exterior to C.

Proof In Fig. 3.4 let AA' be a 'bridge' connecting Γ and Γ', and consider the directed composite contour, C, described by $ABAA'B'A'A$ in Fig. 3.4. Since C contains no equilibrium points, $I_C = 0$ by the Theorem. But by (3.4)

$$0 = I_C = \frac{1}{2\pi} \oint_C d\phi = \frac{1}{2\pi} \left(\oint_\Gamma d\phi + \oint_{AA'} d\phi - \oint_{\Gamma'} d\phi + \oint_{A'A} d\phi \right),$$

where

$$\oint_{\Gamma'} d\phi \quad \text{and} \quad \oint_\Gamma d\phi$$

represent integrals taken in the counterclockwise direction around Γ' and Γ respectively. Since

$$\oint_{A'A} d\phi = - \oint_{AA'} d\phi,$$

we obtain

$$\frac{1}{2\pi} \left(\oint_\Gamma d\phi - \oint_{\Gamma'} d\phi \right) = 0.$$

Therefore

$$I_\Gamma = I_{\Gamma'}. \qquad \blacksquare$$

This theorem shows that the index I_Γ of the vector field (X, Y) with respect to Γ is to a large extent independent of Γ, and enables the index to be associated with special points in the plane rather than with contours. If the smoothness conditions on the field (X, Y) are satisfied in a region containing a single equilibrium point, then any simple closed curve Γ surrounding the point generates the same number I_Γ. We therefore drop the suffix Γ, and say that I is the **index of the equilibrium point**.

Theorem 3.2 *If Γ surrounds n equilibrium points $P_1, P_2, \ldots, P_n$ then*

$$I_\Gamma = \sum_{i=1}^{n} I_i,$$

where I_i is the index of the point P_i, $i = 1, 2, \ldots, n$.

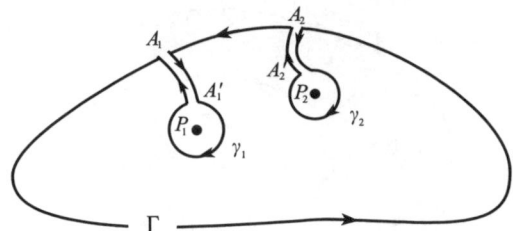

Figure 3.5 The composite simply connected curve $\mathcal{C}$, traversed in the counterclockwise direction. P_1 and P_2 are equilibrium points, exterior to $\mathcal{C}$.

Proof We illustrate the proof for the case of two equilibrium points, at P_1, P_2. Construct a contour $\mathcal{C}$ consisting of Γ, two 'bridges' $A_1 A_1'$ and $A_2 A_2'$, and curves γ_1 and γ_2 surrounding P_1 and P_2 respectively, as shown in Fig. 3.5. The interior of $\mathcal{C}$ does not contain any equilibrium points, so $I_\mathcal{C} = 0$. But also

$$I_\mathcal{C} = \frac{1}{2\pi} \oint_\Gamma \mathrm{d}\phi - \frac{1}{2\pi} \oint_{\gamma_1} \mathrm{d}\phi - \frac{1}{2\pi} \oint_{\gamma_2} \mathrm{d}\phi,$$

since the integrals along the 'bridges' cancel to zero. (Note that the integrals around γ_1 and γ_2 are here taken in the conventional counterclockwise direction.) Therefore

$$I_\Gamma = I_1 + I_2.$$ ∎

A simple way to calculate the index in a particular case is to use the following result.

Theorem 3.3 *Let p be the number of times $Y(x, y)/X(x, y)$ changes from $+\infty$ to $-\infty$, and q the number of times it changes from $-\infty$ to $+\infty$, on Γ. Then $I_\Gamma = \frac{1}{2}(p - q)$.*

Proof Since $\tan\phi = Y/X$, we are simply counting the number of times the direction (X, Y) is vertical (up or down), and associating a direction of rotation. We could instead examine the points where $\tan\phi$ is zero, that is, where Y is zero. Then if P and Q are numbers of changes in $\tan\phi$ from (negative/positive) to (positive/negative) respectively across the zero of Y, $I_\Gamma = \frac{1}{2}(P - Q)$. ∎

Example 3.3 *Find the index of the equilibrium point $(0,0)$ of $\dot{x} = y^3, \dot{y} = x^3$.*
By stretching the theory a little we can let Γ be the square with sides $x = \pm 1$, $y = \pm 1$. Start from $(1, 1)$ and go round counterclockwise. On $y = 1$, $\tan\phi = x^3$, with no infinities; similiarly on $y = -1$. On $x = -1$, $\tan\phi = -y^{-3}$, with a change from $-\infty$ to $+\infty$, and on $x = 1, \tan\phi = y^{-3}$, with a change from $-\infty$ to $+\infty$. Thus $p = 0, q = 2$ and the index is -1. ●

Example 3.4 *Find the index of the equilibrium point at $(0, 0)$ of the system $\dot{x} = y - \frac{1}{2}xy - 3x^2, \dot{y} = -xy - \frac{3}{2}y^2$.*
In using the method of Theorem 3.3, it is quite helpful first to sketch the curves

$$X(x, y) = y - \tfrac{1}{2}xy - 3x^2 = 0 \quad \text{and} \quad Y(x, y) = -xy - \tfrac{3}{2}y^2 = 0,$$

(see Fig. 3.6(a)). Equilibrium points occur where these curves intersect: at $(0, 0)$ and at $(-\frac{1}{4}, \frac{1}{6})$. The origin is a *higher order* equilibrium point (linearization will not classify it), and $(-\frac{1}{4}, \frac{1}{6})$ is a saddle point. Surround $(0, 0)$

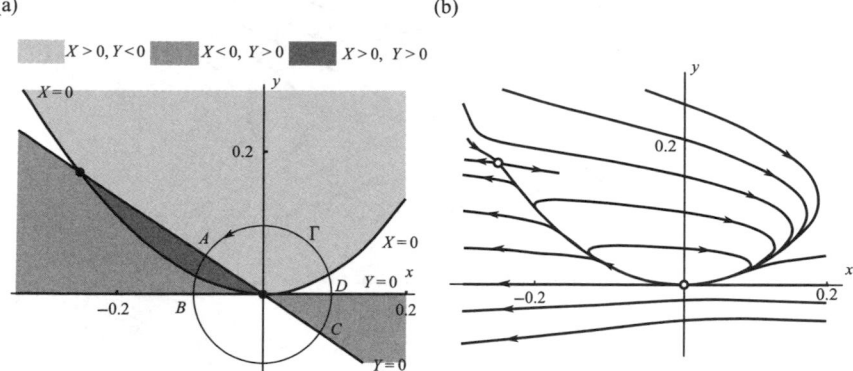

Figure 3.6

by a circle Γ which does not include the saddle point. Then $\tan\phi = Y(x, y)/X(x, y)$ becomes zero at A, B, C, and D. In a counterclockwise circuit of Γ, the sign changes in $\tan\phi$ are:

zero of $Y(x, y)$	A	B	C	D
sign change in $\tan\phi$	$-/+$	$-/+$	$+/-$	$-/+$

Hence $P = 3$ and $Q = 1$, so that the index of $(0, 0)$ is $I = 1$.

The phase diagram of the system is shown in Fig. 3.6(b). Note the peculiarity of the equilibrium point at the origin; it cannot be linearised about the origin. ●

If we already know the nature of the equilibrium point, the index is readily found by simply drawing a figure and following the angle round. The following shows the indices of the elementary types met in Chapter 2.

(i) *A saddle point* (Fig. 3.7) The change in ϕ in a single circuit of the curve Γ surrounding the saddle point is -2π, and the index is therefore -1.

(ii) *A centre* (Fig. 3.8) Γ can be chosen to be a phase path, so that $\Gamma = +1$ irrespective of the direction of the paths.

(iii) *A spiral (stable or unstable)* (Fig. 3.9). The index is $+1$.

(iv) *A node (stable or unstable)* (Fig. 3.10). The index is $+1$.

A simple practical way of finding the index of a curve Γ uses a watch. Set the watch to 12.00 (say) and put the watch at a point on Γ with the minute hand pointing in the direction of the phase path there. Move the watch without rotation around Γ, always turning the direction of the minute hand along the phase path. After one counterclockwise circuit of Γ, the index is given by the gain (-1 for each hour gained) or loss, of the hour hand. For example the 'time' after one circuit of the centre is 11.00, an index of $+1$; for the saddle point the 'time' is 1.00, an index of -1.

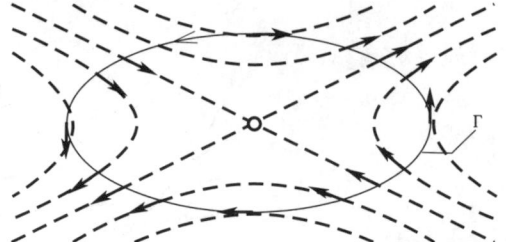

Figure 3.7 Saddle point, index −1.

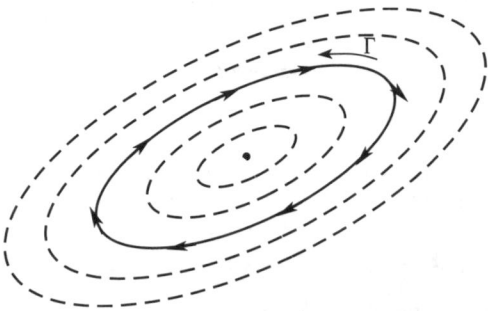

Figure 3.8 Centre, index 1.

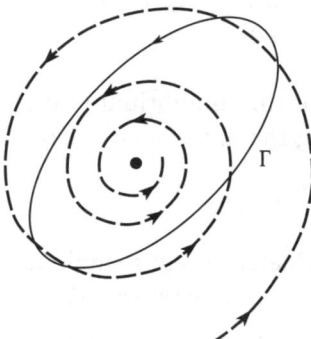

Figure 3.9 Stable spiral, index 1.

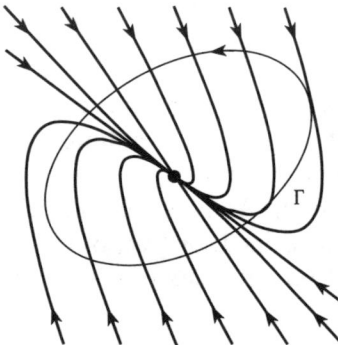

Figure 3.10 Stable node, index 1.

Exercise 3.1
For the system $\dot{x} = 2x^2 - 1$, $\dot{y} = 2xy$, let Γ be defined by $x = a \cos \theta$, $y = b \sin \theta$. Using the results from Example 3.1 deduce that

$$ab \int_0^{2\pi} \frac{a^2 + (a^2 - 1) \cos 2\theta}{(a^2 - 1 + a^2 \cos 2\theta)^2 + a^2 b^2 \sin^2 2\theta} \, d\theta = 2\pi$$

for all $a > 1/\sqrt{2}$, $b > 0$. (Index theory using Theorem 3.1 enables many seemingly complicated integrals to be evaluated by choosing different Γ's.)

Exercise 3.2
Find the index of the higher order saddle point at the origin of $\dot{x} = 2xy$, $\dot{y} = x^2 - y^2$.

3.2   The index at infinity

In the next two sections we show some techniques for getting a more global view of the pattern of trajectories for the system

$$\dot{x} = X(x, y), \qquad \dot{y} = Y(x, y).$$

Definition 3.1   *The index, I_∞, of the point at infinity. Introduce new coordinates x_1, y_1 by the transformation (known as inversion in the origin)*

$$x_1 = \frac{x}{x^2 + y^2}, \qquad y_1 = -\frac{y}{x^2 + y^2}. \tag{3.9}$$

*The index of the origin, $x_1 = y_1 = 0$, for the transformed equation is called the **index at infinity** for the original equation. (In polar coordinates let $x = r \cos \theta$, $y = r \sin \theta$ and $x_1 = r_1 \cos \theta_1$, $y_1 = r_1 \sin \theta_1$. Then the inversion is $r_1 = 1/r$, $\theta_1 = -\theta$.)*

In order to prove a result concerning the index of the point at infinity, it is convenient to use complex variables and write

$$z = x + iy, \qquad z_1 = x_1 + iy_1.$$

The differential equation system (3.1) becomes

$$\frac{dz}{dt} = Z.$$

where $Z = X + iY$ (Z is not, of course, an analytic function of z in general.) Also ϕ (eqn (3.3)) is given by $\phi = \arg Z$.

The transformation in (3.9) is equivalent to the complex transformation

$$z_1 = z^{-1}, \tag{3.10}$$

which maps points at infinity in the Z, or (x, y) plane into a neighbourhood of the origin in the Z_1, or (x_1, y_1) plane and vice versa, Then (3.9) becomes

$$-\frac{1}{z_1^2}\frac{dz_1}{dt} = Z,$$

or, from (3.9)

$$\frac{dz_1}{dt} = -z_1^2 Z = Z_1 \tag{3.11}$$

say, which is the differential equation for the transformed system. Therefore, $I_\infty = $ the index of (3.11) at $Z_1 = 0$. Notice that by eqn (3.11), the transformation maps equilibrium points on any infinite region of the Z plane into equilibrium points on the Z_1 plane.

Theorem 3.4   *The index I_∞ for the system $\dot{x} = X(x, y), \dot{y} = Y(x, y)$, having a finite number n of equilibrium points with indices $I_i, i = 1, 2, \ldots, n$, is given by*

$$I_\infty = 2 - \sum_{i=1}^{n} I_i. \tag{3.12}$$

Proof   To obtain the index at the origin of the transformed plane, $z_1 = 0$, surround $z_1 = 0$ by a simple closed contour C_1 containing no equilibrium points of the new system apart from $Z_1 = 0$. Under the transformation $z = z_1^{-1}$ (eqn (3.10)), C_1 described counterclockwise corresponds to C described clockwise, and the exterior of C_1 corresponds to the interior of C (Fig. 3.11). Thus C embraces all the finite equilibrium points of the original system together with the point at infinity.

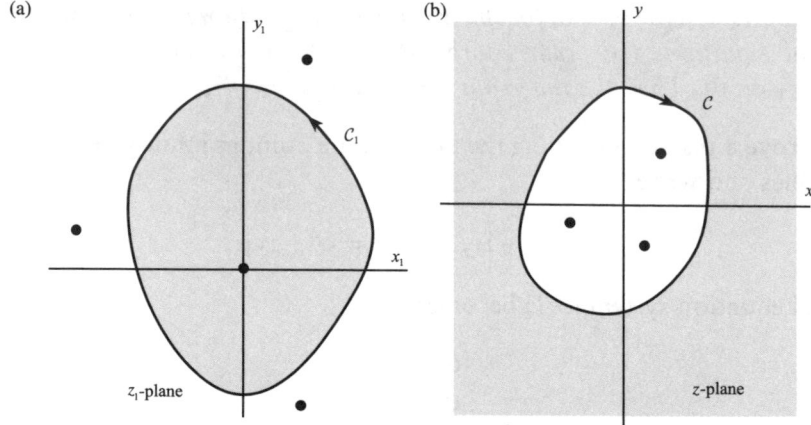

Figure 3.11   (a) The transformed plane and contour C_1 which surrounds the origin, but excludes locations of transformed equilibrium points. (b) The original plane and countour C containing all the equilibrium points.

Now let

$$z_1 = r_1 e^{i\theta_1} \tag{3.13}$$

on $\mathcal{C}_1$; and on $\mathcal{C}_1$ and $\mathcal{C}$, respectively let

$$Z_1 = \rho_1 e^{i\phi_1}, \qquad Z = \rho e^{i\phi},$$

where $\rho_1 = |Z_1|$, $\rho = |Z|$. Then

$$I_\infty = \frac{1}{2\pi}[\phi_1]_{\mathcal{C}_1}.$$

But from (3.11)

$$Z_1 = \rho_1 e^{i\phi_1} = -r_1^2 e^{2i\theta_1} \rho e^{i\phi} = r_1^2 \rho e^{i(2\theta_1 + \phi + \pi)}.$$

Therefore

$$
\begin{aligned}
I_\infty &= \frac{1}{2\pi}[2\theta_1 + \phi + \pi]_{\mathcal{C}_1} \\
&= \frac{1}{2\pi}\{2[\theta_1]_{\mathcal{C}_1} + [\phi]_{\mathcal{C}_1}\} = \frac{1}{2\pi}\{2[\theta_1]_{\mathcal{C}_1} - [\phi]_{\mathcal{C}}\} \\
&= \frac{1}{\pi}\left\{4\pi - 2\pi \sum_{i=1}^{n} I_i\right\}
\end{aligned}
$$

by Theorem 3.2, since also $\mathcal{C}$ is described clockwise. The result follows. ∎

Corollary to Theorem 3.4 *Under the conditions stated in the theorem, the sum of all the indices, including that of the point at infinity, is 2.* ∎

Example 3.5 *Find the index at infinity of the system*

$$\dot{x} = x - y^2, \qquad \dot{y} = x - y.$$

The system has equilibrium points at $(0, 0)$ and $(1, 1)$. We show a different approach from that of Examples 3.3 and 3.4. Putting $x = r \cos\theta$, $y = r \sin\theta$ we have

$$Z = (r\cos\theta - r^2 \sin^2\theta) + i(r\cos\theta - r\sin\theta),$$

and the transformation $z = z_1^{-1}$ gives $r = r_1^{-1}, \theta = -\theta_1$ so from (3.11)

$$Z_1 = -r_1^2(\cos 2\theta_1 + i\sin 2\theta_1)\{ir_1^{-1}(\cos\theta_1 + \sin\theta_1) + r_1^{-1}(\cos\theta_1 - r_1^{-1}\sin^2\theta_1)\}. \tag{3.14}$$

We require the index of eqn (3.11) at the point $r_1 = 0$. To evaluate it choose Γ_1 to be a circle centred at the origin, having arbitrarily small radius equal to r_1. Then we need to consider, in (3.14), only the terms of $O(1)$,

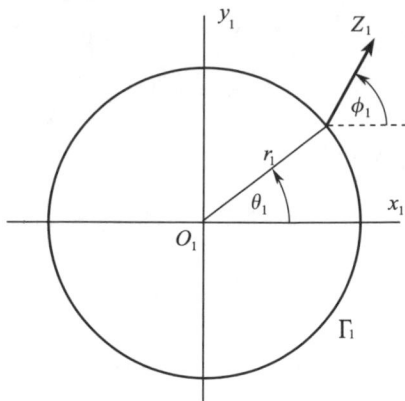

Figure 3.12

the terms of $O(r_1)$ making no contribution in the limit $r_1 \to 0$. Then on Γ

$$Z_1 = (\cos 2\theta_1 + \mathrm{i} \sin 2\theta_1) \sin^2 \theta_1 + O(r_1) = \sin^2 \theta_1 e^{2\mathrm{i}\theta_1} + O(r_1).$$

As shown in Fig. 3.12 the direction, ϕ_1, of Z_1 is equal to $2\theta_1$, and the index at infinity is therefore equal to 2.

It can be confirmed that the other equilibrium points are a spiral or centre at $x = 0$, $y = 0$ with index 1, and a saddle at $x = 1$, $y = 1$ with index -1. The sum of these indices and I_∞ is therefore 2, as required by Theorem 3.4.   ●

Exercise 3.3
Using the inversion $x_1 = x/(x^2 + y^2)$, $y_1 = y/(x^2 + y^2)$, express the equations $\dot{x} = 2xy$, $\dot{y} = x^2 + y^2$ in terms of x_1, y_1. Using the unit circle $x = \cos\theta$, $y = \sin\theta$, find the index at infinity of the system.

3.3 The phase diagram at infinity

Phase diagrams such as we have shown are incomplete since there is always an area outside the picture which we cannot see. When the phase diagram has a simple structure this may not be a serious loss, but in complicated cases the overall structure may be obscure. We therefore show how the whole phase diagram, including the infinitely remote parts, can be displayed.

Projection on to a hemisphere

In Fig. 3.13, $\mathcal{S}$ is a hemisphere of unit radius touching the phase plane $\mathcal{P}$ at O. Its centre is O^*. A representative point P on $\mathcal{P}$ projects from O^* on to P' on the sphere. Paths on $\mathcal{P}$ become paths on $\mathcal{S}$; such features as spirals, nodes, and saddle points on $\mathcal{P}$ appear as such on $\mathcal{S}$. The orientation of curves is also unchanged provided the sphere is viewed from inside. Local distortion occurs increasingly for features far from the origin, radial distance on $\mathcal{P}$ becoming

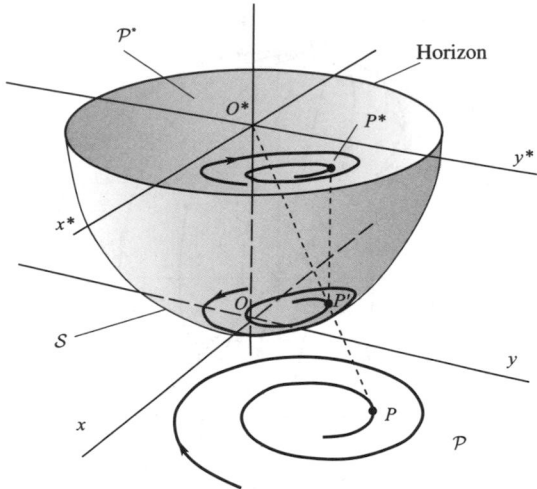

Figure 3.13 Projection of the phase plane $\mathcal{P}$ on to the hemisphere $\mathcal{S}$, and then from $\mathcal{S}$ on to the diametrical plane $\mathcal{P}^*$.

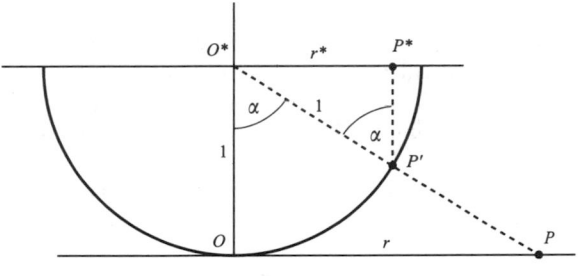

Figure 3.14

greatly scaled down on $\mathcal{S}$. The horizontal great circle is called the **horizon**, and the happenings 'at infinity' on $\mathcal{P}$ occur near the horizon of $\mathcal{S}$. Radial straight lines on $\mathcal{P}$ project on to great circles of $\mathcal{S}$ in a vertical plane, cutting the horizon at right angles. Non-radial straight lines project on to great circles which cut the horizon at other angles.

To represent on another plane the resulting pattern of lines on the hemisphere, we may simply take its orthogonal projection, viewed from above, on to the diametrical plane $\mathcal{P}^*$ containing the horizon, as in the following example. Figure 3.14 shows the vertical section OO^*P of Fig. 3.13. Let $OP = r$ and $O^*P^* = r^*$. Then $r^* = \sin\alpha$ and $r = \tan\alpha$. Consequently

$$r^* = r/\sqrt{(1+r^2)}.$$

Since x, x^* and y, y^* are in the same proportions as r, r^*, it follow that the coordinates in the phase plane (x, y) and disc (x^*, y^*) are related by

$$x^* = x/\sqrt{(1+r^2)}, \qquad y^* = y/\sqrt{(1+r^2)}.$$

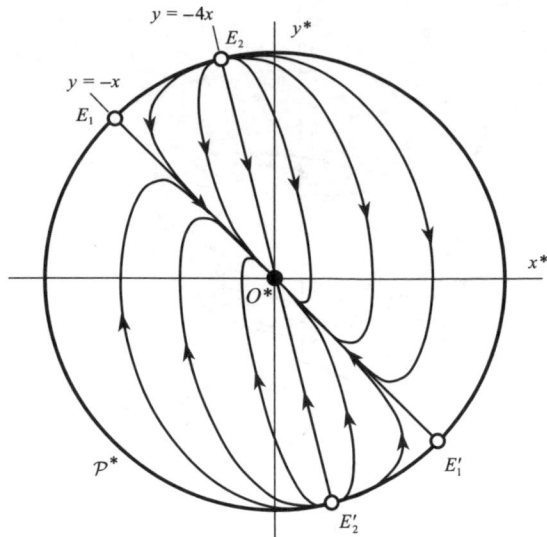

Figure 3.15   Projection of the phase diagram of $\dot{x} = y$, $\dot{y} = -4x - 5y$ on the diametrical plane $\mathcal{P}^*$.

Example 3.6   *Project the phase paths of $\dot{x} = y$, $\dot{y} = -4x - 5y$ on to the diametrical plane*
This represents a heavily damped linear oscillator. On the phase plane $\mathcal{P}$, $(0,0)$ is a stable node. $y = -x$, $y = -4x$ consist of four phase paths, and the results of Chapter 2 show that all other paths are tangent to the first at the origin and parallel to the second at infinity. The appearance of the diametrical plane P^* with axes x^*, y^* is shown in Fig. 3.15.

The paths $y = -x$, $y = -4x$ on the phase plane $\mathcal{P}$ are radial straight lines and project into lines of longitude through O on the sphere. Their projections on to the diametrical plane are therefore again straight lines $E_1 O$, $E_1' O$, $E_2 O$, $E_2' O$ of Fig. 3.15.

Since all other paths become parallel to $y = -4x$ at large distance from the origin, their projections enter the points E_2, E_2' on the sphere as shown. Other mappings of the hemisphere on to a plane can be devised.   ●

Detail on the horizon

In order to study the paths on the hemisphere $\mathcal{S}$ near the horizon we will project the paths found on $\mathcal{S}$ from its centre O^* on to the plane $\mathcal{U}$: $x = 1$. We thus have, simultaneously, a projection of the paths in the phase plane $\mathcal{P}$ on to $\mathcal{U}$. Let u, z be axes with origin at N, where $\mathcal{S}$ touches $\mathcal{U}$: u is parallel to y and z is positive downwards, so u, z is right-handed viewed from $\dot{O}^*$. Points in $\mathcal{P}$ for $x < 0$ go into the left half of the lower hemisphere, then through O^* on to $\mathcal{U}$ for $z < 0$; and $\mathcal{P}$ for $x > 0$ goes into the right half of $\mathcal{S}$, then on to $\mathcal{U}$ for $z > 0$. Points at infinity on $\mathcal{P}$ go to the horizon of $\mathcal{S}$, then to $z = 0$ on $\mathcal{U}$. The points at infinity on $\mathcal{P}$ in direction $\pm y$ go to points at infinity on $\mathcal{U}$, as does the point O. The topological features of $\mathcal{P}$ are preserved on $\mathcal{U}$, except that the orientation of closed paths on the part of $\mathcal{S}$ corresponding to $x < 0$ is reversed (see Fig. 3.16).

It can be shown easily that the transformation from $\mathcal{P}$ to $\mathcal{U}$ is given by

$$u = y/x, \qquad z = 1/x, \tag{3.15}$$

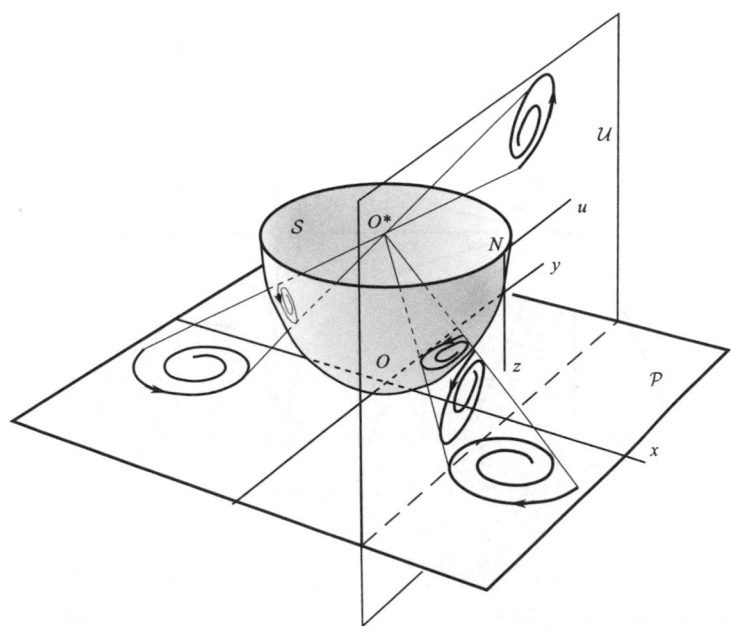

Figure 3.16 Projection of a phase diagram on $\mathcal{P}$ on to the vertical plane $\mathcal{U}$: the hemispherical projection on to $\mathcal{S}$ is also shown. The hemispherical space should be viewed from inside, and the vertical plane from the left.

with inverse

$$x = 1/z, \qquad y = u/z.$$

Example 3.7 *Examine the system $\dot{x} = y$, $\dot{y} = -4x - 5y$ at infinity (excluding the direction $\pm y$).*
(This is the system of Example 3.6.) The transformation (3.15) leads to

$$\dot{u} = x\dot{y} - y\dot{x}/x^2 = -(4x^2 + 5xy + y^2)/x^2 = -4 - 5u - u^2$$

and, similarly,

$$\dot{z} = -\dot{x}/x^2 = -uz.$$

There are equilibrium points at E_1: $u = -1, z = 0$, and at E_2: $u = -4, z = 0$ and it can be confirmed in the usual way that E_1 is a saddle and E_2 a node. The pattern of the paths is shown in Fig. 3.17, and should be compared with Fig. 3.15.

To examine the phase plane at infinity in the directions $\pm y$, $\mathcal{S}$ is projected on another plane $\mathcal{V}$: $y = 1$. The corresponding axes v, z, right-handed viewed from O^*, have their origin at the point of tangency; v points in the direction of $-x$ and z is downward. The required transformation is

$$v = x/y, \qquad z = 1/y, \tag{3.16}$$

or

$$x = v/z, \qquad y = 1/z, \qquad\qquad\qquad\qquad \bullet$$

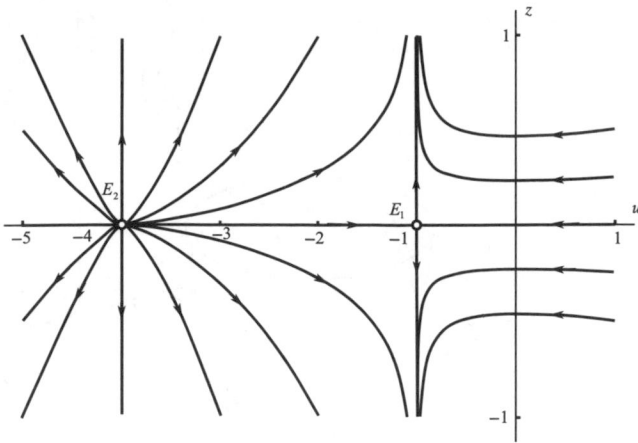

Figure 3.17   Phase paths near the horizon for $\dot{x} = y,\ \dot{y} = -4x - 5y$.

Exercise 3.4
Find the differential equation of

$$\dot{x} = 2xy, \qquad \dot{y} = x^2 - y^2$$

in the diametrical plane of the hemisphere using the transformation

$$x^* = x/\sqrt{(1 + r^2)}, \qquad y^* = y/\sqrt{(1 + r^2)}, \quad r^2 = x^2 + y^2.$$

Where are the equilibrium points at infinity in the diametrical plane?

3.4   Limit cycles and other closed paths

In nonlinear systems there is particular interest in the existence of periodic solutions, and their amplitudes, periods, and phases. If the system is autonomous, and $x(t)$ in any solution, so is $x(t + \tau)$ for *any* value of τ, which means that phase is not significant since the solutions map on to the same phase paths. In the phase plane, periodic solutions appear as *closed* paths. Conservative systems (Section 1.3) and Hamiltonian systems (Section 2.8) often contain nests of closed paths forming centres, which we might expect since they are generally non-dissipative; that is, friction is absent.

As we have seen (Section 1.6), a **limit cycle** is an *isolated* periodic solution of an autonomous system, represented in the phase plane by an isolated closed path. The neighbouring paths are, by definition, not closed, but spiral into or away from the limit cycle $\mathcal{C}$ as shown in Fig. 3.18. In the case illustrated, which is a **stable limit cycle**, the device represented by the system (which might be, for example, an electrical circuit), will spontaneously drift into the corresponding periodic oscillation from a wide range of initial states. The existence of limit cycles is therefore a feature of great practical importance.

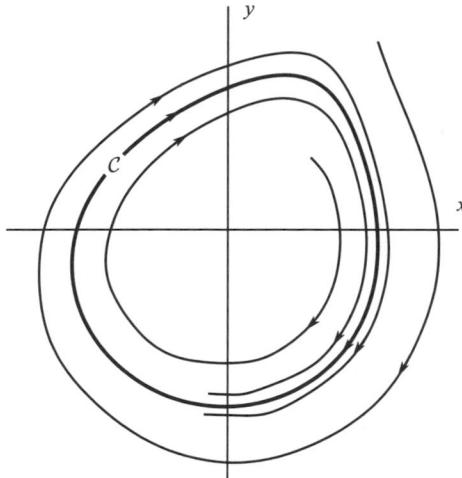

Figure 3.18 An example of a stable limit cycle, showing two phase paths approaching it from the interior and exterior respectively.

Autonomous linear systems (i.e. with constant coefficients) cannot exhibit limit cycles (see Section 2.6(v)). Since normally we cannot solve the nonlinear equations which might have them it is important to be able to establish by indirect means whether a limit cycle is present in the phase diagram. If one exists, then various approximate methods can be used to locate it. In this section we give some simple indications and counter-indications, the question of existence being dealt with in more detail in Chapter 11.

The index of a limit cycle C is 1 since the vector (X, Y) is tangential to C at every point on it, and the change in ϕ around C is 2π. By Theorem 3.2, therefore *if C is a limit cycle then the sum of the indices at the equilibrium points enclosed by C is 1.*

This result applies to any closed path, whether isolated (i.e., a limit cycle) or not, and it provides a **negative criterion**, indicating cases where such a path *cannot* exist. For example, a closed path cannot surround a region containing no equilibrium points, nor one containing only a saddle. If the sum of the indices of a group of equilibrium points does equal unity, the result does *not* allow us to infer the existence of a closed-path surrounding them.

The following result is due to Bendixson (Cesari 1971) and is called Bendixson's Negative Criterion:

Theorem 3.5 (Bendixson's negative criterion)   *There are no closed paths in a simply connected region of the phase plane on which $\partial X/\partial x + \partial Y/\partial y$ is of one sign.*

Proof   We make the usual assumptions about the smoothness of the vector field (X, Y) necessary to justify the application of the divergence theorem. Suppose that there is a closed phase path C in the region $\mathcal{D}$ where $\partial X/\partial x + \partial Y/\partial y$ is of one sign (see Fig. 3.19); we shall show that this assumption leads to a contradiction. By the divergence theorem,

$$\iint_{\mathcal{S}} \left(\frac{\partial X}{\partial x} + \frac{\partial Y}{\partial y} \right) dx \, dy = \int_{C} (X, Y) \cdot \boldsymbol{n} \, ds,$$

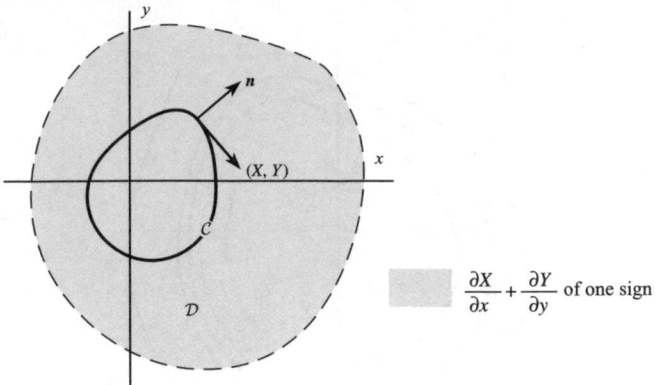

Figure 3.19 C is a closed phase path.

where S is the interior of C, $\boldsymbol{n}$ is the unit outward normal, and $\mathrm{d}s$ is an undirected line element of C. Since on C, (X, Y) is perpendicular to $\boldsymbol{n}$, the integral on the right is zero. But the integrand on the left is of one sign so its integral cannot be zero. Therefore C cannot be a path. ∎

An extension of this result, called Dulac's test, is given in Problem 3.23.

Example 3.8 *Show that the equation $\ddot{x} + f(x)\dot{x} + g(x) = 0$ can have no periodic solution whose phase path lies in a region where f is of one sign. (Such regions have only 'positive damping' or 'negative damping'.)*

The equivalent system is $\dot{x} = y$, $\dot{y} = -f(x)y - g(x)$, so that $(X, Y) = (y, -f(x)y - g(x))$, and

$$\frac{\partial X}{\partial x} + \frac{\partial Y}{\partial y} = -f(x),$$

which is of one sign wherever f is of one sign. ●

Example 3.9 The equations

$$\dot{V}_1 = -\sigma V_1 + f(E - V_2) = X(V_1, V_2),$$
$$\dot{V}_2 = -\sigma V_2 + f(E - V_1) = Y(V_1, V_2),$$

describe the voltages generated over the deflection plates in a simplified triggered sweeping circuit for an oscilloscope (Andronov and Chaikin 1949). E and σ are positive constants, and f has the shape shown in Fig. 3.20 f is a continuous, odd function, tending to a limit as $x \to \infty$, and f' is positive and decreasing on $x > 0$. It can be shown (Chapter 2, Problem 2.33) that there is an equilibrium point at $V_1 = V_2, = V_0$ say. If moreover $f'(E - V_0) > \sigma$, then there are two others, and in this case (V_0, V_0) is a saddle point and the other two are stable nodes. The sum of the indices is 1, so periodic solutions with paths surrounding the three points are not ruled out by this test. However,

$$\frac{\partial X}{\partial V_1} + \frac{\partial Y}{\partial V_2} = -2\sigma,$$

which is of one sign everywhere, so in fact there are no periodic solutions. ●

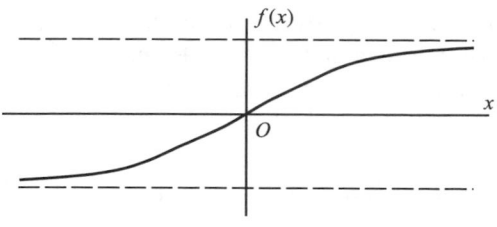

Figure 3.20

Exercise 3.5

Show that the system

$$\dot{x} = (x^2 + y^2 - 1)x + y, \qquad \dot{y} = (x^2 + y^2 - 1)y - x$$

has an unstable limit cycle. (Hint: express the equations in polar coordinates.)

Exercise 3.6

Explain why the following systems can have no periodic solutions:

(a) $\dot{x} = x(y^2 + 1) + y, \qquad \dot{y} = (x^2 + 1)(y^2 + 1)$;

(b) $\dot{x} = x(y^2 + 1) + y, \qquad \dot{y} = x^2 y + x$.

3.5 Computation of the phase diagram

There are many computer packages available which can compute and display phase diagrams and solution curves. Symbolic software such as *Mathematica*™ (Wolfram, 1996) or MAPLE™ can be readily programmed to generate graphical representations. These are also dedicated packages for dynamical systems. (see, e.g. Ermentrout (2002)).

Here we also indicate that, if such facilities are not available, quite unsophisticated methods can work well if they are supported by analytical reasoning, and simple precautions are taken. Some BASIC programs for computing phase diagrams are also given by Acheson (1997).

We consider autonomous equations of the form

$$\dot{x} = P(x, y), \qquad \dot{y} = Q(x, y). \tag{3.17}$$

In plotting a phase path we proceed step-by-step in some way from the initial point x_0, y_0:

$$x(t_0) = x_0, \qquad y(t_0) = y_0. \tag{3.18}$$

The simplest step-by-step method is Euler's, using t as a supplementary variable, which we use as the basis of discussion.

Let the length of the time step be h, assumed small; and let (x_0, y_0), (x_1, y_1), (x_2, y_2), ..., be the points on the phase path at times $t_0, t_0 + h, t_0 + 2h, \dots$. Assume that (x_n, y_n) is known. Then, writing $\dot{x}(t_0 + nh) \simeq (x_{n+1} - x_n)/h$, $\dot{y}(t_0 + nh) \approx (y_{n+1} - y_n)/h$, eqn (3.17) becomes approximately

$$(x_{n+1} - x_n)/h = P(x_n, y_n), \qquad (y_{n+1} - y_n)/h = Q(x_n, y_n),$$

and, after rearrangement,

$$x_{n+1} = x_n + hP(x_n, y_n), \qquad y_{n+1} = y_n + hQ(x_n, y_n). \qquad (3.19)$$

The point (x_{n+1}, y_{n+1}) is recorded and we proceed to the next step. The danger of progressively increasing error is present, even when h is very small so that good accuracy is ostensibly being obtained: the reader should see, for example, Cohen (1973) for a discussion of error.

The point which concerns us here is the general unsuitability of t (time) as the parameter for the step-by-step construction of phase paths. The difficulty is shown schematically in Fig. 3.21, where a path is shown approaching a node, and passing near a saddle. Notionally equal time steps are marked. Since $\dot{x}, \dot{y}$, or P, Q, are small near an equilibrium point, progress is very uneven.

It is preferable to use arc length, s, as the parameter, which gives equally spaced points along the arc. Since $\delta s^2 = \delta x^2 + \delta y^2$, (3.17) with parameter s becomes

$$\frac{\mathrm{d}x}{\mathrm{d}s} = \frac{P}{\sqrt{(P^2 + Q^2)}} = U, \qquad \frac{\mathrm{d}y}{\mathrm{d}s} = \frac{Q}{\sqrt{(P^2 + Q^2)}} = V \qquad (3.20)$$

(s is measured increasing in the direction of t increasing). Letting h now represent the basic step length, (3.20) leads to the iterative scheme

$$x_{n+1} = x_n + hU(x_n, y_y), \qquad y_{n+1} = y_n + hV(x_n, y_y) \qquad (3.21)$$

with the initial values (x_0, y_0) given.

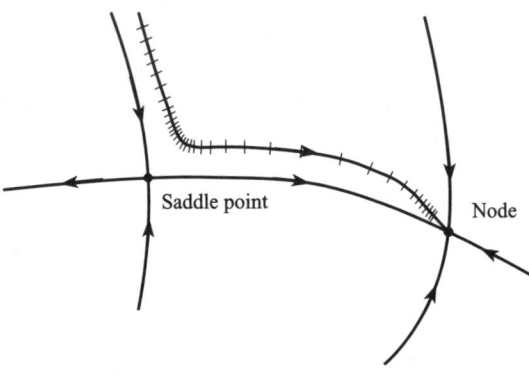

Figure 3.21   Equal time steps near equilibrium points.

Improved processes, such as the Runge–Kutta method (Cohen 1973) are generally to be preferred, and can easily be programmed. It is desirable to have a program which prints out only after a predetermined number of calculation steps. In this way it can be arranged, for example, to produce a point every centimetre of arc along a phase path, the interval being reduced if the path is turning rapidly.

In a practical plot it is helpful, if possible, to locate and classify the equilibrium points and to know at least whether limit cycles are likely to appear. Saddle points in particular are difficult to locate closely by plotting paths. The location of equilibrium points may be obtained by algebraic or numerical solution of the equations $P(x, y) = Q(x, y) = 0$. Classification on the basis of linear approximation may not be possible, for example when P or Q have a zero linear expansion, or when the linear approximation predicts a centre, which may become a spiral in the presence of nonlinear terms (as in Problem 2.7), or in certain other cases. In such cases many exploratory plots may have to be made near the equilibrium point before an adequate characterization is obtained.

By reversing the sign of h in the input to suitable programs, a plot 'backwards' from the starting point is obtained. This is useful to complete a path (Fig. 3.22(a)) or to get close to an unstable node (Fig. 3.22(b)) or spiral.

It may be difficult to distinguish with confidence between certain cases, for example between a node whose paths appear to wind around it and a genuine spiral (Fig. 3.23). A display which appears to represent a spiral may, on scaling up the neighbourhood of the origin, begin to resemble a node, and conversely. This consideration applies, in principle, to all numerical and plotted data: it is impossible to deduce from the data alone that some qualitative deviation does not appear between the plotted points. However, in the case of equilibrium points and limit cycles there may be genuine uncertainty, and it is preferable to establish their nature analytically and to determine the directions of separatrices when this is possible.

A limit cycle is a feature requiring some care. The 'model' of a limit cycle which it is useful to have in mind is a pattern of spirals inside and outside the cycle as in Fig. 3.18. The figure shows a stable limit cycle, for which all nearby paths drift towards the limit cycle. If they all recede (reverse the arrows) the limit cycle is unstable. A semi-stable limit cycle can also occur, in which the paths approach on one side and recede on the other.

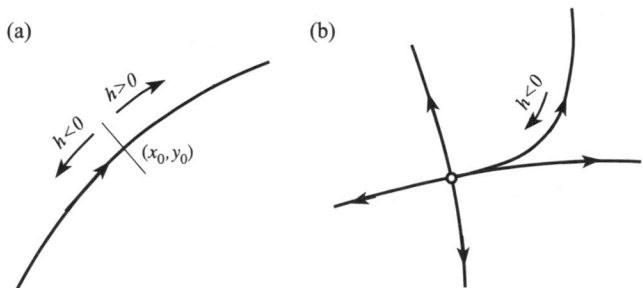

Figure 3.22 Illustrating reversal of step sign.

(a)   (b)

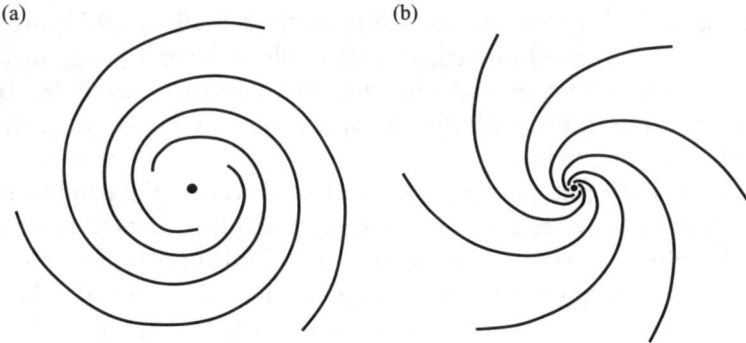

Figure 3.23   (a) Small-scale plot (spiral?); (b) large-scale plot (node?).

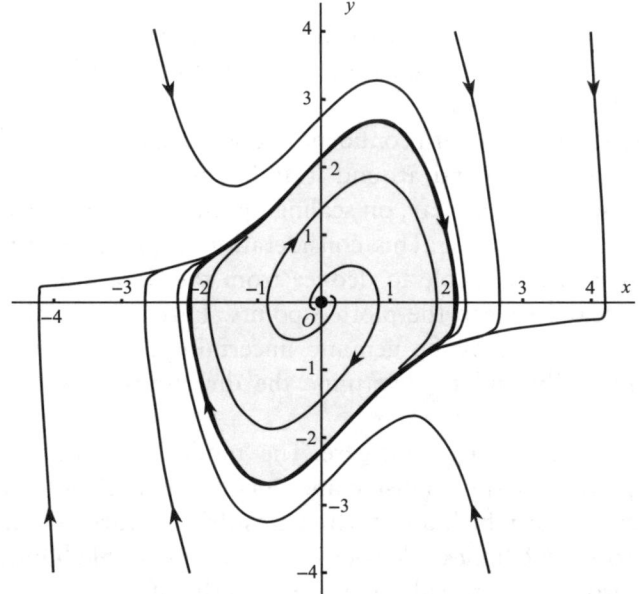

Figure 3.24   Limit cycle for the van der Pol equation $\ddot{x} + (x^2 - 1)\dot{x} + x = 0$.

Both the limit cycle and the neighbouring paths can diverge very considerably from the idealized pattern of Fig. 3.18; for example Fig. 3.24 shows the limit cycle for **van der Pol's** equation

$$\ddot{x} + \varepsilon(x^2 - 1)\dot{x} + x = 0$$

with a moderate value for the parameter ε. In general there is no way of finding a single point, exactly lying on the cycle, from which to start; the cycle has to be located by 'squeezing' it between inner and outer spirals. Clearly it is helpful to reverse the sign of t (or of h in the program input) if necessary, so as to approach rather than recede from the cycle during the process of locating it.

3.6 Homoclinic and heteroclinic paths

A **separatrix** is, generally, a phase path which separates obvious distinct regions in the phase plane. It could be any of the paths which enter or emerge from a saddle point; or a limit cycle; or a path joining two equilibrium points. It is not a precise term. However, certain types of separatrix are important in later work, and we shall define them here in the context of equilibrium points.

Any phase path which joins an equilibrium point to itself is a form of separatrix known as a **homoclinic path**, and any phase path which joins an equilibrium point to another is known as a **heteroclinic path**. For plane autonomous systems this means that homoclinic paths can only be associated with *saddle points* (or some generalized type of saddle) since both outgoing and incoming paths are required of the equilibrium point. On the other hand it is possible for heteroclinic paths to join any two *hyperbolic* equilibrium points, that is (see Section 2.5) any two (including the same type) from the list of a saddle point, a node or a spiral. Phase paths which join the same, or two different, saddles are also known as **saddle connections**. Figure 3.25 shows some examples (dashed lines) of homoclinic and heteroclinic paths.

Example 3.10   Find the equations of the homoclinic phase paths of the system

$$\dot{x} = y, \qquad \dot{y} = x - x^3. \tag{i}$$

Find also the solutions for x in terms of t on the homoclinic paths.

The equilibrium points occur where $y = x - x^3 = 0$ at $(0,0), (\pm 1, 0)$. The origin is a saddle point whilst $x = \pm 1$ are both centres. The phase paths satisfy the separable differential equation

$$\frac{dy}{dx} = \frac{x(1 - x^2)}{y},$$

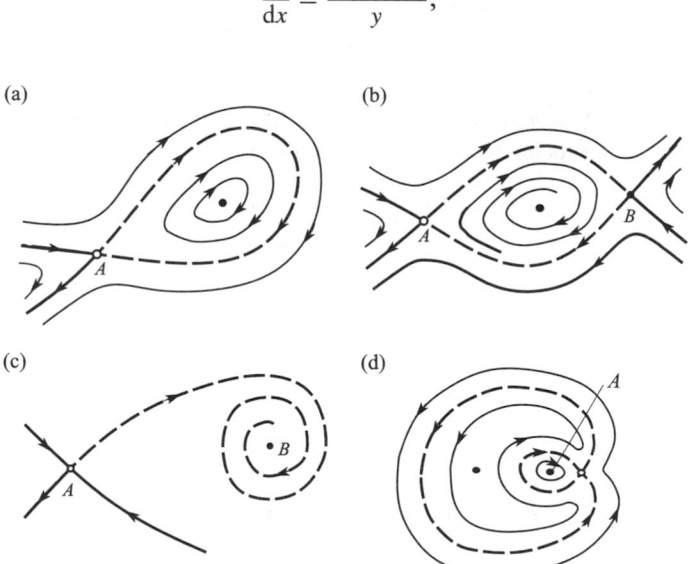

Figure 3.25   Dashed curves show; (a) homoclinic path of A; (b) two heteroclinic paths joining A and B; (c) a saddle-spiral connection; (d) two homoclinic paths of A.

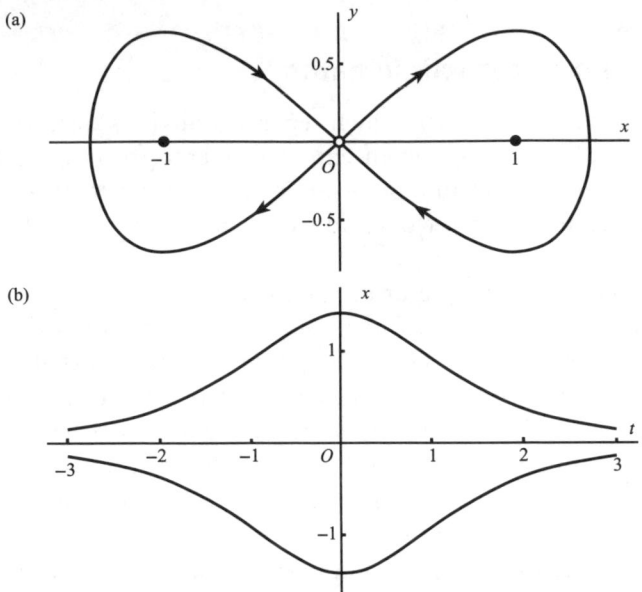

Figure 3.26   (a) Homoclinic paths of $\dot{x} = y$, $\dot{y} = x - x^3$; (b) homoclinic solutions $x = \pm\sqrt{2}\,\mathrm{sech}\,t$ in the (x, t) plane.

which can be integrated to give

$$y^2 = x^2 - \tfrac{1}{2}x^4 + C,$$

where C is a constant. Homoclinic paths can only be associated with the saddle point at the origin. The phase paths approach the origin only if $C = 0$. There are two such paths, both given by

$$y^2 = x^2 - \tfrac{1}{2}x^4,$$

one over the interval $0 \leq x \leq \sqrt{2}$, and one over $-\sqrt{2} \leq x \leq 0$. The homoclinic paths are shown in Fig. 3.26(a).
   The time solutions for the homoclinic paths can be found by integrating

$$\left(\frac{dx}{dt}\right)^2 = x^2 - \frac{1}{2}x^4.$$

Then separating variables

$$\int \frac{dx}{x\sqrt{(1 - \tfrac{1}{2}x^2)}} = \pm(t - t_0),$$

where t_0 is a constant. Using the substitutions $x = \pm\sqrt{2}\,\mathrm{sech}\,u$, it follows that

$$\int du = \pm(t - t_0) \quad \text{or} \quad u = \pm(t - t_0).$$

Hence the homoclinic solutions are $x = \pm 2\,\mathrm{sech}\,(t - t_0)$ for any t_0 since $x \to 0$ as $t \to \pm\infty$. The solutions with $t_0 = 0$ are shown in Fig. 3.26(b): other choices for t_0 will produce translations of these solutions in the t direction.   ●

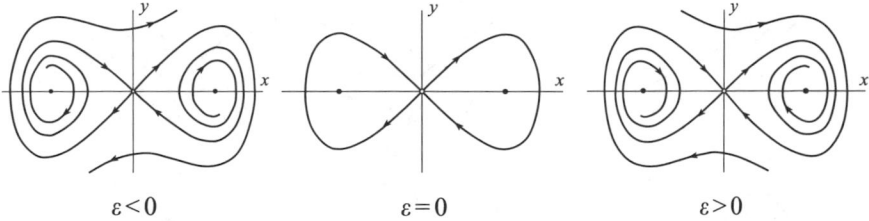

Figure 3.27 Homoclinic bifurcation of $\ddot{x} + \varepsilon\dot{x} - x + x^3 + 0$ at $\varepsilon = 0$.

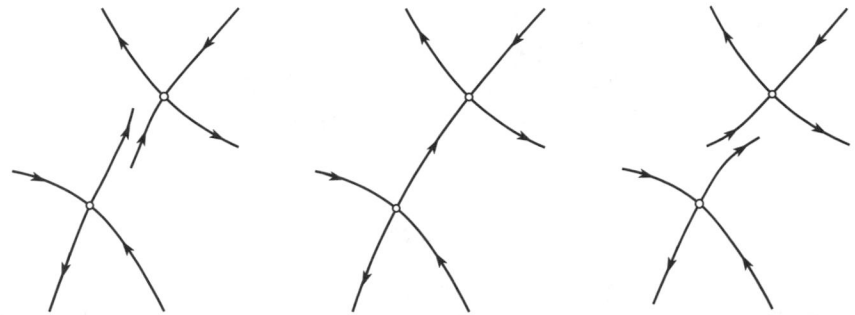

Figure 3.28 An illustration of Heteroclinic bifurcation.

Consider now a perturbation of the system in Example 3.9, namely

$$\ddot{x} + \varepsilon\dot{x} - x + x^3 = 0, \qquad \dot{x} = y,$$

where ε is a small parameter. If $\varepsilon > 0$ then the term $\varepsilon\dot{x}$ indicates damping, and if $\varepsilon < 0$, its effect produces negative damping. The equilibrium points at $(\pm1, 0)$ become stable spirals if $\varepsilon > 0$, and unstable spirals if $\varepsilon < 0$. As ε increases through zero, the path changes from a heteroclinic spiral–saddle connection ($\varepsilon < 0$) to a homoclinic saddle connection ($\varepsilon = 0$) to a heteroclinic saddle–spiral connection ($\varepsilon > 0$), as shown in Fig. 3.27. This transition is known as **homoclinic bifurcation**. A perturbation method for detecting homoclinic bifurcation is given in Section 5.12.

An illustration of transition through heteroclinic bifurcation as a parameter varies, is shown in Fig. 3.28.

Problems

3.1 By considering the variation of path direction on closed curves round the equilibrium points, find the index in each case of Fig. 3.29.

3.2 The motion of a damped pendulum is described by the equations

$$\dot{\theta} = \omega, \qquad \dot{\omega} = -k\omega - \nu^2\sin\theta,$$

where $k\,(>0)$ and ν are constants. Find the indices of all the equilibrium states.

3.3 Find the index of the equilibrium points of the following systems: (i) $\dot{x} = 2xy$, $\dot{y} = 3x^2 - y^2$; (ii) $\dot{x} = y^2 - x^4$, $\dot{y} = x^3 y$; (iii) $\dot{x} = x - y$, $\dot{y} = x - y^2$.

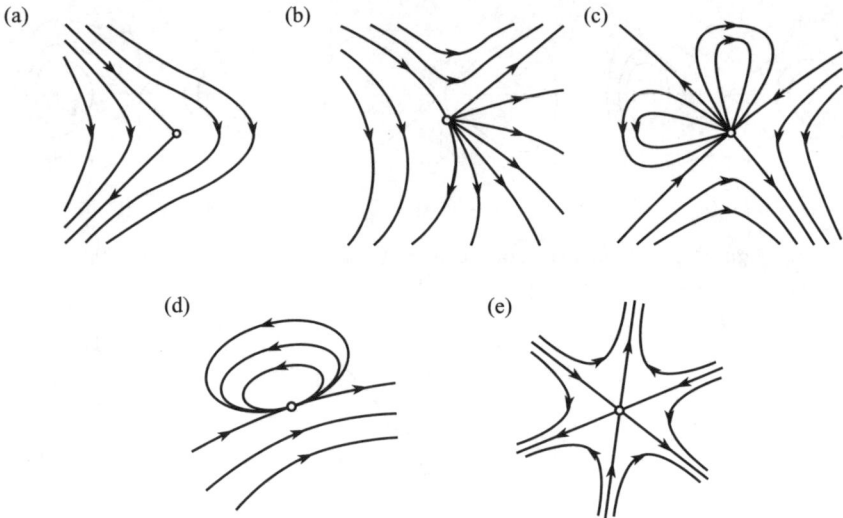

Figure 3.29

3.4 For the linear system $\dot{x} = ax + by, \dot{y} = cx + dy$, where $ad - bc \neq 0$, obtain the index at the origin by evaluating eqn (3.7), showing that it is equal to sgn $(ad - bc)$. (Hint: choose Γ to be the ellipse $(ax + by)^2 + (cx + dy)^2 = 1$.)

3.5 The equation of motion of a bar restrained by springs (see Fig. 3.30)

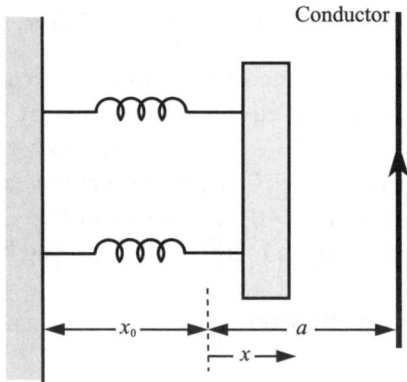

Figure 3.30

and attracted by a parallel current-carrying conductor is

$$\ddot{x} + c\{x - \lambda/(a - x)\} = 0,$$

where c (the stiffness of the spring), a and λ are positive constants. Sketch the phase paths for $-x_0 < x < a$, where x_0 is the unstretched length of each spring, and find the indices of the equilibrium points for all $\lambda > 0$.

3.6 Show that the equation $\ddot{x} - \varepsilon(1 - x^2 - \dot{x}^2)\dot{x} + x = 0$ has an equilibrium point of index 1 at the origin of the phase plane x, y with $\dot{x} = y$. (It also has a limit cycle, $x = \cos t$.) Use eqn (3.7), with Γ a circle of radius a to show that, for all a,

$$\int_0^{2\pi} \frac{d\theta}{1 - 2\varepsilon(1-a^2)\sin\theta\cos\theta + \varepsilon^2(1-a^2)^2\sin^2\theta} = 2\pi.$$

3.7 A limit cycle encloses N nodes, F spirals, C centres, and S saddle points only, all of the linear type of Section 2.5. Show that $N + F + C - S = 1$.

3.8 Given the system

$$\dot{x} = X(x,y)\cos\alpha - Y(x,y)\sin\alpha, \quad \dot{y} = X(x,y)\sin\alpha + Y(x,y)\cos\alpha,$$

where α is a parameter, prove that the index of a simple closed curve which does not meet an equilibrium point is independent of α (See also Problem 2.41).

3.9 Suppose that the system $\dot{x} = X(x)$ has a finite number of equilibrium points, each of which is either a node, a centre, a spiral, or a saddle point, of the elementary types discussed in Section 2.5, and assume that $I_\infty = 0$. Show that the total number of nodes, centres, and spirals is equal to the total number of saddle points plus two.

3.10 Obtain differential equations describing the behaviour of the linear system, $\dot{x} = ax + by$, $\dot{y} = cx + dy$, at infinity. Sketch the phase diagram and analyse the system $\dot{x} = 2x - y$, $\dot{y} = 3x - 2y$ near the horizon.

3.11 A certain system is known to have exactly two equilibrium points, both saddle points. Sketch phase diagrams in which
 (i) a separatrix connects the saddle points;
 (ii) no separatrix connects them.

For example, the system $\dot{x} = 1 - x^2$, $\dot{y} = xy$ has a saddle correction joining the saddle points at $(\pm 1, 0)$. The perturbed system $\dot{x} = 1 - x^2$, $\dot{y} = xy + \varepsilon x^2$ for $0 < \varepsilon \ll 1$ breaks the saddle correction (heteroclinic bifurcation).

3.12 Deduce the index at infinity for the system $\dot{x} = x - y$, $\dot{y} = x - y^2$ by calculating the indices of the equilibrium points.

3.13 Use the geometrical picture of the field (X, Y) in the neighbourhood of an ordinary point to confirm Theorem 3.1.

3.14 Suppose that, for the two plane systems $\dot{x}_1 = X_1(x_1), \dot{x}_2 = X_2(x_2)$, and for a given closed curve Γ, there is no point on Γ at which X_1 and X_2 are opposite in direction. Show that the index of Γ is the same for both systems.

The system $\dot{x} = y$, $\dot{y} = x$ has a saddle point at the origin. Show that the index of the origin for the system $\dot{x} = y + cx^2 y$, $\dot{y} = x - cy^2$ is likewise -1.

3.15 Use Problem 3.14 to show that the index of the equilibrium point $x = 0$, $\dot{x} = 0$ for the equation $\ddot{x} + \sin x = 0$ on the usual phase plane has index 1, by comparing the equation $\ddot{x} + x = 0$.

3.16 The system

$$\dot{x} = ax + by + P(x,y), \qquad \dot{y} = cx + dy + Q(x,y)$$

has an isolated equilibrium point at $(0,0)$, and $P(x,y) = O(r^2)$, $Q(x,y) = O(r^2)$ as $r \to 0$, where $r^2 = x^2 + y^2$. Assuming that $ad - bc \neq 0$, show that the origin has the same index as its linear approximation.

3.17 Show that, on the phase plane with $\dot{x} = y$, $\dot{y} = Y(x,y)$, Y continuous, the index I_Γ of any simple closed curve Γ that encloses all equilibrium points can only be 1, -1, or zero.

Let $\dot{x} = y$, $\dot{y} = f(x, \lambda)$, with f, $\partial f/\partial x$ and $\partial f/\partial \lambda$ continuous, represent a parameter-dependent system with parameter λ. Show that, at a bifurcation point (Section 1.7), where an equilibrium point divides as λ varies, the sum of the indices of the equilibrium points resulting from the splitting is unchanged. (Hint: the integrand in eqn (3.7) is continuous.)

Deduce that the equilibrium points for the system $\dot{x} = y$, $\dot{y} = -\lambda x + x^3$ consist of a saddle point for $\lambda < 0$, and a centre and two saddles for $\lambda > 0$.

3.18 Prove a similar result to that of Problem 3.17 for the system $\dot{x} = y$, $\dot{y} = f(x, y, \lambda)$. Deduce that the system $\dot{x} = y$, $\dot{y} = -\lambda x - ky - x^3$, ($k > 0$), has a saddle point at $(0, 0)$ when $\lambda < 0$, which bifurcates into a stable spiral or node and two saddle points as λ becomes positive.

3.19 A system is known to have three closed paths, C_1, C_2, and C_3, such that C_2, and C_3 are interior to C_1 and such that C_2, C_3 have no interior points in common. Show that there must be at least one equilibrium point in the region bounded by C_1, C_2, and C_3.

3.20 For each of the following systems you are given some information about phase paths and equilibrium points. Sketch phase diagrams consistent with these requirements.
  (i) $x^2 + y^2 = 1$ is a phase path, $(0, 0)$ a saddle point, $(\pm\frac{1}{2}, 0)$ centres.
  (ii) $x^2 + y^2 = 1$ is a phase path, $(-\frac{1}{2}, 0)$ a saddle point, $(0, 0)$ and $(\frac{1}{2}, 0)$ centres.
  (iii) $x^2 + y^2 = 1$, $x^2 + y^2 = 2$ are phase paths, $(0, \pm\frac{3}{2})$ stable spirals, $(\pm\frac{3}{2}, 0)$ saddle points, $(0, 0)$ a stable spiral.

3.21 Consider the system

$$\dot{x} = y(z - 2), \qquad \dot{y} = x(2 - z) + 1, \qquad x^2 + y^2 + z^2 = 1$$

which has exactly two equilibrium points, both of which lie on the unit sphere. Project the phase diagram on to the plane $z = -1$ through the point $(0, 0, 1)$. Deduce that $I_\infty = 0$ on this plane (consider the projection of a small circle on the sphere with its centre on the z axis). Explain, in general terms, why the sum of the indices of the equilibrium points on a sphere is two.

  For a certain problem, the phase diagram on the sphere has centres and saddle points only, and it has exactly two saddle points. How many centres has the phase diagram?

3.22 Show that the following systems have no periodic solutions:
  (i) $\dot{x} = y + x^3$, $\quad \dot{y} = x + y + y^3$;
  (ii) $\dot{x} = y$, $\quad \dot{y} = -(1 + x^2 + x^4)y - x$.

3.23 **(Dulac's test)** For the system $\dot{x} = X(x, y), \dot{y} = Y(x, y)$, show that there are no closed paths in a simply connected region in which $\partial(\rho X)/\partial x + \partial(\rho Y)/\partial y$ is of one sign, where $\rho(x, y)$ is any function having continuous first partial derivatives.

3.24 Explain in general terms how Dulac's test (Problem 3.23) and Bendixson's negative criterion may be extended to cover the cases when $\partial(\rho X)/\partial x + \partial(\rho Y)/\partial y$ is of one sign except on isolated points or curves within a simply connected region.

3.25 For a second-order system $\dot{x} = X(x)$, $\mathbf{curl}(X) = 0$ and $X \neq 0$ in a simply connected region $\mathcal{D}$. Show that the system has no closed paths in $\mathcal{D}$. Deduce that

$$\dot{x} = y + 2xy, \qquad \dot{y} = x + x^2 - y^2$$

has no periodic solutions.

3.26 In Problem 3.25 show that $\mathbf{curl}(X) = 0$ may be replaced by $\mathbf{curl}(\psi X) = 0$, where $\psi(x, y)$ is of one sign in $\mathcal{D}$.

3.27 By using Dulac's test (Problem 3.23) with $\rho = e^{-2x}$, show that

$$\dot{x} = y, \qquad \dot{y} = -x - y + x^2 + y^2$$

has no periodic solutions.

3.28 Use Dulac's test (Problem 3.23) to show that $\dot{x} = x(y - 1)$, $\dot{y} = x + y - 2y^2$, has no periodic solutions.

3.29 Show that the following systems have no periodic solutions:

(i) $\dot{x} = y$, $\dot{y} = 1 + x^2 - (1 - x)y$;

(ii) $\dot{x} = -(1 - x)^3 + xy^2$, $\dot{y} = y + y^3$;

(iii) $\dot{x} = 2xy + x^3$, $\dot{y} = -x^2 + y - y^2 + y^3$;

(iv) $\dot{x} = x$, $\dot{y} = 1 + x + y^2$;

(v) $\dot{x} = y$, $\dot{y} = -1 - x^2$;

(vi) $\dot{x} = 1 - x^3 + y^2$, $\dot{y} = 2xy$;

(vii) $\dot{x} = y$, $\dot{y} = (1 + x^2)y + x^3$;

3.30 Let $\mathcal{D}$ be a doubly connected region in the x, y plane. Show that, if $\rho(x, y)$ has continuous first partial derivatives and $\operatorname{div}(\rho X)$ is of constant sign in $\mathcal{D}$, then the system has not more than one closed path in $\mathcal{D}$. (An extension of Dulac's test Problem 3.23.)

3.31 A system has exactly two limit cycles with one lying interior to the other and with no equilibrium points points between them. Can the limit cycles be described in opposite senses? Obtain the equations of the phase paths of the system

$$\dot{r} = \sin \pi r, \qquad \dot{\theta} = \cos \pi r$$

as described in polar coordinates (r, θ). Sketch the phase diagram.

3.32 Using Bendixson's theorem (Section 3.4) show that the response amplitudes a, b for the van der Pol equation in the 'van der Pol plane' (this will be discussed later in Chapter 7), described by the equations

$$\dot{a} = \frac{1}{2}\varepsilon \left(1 - \frac{1}{4}r^2\right)a - \frac{\omega^2 - 1}{2\omega}b,$$

$$\dot{b} = \frac{1}{2}\varepsilon \left(1 - \frac{1}{4}r^2\right)b + \frac{\omega^2 - 1}{2\omega}a + \frac{\Gamma}{2\pi},$$

$$r = \sqrt{(a^2 + b^2)},$$

have no closed paths in the circle $r < \sqrt{2}$.

3.33 Let C be a closed path for the system $\dot{x} = X(x)$, having $\mathcal{D}$ as its interior. Show that

$$\iint_{\mathcal{D}} \operatorname{div}(X) dx\, dy = 0.$$

3.34 Assume that van der Pol's equation in the phase plane

$$\dot{x} = y, \qquad \dot{y} = -\varepsilon(x^2 - 1)y - x$$

has a single closed path, which, for ε small, is approximately a circle, centre the origin, of radius a. Use the result of Problem 3.33 to show that approximately

$$\int_{-a}^{a} \int_{-\sqrt{(a^2 - x^2)}}^{\sqrt{(a^2 - x^2)}} (x^2 - 1) dy\, dx = 0,$$

and so deduce a.

3.35 Following Problems 3.33 and 3.34, deduce a condition on the amplitudes of periodic solutions of

$$\ddot{x} + \varepsilon h(x, \dot{x})\dot{x} + x = 0, \qquad |\varepsilon| \ll 1.$$

3.36 For the system

$$\ddot{x} + \varepsilon h(x, \dot{x})\dot{x} + g(x) = 0,$$

suppose that $g(0) = 0$ and $g'(x) > 0$. Let C be a closed path in the phase plane (as all paths must be) for the equation

$$\ddot{x} + g(x) = 0$$

having interior C. Use the result of Problem 3.33 to deduce that for small ε, C approximately satisfies

$$\int_{\mathcal{D}} \{h(x, y) + h_y(x, y)y\}dx\, dy = 0.$$

Adapt this result to the equation

$$\ddot{x} + \varepsilon(x^2 - \alpha)\dot{x} + \sin x = 0,$$

with ε small, $0 < \alpha \ll 1$, and $|x| < \frac{1}{2}\pi$. Show that a closed path (a limit cycle) is given by

$$y^2 = 2A + 2\cos x,$$

where A satisfies

$$\int_{-\cos^{-1}(-A)}^{\cos^{-1}(-A)} (x^2 - \alpha)\sqrt{\{2A + 2\cos x\}}dx = 0.$$

3.37 Consider the system

$$\dot{x} = X(x, y) = -(x^2 + y^2)y,$$
$$\dot{y} = Y(x, y) = bx + (1 - x^2 - y^2)y.$$

Let C be the circle $x^2 + y^2 = a^2$ with interior $\mathcal{R}$. Show that

$$\iint_{\mathcal{R}} \operatorname{div}(X, Y)dx\, dy = 0$$

only if $a = 1$. Is C a phase path (compare Problem 3.33)?

3.38 The equation $\ddot{x} + F_0 \tanh k(\dot{x} - 1) + x = 0$, $F_0 > 0$, $k \gg 1$, can be thought of as a plausible continuous representation of the type of Coulomb friction problem of Section 1.6. Show, however, that the only equilibrium point is a stable spiral, and that there are no periodic solutions.

3.39 Show that the third-order system $\dot{x}_1 = x_2$, $\dot{x}_2 = -x_1$, $\dot{x}_3 = 1 - (x_1^2 + x_2^2)$ has no equilibrium points but nevertheless has closed paths (periodic solutions).

3.40 Sketch the phase diagram for the quadratic system $\dot{x} = 2xy$, $\dot{y} = y^2 - x^2$.

3.41 Locate the equilibrium points of the system

$$\dot{x} = x(x^2 + y^2 - 1), \qquad \dot{y} = y(x^2 + y^2 - 1),$$

and sketch the phase diagram.

3.42 Find the equilibrium points of the system

$$\dot{x} = x(1 - y^2), \qquad \dot{y} = x - (1 + e^x)y.$$

Show that the system has no closed paths.

3.43 Show using Bendixson's theorem that the system

$$\dot{x} = x^2 + y^2, \qquad \dot{y} = y^2 + x^2 e^x,$$

has no closed paths in $x + y > 0$ or $x + y < 0$. Explain why the system has no closed paths in the x, y plane.

3.44 Plot a phase diagram, showing the main features of the phase plane, for the equation $\ddot{x} + \varepsilon(1 - x^2 - \dot{x}^2)\dot{x} + x = 0$, using $\dot{x} = y$, $\varepsilon = 0.1$ and $\varepsilon = 5$.

3.45 Plot a phase diagram for the damped pendulum equation $\ddot{x} + 0.15\dot{x} + \sin x = 0$. See Fig. 3.31.

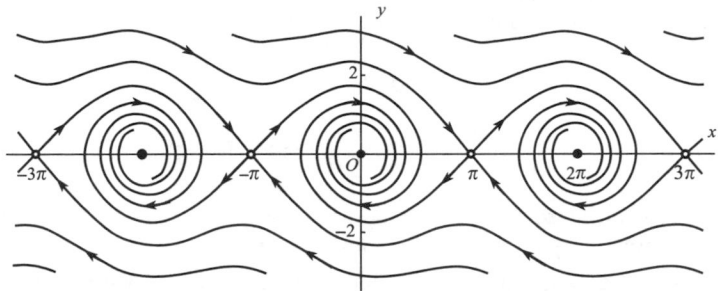

Figure 3.31   Phase diagram for the pendulum equation with small damping.

3.46 The system

$$\dot{x} = -\frac{1}{2\omega} y \left\{ (\omega^2 - 1) - \frac{3}{4}\beta(x^2 + y^2) \right\},$$

$$\dot{y} = \frac{1}{2\omega} x \left\{ (\omega^2 - 1) - \frac{3}{4}\beta(x^2 + y^2) \right\} + \frac{\Gamma}{2\omega}$$

occurs in the theory of the forced oscillations of a pendulum. Obtain the phase diagram when $\omega = 0.975, \Gamma = 0.005, \beta = -\frac{1}{6}$.

3.47 A population of rabbits $R(t)$ and foxes $F(t)$ live together in a certain territory. The combined birth and death rate of the rabbits due to 'natural' causes is $\alpha_1 > 0$, and the additional deaths due to their being eaten by foxes is introduced through an 'encounter factor' β_1, so that

$$\frac{dR}{dt} = \alpha_1 R - \beta_1 RF.$$

The foxes die of old age with death rate $\beta_2 > 0$, and the live birth rate is sustained through an encounter factor α_2, so that (compare Example 2.3)

$$\frac{dF}{dt} = \alpha_2 RF - \beta_2 F.$$

Plot the phase diagram, when $\alpha_1 = 10$, $\beta_1 = 0.2$, $\alpha_2 = 4 \times 10^{-5}$, $\beta_2 = 0.2$. Also plot typical solution curves $R(t)$ and $F(t)$ (these are oscillatory, having the same period but different phase).

3.48 The system

$$\dot{x} = \frac{1}{2}\alpha \left(1 - \frac{1}{4}r^2 \right) x - \frac{\omega^2 - 1}{2\omega} y \quad (r^2 = x^2 + y^2),$$

$$\dot{y} = \frac{\omega^2 - 1}{2\omega} x + \frac{1}{2}\alpha \left(1 - \frac{1}{4}r^2 \right) y + \frac{\Gamma}{2\omega},$$

occurs in the theory of forced oscillations of the van der Pol equation (Section 7.4, and see also Problem 3.32). Plot phase diagrams for the cases

(i) $\alpha = 1, \Gamma = 0.75, \omega = 1.2$;

(ii) $\alpha = 1, \Gamma = 2.0, \omega = 1.6$.

3.49 The equation for a tidal bore on a shallow stream is

$$\varepsilon \frac{\mathrm{d}^2\eta}{\mathrm{d}\xi^2} - \frac{\mathrm{d}\eta}{\mathrm{d}\xi} + \eta^2 - \eta = 0,$$

where (in appropriate dimensions), η is the height of the free surface, and $\xi = x - ct$ where c is the wave speed. For $0 < \varepsilon \ll 1$, find the equilibrium points of the equation and classify them according to their linear approximations.

Plot the phase paths in the plane of η, w, where

$$\frac{\mathrm{d}\eta}{\mathrm{d}\xi} = w, \qquad \varepsilon \frac{\mathrm{d}w}{\mathrm{d}\xi} = \eta + w - \eta^2$$

and show that a separatrix from the saddle point at the origin reaches the other equilibrium point. Interpret this observation in terms of the shape of the wave.

3.50 Determine the nature of the equilibrium point, and compute the phase diagram for the Coulomb friction type problem $\ddot{x} + x = F(\dot{x})$ where

$$F(y) = \begin{cases} -6.0(y - 1), & |y - 1| \le 4 \\ -[1 + 1.4\exp\{-0.5|y - 1| + 0.2\}]\mathrm{sgn}(y - 1), & |y - 1| \ge 0.4. \end{cases}$$

(See Fig. 3.32, and compare the simpler case shown in Section 1.6.)

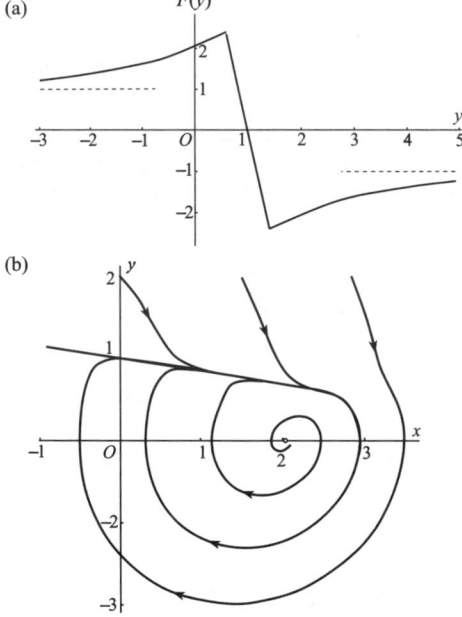

Figure 3.32   (a) Typical Coulomb dry friction force $F(y)$ in terms of slip velocity y; (b) corresponding phase diagram.

3.51 Compute the phase diagram for the system whose polar representation is

$$\dot{r} = r(1-r), \qquad \dot{\theta} = \sin^2(\tfrac{1}{2}\theta).$$

3.52 Compute the phase diagrams for the following systems: (i) $\dot{x} = 2xy$, $\dot{y} = y^2 + x^2$; (ii) $\dot{x} = 2xy$, $\dot{y} = x^2 - y^2$; (iii) $\dot{x} = x^3 - 2x^2y$, $\dot{y} = 2xy^2 - y^3$.

3.53 Obtain the heteroclinic phase paths for the system $\dot{x} = y$, $\dot{y} = -x + x^3$. Show that their time solutions are given by

$$x = \pm \tanh\left[\tfrac{1}{2}\sqrt{2}(t - t_0)\right].$$

3.54 Obtain the heteroclinic phase paths of $\ddot{x} + \sin x = 0$, $\dot{x} = y$. (This is a periodic differential equation in x. If the phase diagram is viewed on a cylinder of circumference 2π, then heteroclinic paths would appear to be homoclinic.)

3.55 Find the homoclinic paths of $\ddot{x} - x + 3x^5 = 0$, $\dot{x} = y$. Show that the time solutions are given by $x = \pm\sqrt{[\mathrm{sech}(t - t_0)]}$.

3.56 Find all heteroclinic phase paths of $\dot{x} = y(1 - x^2)$, $\dot{y} = -x(1 - y^2)$, (see Example 2.1).

3.57 The problem of the bead sliding on the rotating wire was discussed in Example 1.12, where it was shown that the equation of motion of the bead is

$$a\ddot{\theta} = g(\lambda \cos\theta - 1)\sin\theta.$$

Find the equations of all homoclinic and heteroclinic paths carefully distinguishing the cases $0 < \lambda < 1$, $\lambda = 1$ and $\lambda > 1$.

3.58 Consider the equation $\ddot{x} - x(x - a)(x - b) = 0$, $0 < a < b$. Find the equation of its phase paths. Show that a heteroclinic bifurcation occurs in the neighbourhood of $b = 2a$. Draw sketches showing the homoclinic paths for $b < 2a$ and $b > 2a$.
Show that the time solution for the heteroclinic path ($b = 2a$) is

$$x = \frac{2a}{1 + e^{-a\sqrt{2}(t-t_0)}}.$$

3.59 Show that

$$\dot{x} = 4(x^2 + y^2)y - 6xy, \qquad \dot{y} = 3y^2 - 3x^2 - 4x(x^2 + y^2)$$

has a higher-order saddle at the origin (neglect the cubic terms for $\dot{x}$ and $\dot{y}$, and show that near the origin the saddle has solutions in the directions of the straight lines $y = \pm x/3$, $x = 0$. Confirm that the phase paths through the origin are given by

$$(x^2 + y^2)^2 = x(3y^2 - x^2).$$

By plotting this curve, convince yourself that three homoclinic paths are associated with the saddle point at the origin.

3.60 Investigate the equilibrium points of

$$\dot{x} = y[16(2x^2 + 2y^2 - x) - 1], \qquad \dot{y} = x - (2x^2 + 2y^2 - x)(16x - 4),$$

and classify them according to their linear approximations. Show that homoclinic paths through $(0, 0)$ are given by

$$\left(x^2 + y^2 - \tfrac{1}{2}x\right)^2 - \tfrac{1}{16}\left(x^2 + y^2\right) = 0,$$

and that one homoclinic path lies within the other.

3.61 The following model differential equation exhibits two limit cycles bifurcating through homoclinic paths into a single limit cycle of larger amplitude as the parameter ε decreases through zero:

$$\ddot{x} + (\dot{x}^2 - x^2 + \tfrac{1}{2}x^4 + \varepsilon)\dot{x} - x + x^3 = 0.$$

Let $|\varepsilon| < \tfrac{1}{2}$.

(a) Find and classify the equilibrium points of the equation.

(b) Confirm that the equation has phase paths given by

$$\dot{y} = x^2 - \tfrac{1}{2}x^4 - \varepsilon, \qquad y = \dot{x}.$$

Find where the paths cut the x-axis.

(c) As ε decreases through zero what happens to the limit cycles which surround the equilibrium point at $x = \pm 1$? (It could be quite helpful to plot phase paths numerically for a sample of ε values.) Are they all stable?

3.62 Classify the equilibrium points of $\ddot{x} = x - 3x^2$, $\dot{x} = y$. Show that the equation has one homoclinic path given by $y^2 = x^2 - 2x^3$. Solve this equation to obtain the (x, t) solution for the homoclinic path.

3.63 Classify all the equilibrium points of $\dot{x} = y(2y^2 - 1)$, $\dot{y} = x(x^2 - 1)$, according to their linear approximations. Show that the homoclinic paths are given by

$$2y^2(y^2 - 1) = x^2(x^2 - 2),$$

and that the heteroclinic paths lie on the ellipse $x^2 + \sqrt{2}y^2 = \tfrac{1}{2}(2 + \sqrt{2})$, and the hyperbola $x^2 - \sqrt{2}y^2 = \tfrac{1}{2}(2 - \sqrt{2})$. Sketch the phase diagram.

3.64 A dry friction model has the equation of motion $\ddot{x} + x = F(\dot{x})$, where

$$F(y) = \begin{cases} -\mu(y - 1), & |y - 1| \le \varepsilon, \\ -\mu\varepsilon\,\mathrm{sgn}(y - 1) & |y - 1| > \varepsilon, \end{cases}$$

where $0 < \varepsilon < 1$ (see Fig. 3.33). Find the equations of the phase paths in each of the regions $y > 1 + \varepsilon$, $1 - \varepsilon \le y \le 1 + \varepsilon$, $y < 1 - \varepsilon$.

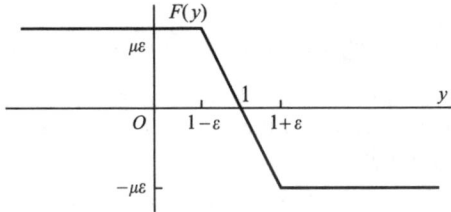

Figure 3.33

3.65 Locate and classify the equilibrium points of

$$\dot{x} = x^2 - 1, \qquad \dot{y} = -xy + \varepsilon(x^2 - 1).$$

Find the equations of all phase paths. Show that the separatrices in $|x| < 1$ which approach $x = \pm 1$ are given by

$$y = \varepsilon\left[\tfrac{1}{2}x\sqrt{(1 - x^2)} + \tfrac{1}{2}\sin^{-1}x \mp \tfrac{1}{4}\pi\right]/\sqrt{(1 - x^2)}.$$

Sketch typical solutions for $\varepsilon >, =, < 0$, and confirm that a heteroclinic bifurcation occurs at $\varepsilon = 0$.

Show that the displacement $d(x)$ in the y direction between the separatrices for $-1 < x < 1$ is given by

$$d(x) = \frac{\pi\varepsilon}{2\sqrt{(1-x^2)}}.$$

(This displacement is zero when $\varepsilon = 0$ which shows that the separatrices become a heteroclinic path joining $(1,0)$ and $(-1,0)$ at this value of ε. This separatrix method is the basis of Melnikov's perturbation method in Chapter 13 for detecting homoclinic and heteroclinic bifurcations.)

3.66 Classify the equilibrium points of the system

$$\dot{x} = y, \qquad \dot{y} = x(1-x^2) + ky^2,$$

according to their linear approximations. Find the equations of the phase paths, and show that, if $k = -\sqrt{(3/2)}$, then there exists a homoclinic path given by

$$y^2 = x^2\left(1 - \sqrt{(\tfrac{2}{3})}x\right)$$

in $x > 0$. Show that the time solution is given by

$$x = \sqrt{(\tfrac{3}{2})}\mathrm{sech}^2\tfrac{1}{2}(t - t_0).$$

3.67 An oscillator has an equation of motion given by $\ddot{x} + f(x) = 0$, where $f(x)$ is a piecewise linear restoring force defined by

$$f(x) = \begin{cases} -x, & |x| \le a, \\ b(x\,\mathrm{sgn}(x) - a) - a, & |x| > a. \end{cases}$$

where $a, b > 0$. Find the equations of the homoclinic paths in the phase plane.

3.68 Consider the system

$$\dot{x} = y\left(2y^2 - 3x^2 + \tfrac{19}{9}x^4\right),$$

$$\dot{y} = y^2\left(3x - \tfrac{38}{9}x^3\right) - \left(4x^3 - \tfrac{28}{3}x^5 + \tfrac{40}{9}x^7\right).$$

Find the locations of its equilibrium points. Verify that the system has four homoclinic paths given by

$$y^2 = x^2 - x^4 \quad \text{and} \quad y^2 = 2x^2 - \tfrac{10}{9}x^4.$$

Show also that the origin is a higher-order saddle with separatrices in the directions with slopes $\pm 1, \pm\sqrt{2}$.

3.69 Find and classify the equilibrium points of

$$\dot{x} = a - x^2, \qquad \dot{y} = -y + (x^2 - a)(1 - 2x)$$

for all a. Show that as a decreases through zero, a saddle point and a node coalesce at $a = 0$ after which the equilibrium points disappear. Using the substitution $y = z + x^2 - a$, determine the equations of the phase paths. Show that the phase path connecting the saddle point and the node is $y = x^2 - a$ for $a > 0$. Compute phase diagrams for $a = 0$ and $a = \pm\tfrac{1}{4}$.

3.70 Locate and classify the equilibrium points of

$$\dot{x} = 1 - x^2, \qquad \dot{y} = -(y + x^2 - 1)x^2 - 2x(1 - x^2)$$

according to their linear approximations. Verify that the phase diagram has a saddle-node connection given by $y = 1 - x^2$. Find the time solutions $x(t), y(t)$ for this connection. Sketch the phase diagram.

3.71 Consider the piecewise linear system

$$\dot{x} = x, \quad \dot{y} = -y, \qquad |x - y| \le 1,$$
$$\dot{x} = y + 1, \quad \dot{y} = 1 - x, \qquad x - y \ge 1,$$
$$\dot{x} = y - 1, \quad \dot{y} = -1 - x, \qquad x - y \le 1.$$

Locate and classify the equilibrium points of the system. By solving the linear equations in each region and matching separatrices, show that the origin has two homoclinic paths.

3.72 Obtain the differential equations for the linear system

$$\dot{x} = ax + by, \quad \dot{y} = cx + dy, \qquad (ad \ne bc),$$

in the $\mathcal{U}$-plane (Fig. 3.16) using the transformation $x = 1/z, y = u/z$.

Under what conditions on $\Delta = p^2 - 4q$, $p = a + d$, $q = ad - bc$ does the system on the $\mathcal{U}$-plane have no equilibrium points?

3.73 Classify all the equilibrium points of the system

$$\dot{x} = X(x, y) = (1 - x^2)(x + 2y), \quad \dot{y} = Y(x, y) = (1 - y^2)(-2x + y).$$

Draw the isoclines $X(x, y) = 0$ and $Y(x, y) = 0$, and sketch the phase diagram for the system. A phase path starts near (but not at) the origin. How does its path evolve as t increases? If, on this path, the system experiences small disturbances which cause it to jump to nearby neighbouring paths, what will eventually happen to the system?

4 Periodic solutions; averaging methods

Consider an equation of the form $\ddot{x} + \varepsilon h(x, \dot{x}) + x = 0$ where ε is small. Such an equation is in a sense close to the simple harmonic equation $\ddot{x} + x = 0$, whose phase diagram consists of circles centred on the origin. It should be possible to take advantage of this fact to construct approximate solutions: the phase paths will be nearly circular for ε small enough. However, the original equation will not, in general, have a centre at the origin. The approximate and the exact solutions may differ only by a little over a single cycle, but the difference may prevent the paths from closing; apart from exceptional paths, which are limit cycles. The phase diagram will generally consist of slowly changing spirals and, possibly, limit cycles, all being close to circular.

We show several methods for estimating the radii of limit cycles and for detecting a centre. Extensions of the methods permit determination of the stability and the period of limit cycles, the shape of the spiral path around limit cycles, and amplitude–frequency relations in the case of a centre.

Similar estimates are successful even in very unpromising cases where ε is not small and the supporting arguments are no longer plausible (see Example 4.12 and Section 4.5). In such cases we can say that we are guided to the appropriate sense of how best to fit a circle to a non-circular path by arguments which are valid when ε is small.

4.1 An energy-balance method for limit cycles

The nonlinear character of isolated periodic oscillations makes their detection and construction difficult. Here we discuss limit cycles and other periodic solutions in the phase plane $\dot{x} = y$, $\dot{y} = Y(x, y)$, which allows the mechanical interpretation of Section 1.6 and elsewhere.

Consider the family of equations of the form

$$\ddot{x} + \varepsilon h(x, \dot{x}) + x = 0 \tag{4.1}$$

(note that the equation $\ddot{x} + \varepsilon h(x, \dot{x}) + \omega^2 x = 0$ can be put into this form by the change of variable $\tau = \omega t$). Then on the phase plane we have

$$\dot{x} = y, \qquad \dot{y} = -\varepsilon h(x, y) - x. \tag{4.2}$$

Assume that $|\varepsilon| \ll 1$, so that the nonlinearity is small, and that $h(0, 0) = 0$, so that the origin is an equilibrium point. Suppose we have reason to think that there is at least one periodic solution with phase path surrounding the origin: an appraisal of the regions of the phase plane in which energy loss and energy gain take place might give grounds for expecting a limit cycle.

When $\varepsilon = 0$, eqn (4.1) becomes

$$\ddot{x} + x = 0, \tag{4.3}$$

called the **linearized equation**. Its general solution is $x(t) = a \cos(t + \alpha)$, where a and α are arbitrary constants. So far as the phase diagram is concerned, we may restrict a and α to the cases

$$a > 0, \qquad \alpha = 0.$$

Since different values of α simply correspond to different time origins, the phase paths and representative points remain unchanged. The family of phase paths for (4.3) is given parametrically by

$$x = a \cos t, \qquad y = -a \sin t,$$

which is the family of circles $x^2 + y^2 = a^2$. The period of all these motions is equal to 2π.

For small enough ε we expect that any limit cycle, or any periodic motion, of (4.1) will be close to one of the circular motions (4.4), and will approach it as $\varepsilon \to 0$. Therefore, for some value of a,

$$x(t) \approx a \cos t, \quad y(t) \approx -a \sin t \quad \text{and} \quad T \approx 2\pi \tag{4.4}$$

on the limit cycle, where T is its period.

From (1.47), (with εh in place of h and $g(x) = x$) the change in energy,

$$\mathcal{E}(t) = \tfrac{1}{2} x^2(t) + \tfrac{1}{2} y^2(t),$$

over one period $0 \leq t \leq T$, is given by

$$\mathcal{E}(T) - \mathcal{E}(0) = -\varepsilon \int_0^T h(x(t), y(t)) y(t) \, dt.$$

Since the path is closed, $\mathcal{E}$ returns to its original value after one circuit. Therefore

$$\int_0^T h(x(t), y(t)) y(t) \, dt = 0$$

on the limit cycle. This relation is exact. Now insert the approximations (4.4) into the integral. We obtain the approximating energy balance equation for the amplitude $a > 0$ of the periodic motion

$$\mathcal{E}(2\pi) - \mathcal{E}(0) = \varepsilon a \int_0^{2\pi} h(a \cos t, -a \sin t) \sin t \, dt = 0 \tag{4.5}$$

or

$$\int_0^{2\pi} h(a \cos t, -a \sin t) \sin t \, dt = 0, \tag{4.6}$$

(after getting rid of the factor $(-\varepsilon a)$). This is an equation which, in principle, can be solved for the unknown amplitude a of a limit cycle. In the case of a centre it is satisfied identically.

Example 4.1 *Find the approximate amplitude of the limit cycle of the van der Pol equation*

$$\ddot{x} + \varepsilon(x^2 - 1)\dot{x} + x = 0 \tag{4.7}$$

when ε is small.

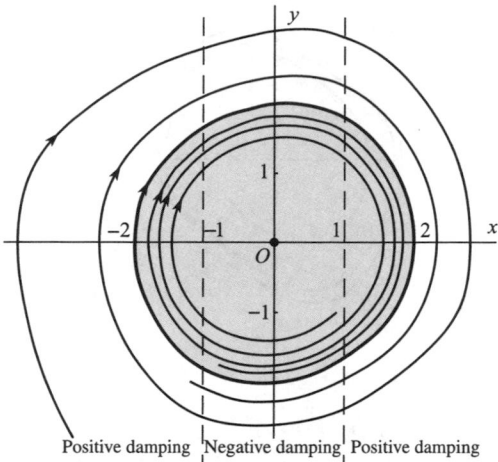

Figure 4.1 Phase diagram for the van der Pol equation $\ddot{x} + \varepsilon(x^2 - 1)\dot{x} + x = 0$ with $\varepsilon = 0.1$. The limit cycle is the outer rim of the shaded region.

Here

$$h(x, y) = (x^2 - 1)y.$$

assuming that $x \approx a \cos t$, the energy balance equation (4.6) becomes

$$\int_0^{2\pi} \left\{ \left(a^2 \cos^2 t - 1 \right) \sin t \right\} \sin t \, dt = 0.$$

This leads to the equation $\frac{1}{4}a^2 - 1 = 0$, with the positive solution $a = 2$. Figure 4.1 shows the limit cycle for $\varepsilon = 0.1$, as obtained numerically.

As ε becomes larger, the shape of the limit cycle becomes significantly different from a circle although the amplitude remains close to 2. This is shown in Fig. 4.2(a) for the case $\varepsilon = 0.5$. The corresponding time solution is shown in Fig. 4.2(b); the period is slightly greater than 2π.   ●

By an extension of this argument the **stability of a limit cycle** can also be determined. Taking the model of a limit cycle as conforming to the type shown in Fig. 4.1, we should expect that *unclosed* paths near enough to the limit cycle, spiralling gradually, will also be given by $x \approx a(t) \cos t$, $y \approx -a(t) \sin t$, where $a(t)$ is nearly constant over a time interval (not now an exact 'period') of $0 \le t \le 2\pi$.

Denote the approximation (4.6) by $g(a)$

$$g(a) = \varepsilon a \int_0^{2\pi} h(a \cos t, -a \sin t) \sin t \, dt; \tag{4.8}$$

and let $a \approx a_0$ (>0) on the limit cycle. Then, by (4.6)

$$g(a_0) = 0$$

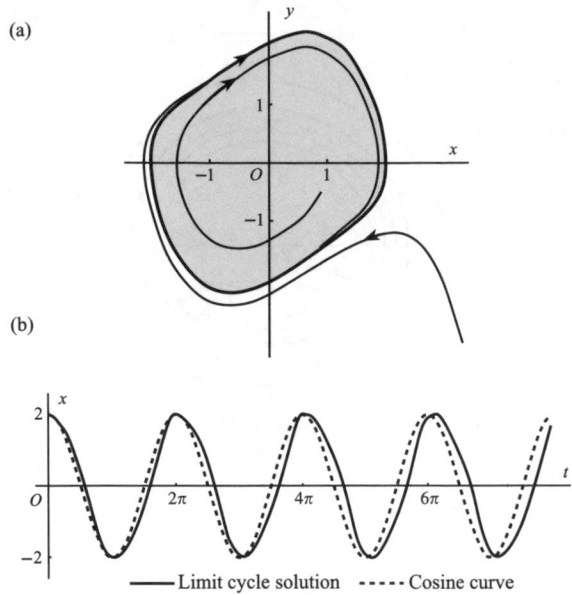

(a)

(b)

Figure 4.2   (a) Phase diagram and (b) time solution for the limit cycle of the van der Pol equation $\ddot{x}+\varepsilon(x^2-1)\dot{x}+x = 0$ with $\varepsilon = 0.5$.

by (4.6). If the limit cycle is stable, then along nearby interior spiral segments ($a < a_0$), energy is gained, and along exterior segments ($a > a_0$), energy is lost. This is to say, for some value of $\delta > 0$,

$$g(a) > 0 \quad \text{when } a_0 - \delta < a < a_0,$$
$$g(a) < 0 \quad \text{when } a_0 < a < a_0 + \delta. \tag{4.9}$$

Similarly if the signs of the inequalities are both reversed, the limit cycle is unstable. Therefore the existence and stability of a limit cycle of amplitude a_0 are determined by the conditions

$$g(a_0) = 0, \quad \textbf{stable if } g'(a_0) < 0, \quad \textbf{unstable if } g'(a_0) > 0. \tag{4.10}$$

Note that the signs of these inequalities are reversed when the sign of ε is reversed. (The case $g(a_0)$ is >0 or <0 on both sides also implies instability, but this is not shown by testing the sign of $g'(x_0)$, because its value is zero in these cases.)

Example 4.2   *Check the stability of the limit cycle in Example 4.1.*
From eqn (4.8)

$$g(a) = -\varepsilon a^2 \int_0^{2\pi} (a^2 \cos^2 t - 1) \sin^2 t \, dt = -\varepsilon a^2 \pi \left(\tfrac{1}{4}a^2 - 1 \right).$$

Therefore,

$$g'(a) = -\varepsilon \pi a(a^2 - 2).$$

Putting $a_0 = 2$ from Example 4.1

$$g'(2) = -4\pi\varepsilon.$$

By (4.10) the cycle is therefore stable when $\varepsilon > 0$, unstable when $\varepsilon < 0$.   ●

Direct differentiation of (4.8) with respect to a gives an alternative criterion for stability. From (4.8),

$$g'(a) = \varepsilon \int_0^{2\pi} h \sin t \, dt + \varepsilon a \int_0^{2\pi} \frac{\partial h}{\partial a} \sin t \, dt$$

$$= \varepsilon \left[-h \cos t \right]_0^{2\pi} + \varepsilon \int_0^{2\pi} \frac{\partial h}{\partial t} \cos t \, dt + \varepsilon a \int_0^{2\pi} \frac{\partial h}{\partial a} \sin t \, dt$$

(integrating the first term by parts). But by the chain rule applied to $h(a \cos t, -a \sin t)$

$$\frac{\partial h}{\partial t} = -ah_1 \sin t - ah_2 \cos t$$

and

$$\frac{\partial h}{\partial a} = h_1 \cos t - h_2 \sin t,$$

where we use the notation

$$h_1(u, v) = \frac{\partial h(u, v)}{\partial u}, \qquad h_2(u, v) = \frac{\partial h(u, v)}{\partial v}.$$

Therefore

$$g'(a) = -\varepsilon a \int_0^{2\pi} h_2(a \cos t, -a \sin t) dt.$$

For **stability** we require $g'(a_0) < 0$ which is equivalent to

$$\varepsilon \int_0^{2\pi} h_2(a_0 \cos t, -a_0 \sin t) dt > 0, \tag{4.11}$$

where a_0 is a positive solution of (4.6).

Example 4.3   *Check the stability of the limit cycle in Example 4.1, using (4.11).*
In this case $h(x, y) = (x^2 - 1)y$ and $h_2(x, y) = x^2 - 1$. The approximate solution for the limit cycle is, from Example 4.1, $x = 2 \cos t$. Equation (4.11), with $a_0 = 2$, gives

$$\varepsilon \int_0^{2\pi} (4 \cos^2 t - 1) dt = 2\pi\varepsilon$$

so that the cycle is stable if $\varepsilon > 0$.   ●

Exercise 4.1
For $|\varepsilon| \ll 1$, find the approximate amplitudes of the limit cycles if

$$\ddot{x} + \varepsilon(x^2 - 1)(x^2 - 6)\dot{x} + x = 0, \quad 0 < \varepsilon \ll 1.$$

Find the function $g(a)$ and the stability of the limit cycles.

4.2 Amplitude and frequency estimates: polar coordinates

Consider again the equation

$$\ddot{x} + \varepsilon h(x, \dot{x}) + x = 0,$$

and the equivalent system

$$\dot{x} = y, \qquad \dot{y} = -\varepsilon h(x, y) - x, \tag{4.12}$$

and suppose that it has at least one periodic time solution, corresponding to a closed path.

Let any phase path of (4.12) be represented parametrically by time-dependent polar coordinates $a(t)$, $\theta(t)$. By eqn (1.58), the polar coordinate form of (4.12) becomes

$$\dot{a} = -\varepsilon h \sin\theta, \tag{4.13}$$

$$\dot{\theta} = -1 - \varepsilon a^{-1} h \cos\theta, \tag{4.14}$$

and the differential equation for the phase paths is therefore

$$\frac{da}{d\theta} = \frac{\varepsilon h \sin\theta}{1 + \varepsilon a^{-1} h \cos\theta}, \tag{4.15}$$

where, for brevity, h stands for $h(a\cos\theta, a\sin\theta)$. These equations hold generally, whether ε is small or not.

Figure 4.3(a) shows a *closed path*, which may be a limit cycle or one of the curves constituting a centre. Let its time period be T. Then $a(t)$, $\theta(t)$ and therefore h all have time period T, meaning that along the closed path $a(t_0 + T) = a(t_0)$ for every t_0, and so for the other variables. Regarded as functions of the angular coordinate θ, all the variables in the problem have θ-period 2π, exactly. As indicated in Fig. 4.3(b), when t increases, θ decreases. The direction along the path for t increasing is clockwise, following the general rule for the phase plane with $y = \dot{x}$; the positive direction for the polar coordinate θ is counterclockwise. A typical cycle of a closed path is therefore described by the end conditions:

$$\begin{aligned} a = a_0, \quad \theta = 2\pi, \quad &\text{at } t = 0, \\ a = a_0, \quad \theta = 0, \quad &\text{at } t = T. \end{aligned} \tag{4.16}$$

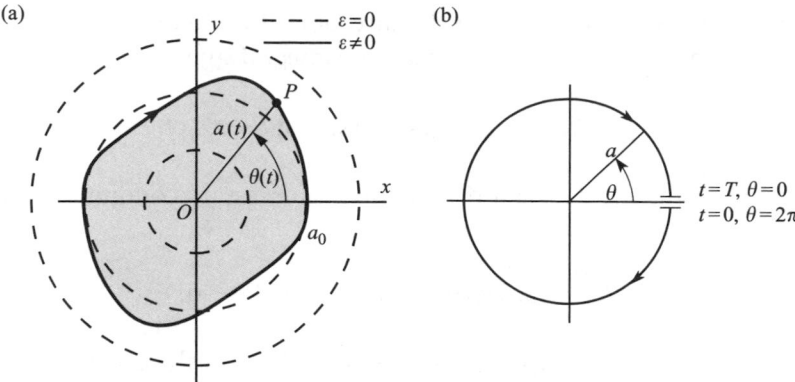

Figure 4.3   (a) A limit cycle and its approximating phase diagram for $\varepsilon = 0$. (b) Showing the relation between θ and t.

Now suppose that ε is small: $|\varepsilon| \ll 1$. As described in the previous section, we expect that the closed path will consist of a small distortion, or **perturbation**, of one of the circular paths of the linearized system $\ddot{x} + x = 0$, or of $\dot{x} = y$, $\dot{y} = -x$, which are indicated as broken lines in Fig. 4.3(a). We shall give more precision to this idea, and also provide estimates of the period T and the circular frequency $\omega = 2\pi/T$, by approximating to eqns (4.13), (4.14), and (4.15) for small values of ε.

By expanding the right-hand side of (4.15) in powers of ε we obtain

$$\frac{\mathrm{d}a}{\mathrm{d}\theta} = \varepsilon h \sin\theta + O(\varepsilon^2). \tag{4.17}$$

Integrate (4.17) with respect to θ, over the range $\theta = 2\pi$ decreasing to $\theta = 0$. We obtain

$$a(\theta) - a(2\pi) = O(\varepsilon), \quad \text{or} \quad a(\theta) = a_0 + O(\varepsilon), \tag{4.18}$$

since $a(2\pi) = a_0$, from (4.16). The deviation from a circular path of radius a_0 is therefore small, of order of magnitude ε.

Integrate (4.17) over the full range of the cycle (4.16), from $\theta = 2\pi$ to $\theta = 0$. We obtain

$$a_0 - a_0 = 0 = \varepsilon \int_{2\pi}^{0} h(a\cos\theta, a\sin\theta)\sin\theta\,\mathrm{d}\theta + O(\varepsilon^2),$$

or

$$0 = -\int_{0}^{2\pi} h(a\cos\theta, a\sin\theta)\sin\theta\,\mathrm{d}\theta + O(\varepsilon),$$

after cancelling the factor ε and reversing the direction of the integral. Now substitute for a from (4.18) and expand $h(a \cos \theta, a \sin \theta)$ under the integral sign: we have

$$\int_0^{2\pi} h(a_0 \cos \theta, a_0 \sin \theta) \sin \theta \, d\theta = O(\varepsilon).$$

Since the integral on the left does not depend on ε, a necessary condition for the phase path to be closed is that

$$\int_0^{2\pi} h(a_0 \cos \theta, a_0 \sin \theta) \sin \theta \, d\theta = 0. \tag{4.19}$$

This serves as an equation for the approximate amplitude a_0 of the cycle. The similar-looking condition (4.7) is recoverable by integrating the expression (4.14) for $\dot\theta$ with respect to t, obtaining

$$\theta = 2\pi - t + O(\varepsilon). \tag{4.20}$$

When this is substituted into (4.19) and the leading term in the expansion of h retained, we have (4.6) again.

To obtain the **period** of the cycle, T, put

$$T = \int_0^T dt = \int_{2\pi}^0 \frac{d\theta}{\dot\theta}.$$

From (4.14), therefore,

$$T = \int_0^{2\pi} \frac{d\theta}{1 + \varepsilon a^{-1} h(a \cos \theta, a \sin \theta) \cos \theta}.$$

Substitute $a = a_0 + O(\varepsilon)$, retain the leading terms, then expand $1/\{1 + \varepsilon a_0^{-1} h \cos \theta\}$ to order ε. We then have

$$T = \int_0^{2\pi} \left\{ 1 - \varepsilon a_0^{-1} h(a_0 \cos \theta, a_0 \sin \theta) \cos \theta + O(\varepsilon^2) \right\} d\theta$$

$$\approx 2\pi - \frac{\varepsilon}{a_0} \int_0^{2\pi} h(a_0 \cos \theta, a_0 \sin \theta) \cos \theta \, d\theta, \tag{4.21}$$

the error being of order ε^2.

The circular frequency of the periodic oscillation is

$$\omega = \frac{2\pi}{T} \approx 1 + \frac{\varepsilon}{2\pi a_0} \int_0^{2\pi} h(a_0 \cos \theta, a_0 \sin \theta) \cos \theta \, d\theta, \tag{4.22}$$

with error of order ε^2.

Example 4.4   *Obtain the frequency of the limit cycle of the van der Pol equation*

$$\ddot{x} + \varepsilon(x^2 - 1)\dot{x} + x = 0$$

correct to order ε.

Since $h(x, \dot{x}) = (x^2 - 1)\dot{x}$ and the amplitude $a_0 = 2$ to order ε (from Example 4.1), eqn (4.22) gives

$$\omega = 1 + \frac{\varepsilon}{4\pi} \int_0^{2\pi} (4\cos^2\theta - 1)(2\sin\theta)\cos\theta \, d\theta$$

$$= 1 + \text{zero}$$

(since the integrand is an odd function about $\theta = \pi$). The frequency is 1, with error $O(\varepsilon^2)$. ●

Example 4.5 *Obtain the relation between frequency and amplitude for moderate-amplitude swings of the pendulum equation* $\ddot{x} + \sin x = 0$.

At first sight this equation seems to have nothing to do with the previous theory. However, write the approximation appropriate to moderate swings:

$$\sin x \approx x - \tfrac{1}{6}x^3$$

(with error about 1% when $x = 1$, corresponding to an angular displacement of $57.3°$). The approximate equation is

$$\ddot{x} - \tfrac{1}{6}x^3 + x = 0.$$

This equation is a member of the family of equations $\ddot{x} + \varepsilon x^3 + x = 0$, with $\varepsilon = -\tfrac{1}{6}$, which is of the right type, with

$$h(x, \dot{x}) = x^3.$$

The equation (4.19) for the amplitude a_0 becomes

$$a_0^3 \int_0^{2\pi} \cos^3\theta \sin\theta \, d\theta = 0.$$

The integrand is an odd function about $\theta = \pi$, so the equation is satisfied identically for all a_0. This is to be expected since the origin $x = 0$, $y = 0$ in the phase plane is a centre.

For the frequency ω, (4.22) becomes

$$\omega = 1 + \frac{\varepsilon}{2\pi a_0} \int_0^{2\pi} a_0^3 \cos^4\theta \, d\theta$$

for $\varepsilon = -\tfrac{1}{6}$ and for all a_0. By carrying out the integration we obtain

$$\omega = 1 - \tfrac{1}{16}a_0^2,$$

the result being closely correct for moderate amplitudes. The frequency decreases (and the period increases) with amplitude for the pendulum. ●

Example 4.6 *Obtain an approximation to the amplitude and frequency of the limit cycle for* **Rayleigh's** *equation* $\ddot{x} + \varepsilon(\tfrac{1}{3}\dot{x}^3 - \dot{x}) + x = 0$.

Here, $h(x, \dot{x}) = \left(\tfrac{1}{3}\dot{x}^2 - 1\right)\dot{x}$. Equation (4.19) becomes

$$\int_0^{2\pi} \left(\tfrac{1}{3}a_0^2 \sin^2\theta - 1\right)\sin^2\theta \, d\theta = 0,$$

which can be verified to have the solution $a_0 = 2$. The frequency equation (4.22) becomes

$$\omega = 1 + \frac{\varepsilon}{4\pi} \int_0^{2\pi} \left(\tfrac{4}{3}\sin^2\theta - 1\right)\sin\theta\cos\theta \, d\theta$$

with error $O(\varepsilon^2)$. But the integral is zero since the integrand is an odd function about $\theta = \pi$. Therefore the frequency is given by

$$\omega = 1 + O(\varepsilon^2).$$

●

Exercise 4.2
Find the approximate amplitude of the limit cycle of the system

$$\ddot{x} + \varepsilon[(x^2 - 1)\dot{x} - x^3] + x = 0, \quad 0 < \varepsilon \ll 1.$$

Show that its frequency ω is given by

$$\omega = 1 - \tfrac{3}{2}\varepsilon + O(\varepsilon^2).$$

4.3   An averaging method for spiral phase paths

The general pattern of the phase diagram for the system

$$\dot{x} = y, \qquad \dot{y} = -\varepsilon h(x, y) - x, \quad |\varepsilon| \ll 1$$

in a bounded region containing the origin is of slowly expanding or contracting spiral phase paths, which may include closed paths, surrounding the origin. This is shown by eqns (4.13) and (4.14). A representative point with polar coordinates $a(t), \theta(t)$ moves round the origin repeatedly at an angular speed $\dot{\theta} \approx -1$, whilst the distance $a(t)$ from the origin changes slowly. We shall obtain an approximate differential equation for the phase paths.

Equation (4.15) for the paths in polar coordinates can be written

$$\frac{\mathrm{d}a}{\mathrm{d}\theta} = \varepsilon h(a(\theta) \cos\theta, a(\theta) \sin\theta) \sin\theta + O(\varepsilon^2). \tag{4.23}$$

The function $h(a(\theta) \cos\theta, a(\theta) \sin\theta) \sin\theta$ is not in general periodic in θ because $a(\theta)$ is not periodic on a spiral. Nevertheless we can construct an expansion of $h \sin\theta$ as a pseudo-Fourier series for all values of θ as follows.

Firstly, treat a as if it were an arbitrary *constant* parameter. Then $h(a \cos\theta, a \sin\theta) \sin\theta$ is periodic with period 2π, so it can be represented by an ordinary Fourier series, valid for all θ and for every *fixed* value of a:

$$h(a \cos\theta, a \sin\theta) \sin\theta = p_0(a) + \sum_{n=1}^{\infty} \{p_n(a) \cos n\theta + q_n(a) \sin n\theta\},$$

in which the coefficients are given by

$$p_0(a) = \frac{1}{2\pi} \int_0^{2\pi} h(a \cos u, a \sin u) \sin u \, \mathrm{d}u; \tag{4.24}$$

and

$$\begin{pmatrix} p_n(a) \\ q_n(a) \end{pmatrix} = \frac{1}{2\pi} \int_0^{2\pi} h(a\cos u, a\sin u)\sin u \begin{pmatrix} \cos nu \\ \sin nu \end{pmatrix} du$$

for $n \geq 1$. Since these formulae are correct for every constant value of a, they still hold good if we put $a(\theta)$ in place of a wherever it appears. The differential equation (4.23) then becomes

$$\frac{da}{d\theta} = \varepsilon p_0(a) + \varepsilon \sum_{n=1}^{\infty}\{p_n(a)\cos n\theta + q_n(a)\sin n\theta\} + O(\varepsilon^2), \tag{4.25}$$

in which a is a function of θ throughout.

Every term on the right of (4.25) contributes something to the solutions $a(\theta)$. However, we shall show that, correct to order ε, the *increment* in $a(\theta)$ over any complete 'loop' $\theta_0 \leq \theta \leq \theta_0 + 2\pi$ of the spiralling phase paths *depends only on the single term* $\varepsilon p_0(a)$.

On one loop of a phase path, $\theta_0 \leq \theta \leq \theta_0 + 2\pi$, put

$$a(\theta_0) = a_1, \qquad a(\theta_0 + 2\pi) = a_2.$$

We require the increment $a_2 - a_1$. By (4.18), $a(\theta) = a_1 + O(\varepsilon)$ on the loop. Substitute this into the terms under the summation sign in (4.25), obtaining $p_n(a) = p_n(a_1) + O(\varepsilon)$, $q_n(a) = q_n(a_1) + O(\varepsilon)$; then (4.25) becomes

$$\frac{da}{d\theta} = \varepsilon p_0(a) + \varepsilon \sum_{n=1}^{\infty}\{p_n(a_1)\cos n\theta + q_n(a_1)\sin n\theta\} + O(\varepsilon^2). \tag{4.26}$$

Integrate (4.26) over the range $\theta_0 \leq \theta \leq \theta_0 + 2\pi$. The integral over each term under the summation sign is zero, and we are left with

$$a_2 - a_1 = \varepsilon \int_{\theta_0}^{\theta_0 + 2\pi} p_0(a(\theta))\, d\theta + O(\varepsilon^2).$$

Therefore, to this order of accuracy, the increment over one loop depends only on $p_0(a)$. The contribution of the higher harmonics in (4.26) to the increment is only $O(\varepsilon^2)$.

Consider now the differential equation that is left over from (4.26) after eliminating the $n \geq 1$ terms which, *on average*, have an effect of order ε^2. It is

$$\frac{da}{d\theta} = \varepsilon p_0(a), \tag{4.27a}$$

where

$$p_0(a) = \frac{1}{2\pi} \int_0^{2\pi} h\{a(\theta)\cos u, a(\theta)\sin u\}\sin u\, du. \tag{4.27b}$$

The integral (4.27b) can, in principle, be evaluated and the approximate differential equation (4.27a) solved to give $a(\theta)$.

To interpret these equations: $p_0(a)$ is nearly equal to the *average value* of $h \sin \theta$ over any loop with polar coordinates $a(\theta), \theta$. Thus the function $h \sin \theta$ in the original differential equation (4.23) has been replaced by this average value. The influence of the higher harmonics in $h \sin \theta$ within any particular loop is disregarded, but the curves generated by (4.27) are still nearly circular for ε small enough, and the *separation between successive loops* of a phase path from point to point is correct to order ε. This process is representative of a group of procedures called **averaging methods** (Krylov and Bogoliubov 1949).

The usual equation for a limit cycle is obtained from (4.27b) by putting $p_0(a) = 0$.

Approximate equations of the same type for the time variation of $a(t), \theta(t)$ may be derived from (4.27). Put

$$\frac{da}{dt} = \frac{da}{d\theta}\frac{d\theta}{dt} = -\frac{da}{d\theta} + O(\varepsilon)$$

from (4.14). From (4.27a), this becomes

$$\frac{da}{dt} = -\varepsilon p_0(a) \tag{4.28}$$

with $p_0(a)$ given by (4.27b).

By a process of averaging similar to that leading up to eqn (4.27a, b) we can also obtain the equation for $\theta(t)$:

$$\frac{d\theta}{dt} = -1 - \varepsilon a^{-1} r_0(a) + O\left(\varepsilon^2\right), \tag{4.29}$$

where

$$r_0(a) = \frac{1}{2\pi} \int_0^{2\pi} h\{a \cos u, a \sin u\} \cos u \, du. \tag{4.30}$$

Example 4.7 *Find approximate time solutions of van der Pol's equation*

$$\ddot{x} + \varepsilon(x^2 - 1)\dot{x} + x = 0,$$

for small positive ε.
From (4.24),

$$p_0(a) = \frac{a}{2\pi} \int_0^{2\pi} (a^2 \cos^2 u - 1) \sin^2 u \, du = \tfrac{1}{2} a \left(\tfrac{1}{4} a^2 - 1\right).$$

We have the approximate equation (4.28), for the radial coordinate:

$$\frac{da}{dt} = -\tfrac{1}{2}\varepsilon a \left(\tfrac{1}{4} a^2 - 1\right).$$

The constant solution $a = 2$ corresponds to the limit cycle. The equation separates, giving

$$\int \frac{da}{a(a^2 - 4)} = -\tfrac{1}{8}\varepsilon(t + C),$$

where C is a constant. By carrying out the integration we obtain

$$-\tfrac{1}{4}\log a + \tfrac{1}{8}\log|a^2 - 4| = -\tfrac{1}{8}\varepsilon(t + C).$$

With the initial condition $a(0) = a_1$, the solution is

$$a(t) = \frac{2}{\{1 - (1 - (4/a_1^2))e^{-\varepsilon t}\}^{1/2}},$$

which tends to 2 as $t \to \infty$. It is easy to verify from eqn (4.30) that $r_0(a) = 0$. Equation (4.29) therefore gives $\theta(t) = -t + \theta_1$ where θ_1 is the initial polar angle. The frequency of the spiral motion is therefore the same as that of the limit cycle to our degree of approximation. Finally, the required approximate time solutions are given by

$$x(t) = a(t)\cos\theta(t) = \frac{2\cos(t - \theta_1)}{\{1 - (1 - (4/a_1^2))e^{-\varepsilon t}\}^{1/2}}.$$

      ●

Example 4.8   *Find the approximate phase paths for the equation*

$$\ddot{x} + \varepsilon(|\dot{x}| - 1)\dot{x} + x = 0.$$

We have $h(x, y) = (|y| - 1)y$. Therefore, from (4.27b),

$$p_0(a) = \frac{a}{2\pi}\int_0^{2\pi} (|a\sin\theta| - 1)\sin^2\theta \, d\theta$$

$$= \frac{a}{2\pi}\left(\int_0^{\pi} (a\sin\theta - 1)\sin^2\theta \, d\theta + \int_{\pi}^{2\pi}(-a\sin\theta - 1)\sin^2\theta \, d\theta\right)$$

$$= \frac{a}{2\pi}2\int_0^{\pi}(a\sin\theta - 1)\sin^2\theta \, d\theta = \frac{1}{\pi}a\left(\frac{4}{3}a - \frac{1}{2}\pi\right).$$

There is a limit cycle when $p_0(a) = 0$; that is, at $a = \tfrac{3}{8}\pi = a_0$, say. Equation (4.27a) becomes

$$\frac{da}{d\theta} = \frac{4\varepsilon}{3\pi}a(a - a_0).$$

The spiral path satisfying $a = a_1$ at $\theta = \theta_1$ is given in polar coordinates by

$$a(\theta) = a_0\bigg/\left(1 - (1 - (a_0/a_1))\exp\left[\varepsilon\frac{4a_0}{3\pi}(\theta - \theta_1)\right]\right).$$

Alternatively, eqn (4.28) gives a as a function t.   ●

Exercise 4.3
Find the limit cycle and approximate spiral paths of

$$\ddot{x} + \varepsilon(\dot{x}^2 - |\dot{x}|)\dot{x} + x = 0, \quad 0 < \varepsilon \ll 1$$

expressed as θ, as a function of a in polars.

> **Exercise 4.4**
> Find the limit cycle and approximate spiral paths of
> $$\ddot{x} + \varepsilon(x^2 + 2\dot{x}^2 - 1)\dot{x} + x = 0, \quad 0 < \varepsilon \ll 1.$$

4.4 Periodic solutions: harmonic balance

One of the most straightforward practical methods for estimating periodic solutions is illustrated by the following examples.

Example 4.9   *Find an approximation to the amplitude and frequency of the limit cycle of van der Pol's equation* $\ddot{x} + \varepsilon(x^2 - 1)\dot{x} + x = 0$.

Assume an approximate solution $x = a \cos \omega t$. (The prescription of zero value for the phase is not a limitation; in the case of an autonomous equation we are, in effect, simply choosing the time origin so that $\dot{x}(0) = 0$.) We expect the angular frequency ω to be close to 1 for small $|\varepsilon|$. Write the equation in the form

$$\ddot{x} + x = -\varepsilon(x^2 - 1)\dot{x}.$$

Upon substituting the assumed form of solution we obtain

$$(-\omega^2 + 1)a \cos \omega t = -\varepsilon(a^2 \cos^2 \omega t - 1)(-a\omega \sin \omega t)$$

$$= \varepsilon a\omega(\tfrac{1}{4}a^2 - 1) \sin \omega t + \tfrac{1}{4}\varepsilon a^3 \omega \sin 3\omega t,$$

after some reduction. The right-hand side is just the Fourier series for $\varepsilon h(x, \dot{x})$: it is easier to get it this way than to work with the equations of the last section. Now, ignore the presence of the higher-harmonic term involving $\sin 3\omega t$, and match terms in $\cos \omega t$, $\sin \omega t$. We find

$$1 - \omega^2 = 0$$

from the $\cos \omega t$ term, and

$$\tfrac{1}{4}a^2 - 1 = 0$$

from the $\sin \omega t$ term. Choosing positive values, the second equation gives $a = 2$, and the first $\omega = 1$ (the signs are equivalent), as in several earlier examples. ●

Example 4.10   *Obtain the amplitude–frequency approximation for the pendulum equation* $\ddot{x} + x - \tfrac{1}{6}x^3 = 0$.

Assume an approximate solution of the form $x = a \cos \omega t$. We obtain, after substituting in the given equation:

$$-\omega^2 a \cos \omega t + a \cos \omega t - \tfrac{1}{6}a^3\left(\tfrac{3}{4} \cos \omega t - \tfrac{1}{4} \cos 3\omega t\right) = 0.$$

Ignore the higher harmonic and collect the coefficients of $\cos \omega t$. We find that

$$a\left[(1 - \omega^2) - \tfrac{1}{8}a^2\right] = 0.$$

Therefore the frequency–amplitude relation becomes

$$\omega = \sqrt{(1 - \tfrac{1}{8}a^2)} \approx 1 - \tfrac{1}{16}a^2$$

for amplitude not too large (compare Example 4.5). The existence of solutions for ω corresponding to a range of values of $a > 0$ indicates that in this case the origin is a centre.   ●

For the general equation

$$\ddot{x} + \varepsilon h(x, \dot{x}) + x = 0, \tag{4.31}$$

suppose that there is a periodic solution close to $a \cos \omega t$ and that $h(a \cos \omega t, -a\omega \sin \omega t)$ has a Fourier series, *the constant term* (which is its mean value over a cycle) *being zero*:

$$h(x, \dot{x}) \approx h(a \cos \omega t, -a\omega \sin \omega t)$$
$$= A_1(a) \cos \omega t + B_1(a) \sin \omega t + \text{ higher harmonics},$$

where

$$A_1(a) = \frac{\omega}{\pi} \int_0^{2\pi/\omega} h(a \cos \omega t, -a\omega \sin \omega t) \cos \omega t \, dt,$$

$$B_1(a) = \frac{\omega}{\pi} \int_0^{2\pi/\omega} h(a \cos \omega t, -a\omega \sin \omega t) \sin \omega t \, dt.$$

Then (4.31) becomes

$$(1 - \omega^2)a \cos \omega t + \varepsilon A_1(a) \cos \omega t + \varepsilon B_1(a) \sin \omega t + \text{higher harmonics} = 0. \tag{4.32}$$

This equation can hold for all t only if

$$(1 - \omega^2)a + \varepsilon A_1(a) = 0, \qquad B_1(a) = 0, \tag{4.33}$$

which determine a and ω.

It will not be possible in general to ensure that full matching is obtained in (4.32): since the solution is not exactly of the form $a \cos \omega t$ we should ideally have represented the solution by a complete Fourier series and matching could then in principle be completed. A way of justifying the non-matching of the higher harmonics is to regard them as additional, neglected input (or forcing terms) to the linear equation $\ddot{x} + x = 0$. The periodic response of this equation to a force $K \cos n\omega t$ is equal to $-K \cos n\omega t / (n^2\omega^2 - 1)$, and is therefore of rapidly diminishing magnitude as n increases.

The method can also be applied to nonlinear equations with a periodic forcing term, and is used extensively in Chapter 7.

Exercise 4.5
Obtain the amplitude and frequency approximation for

$$\ddot{x} + \varepsilon(|x|^3 - 1)\dot{x} + x = 0, \quad 0 < \varepsilon \ll 1.$$

4.5   The equivalent linear equation by harmonic balance

The method of harmonic balance can be adapted to construct a linear substitute for the original nonlinear equation; the process can be described as pseudo-linearization. We illustrate the process by an example.

Example 4.11   *Obtain approximate solutions to the van der Pol equation*

$$\ddot{x} + \varepsilon(x^2 - 1)\dot{x} + x = 0$$

using an equivalent linear equation.

The phase diagram for this by now familiar equation consists of a limit cycle and spirals slowly converging on it when ε is small and positive. We shall approximate to $\varepsilon(x^2 - 1)\dot{x} = \varepsilon h(x, \dot{x})$ on any one 'cycle' of a spiral: this nonlinear term is already small and a suitable approximation retaining the right characteristics should be sufficient. Suppose, then, that we assume (for the purpose of approximating h only) that $x = a\cos\omega t$ and $\dot{x} = -a\omega\sin\omega t$, where a, ω are considered constant on the limit cycle. Then

$$\varepsilon(x^2 - 1)\dot{x} \approx -\varepsilon(a^2\cos^2\omega t - 1)a\omega\sin\omega t$$

$$= -\varepsilon a\omega\left(\tfrac{1}{4}a^2 - 1\right)\sin\omega t - \tfrac{1}{4}\varepsilon a^3\sin 3\omega t. \tag{4.34}$$

(compare Example 4.9). As in the harmonic balance method, we neglect the effect of the higher harmonic, and see that since $-a\omega\sin\omega t = \dot{x}$ on the limit cycle, eqn (4.34) can be written as

$$\varepsilon(x^2 - 1)\dot{x} \approx \varepsilon\left(\tfrac{1}{4}a^2 - 1\right)\dot{x}. \tag{4.35}$$

We now replace this in the differential equation to give the linear equation

$$\ddot{x} + \varepsilon\left(\tfrac{1}{4}a^2 - 1\right)\dot{x} + x = 0. \tag{4.36}$$

This equation is peculiar in that it contains a parameter of its own solutions, namely the amplitude a. If $a = 2$ the damping term vanishes and the solutions are periodic of the form $2\cos t$; hence $\omega = 1$. This is an approximation to the limit cycle.

   The non-periodic solutions are spirals in the phase plane. Consider the motion for which $x(0) = a_0, \dot{x}(0) = 0$: for the next few 'cycles' a_0 will serve as the amplitude used in the above approximation, so (4.40) may be written

$$\ddot{x} + \varepsilon\left(\tfrac{1}{4}a_0^2 - 1\right)\dot{x} + x - 0.$$

With the initial conditions given, the solution is

$$x(t) = \frac{a_0}{\beta}\,e^{\alpha t}[\beta\cos\beta t - \alpha\sin\beta t],$$

where

$$\alpha = \tfrac{1}{2}\varepsilon\left(1 - \tfrac{1}{4}a_0^2\right), \qquad \beta = \tfrac{1}{2}\sqrt{\left[4 - \varepsilon^2(1 - \tfrac{1}{4}a_0^2)\right]}.$$

   It can be confirmed, by expanding both factors in powers of εt, that so long as $\varepsilon t \ll 1$ this 'solution' agrees with the approximate solution found in Example 4.7. Damped or negatively damped behaviour of the solutions occurs according to whether $a_0 > 2$ or $a_0 < 2$; that is, according to whether we start outside or inside the limit cycle.   ●

Example 4.12   *Find the frequency–amplitude relation for the equation* $\ddot{x} + \mathrm{sgn}(x) = 0$ *and compare it with the exact result using an equivalent linear equation.*

It can be shown that the system $\dot{x} = y$, $\dot{y} = -\mathrm{sgn}(x)$ has a centre at the origin: the behaviour is that of a particle on a spring, with a restoring force of constant magnitude as shown in Fig. 4.4(a).

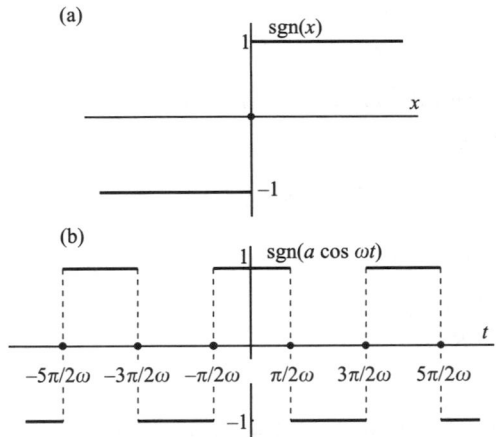

Figure 4.4   (a) Graph of sgn(x); (b) graph of sgn(a cos(ωt)).

Suppose the oscillations have the form $a \cos \omega t$ approximately. Then $\mathrm{sgn}(x) = \mathrm{sgn}(a \cos \omega t)$. This has the shape shown in Fig. 4.4(b). The period is $2\pi/\omega$ and we shall approximate to $\mathrm{sgn}(a \cos \omega t)$ by the first term in its Fourier series on the interval $(0, 2\pi/\omega)$:

$$\mathrm{sgn}(a \cos \omega t) = A_1(a) \cos \omega t + \text{higher harmonics}$$

(the constant term is zero, and there is no sine term since the function is even) where

$$A_1(a) = \frac{\omega}{\pi} \int_0^{2\pi/\omega} \mathrm{sgn}(a \cos \omega t) \cos \omega t \, dt$$

$$= \frac{1}{\pi} \int_0^{2\pi} \mathrm{sgn}(a \cos u) \cos u \, du$$

$$= \frac{1}{\pi} \left(\int_0^{\pi/2} \cos u \, du - \int_{\pi/2}^{3\pi/2} \cos u \, du + \int_{3\pi/2}^{2\pi} \cos u \, du \right) = \frac{4}{\pi}.$$

Therefore $\mathrm{sgn}(x)$ is replaced by $(4/\pi) \cos \omega t$, which in turn is replaced by $(4/\pi)x/a$. The equivalent linear equation is then

$$\ddot{x} + \frac{4}{\pi a} x = 0.$$

The solution, having any amplitude a, of the form $a \cos \omega t$ is

$$x(t) = a \cos \left[\left(\frac{4}{\pi a} \right)^{1/2} t \right].$$

Therefore

$$\omega = \frac{2}{\sqrt{(\pi a)}}$$

and the period T is given by

$$T = \frac{2\pi}{\omega} = \pi^{3/2} a^{1/2} \approx 5.568\sqrt{a}.$$

The exact period can be found as follows. When $x > 0$, $\ddot{x} = -1$. The solution for $x > 0$ is therefore

$$x(t) = -\tfrac{1}{2} t^2 + \alpha t + \beta.$$

The solution for which $x(0) = 0$ and $\dot{x}\left(\tfrac{1}{4} T_e\right) = 0$, where T_e is the exact period, is $\left(-\tfrac{1}{2} t^2 + \tfrac{1}{4} T_e t\right)$.
Further, $x = a$ when $t = \tfrac{1}{4} T_e$, so that

$$T_e = 4\sqrt{(2a)} \approx 5.67\sqrt{a},$$

which compares very favourably with the above estimate.   ●

Example 4.13   *Obtain the period–amplitude relation for the Poisson–Boltzmann-type equation $\ddot{x} + (e^x - e^{-x}) = 0$ using an equivalent linear approximation.*
When $x = a \cos \omega t$,

$$e^x - e^{-x} = e^{a \cos \omega t} - e^{-a \cos \omega t} = A_1(a) \cos \omega t + \text{higher harmonics}$$

(there is no $\sin \omega t$ term since the function is even), where

$$A_1(a) = \frac{\omega}{\pi} \int_0^{2\pi/\omega} (e^{a \cos \omega t} - e^{-a \cos \omega t}) \cos \omega t \, dt$$

$$= \frac{1}{\pi} \int_0^{2\pi} (e^{a \cos u} - e^{-a \cos u}) \cos u \, du.$$

These integrals are expressible in terms of the modified Bessel function I_1 which has the integral representation (Abramowitz and Stegun 1965)

$$I_1(z) = \frac{1}{\pi} \int_0^{\pi} e^{z \cos u} \cos u \, du,$$

and we find that

$$A_1(a) = 4 I_1(a).$$

Neglecting the higher harmonics, we write the equation as

$$\ddot{x} + (4 I_1(a)/a) x = 0.$$

The frequency ω is therefore given by

$$\omega = 2\sqrt{(I_1(a)/a)}$$

and the period T by

$$T = \pi \sqrt{(a/I_1(a))}.$$

Note that a is unrestricted in these equations—it can be verified that $x = 0$ is a centre for the original equation (see Problem 1.21).   ●

Exercise 4.6
Find the equivalent linear equation for $\ddot{x}+(x^2+\dot{x}^2)x=0$, and the resulting frequency-amplitude approximation. What restrictions must be imposed on the amplitude? Find the exact equation of the phase paths, and confirm that $x^2+y^2=1$ is a phase path corresponding to frequency $\omega=1$, amplitude $a=1$.

Problems

4.1 By transforming to polar coordinates, find the limit cycles of the systems

(i) $\dot{x}=y+x(1-x^2-y^2)$, $\dot{y}=-x+y(1-x^2-y^2)$;
(ii) $\dot{x}=(x^2+y^2-1)x-y\sqrt{(x^2+y^2)}$, $\dot{y}=(x^2+y^2-1)y+x\sqrt{(x^2+y^2)}$;

and investigate their stability.

4.2 Consider the system

$$\dot{x}=y+xf(r^2),\qquad \dot{y}=-x+yf(r^2),$$

where $r^2=x^2+y^2$ and $f(u)$ is continuous on $u\geq0$. Show that r satisfies

$$\frac{d(r^2)}{dt}=2r^2f(r^2).$$

If $f(r^2)$ has n zeros, at $r=r_k,\ k=1,2,\ldots,n$, how many periodic solutions has the system? Discuss their stability in terms of the sign of $f'(r_k^2)$.

4.3 Apply the energy balance method of Section 4.1 to each of the following equations assuming $0<\varepsilon\ll1$, and find the amplitude and stability of any limit cycles:

(i) $\ddot{x}+\varepsilon(x^2+\dot{x}^2-1)\dot{x}+x=0$;
(ii) $\ddot{x}+\varepsilon(\frac{1}{3}\dot{x}^3-\dot{x})+x=0$;
(iii) $\ddot{x}+\varepsilon(x^4-1)\dot{x}+x=0$;
(iv) $\ddot{x}+\varepsilon\sin(x^2+\dot{x}^2)\operatorname{sgn}(\dot{x})+x=0$;
(v) $\ddot{x}+\varepsilon(|x|-1)\dot{x}+x=0$;
(vi) $\ddot{x}+\varepsilon(\dot{x}-3)(\dot{x}+1)\dot{x}+x=0$;
(vii) $\dot{x}+\varepsilon(x-3)(x+1)\dot{x}+x=0$.

4.4 For the equation $\dot{x}+\varepsilon(x^2+\dot{x}^2-4)\dot{x}+x=0$, the solution $x=2\cos t$ is a limit cycle. Test its stability, using the method of Section 4.1, and obtain an approximation to the paths close to the limit cycle by the method of Section 4.3.

4.5 For the equation $\ddot{x}+\varepsilon(|x|-1)\dot{x}+x=0$, find approximately the amplitude of the limit cycle and its period, and the polar equations for the phase paths near the limit cycle.

4.6 Repeat Problem 4.5 with Rayleigh's equation, $\ddot{x}+\varepsilon(\frac{1}{3}\dot{x}^3-\dot{x})+x=0$.

4.7 Find approximately the radius of the limit cycle, and its period, for the equation

$$\ddot{x}+\varepsilon(x^2-1)\dot{x}+x-\varepsilon x^3=0,\quad 0<\varepsilon\ll1.$$

4.8 Show that the frequency–amplitude relation for the pendulum equation, $\ddot{x} + \sin x = 0$, is $\omega^2 = 2J_1(a)/a$, using the methods of Section 4.4 or 4.5. (J_1 is the Bessel function of order 1, with representations

$$J_1(a) = \frac{2}{\pi} \int_0^{\pi/2} \sin(a\cos u)\cos u\, du = \sum_{n=0}^{\infty} \frac{(-1)^n (\frac{1}{2}a)^{2n+1}}{n!(n+1)!}.$$

Show that, for small amplitudes, $\omega = 1 - \frac{1}{16}a^2$.

4.9 In the equation

$$\ddot{x} + \varepsilon h(x,\dot{x}) + g(x) = 0$$

suppose that $g(0) = 0$, and that in some interval $|x| < \delta$, g is continuous and strictly increasing. Show that the origin for the equation $\ddot{x} + g(x) = 0$ is a centre. Let $\zeta(t,a)$ represent its periodic solutions near the origin, where a is a parameter which distinguishes the solutions, say the amplitude. Also, let $T(a)$ be the corresponding period.

By using an energy balance argument show that the periodic solutions of the original equation satisfy

$$\int_0^{T(a)} h(\zeta, \dot{\zeta})\dot{\zeta}\, dt = 0.$$

Apply this equation to obtain the amplitude of the limit cycle of the equation

$$\ddot{x} + \varepsilon(x^2 - 1)\dot{x} + v^2 x = 0.$$

4.10 For the following equations, show that, for small ε the amplitude $a(t)$ satisfies approximately the equation given.

(i) $\ddot{x} + \varepsilon(x^4 - 2)\dot{x} + x = 0$, $16\dot{a} = -\varepsilon a(a^4 - 16)$;

(ii) $\ddot{x} + \varepsilon\sin(x^2 + \dot{x}^2)\mathrm{sgn}(\dot{x}) + x = 0$, $\pi\dot{a} = -\varepsilon 2a\sin(a^2)$;

(iii) $\ddot{x} + \varepsilon(x^2 - 1)\dot{x}^3 + x = 0$, $16\dot{a} = -\varepsilon a^3(a^2 - 6)$.

4.11 Verify that the equation

$$\ddot{x} + \varepsilon h(x^2 + \dot{x}^2 - 1)\dot{x} + x = 0,$$

where $h(u)$ is differentiable and strictly increasing for all u, and $h(0) = 0$, has the periodic solutions $x = \cos(t + \alpha)$ for any α. Using the method of slowly varying amplitude show that this solution is a stable limit cycle when $\varepsilon > 0$.

4.12 Find, by the method of Section 4.5 the equivalent linear equation for

$$\ddot{x} + \varepsilon(x^2 + \dot{x}^2 - 1)\dot{x} + x = 0.$$

Show that it gives the limit cycle exactly. Obtain from the linear equation the equations of the nearby spiral paths.

4.13 Use the method of equivalent linearization to find the amplitude and frequency of the limit cycle of the equation

$$\ddot{x} + \varepsilon(x^2 - 1)\dot{x} + x + \varepsilon x^3 = 0, \quad 0 < \varepsilon \ll 1$$

Write down the equivalent linear equation.

4.14 The equation $\ddot{x} + x^3 = 0$ has a centre at the origin of the phase plane with $\dot{x} = y$.

(i) Substitute $x = a\cos\omega t$ to find by the harmonic balance method the frequency–amplitude relation $\omega = \sqrt{3}a/2$.

(ii) Construct, by the method of equivalent linearization, the associated linear equation, and show how the processes (i) and (ii) are equivalent.

4.15 The displacement x of relativistic oscillator satisfies

$$m_0\ddot{x} + k(1 - (\dot{x}/c)^2)^{3/2}x = 0.$$

Show that the equation becomes $\ddot{x} + (\alpha/a)x = 0$ when linearized with respect to the approximate solution $x = a\cos\omega t$ by the method of equivalent linearization, where

$$\alpha = \frac{1}{\pi}\int_0^{2\pi} \frac{ka}{m_0}\cos^2\theta\left(1 - \frac{a^2\omega^2}{c^2}\sin^2\theta\right)^{3/2}d\theta.$$

Confirm that, when $a^2\omega^2/c^2$ is small, the period of the oscillations is given approximately by

$$2\pi\sqrt{\left(\frac{m_0}{k}\right)}\left(1 + \frac{3a^2k}{16m_0c^2}\right).$$

4.16 Show that the phase paths of the equation $\ddot{x} + (x^2 + \dot{x}^2)x = 0$ are given by

$$e^{-x^2}(y^2 + x^2 - 1) = \text{constant}.$$

Show that the surface $e^{-x^2}(y^2 + x^2 - 1) = z$ has a maximum at the origin, and deduce that the origin is a centre.

Use the method of harmonic balance to obtain the frequency–amplitude relation $\omega^2 = 3a^2/(4 - a^2)$ for $a < 2$, assuming solutions of the approximate form $a\cos\omega t$. Verify that $\cos t$ is an exact solution, and that $\omega = 1$, $a = 1$ is predicted by harmonic balance.

Plot some exact phase paths to indicate where the harmonic balance method is likely to be unreliable.

4.17 Show, by the method of harmonic balance, that the frequency–amplitude relation for the periodic solutions of the approximate form $a\cos\omega t$, for

$$\ddot{x} - x + \alpha x^3 = 0, \quad \alpha > 0,$$

is $\omega^2 = \frac{3}{4}\alpha a^2 - 1$.

By analysing the phase diagram, explain the lower bound $2/\sqrt{(3\alpha)}$ for the amplitude of periodic motion. Find the equation of the phase paths, and compare where the separatrix cuts the x-axis, with the amplitude $2/\sqrt{(3\alpha)}$.

4.18 Apply the method of harmonic balance to the equation $\ddot{x} + x - \alpha x^2 = 0$, $\alpha > 0$, using the approximate form of solution $x = c + a\cos\omega t$ to show that,

$$\omega^2 = 1 - 2\alpha c, \quad c = \frac{1}{2\alpha} - \frac{1}{2\alpha}\sqrt{(1 - 2\alpha^2 a^2)}.$$

Deduce the frequency–amplitude relation

$$\omega = (1 - 2\alpha^2 a^2)^{1/4}, \quad a < 1/(\sqrt{2}\alpha).$$

Explain, in general terms, why an upper bound on the amplitude is to be expected.

4.19 Apply the method of harmonic balance to the equation $\ddot{x} - x + x^3 = 0$ in the neighbourhood of the centre at $x = 1$, using the approximate form of solution $x = 1 + c + a\cos\omega t$. Deduce that the mean displacement, frequency, and amplitude are related by

$$\omega^2 = 3c^2 + 6c + 2 + \tfrac{3}{4}a^2, \quad 2c^3 + 6c^2 + c(4 + 3a^2) + 3a^2 = 0.$$

4.20 Consider the van der Pol equation with nonlinear restoring force

$$\ddot{x} + \varepsilon(x^2 - 1)\dot{x} + x - \alpha x^2 = 0,$$

where ε and α are small. By assuming solutions approximately of the form $x = c + a\cos\omega t + b\sin\omega t$, show that the mean displacement, frequency, and amplitude are related by

$$c = 2\alpha, \quad \omega^2 = 1 - 4\alpha^2, \quad a^2 + b^2 = 4(1 - 4\alpha^2)$$

4.21 Suppose that the nonlinear system

$$\dot{x} = p(x), \quad \text{where } x = \begin{bmatrix} x \\ y \end{bmatrix}$$

has an isolated equilibrium point at $x = 0$, and that solutions exist which are approximately of the form

$$\tilde{x} = B \begin{bmatrix} \cos\omega t \\ \sin\omega t \end{bmatrix}, \quad B = \begin{bmatrix} a & b \\ c & d \end{bmatrix}.$$

Adapt the method of equivalent linearization to this problem by approximating $p(\tilde{x})$. by its first harmonic terms:

$$p\{\tilde{x}(t)\} = C \begin{bmatrix} \cos\omega t \\ \sin\omega t \end{bmatrix},$$

where C is a matrix of the Fourier coefficients. It is assumed that

$$\int_0^{2\pi/\omega} p\{\tilde{x}(t)\}dt = 0.$$

Substitute in the system to show that

$$BU = C, \quad \text{where } U = \begin{bmatrix} 0 & -\omega \\ \omega & 0 \end{bmatrix}.$$

Deduce that the equivalent linear system is

$$\dot{\tilde{x}} = BUB^{-1}\tilde{x} = CUC^{-1}\tilde{x},$$

when B and C are nonsingular.

4.22 Use the method of Problem 4.21 to construct a linear system equivalent to the van der Pol equation

$$\dot{x} = y, \quad \dot{y} = -x - \varepsilon(x^2 - 1)y.$$

4.23 Apply the method of Problem 4.21 to construct a linear system equivalent to

$$\begin{bmatrix} \dot{x} \\ \dot{y} \end{bmatrix} = \begin{bmatrix} \varepsilon & 1 \\ -1 & \varepsilon \end{bmatrix}\begin{bmatrix} x \\ y \end{bmatrix} + \begin{bmatrix} 0 \\ -\varepsilon x^2 y \end{bmatrix},$$

and show that the limit cycle has frequency given by $\omega^2 = 1 - 5\varepsilon^2$ for ε small.

4.24 Apply the method of Problem 4.21 to the predator–prey equation (see Section 2.2)

$$\dot{x} = x - xy, \quad \dot{y} = -y + xy,$$

in the neighbourhood of the equilibrium point $(1, 1)$, by using the displaced approximations

$$x = m + a\cos\omega t + b\sin\omega t, \quad y = n + c\cos\omega t + d\sin\omega t.$$

Show that

$$m = n, \quad \omega^2 = 2m - 1 \quad \text{and} \quad a^2 + b^2 = c^2 + d^2.$$

4.25 Show that the approximate solution for oscillations of the equation

$$\ddot{x} = x^2 - x^3$$

in the neighbourhood of $x = 1$ is $x = c + a \cos \omega t$ where

$$\omega^2 = \frac{c(15c^2 - 15c + 4)}{2(3c - 1)}, \quad a^2 = \frac{2c^2(1 - c)}{3c - 1}.$$

4.26 Use the method of Section 4.2 to obtain approximate solutions of the equation

$$\ddot{x} + \varepsilon \dot{x}^3 + x = 0, \quad |\varepsilon| \ll 1.$$

4.27 Suppose that the equation $\ddot{x} + f(x)\dot{x} + g(x) = 0$ has a periodic solution. Represent the equation in the (x, y) phase plane given by $\dot{x} = y - F(x)$, $\dot{y} = -g(x)$, where

$$F(x) = \int_0^x f(u) \mathrm{d}u$$

(this particular phase plane is known as the **Liénard plane**). Let

$$v(x, y) = \tfrac{1}{2} y^2 + \int_0^x g(u) \mathrm{d}u,$$

and by considering $\mathrm{d}v/\mathrm{d}t$ on the closed phase path C show that

$$\int_C F(x) \mathrm{d}y = 0.$$

On the assumption that van der Pol's equation $\ddot{x} + \varepsilon(x^2 - 1)\dot{x} + x = 0$ has a periodic solution approximately of the form $x = A \cos \omega t$, deduce that $A \approx 2$ and $\omega \approx 1$.

4.28 Apply the slowly varying amplitude method of Section 4.3 to

$$\ddot{x} - \varepsilon \sin \dot{x} + x = 0 \quad (0 < \varepsilon \ll 1),$$

and show that the amplitude a satisfies

$$\dot{a} = \varepsilon J_1(a)$$

approximately. [Use the formula

$$J_1(a) = \frac{1}{\pi} \int_0^\pi \sin(a \sin u) \sin u \, \mathrm{d}u$$

for the Bessel function $J_1(a)$: see Abramowitz and Stegun (1965, p. 360).

Find also the approximate differential equation for θ. Using a graph of $J_1(a)$ decide how many limit cycles the system has. Which are stable?

5 Perturbation methods

This chapter describes techniques for obtaining approximations to periodic time solutions of nearly linear second-order differential equations subject to a harmonic forcing term, and to limit cycles of autonomous equations. The approximations take the form of an expansion in integer powers of a small parameter, having coefficients that are functions of time. There is some freedom in assigning the time-dependent coefficients; this is utilized to produce uniformly valid approximations under different circumstances, depending on the values of the main parameters of the equation. It is also shown how a homo-clinic path can be approximated by using similar methods. The processes reveal physical features having no analogue in linear theory, despite the fact that the equations considered have small nonlinearities.

5.1 Nonautonomous systems: forced oscillations

Up to the present point, apart from a brief reference in Section 1.2, we have considered only autonomous systems. Differential equations of the form

$$\ddot{x} = f(x, \dot{x}, t), \tag{5.1}$$

and first-order systems having the general form

$$\dot{x} = X(x, y, t), \quad \dot{y} = Y(x, y, t), \tag{5.2}$$

in which the clock time t appears explicitly, are called **nonautonomous**, and are the main subject of this chapter.

The **states** of the system, given by the pair of values $(x, \dot{x})$ or (x, y), have the same meaning as before (for the equation (5.1) we usually define $y = \dot{x}$). If an autonomous system passes through a particular state (x_0, y_0) at time $t = t_0$, these initial conditions determine the succession of states in the motion uniquely independently of t_0 and they constitute a unique phase path through the point (x_0, y_0). However, *nonautonomous* equations generate an infinite number of phase paths through every point (x_0, y_0); a different one, in general, for every value of t_0. This can be seen by evaluating the slope of a particular phase path passing through (x_0, y_0) at time t_0. From (5.2)

$$\left(\frac{dy}{dx}\right)_{(x_0, y_0, t_0)} = \frac{\dot{y}}{\dot{x}} = \frac{Y(x_0, y_0, t_0)}{X(x_0, y_0, t_0)}.$$

the slope depends on the value of t_0, whereas for autonomous equations it is independent of t_0. This greatly reduces the usefulness of a phase-plane representation, which no longer consists of a pattern of distinct, non-inter-secting curves whose structure can easily be grasped.

In this chapter we shall obtain approximations to periodic solutions of nonautonomous equations of a particular type. Two examples are the forced pendulum equation and the forced van der Pol equation, given by

$$\ddot{x} + k\dot{x} + \omega_0^2 x + \varepsilon x^3 = F \cos \omega t$$

and

$$\ddot{x} + \varepsilon(x^2 - 1)\dot{x} + \omega_0^2 x = F \cos \omega t,$$

where ε is a small parameter. When $\varepsilon = 0$ these equations become linear. The term on the right, $F \cos \omega t$, is called a harmonic **forcing term**, and can be thought of as an external force of amplitude F and circular frequency ω applied to a unit particle on a nonlinear spring, as shown in Fig. 1.10. The forcing term tries to drive the system at frequency ω against its natural tendency to perform **free oscillations** or other motions described by the homogeneous equations

$$\ddot{x} + k\dot{x} + \omega_0^2 x + \varepsilon x^3 = 0 \quad \text{or} \quad \ddot{x} + \varepsilon(x^2 - 1)\dot{x} + \omega_0^2 x = 0.$$

It will be shown that although these systems are close to linear by virtue of the smallness of the nonlinear terms, which have a factor ε, large-scale phenomena are produced which are quite unlike those associated with linear equations. With nonlinear equations small causes do not necessarily produce small effects on a global scale.

For convenience of reference we summarize the properties of the solutions of the equation that represents a damped **linear oscillator** with a harmonic forcing term:

$$\ddot{x} + k\dot{x} + \omega_0^2 x = F \cos \omega t, \tag{5.3a}$$

where $k > 0$, $\omega_0 > 0$, $\omega > 0$ and F are constants, and the damping coefficient k is not too large:

$$0 < k < 2\omega_0. \tag{5.3b}$$

This is the equation of motion of the mechanical system in Fig. 5.1. The block has unit mass, the stiffness of the spring is equal to ω_0^2, and the friction is proportional to the velocity through the constant k. The coordinate x is measured from the position of the mass when the spring has its natural length.

The general solution of (5.3a) in the standard form:

Particular solution + the complementary functions

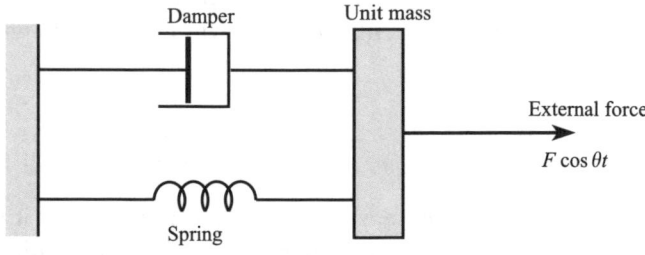

Figure 5.1

(the complementary functions consisting of the general solution of the corresponding homogeneous equation $\ddot{x} + k\dot{x} + \omega_0^2 x = 0$) is

$$x(t) = \frac{F\cos(\omega t - \gamma)}{[(\omega_0^2 - \omega^2)^2 + k^2\omega^2]^{\frac{1}{2}}} + Ce^{-\frac{1}{2}kt}\cos\left[\left(\omega_0^2 - \frac{1}{4}k^2\right)^{\frac{1}{2}}t - \phi\right]. \tag{5.4}$$

In the first term, γ is equal to the polar angle of the complex number $(\omega_0^2 - \omega^2) + ik\omega$ on an Argand diagram. In the second term (the complementary functions) C and ϕ are arbitrary constants, adjustable so as to conform with any given initial conditions $x(t_0) = x_0$, $\dot{x}(t_0) = y_0$. This term always tends to zero as $t \to \infty$ due to the factor $e^{-(1/2)kt}$; it is described as a **transient** term. All solutions of (5.3) therefore drift into the same state of steady oscillation, described by the particular solution $x_p(t)$ where

$$x_p(t) = F\cos(\omega t - \gamma)/[(\omega_0^2 - \omega^2)^2 + k^2\omega^2]^{\frac{1}{2}}, \tag{5.5}$$

which has the same frequency as the forcing term $F\cos\omega t$, but is out of phase with it by an angle γ. This term of (5.4) is also called the **forced oscillation**, the second term the **free oscillation**, when it is necessary to distinguish them.

The amplitude A of (5.5) is given by

$$A = |F|/[(\omega_0^2 - \omega^2)^2 + k^2\omega^2]^{\frac{1}{2}}.$$

Suppose we conduct experiments on the system in Fig. 5.1. by varying the forcing frequency ω, in order to find how the amplitude A varies in response. For a fixed value of k, A attains its maximum value when $(\omega_0^2 - \omega^2)^2 + k^2\omega^2$ takes its minimum value with respect to ω^2, which occurs when

$$\omega^2 = \omega_0^2 - \tfrac{1}{2}k^2,$$

and then

$$A = A_{max} = |F|/\left[k(\omega_0^2 - \tfrac{1}{4}k^2)^{\frac{1}{2}}\right].$$

Suppose now that k is very small. These expressions show that when ω takes values close to ω_0, A_{max} becomes very large, and the system is said to be in a state of **resonance**. A more detailed analysis shows that under these conditions the phase angle γ in (5.4) is such that the applied force $F\cos\omega t$, with $\omega^2 = \omega_0^2 - \tfrac{1}{2}k^2$, is timed so that it always gives the maximum reinforcement to the tendency of the system to follow its free oscillations. In this way the amplitude builds up to a large value.

The case when $k = 0$ in (5.3a) is special in that the second term in the general solution (5.4) is not a transient; it does not decay with time. The solutions are periodic only for exceptional values of ω and ω_0 (specifically, when $\omega = (p/q)\omega_0$, where p and q are integers). There is a permanent, but erratic state of oscillation which depends on the initial conditions. The general solution of (5.3a) when $k = 0$ is

$$x(t) = \frac{F}{\omega_0^2 - \omega^2}\cos\omega t + C\cos(\omega_0 t + \phi)$$

If $k = 0$ and $\omega = \omega_0$ then (5.4) becomes meaningless, so we must revert to the differential equation (5.3), which in this case takes the form

$$\ddot{x} + \omega_0^2 x = F \cos \omega_0 t.$$

The general solution is

$$x(t) = \frac{F}{2\omega_0} t \sin \omega_0 t + C \cos(\omega_0 t - \phi), \tag{5.6}$$

the first term having different form from that in (5.4). The solutions consist of oscillations of frequency ω_0 whose amplitude increases to infinity as $t \to \pm\infty$, which amounts to an extreme case of resonance, uncontrolled by damping.

The equation of motion of a damped pendulum with a harmonic forcing term is easy to visualize, and leads to the consideration of an important family of nonlinear equations called in general **Duffing's equations**. In a standard form the pendulum equation is written

$$\ddot{x} + k\dot{x} + \omega_0^2 \sin x = F \cos \omega t, \tag{5.7}$$

where, in the usual notation, x is the angular displacement from the vertical; $k = \alpha/(ma^2)$ where m is the mass and a the length, and $\alpha\dot{x}$ is the moment of the friction about the support; $\omega_0^2 = g/a$; $F = M/(ma^2)$ where M is the amplitude of the driving moment about the support. The physical dimensions of the coefficients in (5.7) are $[T^{-2}]$.

To envisage how a regular forcing moment might be applied to a pendulum at its support, consider Fig. 5.2. The pendulum is constructed of a spindle (the support) rigidly attached is a rod which carries the bob. The spindle is friction-driven by a close-fitting sleeve which is caused to rotate independently about the spindle with angular velocity $A \cos \omega t$. Supposing that the contact area is lubricated by a viscous liquid, the driving moment M is proportional to the

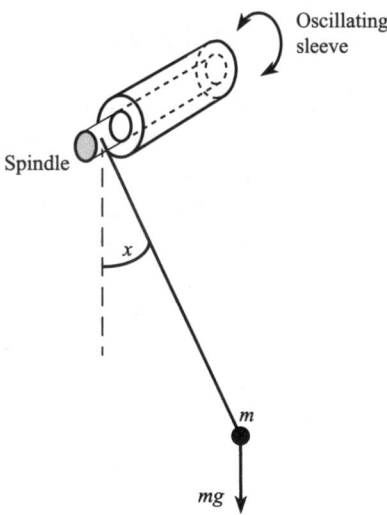

Figure 5.2   Friction-driven pendulum.

relative angular velocity between the spindle and the sleeve:

$$M = \alpha(A \cos \omega t - \dot{x}),$$

where A is a constant. By adjusting (in imagination) the constants α and A, the equation (5.7) may be generated for any value of F and any value of $k > 0$.

5.2 The direct perturbation method for the undamped Duffing's equation

Consider first the forced oscillation of an undamped pendulum,

$$\ddot{x} + \omega_0^2 \sin x = F \cos \omega t, \qquad (5.8)$$

in which, without loss of generality, we can suppose that $\omega_0 > 0$, $\omega > 0$, and $F > 0$ (since $F < 0$ implies a phase difference that can be eliminated by a change of time origin and corresponding modification of initial conditions).

Put

$$\sin x \approx x - \tfrac{1}{6}x^3$$

to allow for moderately large swings, which is accurate to 1% for $|x| < 1$ radian (57°). Then eqn (5.8) becomes, approximately,

$$\ddot{x} + \omega_0^2 x - \tfrac{1}{6}\omega_0^2 x^3 = F \cos \omega t. \qquad (5.9)$$

Standardize the form of (5.9) by putting

$$\tau = \omega t, \quad \Omega^2 = \omega_0^2/\omega^2 \ (\Omega > 0), \quad \Gamma = F/\omega^2. \qquad (5.10)$$

Then we obtain

$$x'' + \Omega^2 x - \tfrac{1}{6}\Omega^2 x^3 = \Gamma \cos \tau, \qquad (5.11)$$

where dashes represent differentiation with respect to τ. This is a special case of Duffing's equation, which is characterized by a cubic nonlinear term. If this eqn (5.11) actually arises by consideration of a pendulum, the coefficients and variables are all dimensionless.

The methods to be described depend on the nonlinear terms being small. Here we assume that $\tfrac{1}{6}\Omega^2$ is small and write

$$\tfrac{1}{6}\Omega^2 = \varepsilon_0. \qquad (5.12)$$

Then (5.11) becomes

$$x'' + \Omega^2 x - \varepsilon_0 x^3 = \Gamma \cos \tau. \qquad (5.13)$$

Instead of taking (5.13) as it stands, with Ω, Γ, ε_0 as constants, we consider the *family of differential equations*

$$x'' + \Omega^2 x - \varepsilon x^3 = \Gamma \cos \tau, \qquad (5.14)$$

where ε is a parameter occupying an interval I_ε which includes $\varepsilon = 0$. When $\varepsilon = \varepsilon_0$ we recover (5.13), and when $\varepsilon = 0$ we obtain the linearized equation corresponding to the family (5.14):

$$x'' + \Omega^2 x = \Gamma \cos \tau. \tag{5.15}$$

The solutions of (5.14) are now to be thought of as functions of both ε and τ and we will write $x(\varepsilon, \tau)$.

The most elementary version of the **perturbation method** is to attempt a representation of the solutions of (5.14) in the form of a power series in ε:

$$x(\varepsilon, \tau) = x_0(\tau) + \varepsilon x_1(\tau) + \varepsilon^2 x_2(\tau) + \cdots, \tag{5.16}$$

whose coefficients $x_i(\tau)$ are functions only of τ. To form equations for $x_i(\tau)$, $i = 0, 1, 2, \ldots$, substitute the series (5.16) into eqn (5.14):

$$(x_0'' + \varepsilon x_1'' + \cdots) + \Omega^2(x_0 + \varepsilon x_1 + \cdots) - \varepsilon(x_0 + \varepsilon x_1 + \cdots)^3 = \Gamma \cos \tau.$$

Since this is assumed to hold for *every member of the family* (5.14), that is for *every* ε on I_ε the coefficients of powers of ε must balance and we obtain

$$x_0'' + \Omega^2 x_0 = \Gamma \cos \tau, \tag{5.17a}$$

$$x_1'' + \Omega^2 x_1 = x_0^3, \tag{5.17b}$$

$$x_2'' + \Omega^2 x_2 = 3x_0^2 x_1, \tag{5.17c}$$

and so on.

We shall be concerned only with *periodic solutions having the period, 2π, of the forcing term.* (There are periodic solutions having other periods: see Chapter 7; also, by a modification of the present procedure as in Section 5.4, we shall find yet more solutions having period 2π.) Then, for all ε on I_ε and for all τ,

$$x(\varepsilon, \tau + 2\pi) = x(\varepsilon, \tau). \tag{5.18}$$

By (5.16), it is sufficient that for all τ

$$x_i(\tau + 2\pi) = x_i(\tau), \quad i = 0, 1, 2, \ldots. \tag{5.19}$$

Equations (5.17) together with the conditions (5.19) are sufficient to provide the solutions required. The details will be worked out in Section 5.3; for the present, note that (5.17a) is the same as the 'linearized equation' (5.15): necessarily so, since putting $\varepsilon_0 = 0$ in (5.13) implies putting $\varepsilon = 0$ in (5.16). The major term in (5.16) is therefore a periodic solution of the linearized equation (5.15). It is therefore clear that this process restricts us to finding the solutions of the nonlinear equation which are close to (or branch from, or **bifurcate** from) the solution of the linearized equation. The method will not expose any other periodic solutions. The zero-order solution $x_0(\tau)$ is known as a **generating solution** for the family of eqns (5.14).

Exercise 5.1
Substitute the perturbation series

$$x(\varepsilon, \tau) = x_0(\tau) + \varepsilon x_1(\tau) + \varepsilon^2 x_2(\tau) + \cdots$$

into the equation

$$x'' + \Omega^2 x + \varepsilon f(x) = \Gamma \cos \tau, \quad 0 < \varepsilon \ll 1,$$

where $f(x)$ is a smooth odd function of x. Obtain the differential equation for $x_0, x_1, x_2,$ and x_3.

5.3 Forced oscillations far from resonance

Suppose that

$$\Omega \neq \text{ an integer.} \tag{5.20}$$

We now solve (5.17) subject to the periodicity condition (5.19). The solutions of (5.17a) are

$$a_0 \cos \Omega \tau + b_0 \sin \Omega \tau + \frac{\Gamma}{\Omega^2 - 1} \cos \tau, \tag{5.21}$$

where a_0, b_0 are constants. Since Ω is not an integer the only solution having period 2π is obtained by putting $a_0 = b_0 = 0$ in (5.21); that is,

$$x_0(\tau) = \frac{\Gamma}{\Omega^2 - 1} \cos \tau. \tag{5.22}$$

Equation (5.17b) then becomes

$$x_1'' + \Omega^2 x_1 = \left(\frac{\Gamma}{\Omega^2 - 1} \right)^3 \cos^3 \tau = \frac{\Gamma^3}{(\Omega^2 - 1)^3} \left(\frac{3}{4} \cos \tau + \frac{1}{4} \cos 3\tau \right).$$

The only solution with period 2π is given by

$$x_1(\tau) = \frac{3}{4} \frac{\Gamma^3}{(\Omega^2 - 1)^4} \cos \tau + \frac{1}{4} \frac{\Gamma^3}{(\Omega^2 - 1)^3 (\Omega^2 - 9)} \cos 3\tau$$

by eqn (5.20).

The first two terms of the expansion (5.16) provide the approximation

$$x(\varepsilon, \tau) = \frac{\Gamma}{\Omega^2 - 1} \cos \tau + \varepsilon \left(\frac{3}{4} \frac{\Gamma^3}{(\Omega^2 - 1)^4} \cos \tau \right.$$

$$\left. + \frac{1}{4} \frac{\Gamma^3}{(\Omega^2 - 1)^3 (\Omega^2 - 9)} \cos 3\tau \right) + O(\varepsilon^2). \tag{5.23}$$

The series continues with terms of order ε^2, ε^3 and involves the harmonics $\cos 5\tau$, $\cos 7\tau$, and so on. For the pendulum, $\varepsilon = \varepsilon_0 = \frac{1}{6}\Omega^2$, by (5.12).

The method obviously fails if Ω^2 takes one of the values $1, 9, 25, \ldots$, since certain terms would then be infinite, and this possibility is averted by condition (5.20). However, the series would not converge well if Ω^2 were even close to one of these values, and so a few terms of the series would describe $x(\varepsilon, \tau)$ only poorly. Such values of Ω correspond to conditions of **near-resonance**. $\Omega \approx 1$ is such a case; not surprisingly, since it is a value close to resonance for the linearized equation. The other odd-number values of Ω correspond to **nonlinear resonances** caused by the higher harmonics present in $x^3(\varepsilon, \tau)$, which can be regarded as 'feeding-back' into the linear equation like forcing terms. This idea is exploited in Chapter 7.

If Ω is an even integer the straightforward procedure fails because (5.21) has period 2π for all values of a_0, b_0. Their required values can only be established by carrying the terms forward into the second stage (5.17b), as is done in Section 5.4 for a similar situation.

Example 5.1   *Obtain an approximation to the forced response, of period 2π, for the equation* $x'' + \frac{1}{4}x + 0.1x^3 = \cos \tau$.

Consider the family $x'' + \frac{1}{4}x + \varepsilon x^3 = \cos \tau$: this is eqn (5.13) with $\Omega = \frac{1}{2}$ and $\varepsilon_0 = -0.1$. Assume that $x(\varepsilon, \tau) = x_0(\tau) + \varepsilon x_1(\tau) + \cdots$. The periodicity condition is met if $x_0(\tau), x_1(\tau), \ldots$, have period 2π. The equations for x_0, x_1 are (see (5.17a,b))

$$x_0'' + \tfrac{1}{4}x_0 = \cos \tau,$$
$$x_1'' + \tfrac{1}{4}x_1 = -x_0^3.$$

The only 2π-periodic solution of the first equation is

$$x_0(\tau) = -\tfrac{4}{3}\cos \tau,$$

and the second becomes

$$x_1'' + \tfrac{1}{4}x_1 = \tfrac{16}{9}\cos \tau + \tfrac{16}{27}\cos 3\tau$$

The 2π-periodic solution is

$$x_1(\tau) = -\tfrac{64}{27}\cos \tau - \tfrac{64}{945}\cos 3\tau.$$

Therefore,

$$x(\varepsilon, \tau) = -\tfrac{4}{3}\cos \tau - \varepsilon\left(\tfrac{64}{27}\cos \tau + \tfrac{64}{945}\cos 3\tau\right) + O(\varepsilon^2).$$

With $\varepsilon = 0.1$,

$$x(\varepsilon, \tau) \approx -1.570\cos \tau - 0.007\cos 3\tau.$$

●

> **Exercise 5.2**
> The perturbation procedure described above is applied to the equation
> $$x'' + \tfrac{2}{3}x - \tfrac{1}{6}x^3 = \Gamma \cos \tau$$
> to find the 2π-period solution. What is the maximum value that Γ can take in order that the leading coefficient of $\cos \tau$ in (5.23) is accurate to within 5% to order ε^2?

5.4 Forced oscillations near resonance with weak excitation

Not only does the method of the previous sections fail to give a good representation near certain critical values of Ω, but it does not reveal all the solutions having the period 2π of the forcing term. The only solution it can detect is the one which bifurcates from the one-and-only solution having period 2π of the linearized eqn (5.15). More solutions can be found by, so to speak, offering the nonlinear equation a greater range of choice of possible generating solutions by modifying the structure so that a different linearized equation is obtained.

We will include damping in the approximate pendulum equation:

$$\ddot{x} + k\dot{x} + \omega_0^2 x - \tfrac{1}{6}\omega_0^2 x^3 = F \cos \omega t. \tag{5.24}$$

Corresponding to eqn (5.13) we have

$$x'' + Kx' + \Omega^2 x - \varepsilon_0 x^3 = \Gamma \cos \tau, \tag{5.25a}$$

where

$$\tau = \omega t, \quad \Omega^2 = \omega_0^2/\omega^2, \quad \varepsilon_0 = \tfrac{1}{6}\Omega^2, \quad K = k/\omega, \quad \Gamma = F/\omega^2. \tag{5.25b}$$

Assume that Γ is small (weak excitation), and K is small (small damping) and therefore put

$$\Gamma = \varepsilon_0 \gamma, \quad K = \varepsilon_0 \kappa \quad (\gamma, \kappa > 0). \tag{5.26}$$

Suppose also that Ω is close to one of the critical resonance values 1, 3, 5, ... Consider the simplest case $\Omega \approx 1$, which corresponds to near-resonance of the linearized equation, and write

$$\Omega^2 = 1 + \varepsilon_0/\beta. \tag{5.27}$$

Equation (5.25) becomes

$$x'' + x = \varepsilon_0(\gamma \cos \tau - \kappa x' - \beta x + x^3). \tag{5.28}$$

Now consider the *family of equations*

$$x'' + x = \varepsilon(\gamma \cos \tau - \kappa x' - \beta x + x^3), \tag{5.29}$$

with parameter ε, in which γ, κ, β retain the constant values given by (5.26) and (5.27). When $\varepsilon = \varepsilon_0$ we return to (5.28). When $\varepsilon = 0$ we obtain the new linearized equation $x'' + x = 0$. This has *infinitely many solutions with period* 2π, offering a wide choice of generating solutions (compare (5.15) in which only one presents itself).

Now assume that $x(\varepsilon, \tau)$ may be expanded in the form

$$x(\varepsilon, \tau) = x_0(\tau) + \varepsilon x_1(\tau) + \varepsilon^2 x_2(\tau) + \cdots, \tag{5.30}$$

where (by the same argument as led up to (5.19)), for all τ

$$x_i(\tau + 2\pi) = x_i(\tau), \quad i = 0, 1, 2, \ldots. \tag{5.31}$$

Substitute (5.30) into (5.29). By the argument leading to (5.17) we obtain

$$x_0'' + x_0 = 0, \tag{5.32a}$$

$$x_1'' + x_1 = \gamma \cos \tau - \kappa x_0' - \beta x_0 + x_0^3, \tag{5.32b}$$

$$x_2'' + x_2 = -\kappa x_1' - \beta x_1 + 3x_0^2 x_1, \tag{5.32c}$$

and so on.

The solution of (5.32a) is

$$x_0(\tau) = a_0 \cos \tau + b_0 \sin \tau \tag{5.33}$$

for every a_0, b_0. Now put (5.33) into (5.32b). After writing (see Appendix E)

$$\cos^3 \tau = \tfrac{3}{4} \cos \tau + \tfrac{1}{4} \cos 3\tau, \quad \sin^3 \tau = \tfrac{3}{4} \sin \tau - \tfrac{1}{4} \sin 3\tau$$

and making other similar reductions[†] we have

$$x_1'' + x_1 = \{\gamma - \kappa b_0 + a_0[-\beta + \tfrac{3}{4}(a_0^2 + b_0^2)]\} \cos \tau$$

$$+ \{\kappa a_0 + b_0[-\beta + \tfrac{3}{4}(a_0^2 + b_0^2)]\} \sin \tau$$

$$+ \tfrac{1}{4} a_0(a_0^2 - 3b_0^2) \cos 3\tau + \tfrac{1}{4} b_0(3a_0^2 - b_0^2) \sin 3\tau. \tag{5.34}$$

The solution $x_1(\tau)$ is required to have period 2π, but unless the coefficients of $\cos \tau$ and $\sin \tau$ in (5.34) are zero, there are no periodic solutions, since any solution would contain terms of the form $\tau \cos \tau$, $\tau \sin \tau$ (compare eqn (5.6)). Such non-periodic or otherwise undesirable constituents of a solution are often called **secular terms**. We eliminate the secular terms by

---

[†] Often, the easiest way to carry out such calculations is to write the solution of (5.32a) as $A_0 e^{i\tau} + \bar{A}_0 e^{-i\tau}$. Then

$$x_0^3 = (A_0 e^{i\tau} + \bar{A}_0 e^{-i\tau})^3 = A_0^3 e^{3i\tau} + 3A_0^2 \bar{A}_0 e^{i\tau} + \text{complete conjugate}.$$

The substitution $A_0 = \tfrac{1}{2}(a_0 - ib_0)$, a_0 real, b_0 real, is left to the very end.

requiring the coefficients of $\cos \tau$ and $\sin \tau$ to be zero:

$$\kappa a_0 - b_0\{\beta - \tfrac{3}{4}(a_0^2 + b_0^2)\} = 0, \qquad (5.35a)$$

$$\kappa b_0 + a_0\{\beta - \tfrac{3}{4}(a_0^2 + b_0^2)\} = \gamma. \qquad (5.35b)$$

These equations settle the values of a_0 and b_0. To solve them, let r_0 be the amplitude of the generating solution:

$$r_0 = \sqrt{(a_0^2 + b_0^2)} > 0. \qquad (5.36)$$

By squaring and adding eqns (5.36a) and (5.36b), we obtain the following cubic **amplitude equation** for r_0^2:

$$r_0^2\left\{\kappa^2 + \left(\beta - \tfrac{3}{4}r_0^2\right)^2\right\} = \gamma^2. \qquad (5.37)$$

When (5.37) is solved for r_0, a_0 and b_0 can be obtained from (5.35) and (5.36).

Equation (5.37) will be analysed in the next section. For the present, note that there may be as many as three positive values of r_0^2 (hence of $r_0 > 0$) satisfying (5.37). This indicates that for certain ranges of κ, β, γ there may be three distinct 2π-periodic solutions of (5.24), (5.28), or (5.29), each bifurcating from one of three distinct generating solutions.

Having selected a pair of values a_0 and b_0 satisfying (5.35), we solve (5.28) to give

$$x_1(\tau) = a_1 \cos \tau + b_1 \sin \tau - \tfrac{1}{32}a_0(a_0^2 - 3b_0^2)\cos 3\tau - \tfrac{1}{32}b_0(3a_0^2 - b_0^2)\sin 3\tau,$$

where a_1 and b_1 are any constants. This expression is substituted into eqn (5.32c); the requirement that $x_2(\tau)$ should have period 2π provides equations determining a_1, b_1, as before. In this and subsequent stages the determining equations for a_i, b_i are linear equations, so no further multiplicity of solutions is introduced and the process is much less complicated.

Exercise 5.3
Establish that the latter assertion is true for a_1 and b_1, that is, that they are given by linear equations: the differential equation for x_2 is given by (5.32c). [Hints: many higher harmonics can be discarded since interest is only in the coefficients of $\cos \tau$ and $\sin \tau$; symbolic computation using trigonometric reduction has to be used to handle the algebra.]

5.5   The amplitude equation for the undamped pendulum

Suppose that the damping coefficient is zero; then in (5.24), (5.25), (5.29)

$$k = K = \kappa = 0. \qquad (5.38)$$

Instead of seeking r_0 through (5.37), the coefficients a_0, b_0 can be found directly from (5.35): the only solutions are given by

$$b_0 = 0, \tag{5.39a}$$

$$a_0(\beta - \tfrac{3}{4}a_0^2) = \gamma. \tag{5.39b}$$

We shall consider in detail only the pendulum case with $\varepsilon = \varepsilon_0 = \tfrac{1}{6}\Omega^2$. The original parameters ω, ω_0 and F of eqn (5.24) can be restored through (5.25) to (5.27). Equation (5.39b) becomes

$$a_0(\omega^2 - \omega_0^2 + \tfrac{1}{8}\omega_0^2 a_0^2) = -F. \tag{5.40}$$

The solutions a_0 can be obtained by drawing a cubic curve

$$z = a_0(\omega^2 - \omega_0^2 + \tfrac{1}{8}\omega_0^2 a_0^2) = f(a_0) \tag{5.41}$$

on a graph with axes a_0, z for any fixed value of ω and ω_0, then finding the intersections with the lines $z = -F$ for various $F > 0$, as in Fig. 5.3.

The main features revealed are as follows:

(i) When $\omega^2 > \omega_0^2$ (Fig. 5.3(a)), there is exactly one periodic oscillation possible. When F is small, the amplitude approximates to the linear response (5.22), and to the 'corrected linear response' (5.23), unless $\omega \approx \omega_0^2$ (very near to resonance) in which case it is considerably different (Fig. 5.3(b)). These responses are 180° out of phase with the forcing term.

(ii) When $\omega^2 < \omega_0^2$ (Fig. 5.3(c)), there is a single response, 180° out of phase, when F is relatively large. When F is smaller there are three distinct periodic responses, two in phase and one out of phase with the forcing term. The response marked 'A' in Fig. 5.3(c) corresponds to the response (5.22) of the linearized equation, and to the corrected linear response (5.23).

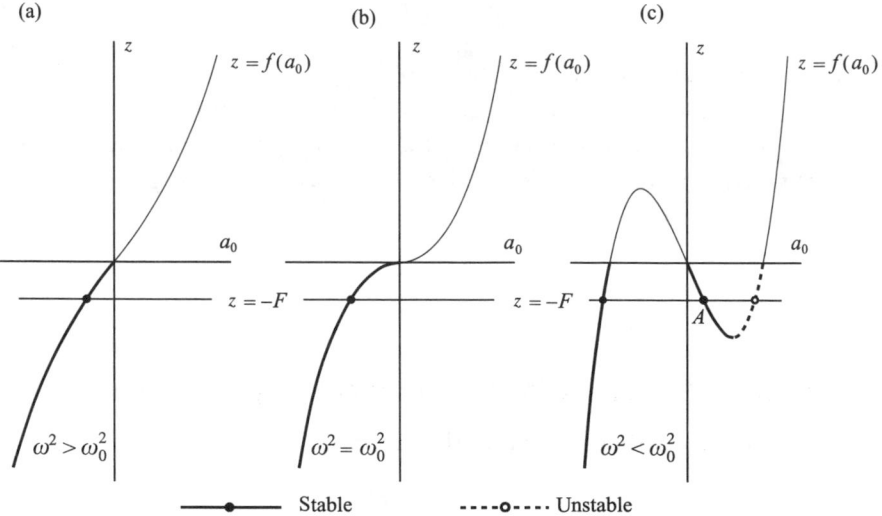

Figure 5.3   Illustrating eqn (5.41).

(iii) All three of the responses described in (ii) will be of small amplitude if the intersections of the curve $z = f(a_0)$ with the a_0 axis (Fig. 5.3(c)) occur close to the origin. These intersections are at $a_0 = 0$, $a_0 = \pm 2\sqrt{2}(1 - \omega^2/\omega_0^2)^{1/2}$. Three intersections are actually achieved provided

$$F < \tfrac{4}{3}\sqrt{\tfrac{2}{3}}\omega_0^2(1 - \omega^2/\omega_0^2)^{3/2}$$

(recalling that F is positive). This result can also be obtained from the properties of the cubic equation (5.40) in a_0 (see Appendix E). Therefore, by choosing ω^2/ω_0^2 sufficiently near to 1 (near to resonance) and F correspondingly small, the amplitudes of *all the three responses can be made as small as we wish*. In particular, they can all be confined in what we should normally regard as the linear range of amplitude for the pendulum.

(iv) Despite there being no damping, there are steady, bounded oscillations even when $\omega = \omega_0$ (unlike the linearized case). The nonlinearity controls the amplitude in the following way. The amplitude increases indefinitely if the forcing term remains in step with the natural oscillation and reinforces it cycle by cycle. However (e.g., Example 4.10), the frequency of the *natural* (nonlinear) oscillation varies with amplitude due to the nonlinearity and therefore does not remain in step with the forcing term.

(v) Whether a steady oscillation is set up or approached at all, and if it is, which of the possible modes is adopted, depends on the initial conditions of the problem, which are not considered at all here (see Chapter 7).

(vi) Whether a particular mode can be sustained in practice depends on its stability, of which an indication can be got as follows. If, in the neighbourhood of amplitude a_0, the forcing amplitude F required to sustain a_0 increases / decreases as a_0 increases / decreases, we expect a stable solution, in which an accidental small disturbance of amplitude cannot be sustained and amplified. If, however, F increases / decreases as a_0 decreases / increases, the conditions are right for growth of the disturbance and instability results. Further justification is given for this argument in Section 9.5. In anticipation of the analysis of Chapters 7 and 9, the stable and unstable branches are indicated in Figs 5.3 and 5.4.

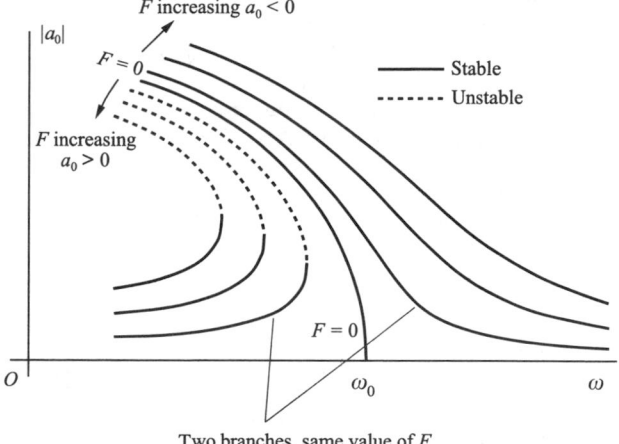

Figure 5.4 Amplitude–frequency curves for the undamped pendulum (eqn (5.40)).

The nature of the solutions of (5.40) as a function of the parameters ω, ω_0, with F given, can be exhibited on a single 'response diagram', Fig. 5.4. The figure can be plotted directly by writing (5.40) in the form

$$\omega = \sqrt{\{\omega_0^2(1 - \tfrac{1}{8}a_0^2) - F/a_0\}}.$$

For each value of $F > 0$ the amplitudes lie on two smooth branches; a typical pair is indicated on the figure. F increases on 'contours' increasingly distant from the curve $F = 0$ (which gives the amplitude–frequency curve for free oscillations, and is part of an ellipse) on both sides of it. Note that

$$\frac{d\omega}{da_0} = \frac{1}{2\omega}\left(-\frac{1}{4}a_0\omega_0^2 + \frac{F}{a_0^2}\right).$$

Therefore, if $a_0 < 0$ (and $F > 0$), $d\omega/da_0$ is never zero, whilst if $a_0 > 0$ (and $F > 0$), $d\omega/da_0$ is zero for a value of a_0 and $\omega < \omega_0$ as shown in Fig. 5.4.

Example 5.2 *Investigate the forced periodic solutions of the equation*

$$x'' + (9 + \varepsilon\beta)x - \varepsilon x^3 = \Gamma\cos\tau,$$

where ε is small and β, Γ are not too large.

This is a case where $\Omega^2 = 9 + \varepsilon\beta$ (eqn (5.14)) has a value which causes the direct expansion (5.23) to fail. The given equation may be rewritten as

$$x'' + 9x = \Gamma\cos\tau + \varepsilon(x^3 - \beta x).$$

Write $x(\varepsilon, \tau) = x_0(\tau) + \varepsilon x_1(\tau) + \cdots$, where $x_0(\tau), x_1(\tau), \ldots$ have period 2π. Then

$$x_0'' + 9x_0 = \Gamma\cos\tau,$$

$$x_1'' + 9x_1 = x_0^3 - \beta x_0,$$

and so on. These have the same form as (5.17), but since 9 is the square of an integer, the first equation has solutions of period 2π of the form

$$x_0(\tau) = a_0\cos 3\tau + b_0\sin 3\tau + \tfrac{1}{8}\Gamma\cos\tau.$$

When this is substituted into the equation for x_1, terms in $\cos 3\tau$, $\sin 3\tau$ emerge on the right-hand side, preventing periodic solutions unless their coefficents are zero. The simplest way to make the substitution is to write.

$$x_0(\tau) = A_0 e^{3i\tau} + \bar{A}_0 e^{-3i\tau} + \tfrac{1}{16}\Gamma e^{i\tau} + \tfrac{1}{16}\Gamma e^{-i\tau},$$

where $A_0 = \tfrac{1}{2}a_0 - \tfrac{1}{2}ib_0$. We find

$$x_0^3 - \beta x_0 = \left[\left(\frac{\Gamma^3}{16^3} + \frac{6\Gamma^2}{16^2}A_0 + 3A_0^2\bar{A}_0 - \beta A_0\right)e^{3i\tau} + \text{complete conjugate}\right]$$

$$+ \text{ other harmonics.}$$

Therefore, we require

$$A_0 \left(3A_0 \bar{A}_0 - \beta + \frac{6\Gamma^2}{16^2} \right) = -\frac{\Gamma^3}{16^3}.$$

This implies that A_0 is real: $b_0 = 0$, $A_0 = \frac{1}{2}a_0$, and the equation for a_0 is

$$\frac{1}{2}a_0 \left(\frac{3}{4}a_0^2 - \beta + \frac{6\Gamma^2}{16^2} \right) + \frac{\Gamma^3}{16^3} = 0.$$

●

Exercise 5.4
Find the leading approximation of the undamped system

$$x'' + x = \varepsilon(\tfrac{5}{4} \cos \tau - 2x + x^3).$$

Show that there are three possible amplitudes.

Exercise 5.5
Find the leading approximation of

$$x'' + x = \varepsilon(\gamma \sin \tau - \beta x + x^3).$$

5.6 The amplitude equation for a damped pendulum

The amplitude equation (5.37) translated into the parameters of eqn (5.24) by (5.25), (5.26), and (5.27), becomes

$$r_0^2 \left\{ k^2\omega^2 + \left(\omega^2 - \omega_0^2 + \tfrac{1}{8}\omega_0^2 r_0^2 \right)^2 \right\} = F^2. \tag{5.42}$$

Only solutions with $r_0 > 0$ are valid (see eqn (5.36)). Solving the quadratic equation for ω^2 given by (5.42) we find that

$$\omega^2 = \tfrac{1}{2} \left\{ 2\omega_0^2 \left(1 - \tfrac{1}{8}r_0^2 \right) - k^2 \pm \sqrt{ \left\{ k^4 - 4\omega_0^2 k^2 \left(1 - \tfrac{1}{8}r_0^2 \right) + 4F^2/r_0^2 \right\} } \right\}.$$

Typical response curves are shown in Fig. 5.5 for fixed values of k and ω_0, and selected values of $F > 0$. k was chosen fairly small so that comparison might be made with the undamped case shown in Fig. 5.4. The figure is similar to Fig. 5.4, with some important differences. There are no longer two distant branches corresponding to each value of F: the branches join up to make continuous curves. In the near neighbourhood of the point $\omega = \omega_0$, $r_0 = 0$ the curves do not turn over: that is to say, a certain minimum value of F is required before it is possible

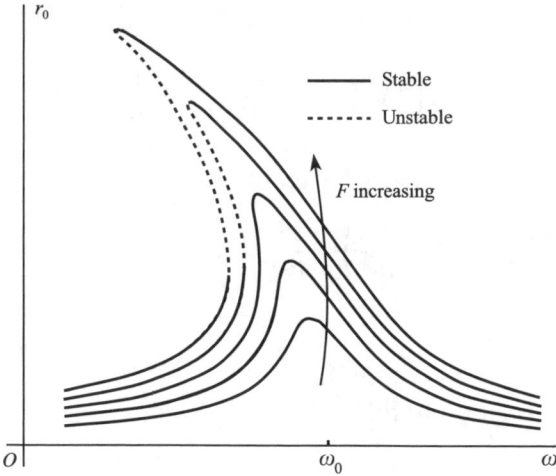

Figure 5.5 Amplitude (r_0)–frequency (ω) curves for the damped pendulum (eqn (5.42)): on each curve F is constant.

to have three alternative forced responses (see Problem 5.10). The response curves for small F represent the approximately linear response, as can be seen from (5.35).

The 'jump effect' associated with the response curves is presented in detail in Section 7.3.

5.7   Soft and hard springs

We have so far been concerned with equations of the type

$$\ddot{x} + k\dot{x} + cx + \varepsilon x^3 = F\cos\omega t,$$

which arise from an approximation to the pendulum equation. In the general case this is called a Duffing's equation with a forcing term. Now consider

$$\ddot{x} + k\dot{x} + cx + \varepsilon g(x) = F\cos\omega t, \tag{5.43}$$

where $k > 0$, $c > 0$, and $|\varepsilon| \ll 1$. This can be interpreted as the equation of forced motion with damping of a particle on a spring, the spring providing a restoring force which is almost but not quite linear. This is a useful physical model, since it leads us to expect that the features of linear motion will be to some extent preserved; for example, if $k = 0$ and $F = 0$ we guess that solutions for small enough amplitude will be oscillatory, and that the effect of small damping is to reduce the oscillations steadily to zero.

Important classes of cases are represented by the example $g(x) = x^3$, with $\varepsilon > 0$ and $\varepsilon < 0$. The corresponding (symmetrical) restoring forces are illustrated in Fig. 5.6(a) and (b). When ε is negative the restoring force become progressively weaker in extension than for the linear spring: such a spring is called **soft**. A pendulum is modelled by a soft spring system. When ε is positive the spring becomes stiffer as it extends and is called a **hard spring**.

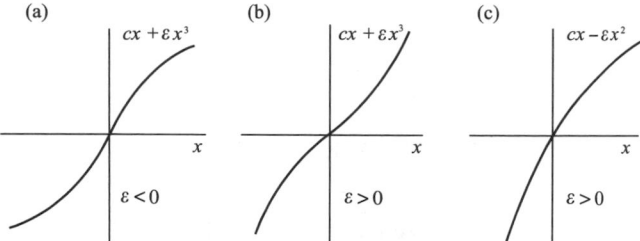

Figure 5.6   Restoring force functions for (a) a soft spring; (b) a hard spring; (c) an unsymmetrical case.

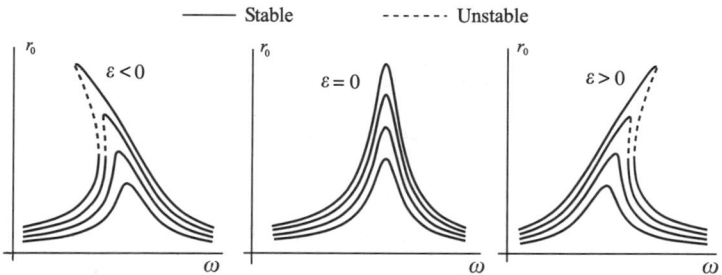

Figure 5.7   Amplitude–frequency curves for Duffing's equation $\ddot{x} + k\dot{x} + cx + \varepsilon x^3 = F \cos \omega t$, $k > 0$, $c > 0$.

The nature of the response diagrams for the forced oscillations, period $2\pi/\omega$, of (5.43) in the case of a soft spring ($\varepsilon < 0$), a linear spring ($\varepsilon = 0$), and a hard spring ($\varepsilon > 0$) are shown in Fig. 5.7.

There are various ways in which the restoring force can be **unsymmetrical**: an important case is shown in Fig. 5.6(c) where $g(x) = -x^2$ and $\varepsilon > 0$. The spring is 'soft' for $x > 0$ and 'hard' for $x < 0$. Problem 5.22 shows that for free oscillations ($F = 0$ in (5.43)) the effect is to shift the centre of oscillation to small positive x values, as the diagram suggests; the same occurs when the oscillation is forced, as is shown in the following example.

Example 5.3   *Obtain approximately the solutions of period 2π of the equation $\ddot{x} + \Omega^2 x - \varepsilon x^2 = \Gamma \cos t$, $\varepsilon > 0$.*
First, suppose that Ω is not close to an integer (i.e., it is far from resonance). Writing $x(\varepsilon, t) = x_0(t) + \varepsilon x_1(t) + \cdots$, the required sequence of equations is

$$\ddot{x}_0 + \Omega^2 x_0 = \Gamma \cos t,$$
$$\ddot{x}_1 + \Omega^2 x_1 = x_0^2,$$

and so on. The periodic solution of the first is

$$x_0(t) = \frac{\Gamma}{\Omega^2 - 1} \cos t,$$

and the second becomes

$$\ddot{x}_1 + \Omega^2 x_1 = \frac{\Gamma^2}{2(\Omega^2 - 1)^2}(1 + \cos 2t)$$

whose solutions are

$$x_1(t) = \frac{\Gamma^2}{2\Omega^2(\Omega^2-1)^2} + \frac{\Gamma^2}{2(\Omega^2-1)^2(\Omega^2-4)} \cos 2t + a_1 \cos \Omega t + b_1 \sin \Omega t.$$

For $x_1(t)$ to have period 2π, $a_1 = b_1 = 0$: therefore

$$x(\varepsilon, t) = \frac{\Gamma}{\Omega^2-1} \cos t + \frac{\varepsilon \Gamma^2}{2\Omega^2(\Omega^2-1)^2} + \frac{\varepsilon \Gamma^2}{2(\Omega^2-1)^2(\Omega^2-4)} \cos 2t.$$

Now suppose $\Omega \approx 1$ (i.e., close to resonance), and Γ small. Put

$$\Omega^2 = 1 + \varepsilon \beta, \qquad \Gamma = \varepsilon \gamma.$$

The equation becomes

$$\ddot{x} + x = \varepsilon(\gamma \cos t + x^2 - \beta x).$$

Assume that $x(\varepsilon, \tau) = x_0(\tau) + \varepsilon x_1(\tau) + \cdots$. The first two equations are

$$\ddot{x}_0 + x_0 = 0,$$

$$\ddot{x}_1 + x_1 = \gamma \cos t + x_0^2 - \beta x_0.$$

The solution of the first equation, of period 2π, is $a_0 \cos t + b_0 \sin t$, where a_0, b_0 are to be determined. The second equation becomes

$$\ddot{x}_1 + x_1 = \tfrac{1}{2}(a_0^2 + b_0^2) + (\gamma - \beta a_0) \cos t - \beta b_0 \sin t$$
$$+ \tfrac{1}{2}(a_0^2 - b_0^2) \cos 2t + a_0 b_0 \sin 2t.$$

Secular terms vanish only if

$$b_0 = 0, \qquad a_0 = \gamma/\beta.$$

Then the equation for x_1 simplifies to

$$\ddot{x}_1 + x_1 = \frac{\gamma^2}{2\beta^2}(1 + \cos 2t),$$

which has the 2π-period solution

$$x_1 = \frac{\gamma^2}{2\beta^2} - \frac{\gamma^2}{6\beta^2} \cos 2t + a_1 \cos t + b_1 \sin t.$$

To order ε,

$$x(\varepsilon, t) \approx \frac{\gamma}{\beta} \cos t + \varepsilon \left(\frac{1}{2} \frac{\gamma^2}{\beta^2} - \frac{1}{6} \frac{\gamma^2}{\beta^2} \cos 2t + a_1 \cos t + b_1 \sin t \right).$$

The constants a_1 and b_1 can be determined by eliminating secular terms in the equation for x_2. ●

Exercise 5.6
In Example 5.3, write down the equation for x_2. By eliminating secular terms in this equation, find a_1 and b_1 which arise in the solution for x_1.

5.8 Amplitude–phase perturbation for the pendulum equation

In Section 5.4, we stopped at the first approximation to a solution because of the rapidly developing complexity of the algebra. The following method allows a higher approximation to be obtained rather more efficiently.

Consider again the family (5.29):

$$x'' + x = \varepsilon(\gamma \cos \tau - \kappa x' - \beta x + x^3). \tag{5.44}$$

Instead of seeking, as in Section 5.4, solutions in effect of the form

$$(a_0 + \varepsilon a_1 + \cdots) \cos \tau + (b_0 + \varepsilon b_1 + \cdots) \sin \tau + \text{higher harmonics}$$

We shall arrange for them to appear in the form

$$x(\varepsilon, \tau) = (r_0 + \varepsilon r_1 + \cdots) \cos(\tau + \alpha_0 + \varepsilon \alpha_1 + \cdots) + \text{higher harmonics}, \tag{5.45}$$

where

$$\alpha = \alpha_0 + \varepsilon \alpha_1 + \cdots$$

is the phase difference between the response and forcing term. α_0 is expected to be the phase of the generating solution and r_0 its amplitude.

It is more convenient for manipulative reasons to have the unknown phase, α, appear in the equation itself. Therefore we discuss instead the equation

$$X'' + X = \varepsilon(\gamma \cos(s - \alpha) - \kappa X' - \beta X + X^3), \tag{5.46}$$

where we have put $s = \tau + \alpha$ and $X(\varepsilon, s) = x(\varepsilon, \tau) = x(\varepsilon, \ s - \alpha)$ into (5.44) with differentiation now with respect to s.

Assume that for all small enough ε

$$X(\varepsilon, s) = X_0(s) + \varepsilon X_1(s) + \cdots,$$
$$\alpha = \alpha_0 + \varepsilon \alpha_1 + \cdots. \tag{5.47}$$

We require solutions $X(\varepsilon, s)$ having the period, 2π, of the forcing term, which implies that for all s,

$$X_i(s + 2\pi) = X_i(s), \quad i = 1, 2, \ldots. \tag{5.48}$$

Finally we shall impose the extra condition $X'(\varepsilon, 0) = 0$. This is not a real restriction: we simply adjust the time origin, and hence the phase, so that it is so. Therefore

$$X_i'(0) = 0, \quad i = 0, 1, \ldots. \tag{5.49}$$

Substitute (5.47) into (5.46), using the Taylor series

$$\cos(s - \alpha) = \cos(s - \alpha_0) + \varepsilon \alpha_1 \sin(s - \alpha_0) + \cdots.$$

By assembling powers of ε and equating their coefficients to zero we obtain

$$X_0'' + X_0 = 0, \tag{5.50a}$$

$$X_1'' + X_1 = \gamma \cos(s - \alpha_0) - \kappa X_0' - \beta X_0 + X_0^3, \tag{5.50b}$$

$$X_2'' + X_2 = \gamma \alpha_1 \sin(s - \alpha_0) - \kappa X_1' - \beta X_1 + 3 X_0^2 X_1, \tag{5.50c}$$

and so on.

The periodic solutions of (5.50a) satisfying (5.49) are

$$X_0(s) = r_0 \cos s, \quad r_0 > 0 \tag{5.51}$$

where we choose $r_0 > 0$ by later adjusting the phase α_0. From (5.50b)

$$X_1'' + X_1 = (\gamma \cos \alpha_0 - \beta r_0 + \tfrac{3}{4} r_0^3) \cos s$$

$$+ (\kappa r_0 + \gamma \sin \alpha_0) \sin s + \tfrac{1}{4} r_0^3 \cos 3s, \tag{5.52}$$

since $\cos^3 s = \tfrac{3}{4} \cos s + \tfrac{1}{4} \cos 3s$. For there to be a periodic solution the secular terms (in $\cos s$ and $\sin s$) must be eliminated, so

$$\beta r_0 - \tfrac{3}{4} r_0^3 = \gamma \cos \alpha_0, \tag{5.53a}$$

$$\kappa r_0 = -\gamma \sin \alpha_0. \tag{5.53b}$$

By squaring and adding we obtain eqn (5.37) again:

$$r_0^2 \left\{ \kappa^2 + \left(\beta - \tfrac{3}{4} r_0^2 \right)^2 \right\} = \gamma^2. \tag{5.54}$$

α_0 is then obtainable from (5.53). Considering only $-\tfrac{1}{2}\pi \le \alpha_0 \le \tfrac{1}{2}\pi$

$$\alpha_0 = -\sin^{-1}(\kappa r_0 / \gamma). \tag{5.55}$$

Equation (5.52) becomes

$$X_1'' + X_1 = \tfrac{1}{4} r_0^3 \cos 3s,$$

with solutions

$$X_1(s) = r_1 \cos s - \tfrac{1}{32} r_0^3 \cos 3s, \quad r_1 > 0, \tag{5.56}$$

satisfying (5.48) and (5.49). Substitute (5.56) into (5.50c):

$$X_2'' + X_2 = \left(-\gamma \alpha_1 \sin \alpha_0 - \beta r_1 + \tfrac{9}{4} r_0^2 r_1 - \tfrac{3}{128} r_0^5 \right) \cos s$$

$$+ (\gamma \alpha_1 \cos \alpha_0 + \kappa r_1) \sin s + \left(\tfrac{3}{4} r_0^2 r_1 - \tfrac{3}{64} r_0^5 + \tfrac{1}{32} \beta r_0^3 \right) \cos 3s$$

$$- \tfrac{3}{32} \kappa r_0^3 \sin 3s - \tfrac{3}{128} r_0^5 \sin 5s, \tag{5.57}$$

using the identity $\cos^2 s \cos 3s = \frac{1}{4}\cos s + \frac{1}{2}\cos 3s + \frac{1}{4}\cos 5s$. Periodicity requires that the coefficients of $\cos s$ and $\sin s$ should be zero, which leads to the results

$$r_1 = \tfrac{3}{128} r_0^5 / \left(\tfrac{9}{4} r_0^2 - \beta + \kappa \tan \alpha_0 \right), \tag{5.58a}$$

$$\alpha_1 = -\frac{3}{128} \frac{\kappa r_0^5}{\gamma \cos \alpha_0} \Big/ \left(\frac{9}{4} r_0^2 - \beta + \kappa \tan \alpha_0 \right). \tag{5.58b}$$

The solution to eqn (5.44) is then

$$x(\varepsilon, \tau) = (r_0 + \varepsilon r_1)\cos(\tau + \alpha_0 + \varepsilon\alpha_1) - \varepsilon r_0^2 \cos 3(\tau + \alpha_0) + O(\varepsilon^2). \tag{5.59}$$

5.9 Periodic solutions of autonomous equations (Lindstedt's method)

Consider the oscillations of the autonomous pendulum-type equation (a form of Duffing's equation)

$$\frac{\mathrm{d}^2 x}{\mathrm{d}t^2} + x - \varepsilon x^3 = 0. \tag{5.60}$$

For a soft spring $\varepsilon > 0$, and for a hard spring $\varepsilon < 0$. The system is conservative, and the method of Section 1.3 can be used to show that all motions of small enough amplitude are periodic.

In this case of unforced vibration, the frequency, ω, is not known in advance, but depends on the amplitude. It reduces to 1 when $\varepsilon = 0$. Assume that

$$\omega = 1 + \varepsilon\omega_1 + \cdots, \tag{5.61}$$

$$x(\varepsilon, t) = x_0(t) + \varepsilon x_1(t) + \cdots. \tag{5.62}$$

We could substitute these into (5.60) and look for solutions of period $2\pi/\omega$, but it is mechanically simpler to cause ω to appear as a factor in the differential equation by writing

$$\omega t = \tau. \tag{5.63}$$

Then (5.60) becomes

$$\omega^2 x'' + x - \varepsilon x^3 = 0. \tag{5.64}$$

By this substitution we have replaced eqn (5.60), which has unknown period, by eqn (5.64) which has known period 2π. Therefore, as before, for all τ,

$$x_i(\tau + 2\pi) = x_i(\tau), \quad i = 0, 1, \ldots. \tag{5.65}$$

Equation (5.64) becomes

$$(1 + \varepsilon 2\omega_1 + \cdots)(x_0'' + \varepsilon x_1'' + \cdots) + (x_0 + \varepsilon x_1 + \cdots) = \varepsilon(x_0 + \varepsilon x_1 + \cdots)^3,$$

and by assembling powers of ε we obtain

$$x_0'' + x_0 = 0, \qquad (5.66a)$$

$$x_1'' + x_1 = -2\omega_1 x_0'' + x_0^3, \qquad (5.66b)$$

and so on.

To simplify the calculations we can impose the conditions

$$x(\varepsilon, 0) = a_0, \qquad x'(\varepsilon, 0) = 0 \qquad (5.67)$$

without loss of generality (only in the autonomous case!). This implies that

$$x_0(0) = a_0, \qquad x_0'(0) = 0, \qquad (5.68a)$$

and

$$x_i(0) = 0, \quad x_i'(0) = 0, \qquad i = 1, 2, \dots . \qquad (5.68b)$$

The solution of (5.66a) satisfying (5.68a) is

$$x_0 = a_0 \cos \tau. \qquad (5.69)$$

Equation (5.66b) then becomes

$$x_1'' + x_1 = (2\omega_1 a_0 + \tfrac{3}{4}a_0^3) \cos \tau + \tfrac{1}{4}a_0^3 \cos 3\tau. \qquad (5.70)$$

The solutions will be periodic only if

$$\omega_1 = -\tfrac{3}{8}a_0^2. \qquad (5.71)$$

From (5.70), (5.71),

$$x_1(\tau) = a_1 \cos \tau + b_1 \sin \tau - \tfrac{1}{32}a_0^3 \cos 3\tau,$$

and (5.68b) implies

$$b_1 = 0, \qquad a_1 = \tfrac{1}{32}a_0^3.$$

Therefore,

$$x_1(\tau) = \tfrac{1}{32}a_0^3(\cos \tau - \cos 3\tau). \qquad (5.72)$$

Finally, from (5.69) and (5.72),

$$x(\varepsilon, \tau) \approx a_0 \cos \tau + \tfrac{1}{32}\varepsilon a_0^3(\cos \tau - \cos 3\tau) + O(\varepsilon^2). \qquad (5.73)$$

Returning to the variable t (eqn (5.63)), we have the approximation

$$x(\varepsilon, t) \approx a_0 \cos \omega t + \tfrac{1}{32}\varepsilon a_0^3(\cos \omega t - \cos 3\omega t), \qquad (5.74a)$$

where

$$\omega \approx 1 - \tfrac{3}{8}\varepsilon a_0^2; \qquad (5.74b)$$

this gives the dependence of frequency on amplitude.

> **Exercise 5.7**
> Find the frequency–amplitude relation for periodic solutions of
> $$\ddot{x} + x - \varepsilon(x^3 + x^5) = 0, \quad x(\varepsilon, 0) = a_0, \quad \dot{x}(\varepsilon, 0) = 0,$$
> using Lindstedt's method. (Use the trigonometric identities in Appendix E.)

5.10 Forced oscillation of a self-excited equation

Consider the van der Pol equation with a forcing term:

$$\ddot{x} + \varepsilon(x^2 - 1)\dot{x} + x = F \cos \omega t. \tag{5.75}$$

The unforced equation has a limit cycle with radius approximately 2 and period approximately 2π (see Sections 4.1 and 4.2). The limit cycle is generated by the balance between internal energy loss and energy generation (see Section 1.5), and the forcing term will alter this balance. If F is 'small' (**weak excitation**), its effect depends on whether or not ω is close to the natural frequency. If it is, it appears that an oscillation might be generated which is a perturbation of the limit cycle. If F is not small (**hard excitation**) or if the natural and imposed frequency are not closely similar, we should expect that the 'natural oscillation' might be extinguished, as occurs with the corresponding linear equation.

Firstly, write

$$\omega t = \tau; \tag{5.76}$$

then (5.75) becomes

$$\omega^2 x'' + \varepsilon\omega(x^2 - 1)x' + x = F \cos \tau, \tag{5.77}$$

where the dashes signify differentiation with respect to τ.

Hard excitation, far from resonance

Assume that ω is not close to an integer. In (5.77), let

$$x(\varepsilon, \tau) = x_0(\tau) + \varepsilon x_1(\tau) + \cdots. \tag{5.78}$$

The sequence of equations for x_0, x_1, ... begins

$$\omega^2 x_0'' + x_0 = F \cos \tau, \tag{5.79a}$$

$$\omega^2 x_1'' + x_1 = -(x_0^2 - 1)x_0', \tag{5.79b}$$

$x_0(\tau)$, $x_1(\tau)$ having period 2π.

The only solution of (5.79a) having period 2π is

$$x_0(\tau) = \frac{F}{1 - \omega^2} \cos \tau,$$

and therefore

$$x(\varepsilon, \tau) = \frac{F}{1 - \omega^2} \cos \tau + O(\varepsilon). \tag{5.80}$$

The solution is therefore a perturbation of the ordinary linear response and the limit cycle is suppressed as expected.

Soft excitation, far from resonance

This case is similar to hard excitation above but with $F = \varepsilon F_0$, and is left as Problem 5.20. However, this solution is normally unstable (see Section 7.4), and there is no limit cycle.

Soft excitation, near-resonance

For soft excitation write in (5.77)

$$F = \varepsilon \gamma \tag{5.81}$$

and for near-resonance

$$\omega = 1 + \varepsilon \omega_1. \tag{5.82}$$

The expansion is assumed to be

$$x(\varepsilon, \tau) = x_0(\tau) + \varepsilon x_1(\tau) + \cdots . \tag{5.83}$$

Equations (5.80), (5.81), and (5.82) lead to the sequence

$$x_0'' + x_0 = 0, \tag{5.84a}$$

$$x_1'' + x_1 = -2\omega_1 x_0'' - (x_0^2 - 1)x_0' + \gamma \cos \tau, \tag{5.84b}$$

and so on. We require solutions with period 2π. Equation (5.84a) has the solutions

$$x_0(\tau) = a_0 \cos \tau + b_0 \sin \tau. \tag{5.85}$$

After some manipulation, (5.84b) becomes

$$x_1'' + x_1 = \{\gamma + 2\omega_1 a_0 - b_0(\tfrac{1}{4}r_0^2 - 1)\} \cos \tau$$
$$+ \{2\omega_1 b_0 + a_0(\tfrac{1}{4}r_0^2 - 1)\} \sin \tau + \text{higher harmonics}, \tag{5.86}$$

where

$$r_0 = \sqrt{(a_0^2 + b_0^2)} > 0. \tag{5.87}$$

For a periodic solution we require

$$2\omega_1 a_0 - b_0(\tfrac{1}{4}r_0^2 - 1) = -\gamma, \tag{5.88a}$$

$$2\omega_1 b_0 - a_0(\tfrac{1}{4}r_0^2 - 1) = 0. \tag{5.88b}$$

By squaring and adding these two equations we obtain

$$r_0^2 \left\{ 4\omega_1^2 + \left(\tfrac{1}{4}r_0^2 - 1\right)^2 \right\} = \gamma^2 \tag{5.89}$$

which give the possible amplitudes r_0 of the response for given ω_1 and γ. The structure of this equation is examined in Chapter 7 in a different connection: it is sufficient to notice here its family resemblance to (5.37) for the pendulum equation. Like (5.37), it may have as many as three real solutions for $r_0 > 0$. The limit cycle is extinguished.

Exercise 5.8
The frequency–amplitude equation (5.89) for the forced van der Pol equation under near-resonance soft excitation is

$$r_0^2\{4\omega_1^2 + (\tfrac{1}{4}r_0^2 - 1)^2\} = \gamma^2.$$

If $\omega_1 = \tfrac{1}{4}r > 0$, find all a_0, b_0 in the leading term for the periodic solution.

5.11 The perturbation method and Fourier series

In the examples given in Sections 5.3 and 5.4 the solutions emerge as series of sines and cosines with frequencies which are integer multiples of the forcing frequency. These appeared as a result of reorganizing terms like x^3, but by making a direct attack using Fourier series we can show that this form always occurs, even when we are not involved with polynomial terms, or harmonic forcing.

Consider the more general forced equation

$$x'' + \Omega^2 x = F(\tau) - \varepsilon h(x, x'), \tag{5.90}$$

where ε is a small parameter. Suppose that F is periodic, with the time variable already scaled to give it period 2π, and that its mean value is zero so that there is zero constant term in its Fourier series representation (meaning that its time-averaged value of F is zero over a period) so that we may expand F is the Fourier series

$$F(\tau) = \sum_{n=1}^{\infty} (A_n \cos n\tau + B_n \sin n\tau), \tag{5.91}$$

in which the Fourier coefficients are given by

$$A_n = \frac{1}{\pi} \int_0^{2\pi} F(\tau) \cos n\tau \, d\tau, \qquad B_n = \frac{1}{\pi} \int_0^{2\pi} F(\tau) \sin n\tau \, d\tau.$$

We shall allow the possibility that Ω *is close to some integer* N by writing

$$\Omega^2 = N^2 + \varepsilon\beta \tag{5.92}$$

($N = 1$ in Section 5.4). The perturbation method requires that the periodic solutions emerge from periodic solutions of some appropriate linear equation. If (5.91) has a nonzero term of order N then (5.90), with $\varepsilon = 0$, is clearly not an appropriate linearization, since the forcing term has a component equal to the natural frequency N and there will be no periodic solutions. However, if we write

$$A_N = \varepsilon A, \qquad B_N = \varepsilon B, \tag{5.93}$$

the term in F giving resonance is removed from the *linearized* equation and we have a possible family of generating solutions. Now rearrange (5.90), isolating the troublesome term in F by writing

$$f(\tau) = F(\tau) - \varepsilon A \cos N\tau - \varepsilon B \sin N\tau = \sum_{n \neq N} (A_n \cos n\tau + B_n \sin n\tau). \tag{5.94}$$

Equation (5.90) becomes

$$x'' + N^2 x = f(\tau) + \varepsilon\{-h(x, x') - \beta x + A \cos N\tau + B \sin N\tau\}. \tag{5.95}$$

The linearized equation is now $x'' + N^2 x = f(\tau)$, with no resonance.
   Now write as usual

$$x(\varepsilon, \tau) = x_0(\tau) + \varepsilon x_1(\tau) + \cdots, \tag{5.96}$$

where $x_0, x_1, \ldots$ have period 2π. By expanding h in (5.90) in powers of ε we have

$$h(x, x') = h(x_0, x_0') + \varepsilon h_1(x_0, x_0', x_1, x_1') + \cdots, \tag{5.97}$$

where h_1 can be calculated, and by substituting (5.96) and (5.97) into (5.95) we obtain the sequence

$$x_0'' + N^2 x_0 = \sum_{n \neq N} (A_n \cos n\tau + B_n \sin n\tau), \tag{5.98a}$$

$$x_1'' + N^2 x_1 = -h(x_0, x_0') - \beta x_0 + A \cos N\tau + B \sin N\tau, \tag{5.98b}$$

$$x_2'' + N^2 x_2 = -h_1(x_0, x_0', x_1, x_1') - \beta x_1, \tag{5.98c}$$

and so on. The solution of (5.98a) is

$$x_0(\tau) = a_0 \cos N\tau + b_0 \sin N\tau + \sum_{n \neq N} \frac{A_n \cos n\tau + B_n \sin n\tau}{N^2 - n^2}$$

$$= a_0 \cos N\tau + b_0 \sin N\tau + \phi(\tau), \tag{5.99}$$

say; a_0, b_0 are constants to be determined at the next stage.

We require x_0, as determined by (5.99), to be such that (5.98b) has 2π-periodic solutions. This is equivalent to requiring that its right side has no Fourier term of order N, since such a term would lead to resonance. We required therefore that

$$\beta a_0 = -\frac{1}{\pi} \int_0^{2\pi} h(a_0 \cos N\tau + b_0 \sin N\tau + \phi(\tau),$$

$$- a_0 N \sin N\tau + b_0 N \cos N\tau + \phi'(\tau)) \cos N\tau \, \mathrm{d}\tau + A, \tag{5.100a}$$

$$\beta b_0 = -\frac{1}{\pi} \int_0^{2\pi} h(a_0 \cos N\tau + b_0 \sin N\tau + \phi(\tau),$$

$$- a_0 N \sin N\tau + b_0 N \cos N\tau + \phi'(\tau)) \sin N\tau \, \mathrm{d}\tau + B, \tag{5.100b}$$

which constitute two equations for the unknowns a_0, b_0. The reader should confirm that the resulting equations are the same as those for the first order approximation, with $N = 1$, obtained in Section 5.4.

Each equation in the sequence (5.98) has solutions containing constants a_1, b_1, a_2, $b_2, \ldots$, whose values are similarly established at the succeeding step. However, the equations for subsequent constants are linear: the pair (5.100) are the only ones which may have several solutions (compare the remark at the end of Section 5.4).

5.12 Homoclinic bifurcation: an example

For autonomous systems a **homoclinic path** is a phase path which joins a saddle point to itself. An example is given in Section 3.6. **Homoclinic bifurcation** takes place when a change of a parameter in a system causes a homoclinic path to appear for a specific value of the parameter, and then disappear again as illustrated in Fig. 3.27.

Consider the equation

$$\ddot{x} + \varepsilon(\alpha x^2 - \beta)\dot{x} - x + x^3 = 0 \qquad (\alpha > \beta > 0), \tag{5.101}$$

which has van der Pol 'damping' and a negative restoring force for x small. It is assumed that $0 < \varepsilon \ll 1$. The equation has three equilibrium points, at $y = 0$, $x = 0$, ± 1. Near $x = 0$, x satisfies

$$\ddot{x} - \varepsilon\beta\dot{x} - x = 0,$$

which has the characteristic equation

$$m^2 - \varepsilon\beta m - 1 = 0.$$

Both roots are real but of opposite sign, indicating a saddle point at the origin in the phase plane. Close to $x = 1$, let $x = 1 + \xi$ so that, approximately,

$$\ddot{\xi} + \varepsilon(\alpha - \beta)\dot{\xi} + 2\xi = 0.$$

For $\alpha > \beta$, this equilibrium point is a stable spiral for ε sufficiently small. Similarly it can be shown that the equilibrium point at $x = -1$ is also a stable spiral.

For $\varepsilon = 0$, the unperturbed system

$$\ddot{x} - x + x^3 = 0$$

has homoclinic paths meeting at the origin given by

$$y^2 = x^2 - \tfrac{1}{2}x^4,$$

and a time solution for $x > 0$ of

$$x = x_0(t) = \sqrt{2}\operatorname{sech} t, \quad -\infty < t < \infty \tag{5.102}$$

(see Example 3.9 and Fig. 3.26). The intention here is to investigate, using a perturbation method, whether there are any values of the parameters α and β for which homoclinic paths persist. Physically this seems likely since the damping term in (5.101) will be negative for small x, but become positive for larger x so that perhaps a balance will be achieved over the range of a homoclinic path.

Consider the following perturbation scheme:

$$x = x_0 + \varepsilon x_1 + \cdots \quad (-\infty < t < \infty).$$

Substitute this series into (5.101), and collect terms in like powers of ε. The first two equations are

$$\ddot{x}_0 - x_0 + x_0^3 = 0, \tag{5.103}$$

$$\ddot{x}_1 + (3x_0^2 - 1)x_1 = (\beta - \alpha x_0^2)\dot{x}_0, \tag{5.104}$$

where the solution of (5.103) is given by (5.102). For a homoclinic solution the required boundary conditions are that $x, \dot{x} \to 0$, $t \to \pm\infty$, which means that for each term in the series $x_n, \dot{x}_n \to 0$ ($n = 0, 1, 2, \ldots$) as $t \to \pm\infty$. This condition is already satisfied by x_0.

Equation (5.104) is a forced linear second-order equation for $x_1(t)$. A direct solution of (5.104) can be obtained, at least as an integral, but what we require from (5.104) is the condition that the phase path is homoclinic to order ε. Multiply both sides of (5.104) by $\dot{x}_0$, giving

$$\ddot{x}_1\dot{x}_0 + (3x_0^2 - 1)x_1\dot{x}_0 = (\beta - \alpha x_0^2)\dot{x}_0^2.$$

From (5.103), this may be written as

$$\frac{d}{dt}(\dot{x}_1\dot{x}_0) - \dot{x}_1\ddot{x}_0 + x_1\frac{d}{dt}(x_0^3 - x_0) = (\beta - \alpha x_0^2)\dot{x}_0^2,$$

or, from (5.103), as

$$\frac{d}{dt}(\dot{x}_1\dot{x}_0) + \dot{x}_1(x_0^3 - x_0) + x_1\frac{d}{dt}(x_0^3 - x_0) = (\beta - \alpha x_0^2)\dot{x}_0^2,$$

or

$$\frac{d}{dt}[\dot{x}_1\dot{x}_0 + x_1(x_0^3 - x_0)] = (\beta - \alpha x_0^2)\dot{x}_0^2,$$

Now integrate from $t = -\infty$ to $t = \tau$:

$$\dot{x}_1(\tau)\dot{x}_0(\tau) + x_1(\tau)[x_0^3(\tau) - x_0(\tau)] = \int_{-\infty}^{\tau} [\beta - \alpha x_0^2(t)]\dot{x}_0^2(t)\,dt. \tag{5.105}$$

Since $x_0(t) = \sqrt{2}\,\mathrm{sech}\,t$, then $x_0(\tau) = O(e^{-\tau})$ and $x_0^3(\tau) - x_0(\tau) = O(e^{-\tau})$ as $\tau \to \infty$. It follows also that

$$\int_{-\infty}^{\tau} [\beta - \alpha x_0^2(t)]\dot{x}_0^2(t)\,dt = I(\alpha, \beta) - \int_{\tau}^{\infty} [\beta - \alpha x_0^2(t)]\dot{x}_0^2(t)\,dt$$

$$= I(\alpha, \beta) + O(e^{-2\tau}),$$

where

$$I(\alpha, \beta) = \int_{-\infty}^{\infty} [\beta - \alpha x_0^2(t)]\dot{x}_0^2(t)\,dt.$$

Multiply (5.105) through by e^T, and examine the order of the terms in x_0, so that

$$\dot{x}_1(\tau)(2\sqrt{2} + O(e^{-\tau})) + x_1(\tau)(-2\sqrt{2} + O(e^{-\tau})) = e^{\tau} I(\alpha, \beta) + O(e^{-\tau}). \tag{5.106}$$

If $I(\alpha, \beta) = 0$, then the right-hand side of (5.105) is $O(e^{-\tau})$. On the left-hand side, $x_1(\tau) = e^{\tau} + O(1)$ is not possible since it would then be $O(1)$, nor is $x_1(\tau) = A + O(e^{-\tau})$ possible with a non-zero constant. We conclude that $x_1(\tau) = O(e^{-\tau})$, which ensures that

$$I(\alpha, \beta) = \int_{-\infty}^{\infty} [\beta - \alpha x_0^2(t)]\dot{x}_0^2(t)\,dt = 0$$

is a necessary and sufficient condition for a homoclinic path to $O(\varepsilon^2)$.

In our example, it follows from (5.102) (switching back to the variable t instead of τ),

$$\dot{x}_0(t) = -\sqrt{2}\,\mathrm{sech}^2 t \sinh t.$$

Hence (5.105) becomes

$$2\beta \int_{-\infty}^{\infty} \text{sech}^4 t\ \sinh^2 t\ dt - 4\alpha \int_{-\infty}^{\infty} \text{sech}^6 t\ \sinh^2 t\ dt = 0. \qquad (5.107)$$

The last two integrals can be evaluated as follows. Using the substitution $u = \tanh t$ in both cases,

$$\int_{-\infty}^{\infty} \text{sech}^4 t\ \sinh^2 t\ dt = \int_{-1}^{1} u^2 du = \tfrac{2}{3},$$

$$\int_{-\infty}^{\infty} \text{sech}^6 t\ \sinh^2 t\ dt = \int_{-1}^{1} u^2 (1 - u^2) du = \tfrac{2}{3} - \tfrac{2}{5} = \tfrac{4}{15}.$$

Hence, to order ε, a homoclinic path in $x > 0$ exists if

$$\beta = \tfrac{4}{5}\alpha \qquad (5.108)$$

from (5.107). It can be shown also that a homoclinic path in $x < 0$ exists for the same parameter ratio.

Figure 5.8 shows a computed phase diagram for eqn (5.101) for $\varepsilon = 0.3$, $\alpha = 1$ and $\beta = 0.801$. This value of β should be compared with the value of $\beta = 0.8$ predicted by eqn (5.108). The agreement is good for $0 < \varepsilon < 0.3$ but the perturbation method will become less accurate as ε increases further. As β increases through the critical value $\tfrac{4}{5}\alpha$ for any fixed α, a homoclinic bifurcation takes place (see section 3.6).

A more general perturbation method valid also for forced systems will be developed in Section 13.7.

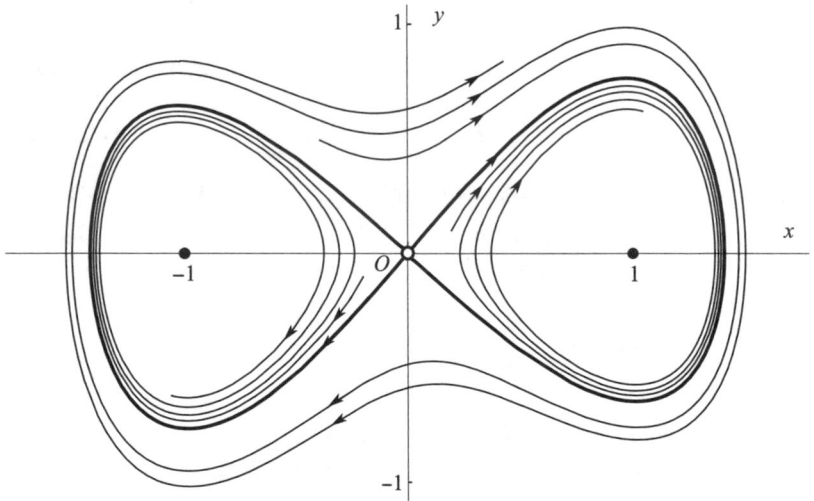

Figure 5.8   Homoclinic paths for $\ddot{x} + \varepsilon(\alpha x^2 - \beta)\dot{x} - x + x^3 = 0$ for $\varepsilon = 0.3$, $\alpha = 1$, $\beta = 0.801$.

Exercise 5.9
Verify that the system $\ddot{x} + x - x^3 = 0$ has the heteroclinic solutions $x = \pm\tanh(t/\sqrt{2})$ connecting the saddle points at $x = \pm 1$. In the equation

$$\ddot{x} + \varepsilon(\alpha x^2 - \beta)\dot{x} + x - x^3 = 0, \quad 0 < \varepsilon \ll 1,$$

let $x = x_0 + \varepsilon x_1 + \cdots$. Show that a heteroclinic bifurcation occurs for $\alpha \approx 3\beta$.

Problems

5.1 Find all the periodic solutions of $\ddot{x} + \Omega^2 x = \Gamma \cos t$ for all values of Ω^2.

5.2 Find the first two harmonics of the solutions of period 2π of the following
   (i) $\ddot{x} - 0.5x^3 + 0.25x = \cos t$;
   (ii) $\ddot{x} - 0.1x^3 + 0.6x = \cos t$;
   (iii) $\ddot{x} - 0.1\dot{x}^2 + 0.5x = \cos t$.

5.3 Find a first approximation to the limit cycle for Rayleigh's equation

$$\ddot{x} + \varepsilon(\tfrac{1}{3}\dot{x}^3 - \dot{x}) + x = 0, \quad |\varepsilon| \ll 1,$$

using the method of Section 5.9.

5.4 Use the method of Section 5.9 to order ε to obtain solutions of period 2π, and the amplitude–frequency relation, for
   (i) $\ddot{x} - \varepsilon x\dot{x} + x = 0$;
   (ii) $(1 + \varepsilon\dot{x})\ddot{x} + x = 0$.

5.5 Apply the perturbation method to the equation $\ddot{x} + \Omega^2 \sin x = \cos t$ by considering $\ddot{x} + \Omega^2 x + \varepsilon\Omega^2 (\sin x - x) = \cos t$, with $\varepsilon = 1$, and assuming that Ω is not close to an odd integer to find 2π-periodic solutions. Use the Fourier expansion

$$\sin(a \cos t) = 2\sum_{n=0}^{\infty}(-1)^n J_{2n+1}(a)\cos\{(2n+1)t\},$$

where J_{2n+1} is the Bessel function of order $2n+1$. Confirm that the leading terms are given by

$$x = \frac{1}{\Omega^2 - 1}[1 + \Omega^2 - 2\Omega^2 J_1\{1/(\Omega^2 - 1)\}]\cos t + \frac{2}{\Omega^2 - 9}J_3\{1/(\Omega^2 - 1)\}\cos 3t.$$

5.6 For the equation $\ddot{x} + \Omega^2 x - 0.1x^3 = \cos t$, where Ω is not near $1, 3, 5, \ldots$, find to order ε the ratio of the magnitudes of the first two harmonics.

5.7 In the equation $\ddot{x} + \Omega^2 x + \varepsilon f(x) = \Gamma \cos t$ Ω is not close to an odd integer, and $f(x)$ is an odd function of x, with expansion

$$f(a \cos t) = -a_1(a)\cos t - a_3(a)\cos 3t - \cdots.$$

Derive a perturbation solution of period 2π, to order ε.

5.8   The Duffing equation near resonance at $\Omega = 3$, with weak excitation, is

$$\ddot{x} + 9x = \varepsilon(\gamma \cos t - \beta x + x^3).$$

Show that there are solutions of period 2π if the amplitude of the zero-order solution is 0 or $2\sqrt{(\beta/3)}$.

5.9   From eqn (5.40), the amplitude equation for an undamped pendulum is

$$-F = a_0(\omega^2 - \omega_0^2 + \tfrac{1}{8}\omega_0^2 a_0^2).$$

When ω_0 is given, find for what values of ω there are three possible responses. (Find the stationary values of F with respect to a_0, with ω fixed. These are the points where the response curves of Fig. 5.4 turn over.)

5.10 From eqn (5.42), the amplitude equation for the damped pendulum is $F^2 = r_0^2\{k^2\omega^2 + (\omega^2 - \omega_0^2 + \tfrac{1}{8}\omega_0^2 r_0^2)\}$. By considering $\mathrm{d}(F^2)/\mathrm{d}(r_0^2)$, show that if $(\omega^2 - \omega_0^2)^2 \leq 3k^2\omega^2$, then the amplitude equation has only one real root r_0 for all F, and three real roots if $(\omega^2 - \omega_0^2)^2 > 3k^2\omega^2$.

5.11 Find the equivalent linear form (Section 4.5) of the expression $\ddot{x} + \Omega^2 x - \varepsilon x^3$, with respect to the periodic form $x = a\cos t$. Use the linear form to obtain the frequency–amplitude relation for the equation

$$\ddot{x} + \Omega^2 x - \varepsilon x^3 = \Gamma \cos t.$$

Solve the equation approximately by assuming that $a = a_0 + \varepsilon a_1$, and show that this agrees with the first harmonic in eqn (5.23). (Note that there may be three solutions, but that this method of solution shows only the one close to $\{\Gamma/(1 - \Omega^2)\}\cos t$.)

5.12 Generalize the method of Problem 5.11 for the same equation by putting $x = x^{(0)} + x^{(1)} + \cdots$, where $x^{(0)}$ and $x^{(1)}$ are the first two harmonics to order ε, $a\cos t$ and $b\cos 3t$, say, in the expansion of the solution. Show that the linear form equivalent to x^3 is

$$\left(\tfrac{3}{4}a^2 + \tfrac{3}{4}ab + \tfrac{3}{2}b^2\right)x^{(0)} + \left(\tfrac{1}{4}a^3 + \tfrac{3}{2}a^2 b + \tfrac{3}{4}b^3\right)x^{(1)}/b.$$

Split the pendulum equation into the two equations

$$\ddot{x}^{(0)} + \left\{\Omega^2 - \varepsilon\left(\tfrac{3}{4}a^2 + \tfrac{3}{4}ab + \tfrac{3}{2}b^2\right)\right\}x^{(0)} = \Gamma \cos t,$$

$$\ddot{x}^{(1)} + \left\{\Omega^2 - \varepsilon\left(\tfrac{1}{4}a^3 + \tfrac{3}{2}a^2 b + \tfrac{3}{4}b^3\right)/b\right\}x^{(1)} = 0.$$

Deduce that a and b must satisfy

$$a\left\{\Omega^2 - 1 - \varepsilon\left(\tfrac{3}{4}a^2 + \tfrac{3}{4}ab + \tfrac{3}{2}b^2\right)\right\} = \Gamma,$$

$$b\left\{\Omega^2 - 9 - \varepsilon\left(\tfrac{1}{4}a^3 + \tfrac{3}{2}a^2 b + \tfrac{3}{4}b^3\right)\right\} = 0.$$

Assume that $a \approx a_0 + \varepsilon a_1$, $b \approx \varepsilon b_1$ and obtain a_0, a_1, and b_1 (giving the perturbation solution (5.23)).

5.13 Apply the Lindstedt method, Section 5.9, to van der Pol's equation $\ddot{x} + \varepsilon(x^2 - 1)\dot{x} + x = 0$, $|\varepsilon| \ll 1$. Show that the frequency of the limit cycle is given by $\omega = 1 - \tfrac{1}{16}\varepsilon^2 + 0(\varepsilon^3)$.

5.14 Investigate the forced periodic solutions of period $\tfrac{2}{3}\pi$ for the Duffing equation in the form $\ddot{x} + (1 + \varepsilon\beta)x - \varepsilon x^3 = \Gamma \cos 3t$.

5.15 For the equation $\ddot{x} + x + \varepsilon x^3 = 0$, $|\varepsilon| \ll 1$, with $x(0) = a$, $\dot{x}(0) = 0$, assume an expansion of the form $x(t) = x_0(t) + \varepsilon x_1(t) + \ldots$, and carry out the perturbation process without assuming periodicity of the solution. Show that

$$x(t) = a\cos t + \varepsilon a^3\left\{-\tfrac{3}{8}t\sin t + \tfrac{1}{32}(\cos 3t - \cos t)\right\} + O(\varepsilon^2).$$

(This expansion is valid, so far as it goes. Why is it not so suitable as those already obtained for describing the solutions?)

5.16 Find the first few harmonics in the solution, period 2π, of $\ddot{x} + \Omega^2 x + \varepsilon x^2 = \Gamma \cos t$, by the direct method of Section 5.2. Explain the presence of a constant term in the expansion.

   For what values of Ω does the expansion fail? Show how, for small values of Γ, an expansion valid near $\Omega = 1$ can be obtained.

5.17 Use the method of amplitude–phase perturbation (Section 5.8) to approximate to the solutions, period 2π, of $\ddot{x} + x = \varepsilon(\gamma \cos t - x\dot{x} - \beta x)$.

5.18 Investigate the solutions, period 2π, of $\ddot{x} + 9x + \varepsilon x^2 = \Gamma \cos t$ obtained by using the direct method of Section 5.2. If $x = x_0 + \varepsilon x_1 + \cdots$, show that secular terms first appear in x_2.

5.19 For the damped pendulum equation with a forcing term,

$$\ddot{x} + k\dot{x} + \omega_0^2 x - \tfrac{1}{6}\omega_0^2 x^3 = F \cos \omega t,$$

show that the amplitude–frequency curves have their maxima on

$$\omega^2 = \omega_0^2(1 - \tfrac{1}{8}r_0^2) - \tfrac{1}{2}k^2.$$

5.20 Show that the first harmonic for the forced van der Pol equation $\ddot{x} + \varepsilon(x^2 - 1)\dot{x} + x = F \cos \omega t$ is the same for both weak and hard excitation, far from resonance.

5.21 The orbital equation of a planet about the sun is

$$\frac{d^2 u}{d\theta^2} + u = k(1 + \varepsilon u^2),$$

where $u = r^{-1}$ and r, θ, are polar coordinates, $k = \gamma m/h^2$, γ is the gravitational constant, m is the mass of the planet and h is its moment of momentum, a constant. $\varepsilon k u^2$ is the relativistic correction term.

   Obtain a perturbation expansion for the solution with initial conditions $u(0) = k(e+1)$, $\dot{u}(0) = 0$. (e is the eccentricity of the unperturbed orbit, and these are initial conditions at the perihelion: the nearest point to the sun on the unperturbed orbit.) Note that the solution of the perturbed equation is not periodic, and that 'secular' terms cannot be eliminated. Show that the expansion to order ε predicts that in each orbit the perihelion advances by $2k^2\pi\varepsilon$.

5.22 Use the Lindstedt procedure (Section 5.9) to find the first few terms in the expansion of the periodic solutions of $\ddot{x} + x + \varepsilon x^2 = 0$. Explain the presence of a constant term in the expansion.

5.23 Investigate the forced periodic solutions of period 2π of the equation

$$\ddot{x} + (4 + \varepsilon\beta)x - \varepsilon x^3 = \Gamma \cos t,$$

where ε is small and β and Γ are not too large. Confirm that there is always a periodic solution of the form $a_0 \cos 2t + b_0 \sin 2t + \tfrac{1}{3}\Gamma \cos t$ where

$$a_0(\tfrac{3}{4}r_0^2 + \tfrac{1}{6}\Gamma^2 - \beta) = b_0(\tfrac{3}{4}r_0^2 + \tfrac{1}{6}\Gamma^2 - \beta) = 0.$$

5.24 Investigate the equilibrium points of $\ddot{x} + \varepsilon(\alpha x^4 - \beta)\dot{x} - x + x^3 = 0$, $(\alpha > \beta > 0)$ for $0 < \varepsilon \ll 1$. Use the perturbation method of Section 5.12 to find the approximate value of β/α at which homoclinic paths exist.

5.25 Investigate the equilibrium points of

$$\ddot{x} + \varepsilon(\alpha x^2 - \beta)\dot{x} - x + 3x^5 = 0 \quad (\alpha, \beta > 0)$$

for $0 < \varepsilon \ll 1$. Confirm that the equation has an unperturbed time solution

$$x_0 = \sqrt{[\text{sech } 2t]}.$$

(see Problem 3.55). Use the perturbation method of Section 5.12 to show that a homoclinic bifurcation takes place for $\beta \approx 4\alpha/(3\pi)$.

5.26 The equation $\ddot{x} + \varepsilon g(x, \dot{x})\dot{x} + f(x) = 0$, $\dot{x} = y$, $0 < \varepsilon \ll 1$, is known to have a saddle point at $(0, 0)$ with an associated homoclinic trajectory with solution $x = x_0(t)$ for $\varepsilon = 0$. Work through the perturbation method of Section 5.12, and show that any homoclinic paths of the perturbed system occur where

$$\int_{-\infty}^{\infty} g(x_0, \dot{x}_0)\,\dot{x}_0^2\,dt = 0.$$

If $g(x, \dot{x}) = \beta - \alpha x^2 - \gamma \dot{x}^2$ and $f(x) = -x + x^3$, show that homoclinic bifurcation occurs approximately where

$$\beta = \frac{28\alpha + 12\gamma}{35}$$

for small ε.

5.27 Apply Lindstedt's method to $\ddot{x} + \varepsilon x \dot{x} + x = 0$, $0 < \varepsilon \ll 1$, where $x(0) = a_0$, $\dot{x}(0) = 0$. Show that the frequency–amplitude relation for periodic solutions is given by $\omega = 1 - \frac{1}{24}a_0^3\varepsilon^2 + O(\varepsilon^3)$.

5.28 Find the first three terms in a direct expansion for x in powers of ε for the 2π-period solutions of equation

$$\ddot{x} + \Omega^2 x - \varepsilon \dot{x}^2 = \cos t,$$

where $0 < \varepsilon \ll 1$ and $\Omega \neq$ an integer.

6 Singular perturbation methods

This chapter is concerned with approximating to the solutions of differential equations containing a small parameter ε in what might be called difficult cases where, for one reason or another, a straightforward expansion of the solution in powers of ε is unobtainable or unusable. Often in such cases it is possible to modify the method so that the essential features of such expansions are redeemed. Ideally, we wish to be able to take a few terms of an expansion, and to be able to say that for some small fixed numerical value of ε supplied in a practical problem the truncated series is close to the required solution *for the whole range of the independent variable* in the differential equation. Failure of this condition is the rule rather than the exception. In fact the examples of Chapter 5 do not give useful approximations if we work solely from the initial conditions (see Problem 5.15): for satisfactory approximation we must use the technically redundant information that the solutions are periodic.

This chapter illustrates several other methods which have been used in such cases. If one method fails, another may work, or perhaps a combination of methods may work, but generally speaking the approach is tentative. For a treatment of the whole topic, not restricted to ordinary differential equations, see, for example, Nayfeh (1973), van Dyke (1964), O'Malley (1974), and Kevorkian and Cole (1996).

6.1 Non-uniform approximations to functions on an interval

The solutions of the differential equations we are considering,

$$\ddot{x} = f(x, \dot{x}, t, \varepsilon),$$

are functions of t and ε. Some of the problems which may arise in approximating to them can be illustrated by considering simple functions which are not necessarily associated with any particular differential equation. For example, consider the function

$$x(\varepsilon, t) = e^{-\varepsilon t} \tag{6.1}$$

on $t \geq 0$, where ε lies in a neighbourhood of zero. The first three terms of the Taylor expansion in powers of ε are

$$1 - \varepsilon t + \tfrac{1}{2}\varepsilon^2 t^2, \tag{6.2}$$

where the error is

$$O(\varepsilon^3). \tag{6.3}$$

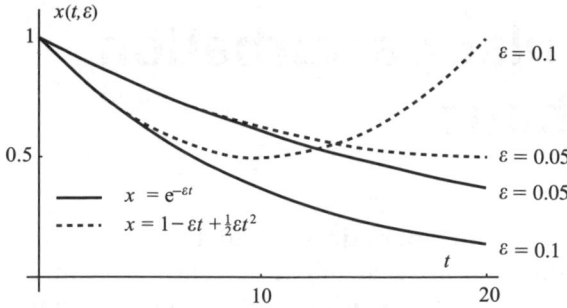

Figure 6.1   The function $e^{-\varepsilon t}$ (6.1) and its three-term Taylor series (6.2) compared for $\varepsilon = 0.05$ and $\varepsilon = 0.1$.

The error estimate (6.3) implies that for any *fixed* t, however large, we can choose ε small enough for the error to be as small as we please, and, indeed, smaller than the smallest term in the approximation, $\frac{1}{2}\varepsilon^2 t^2$. However, the trend of the terms shows clearly that *if ε is fixed*, at however small a value (which is the usual situation), t may be chosen large enough to destroy the approximation completely (see Fig. 6.1).

Consider now a function which appears in Section 6.4 as part of the solution to a differential equation:

$$x(\varepsilon, t) = \cos\left\{(1 - \varepsilon)^{1/2} t\right\}, \quad 0 \le t < \infty, \tag{6.4}$$

where ε lies in a neighbourhood of zero. The first three terms of the Taylor series for small ε give

$$\cos t + \tfrac{1}{2}\varepsilon t \sin t + \tfrac{1}{8}\varepsilon^2(t \sin t - t^2 \cos t) \tag{6.5}$$

with error $O(\varepsilon^3)$. Once again, for t large enough the approximation fails. We can see that it fails when t is so large that εt ceases to be small. The condition that (6.5) should be an approximation *for fixed ε* is that

$$t \ll \varepsilon^{-1}. \tag{6.6}$$

If this is satisfied, then the error, which is dominated by a term like $\varepsilon^3 t^3$, is small. It is also small compared with each term in (6.5), unless t has a fortuitous value making one of these terms very small or zero. The same conclusion is reached no matter how many terms of the Taylor series we take: if more terms are taken the approximation may be better while it lasts, but is still fails when t becomes comparable with ε^{-1}. Note that if the series is asymptotic rather than convergent as $\varepsilon \to 0$, the approximation does not necessarily improve by taking more terms (Copson 1965).

We say that (6.5) *does not provide an approximation that is uniformly valid on $t \ge 0$*. Failure of the ordinary perturbation method to produce a **uniform approximation** to the solution of a differential equation is very common, and the occasions of failure are called **singular perturbation problems**.

Formally, we require an approximation $x^*(\varepsilon, t)$ to a function $x(\varepsilon, t)$, usually to be valid for an infinite range of t:

$$x(\varepsilon, t) = x^*(\varepsilon, t) + E(\varepsilon, t), \quad t_0 \leq t < \infty,$$

with error $E(\varepsilon, t)$ where $\lim_{\varepsilon \to 0} E(\varepsilon, t) = 0$ *uniformly* on $t_0 \leq t < \infty$. That is to say, given any $\delta > 0$, there exists $\eta > 0$ independent of t such that

$$|\varepsilon| < \eta \;\; \Rightarrow \;\; |E(\varepsilon, t)| < \delta.$$

The general question of uniform approximation to functions is further illustrated in Section 6.6.

6.2 Coordinate perturbation

Consider the family of Duffing equations with parameter ε:

$$\ddot{x} + x = \varepsilon x^3. \tag{6.7}$$

The expansion

$$x(\varepsilon, t) = x_0(t) + \varepsilon x_1(t) + \cdots$$

leads to

$$\ddot{x}_0 + x_0 = 0,$$

$$\ddot{x}_1 + x_1 = x_0^3,$$

and so on. The general solution of the first equation can be written

$$x_0(t) = A \cos(t - \alpha),$$

where A and α are arbitrary constants. For the second equation

$$\ddot{x}_1 + x_1 = x_0^3 = \tfrac{3}{4}A^3 \cos(t - \alpha) + \tfrac{1}{4}A^3 \cos 3(t - \alpha), \tag{6.8}$$

so that secular terms of the form $t \cos(t - \alpha)$ (Section 5.4) begin to appear in the solution. They cannot be eliminated: only if $A = 0$ are they absent. Therefore a series similar in form to (6.5) emerges, and the truncated series does not approximate $x(\varepsilon, t)$ uniformly on $t \geq 0$.

This problem was treated in Section 5.9. There, the difficulty was avoided by anticipating a *periodic solution* of period $2\pi/\omega$ in t, ω being unknown; we put

$$\omega = 1 + \varepsilon \omega_1 + \varepsilon^2 \omega_2 + \cdots$$

where $\omega_1, \omega_2, \ldots$, are unknown constants, and then changed the variable from t to τ

$$\tau = \omega t, \tag{6.9}$$

so that the equation in τ had known period 2π (the Lindstedt procedure). Equation (6.9) introduced sufficient free constants ω_i to allow the secular terms to be eliminated. This method can be looked on as a device for obtaining uniform approximations by adopting prior knowledge of the periodicity of the solutions.

We can look at this procedure in another way. Write (6.9) in the form

$$t = \tau/\omega = \tau/(1 + \varepsilon\omega_1 + \varepsilon^2\omega_2 + \cdots) = \tau(1 + \varepsilon\tau_1 + \varepsilon^2\tau_2 + \cdots), \qquad (6.10)$$

where $\tau_1, \tau_2, \ldots$, are unknown constant coefficients. Also put, as in the earlier method,

$$x(\varepsilon, t) = X(\varepsilon, \tau) = X_0(\tau) + \varepsilon X_1(\tau) + \varepsilon^2 X_2(\tau) + \cdots, \qquad (6.11)$$

and substitute (6.10) and (6.11) into (6.7). We know this leads to an expansion uniform on $t \geq 0$ since it is equivalent to the Lindstedt procedure. However, the interpretation is now different: it appears that by (6.10) and (6.11) we have made *a simultaneous expansion in powers of ε of both the dependent and the independent variables*, generating an implicit relation between x and t through a parameter τ. The coefficients can be adjusted to eliminate terms which would give non-uniformity (in fact, the 'secular terms'), and so a uniformly valid (and, as it turns out, a periodic) solution is obtained.

We were guided to the form (6.10) by the prior success of the Lindstedt procedure, which suggests its appropriateness. If however we assume no prior experience, we have to abandon the assumption of constant coefficients in (6.10). Therefore let us take, instead of the constants $\tau_1, \tau_2, \ldots$, a set of unknown functions $T_1, T_2, \ldots$, of τ, and see what happens. We have

$$x(\varepsilon, t) = X(\varepsilon, \tau) = X_0(\tau) + \varepsilon X_1(\tau) + \varepsilon^2 X_2(\tau) + \cdots \qquad (6.12a)$$

and

$$t = T(\varepsilon, \tau) = \tau + \varepsilon T_1(\tau) + \varepsilon^2 T_2(\tau) + \cdots. \qquad (6.12b)$$

The first term in the expansion of t remains as τ, since this is appropriate when $\varepsilon \to 0$: τ is called a **strained coordinate**, or a **perturbed coordinate**. The technique is also known as **Poincaré's method of strained coordinates**.

Equations (6.12a) and (6.12b) are the basis of Lighthill's method (see Section 6.3), in which the expansions are substituted directly into the differential equation. Here we show a different approach. The ordinary perturbation process is fairly easy to apply, and gives a series of the form

$$x(\varepsilon, t) = x_0(t) + \varepsilon x_1(t) + \varepsilon^2 x_2(t) + \cdots. \qquad (6.13)$$

A finite number of terms of this series will not generally give an approximation of $x(\varepsilon, t)$ holding uniformly for all t, but when (6.12b) is substituted into (6.13), it may be possible to choose $T_1(\tau), T_2(\tau)$, so as to force (6.13) into a form which does give a uniform approximation. This process is called **coordinate perturbation** (Crocco 1972), and will be carried out in the following two examples.

Example 6.1   *Obtain an approximate solution of the family of autonomous equations*

$$\ddot{x} + x = \varepsilon x^3,$$

with $x(\varepsilon, 0) = 1$, $\dot{x}(\varepsilon, 0) = 0$, and error $O(\varepsilon^3)$ uniformly on $t \geq 0$, by the method of coordinate pertubation.

The expansion (6.13), together with the initial conditions, gives

$$\ddot{x}_0 + x_0 = 0, \qquad x_0(0) = 1, \qquad \dot{x}_0(0) = 0;$$

$$\ddot{x}_1 + x_1 = x_0^3, \qquad x_1(0) = 0, \qquad \dot{x}_1(0) = 0;$$

$$\ddot{x}_2 + x_2 = 3x_0^2 x_1, \qquad x_2(0) = 0, \qquad \dot{x}_2(0) = 0;$$

and so on. Then

$$x(\varepsilon, t) = \cos t + \varepsilon \left(\tfrac{1}{32} \cos t + \tfrac{3}{8} t \sin t - \tfrac{1}{32} \cos 3t \right)$$

$$+ \varepsilon^2 \left(\tfrac{23}{1024} \cos t + \tfrac{3}{32} t \sin t - \tfrac{9}{128} t^2 \cos t - \tfrac{3}{128} \cos 3t \right.$$

$$\left. - \tfrac{9}{256} t \sin 3t + \tfrac{1}{1024} \cos 5t \right) + O(\varepsilon^3). \tag{6.14}$$

The expansion is clearly non-uniform on $t \geq 0$.

Now put

$$t = \tau + \varepsilon T_1(\tau) + \varepsilon^2 T_2(\tau) + \cdots \tag{6.15}$$

into (6.14), expand the terms in powers of ε, and rearrange:

$$X(\varepsilon, \tau) = \cos \tau + \varepsilon \left(\tfrac{1}{32} \cos \tau - T_1 \sin \tau + \tfrac{3}{8} \tau \sin \tau - \tfrac{1}{32} \cos 3\tau \right)$$

$$+ \varepsilon^2 \left(\tfrac{23}{1024} \cos \tau - \tfrac{1}{2} T_1^2 \cos \tau + \tfrac{11}{32} T_1 \sin \tau - T_2 \sin \tau \right.$$

$$+ \tfrac{3}{8} \tau T_1 \cos \tau + \tfrac{3}{32} \tau \sin \tau - \tfrac{9}{128} \tau^2 \cos \tau - \tfrac{3}{128} \cos 3\tau$$

$$+ \tfrac{3}{32} T_1 \sin \tau - \tfrac{9}{128} \tau^2 \cos \tau - \tfrac{3}{128} \cos 3\tau + \tfrac{3}{32} T_1 \sin 3\tau$$

$$\left. - \tfrac{9}{256} \tau \sin 3\tau + \tfrac{1}{1024} \cos 5\tau \right) + O(\varepsilon^3).$$

To avoid the form $\tau \sin \tau$ in the ε term, which would given an obvious non-uniformity, define T_1 by

$$T_1(\tau) = \tfrac{3}{8} \tau. \tag{6.16}$$

The ε^2 coefficient then becomes

$$\tfrac{23}{1024} \cos \tau - T_2 \sin \tau + \tfrac{57}{256} \tau \sin \tau - \tfrac{3}{128} \cos 3\tau + \tfrac{1}{1024} \cos 5\tau,$$

and we must have

$$T_2(\tau) = \tfrac{57}{256} \tau \tag{6.17}$$

to eliminate the non-uniformity $\tau \sin \tau$. On the assumption that this step-by-step process could in principle continue indefinitely we have

$$x = X(\varepsilon, \tau) = \cos \tau + \tfrac{1}{32} \varepsilon (\cos \tau - \cos 3\tau) + \varepsilon^2$$

$$\left(\tfrac{23}{1024} \cos \tau - \tfrac{3}{128} \cos 3\tau + \tfrac{1}{1024} \cos 5\tau \right) + O(\varepsilon^3), \tag{6.18a}$$

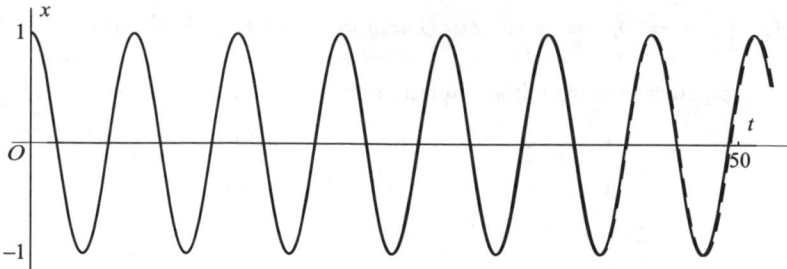

Figure 6.2   Comparison between the numerically computed solution of $\ddot{x}+x = \frac{1}{4}x^3(\varepsilon = 1$ in (6.7)) and the coordinate perturbation given by (6.18): a slight divergence in the periods is beginning to show at $t \approx 50$.

where

$$t = \tau \left(1 + \tfrac{3}{8}\varepsilon + \tfrac{57}{256}\varepsilon^2\right) + O(\varepsilon^3). \tag{6.18b}$$

In Fig. 6.2 the numerical solution of the differential equation with $\varepsilon = 1$ for $0 \leq t \leq 50$ is shown. On this scale the error in the approximation (6.18a) is almost imperceptible.   ●

Example 6.2 (*Lighthill's equation*)   *Find an approximation, uniform on $0 \leq t \leq 1$, to the solution $x(\varepsilon, t)$ of*

$$(\varepsilon x + t)\dot{x} + (2 + t)x = 0, \quad \varepsilon \geq 0, \tag{6.19}$$

satisfying $x(\varepsilon, 1) = e^{-1}$.

Begin by noting certain features of the solution curve in the interval $0 \leq t \leq 1$ (see Fig. 6.3). Since $dx/dt = -(2+t)x/(\varepsilon x + t)$ and $x(\varepsilon, 1) = e^{-1}$, we have

$$x > 0 \quad \text{and} \quad \dot{x} < 0 \qquad \text{at } t = 1.$$

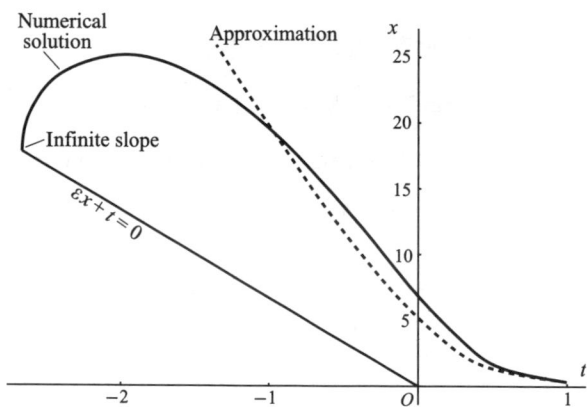

Figure 6.3   Diagram showing the numerical solution of the equation in Example 6.2 and the approximation for $\varepsilon = 0.15$. Note that the continuation of the solution terminates with infinite slope at the point where $\varepsilon x + t = 0$.

We show that x is bounded on the interval $[0, 1]$ by proving that dx/dt is bounded on $[0, 1]$. An infinity can only occur when $\varepsilon x + t = 0$, or at $t = -\varepsilon x$. Moving from $t = 1$ towards the *left*, let $t = t_0$ be the first point at which the solution curve crosses the t axis and becomes negative. To the right of this point x is continuously differentiable. However, x is non-decreasing and positive on $t_0 < t \le 1$ (towards the left) since it is increasing and positive at $t = 1$, and the first turning point is where $dx/dt = 0$, that is, at $t = -2$, which is out of the range $[0, 1]$. Therefore t_0 is out of range. Therefore x is bounded on $0 \le t \le 1$, and non-increasing as we move from $t = 0$ to $t = 1$.

The direct approach, writing

$$x(\varepsilon, t) = x_0(t) + \varepsilon x_1(t) + \varepsilon^2 x_2(t) + \cdots \tag{6.20}$$

leads to the sequence of linear equations

$$t\dot{x}_0 + (2 + t)x_0 = 0, \quad x_0(1) = e^{-1}; \tag{6.21a}$$

$$t\dot{x}_1 + (2 + t)x_1 = -x_0\dot{x}_0, \quad x_1(1) = 0; \tag{6.21b}$$

$$t\dot{x}_2 + (2 + t)x_2 = -\dot{x}_0 x_1 - x_0\dot{x}_1, \quad x_2(1) = 0; \tag{6.21c}$$

and so on. The solution of (6.21a) is

$$x_0(t) = e^{-t}/t^2 \tag{6.22}$$

(predicting, incorrectly, that $x(0) = \infty$). Equation (6.21b) becomes

$$\dot{x}_1 + \left(\frac{2}{t} + 1\right)x_1 = e^{-2t}\left(\frac{2}{t^6} + \frac{1}{t^5}\right).$$

Therefore,

$$x_1(t) = \frac{e^{-t}}{t^2}\int_1^t e^{-u}\left(\frac{2}{u^4} + \frac{1}{u^3}\right)du. \tag{6.23}$$

This is of order $1/t^5$ at $t = 0$, and is even more singular than (6.22). A similar treatment of (6.21c) produces a singularity $O(1/t^8)$ as $t \to 0$. The approximation (6.20):

$$x(\varepsilon, t) \approx x_0(t) + \varepsilon x_1(t),$$

with error $O(\varepsilon^2)$ for fixed t, is clearly not uniform on $0 < t \le 1$ and breaks down completely at $t = 0$.

As in the last example, write the near-identity transformation

$$t = T(\varepsilon, \tau) = \tau + \varepsilon T_1(\tau) + \varepsilon^2 T_2(\tau) + \cdots. \tag{6.24}$$

Then

$$\frac{e^{-t}}{t^2} = \frac{e^{-\tau}}{\tau^2}\left\{1 - \varepsilon T_1(\tau)\left(\frac{2}{\tau} + 1\right)\right\} + O(\varepsilon^2) \tag{6.25}$$

($O(\varepsilon^2)$ refers to fixed τ, not fixed t) and

$$\int_1^t e^{-u}\left(\frac{2}{u^4} + \frac{1}{u^3}\right)du = \int_1^\tau e^{-u}\left(\frac{2}{u^4} + \frac{1}{u^3}\right)du$$

$$+ \int_\tau^{\tau + \varepsilon T_1(\tau) + \cdots} e^{-u}\left(\frac{2}{u^4} + \frac{1}{u^3}\right)du$$

$$= \int_1^\tau e^{-u}\left(\frac{2}{u^4} + \frac{1}{u^3}\right)du + O(\varepsilon). \tag{6.26}$$

From (6.22), (6.23), (6.25), and (6.26) we find, to $O(\varepsilon^2)$ (referring to fixed τ),

$$x = \frac{e^{-\tau}}{\tau^2} \left\{ 1 + \varepsilon \left[\int_1^\tau e^{-u} \left(\frac{2}{u^4} + \frac{1}{u^3} \right) du - T_1(\tau) \left(\frac{2}{\tau} + 1 \right) \right] \right\} + O(\varepsilon^2). \qquad (6.27)$$

The best thing we can do is to choose T_1 to eliminate the worst singularity for small τ arising from the integral. This comes from the term

$$2 \int_1^\tau \frac{e^{-u}}{u^4} du = -\frac{2}{3} \int_1^\tau e^{-u} \frac{d}{du} \left(\frac{1}{u^3} \right) du$$

$$= -\frac{2}{3} \left[\frac{e^{-u}}{u^3} \right]_1^\tau + \frac{2}{3} \int_1^\tau \frac{e^{-u}}{u^3} du$$

$$= -\frac{2e^{-\tau}}{3\tau^3} + O \left(\frac{1}{\tau^2} \right)$$

as $\tau \to 0$. The choice

$$T_1(\tau) = -\frac{1}{3\tau^2} \qquad (6.28)$$

eliminates this singularity (of course, we are left with one of order τ^{-2}).

Finally, from (6.27), (6.24), and (6.28)

$$x = e^{-\tau}/\tau^2 \qquad (6.29a)$$

together with

$$t = \tau - \frac{\varepsilon}{3\tau^2} \qquad (6.29b)$$

give a uniform approximation for $0 \le t \le 1$ (Fig. 6.3).

Note that there is now no infinity predicted at $t = 0$: from (6.29b), $t = 0$ corresponds to $\tau = (\varepsilon/3)^{1/3}$, and $x(0)$ therefore has the value approximately $(\varepsilon/3)^{-2/3}$. The error in (6.29a,b) is $O(\varepsilon^2)$ for fixed τ, but only $O(\varepsilon^{1/3})$ for fixed t. ●

Exercise 6.1

Obtain an approximation solution of

$$\ddot{x} + x = \varepsilon x \dot{x}^2, \quad x(\varepsilon, 0) = 1, \quad \dot{x}(\varepsilon, 0) = 0$$

with error $O(\varepsilon^2)$ uniformly on $t \ge 0$, by the method of coordinate perturbation.

6.3   Lighthill's method

We reconsider the equation (see Example 6.2)

$$(\varepsilon x + t)\dot{x} + (2 + t)x = 0, \qquad (6.30a)$$

$$x(\varepsilon, 1) = e^{-1}, \qquad (6.30b)$$

on $0 \leq t \leq 1$, using the direct attack suggested in the previous section. A systematic procedure known as **Lighthill's method** is developed. Substitute

$$x = X(\varepsilon, \tau) = X_0(\tau) + \varepsilon X_1(\tau) + \cdots \tag{6.31a}$$

and

$$t = T(\varepsilon, \tau) = \tau + \varepsilon T_1(\tau) + \cdots \tag{6.31b}$$

into the differential equation and boundary condition.

First, we do not expect that $t = 1$ will correspond with $\tau = 1$. Suppose that $t = 1$ corresponds to $\tau = \tau^*(\varepsilon)$. Then we must solve (6.31b) for τ^*; that is

$$1 = \tau^* + \varepsilon T_1(\tau^*) + \cdots. \tag{6.32}$$

Then (6.30b), (6.31a) give the transformed boundary condition

$$e^{-1} = X_0(\tau^*) + \varepsilon X_1(\tau^*) + \cdots. \tag{6.33}$$

To solve (6.32), assume τ^* is close to 1 for ε small and has the expansion

$$\tau^* = 1 + \varepsilon \tau_1 + \cdots, \tag{6.34}$$

where $\tau_1, \ldots$ are constants. Equation (6.32) becomes (writing $T_1(\tau^*) = T_1(1) + \varepsilon \tau_1 T_1'(1) + \cdots$)

$$1 = 1 + \varepsilon(\tau_1 + T_1(1)) + \cdots$$

so $\tau_1 = -T_1(1)$, and from (6.34), the boundary $t = 1$ corresponds to $\tau = \tau^*$, where

$$\tau^* = 1 - \varepsilon T_1(1) + \cdots.$$

Therefore, by expanding $X_0(\tau^*)$ and $X_1(\tau^*)$ in (6.33), the boundary condition (6.33) becomes

$$e^{-1} = X_0(1) + \varepsilon X_1(1) - X_0'(1)T_1(1)) + \cdots. \tag{6.35}$$

Next, the derivative in (6.30a) is transformed by

$$\frac{dx}{dt} = \frac{dX}{d\tau} \bigg/ \frac{dT}{d\tau} = \frac{X_0' + \varepsilon X_1' + \cdots}{1 + \varepsilon T_1' + \cdots}$$

$$= X_0' + \varepsilon(X_1' - X_0'T_1') + \cdots. \tag{6.36}$$

Thus eqn (6.30a) becomes, to order ε,

$$(\varepsilon X_0 + \tau + \varepsilon T_1)(X_0' + \varepsilon\{X_1' - X_0'T_1'\}) + (2 + \tau + \varepsilon T_1)(X_0 + \varepsilon X_1) = 0. \tag{6.37}$$

Equations (6.35) and (6.37) give

$$\tau X_0' + (2 + \tau)X_0 = 0, \quad X_0(1) = e^{-1}; \tag{6.38a}$$

$$\left.\begin{array}{l} \tau X_1' + (2 + \tau)X_1 = -T_1(X_0' + X_0) + \tau X_0'T_1' - X_0X_0', \\ X_1(1) = X_0'(1)T_1(1). \end{array}\right\} \tag{6.38b}$$

From (6.38a) the zero-order approximation is

$$X_0(\tau) = e^{-\tau}/\tau^2. \tag{6.39}$$

Then (6.38b) becomes

$$\tau X_1' + (2+\tau)X_1 = \frac{2e^{-\tau}}{\tau^3}T_1 - e^{-\tau}\left(\frac{2}{\tau^2} - \frac{1}{\tau}\right)T_1' + e^{-2\tau}\left(\frac{2}{\tau^5} - \frac{1}{\tau^4}\right), \tag{6.40}$$

with initial condition

$$X_1(1) = -3e^{-1}T_1(1).$$

We now have a free choice of $T_1(\tau)$: it could be chosen to make the right-hand side of (6.40) zero, for example; this would lead to a solution for (6.40) of the type (6.39) again, but is in any case impracticable. We shall choose T_1 to nullify the worst visible singularity on the right of (6.40), which is of order $1/\tau^5$. We attempt this by writing $e^{-\tau}$, $e^{-2\tau} \approx 1$ for small τ and solving

$$\frac{2}{\tau^3}T_1 - \frac{2}{\tau^2}T_1' + \frac{2}{\tau^5} = 0.$$

The simplest solution is

$$T_1(\tau) = -\frac{1}{3\tau^2} \tag{6.41}$$

(compare (6.28)). We have therefore achieved the same result as in the last section; though, in this example, at considerably more effort. Note that throughout the argument we have regarded the equation as a member of a family of equations with parameter ε, as in Chapter 5.

Exercise 6.2
Find the general solution of

$$(\varepsilon x + t)\dot{x} + (2+t)x = 0, \quad x(\varepsilon, 1) = e^{-1}$$

when $\varepsilon = 0$. How does the solution behave near $t = 0$?

6.4 Time-scaling for series solutions of autonomous equations

Consider, for illustration, the family of linear differential equations with parameter ε,

$$\ddot{x} + \varepsilon x + x = 0, \quad t \geq 0, \tag{6.42a}$$

and initial conditions

$$x(0) = 1, \quad \dot{x}(0) = 0, \tag{6.42b}$$

where $|\varepsilon| \ll 1$. The exact solution is

$$x(t, \varepsilon) = \cos(1 + \varepsilon)^{1/2}t. \tag{6.43}$$

Equation (6.42a) is of the type $\ddot{x} + \varepsilon h(x, \dot{x}) + x = 0$ (where $h(x, \dot{x}) = x$) and the discussion is to a large extent representative of equations of this type.

We may carry out a perturbation process for small ε to obtain the solution (6.43) as a power series in ε by substituting the series

$$x(t, \varepsilon) = x_0(t) + \varepsilon x_1(t) + \cdots$$

into the differential equation and the initial conditions and matching the coefficients of $\varepsilon^n, n = 0, 1, \ldots$ successively to generate a sequence of differential equations for $x_0(t), x_1(t), \ldots$. This process gives the same result as we should get by expanding the exact solution (6.43) as a Taylor series in powers of ε. Up to the term in ε^2 we have

$$x(t, \varepsilon) = \cos(1 + \varepsilon)^{1/2}t \approx \cos t - \tfrac{1}{2}\varepsilon t \sin t + \tfrac{1}{8}\varepsilon^2(t \sin t - t^2 \cos t). \tag{6.44}$$

The accuracy of this approximation depends upon the balance between the range of ε, close to zero, and the range of t, which we should like to be as large as possible. The indications are that it will be useless for any *fixed* value of ε as soon as t becomes so large that the three terms cease to diminish rapidly in magnitude. The combinations of ε and t that occur in (6.44) are εt, $\varepsilon^2 t$ and $\varepsilon^2 t^2$. Failure occurs as soon as εt ceases to be small, so the approximation (6.44) is useful only so long as

$$\varepsilon t \ll 1. \tag{6.45}$$

If ε is reduced the breakdown is delayed, but there is no fixed number of terms in the series and no fixed value of ε which will deliver a good approximation for all $t \geq 0$.

However, there exist other representations of $x(t, \varepsilon)$ in series form that are satisfactory for a much larger range of t. The Taylor expansion for $(1 + \varepsilon)^{1/2}$ can be expressed as

$$(1 + \varepsilon)^{1/2} = 1 + \tfrac{1}{2}\varepsilon - \tfrac{1}{8}\varepsilon^2 + \tfrac{1}{16}\varepsilon^3 + O(\varepsilon^4).$$

Use the leading terms to expand $x(t, \varepsilon)$ in a Taylor series centred on the value $(1 + \tfrac{1}{2}\varepsilon)t$; then

$$x(t, \varepsilon) = \cos\left\{\left(1 + \tfrac{1}{2}\varepsilon\right)t - \tfrac{1}{8}\varepsilon^2 t + \tfrac{1}{16}\varepsilon^3 t^2 + O(\varepsilon^4 t)\right\}$$

$$= \cos\left(1 + \tfrac{1}{2}\varepsilon\right)t + \tfrac{1}{8}\varepsilon^2 t \sin\left(1 + \tfrac{1}{2}\varepsilon\right)t - \tfrac{1}{16}\varepsilon^3 t \sin(1 + \varepsilon t)$$

$$- \tfrac{1}{128}\varepsilon^4 t^2 \cos\left(1 + \tfrac{1}{2}\varepsilon\right)t + O(\varepsilon^4 t). \tag{6.46}$$

For given values of ε in the range $|\varepsilon| \ll 1$, consider values of t which may be very large, but are still restricted to the finite range

$$0 \leq t \leq \frac{\kappa}{\varepsilon}, \tag{6.47a}$$

where κ is a constant, so that we may write

$$t = O(1/\varepsilon), \quad \text{or} \quad \varepsilon t = O(1), \quad \text{as } \varepsilon \to 0. \tag{6.47b}$$

Now introduce a new parameter η, called **slow time**, defined by

$$\eta = \varepsilon t = O(1) \tag{6.47c}$$

when $t = O(1/\varepsilon)$, by (6.47b). Then the series (6.46) may be written in the form

$$x(t, \varepsilon) = \cos\left(t + \tfrac{1}{2}\eta\right) + \tfrac{1}{8}\varepsilon\eta \sin\left(t + \tfrac{1}{2}\eta\right)$$
$$- \tfrac{1}{16}\varepsilon^2 \left\{\eta \sin\left(t + \tfrac{1}{2}\eta\right) - \tfrac{1}{8}\eta^2 \cos\left(t + \tfrac{1}{2}\eta\right)\right\} + O(\varepsilon^3). \tag{6.48}$$

A series in powers of ε is emerging whose coefficients are all $O(1)$ when the range of t is $O(\varepsilon^{-1})$. This range is an order of magnitude greater than that permitted in the more elementary representation, (6.44).

We have shown the possibility of an improved representation of a solution of a differential equation by choosing one for which we know the exact solution. This, of course, will not usually be available. We now show how to obtain such a form for the solution by working directly from a differential equation without referring to the exact solution. For the purpose of illustration, however, we shall use the same differential equation, (6.42a), as before.

Consider again the initial value problem

$$\ddot{x} + \varepsilon x + x = 0; \quad x(0, \varepsilon) = 1, \quad \dot{x}(0, \varepsilon) = 0, \tag{6.49}$$

where this is regarded as a family of problems with parameter ε, where $|\varepsilon| \ll 1$. Taking the lead from the previous discussion, we shall aim at obtaining approximations valid over a range of t given by

$$t = O(\varepsilon^{-1}). \tag{6.49a}$$

Introduce a slow-time parameter η:

$$\eta = \varepsilon t = O(1). \tag{6.49b}$$

Then we shall seek a form of solution

$$x(t, \varepsilon) = X(t, \eta, \varepsilon)$$
$$= X_0(t, \eta) + \varepsilon X_1(t, \eta) + \varepsilon^2 X_2(t, \eta) + O(\varepsilon^3), \tag{6.50a}$$

(and assume it may be continued in the same way) in which

$$X_0(t, \eta), X_1(t, \eta), X_2(t, \eta), \ldots = O(1) \tag{6.50b}$$

provided that $t = O(\varepsilon^{-1})$ as $\varepsilon \to 0$.

Since $\eta = \varepsilon t$, the variables η and t are not independent in the original problem. However, $X(t, \eta, \varepsilon)$ is certainly a *particular solution of some partial differential equation in which t and*

η *figure as independent variables.* We obtain this equation in the following way. We have, for all t and ε

$$x(t,\varepsilon) \equiv X(t,\varepsilon t,\varepsilon) \tag{6.51a}$$

when $x(t,\varepsilon)$ is the required solution. Therefore,

$$\frac{\mathrm{d}x}{\mathrm{d}t} \equiv \frac{\mathrm{d}}{\mathrm{d}t}X(t,\varepsilon t,\varepsilon) = \frac{\partial X}{\partial t} + \varepsilon \frac{\partial X}{\partial \eta}, \tag{6.51b}$$

and

$$\frac{\mathrm{d}^2 x}{\mathrm{d}t^2} = \frac{\partial^2 X}{\partial t^2} + 2\varepsilon \frac{\partial^2 X}{\partial \eta \partial t} + \varepsilon^2 \frac{\partial^2 X}{\partial \eta^2}. \tag{6.51c}$$

Since $x(t,\varepsilon)$ satisfies (6.49), $X(t,\eta,\varepsilon)$ satisfies

$$\frac{\partial^2 X}{\partial t^2} + 2\varepsilon \frac{\partial^2 X}{\partial t \partial \eta} + \varepsilon^2 \frac{\partial^2 X}{\varepsilon \eta^2} + (1+\varepsilon)X = 0. \tag{6.52}$$

This is the required partial differential equation. The initial conditions in (6.49) become, using (6.51a), (6.51b),

$$X(0,0,\varepsilon) = 1, \quad \frac{\partial X}{\partial t}(0,0,\varepsilon) + \varepsilon \frac{\partial X}{\partial \eta}(0,0,\varepsilon) = 0. \tag{6.53}$$

The initial conditions (6.53), together with the requirements (6.50a) and (6.50b), are sufficient to isolate the required solution of the partial differential equation (6.52) for all $|\varepsilon| \ll 1$. We shall use an abbreviated notation for the partial derivatives:

$$X^{(t)} = \frac{\partial X}{\partial t}, \quad X^{(t,\eta)} = \frac{\partial^2 X}{\partial t \partial x}, \quad X^{(t,t)} = \frac{\partial^2 X}{\partial t^2},$$

and so on.

Substitute the series (6.50a) into the differential equation (6.52); this must hold good for all ε, so the coefficients of the powers of ε are zero. For terms up to ε^2 we obtain the sequence

$$X_0^{(t,t)} + X_0 = 0; \tag{6.54a}$$

$$X_1^{(t,t)} + X_1 = -2X_0^{(t,\eta)} - X_0; \tag{6.54b}$$

$$X_2^{(t,t)} + X_2 = -2X_1^{(t,\eta)} - X_1 - X_0^{(\eta,\eta)}. \tag{6.54c}$$

From (6.53) we have, similarly, the sequence

$$X_0(0,0) = 1, \quad X_0^{(t)}(0,0) = 0; \tag{6.55a}$$

$$X_1(0,0) = 0, \quad X_1^{(t)}(0,0) + X_0^{\eta}(0,0) = 0; \tag{6.55b}$$

$$X_2(0,0) = 0, \quad X_2^{(t)}(0,0) + X_1^{(\eta)}(0,0) = 0. \tag{6.55c}$$

We further require, by (6.50b), that X_0, X_1, X_2 should be $O(1)$ when $t = O(\varepsilon^{-1})$. This will produce a further condition: that the terms on the right of (6.54b) and (6.54c) (and in any subsequent equations) are zero.

Start with (6.54a). The general solution is

$$X_0(t, \eta) = a_0(\eta) \cos t + b_0(\eta) \sin t, \qquad (6.56a)$$

where a_0 and b_0 are arbitrary functions of η. The initial conditions (6.55a) are satisfied if

$$a_0(0) = 1, \quad b_0(0) = 0. \qquad (6.56b)$$

Substitute (6.56a) into the right-hand side of (6.54b); this becomes

$$X_1^{(t,t)} + X_1 = (2a_0' - b_0) \sin t - (2b_0' + a_0) \cos t.$$

Unless the right-hand side is zero for all t, the solutions X_1 will contain terms behaving like $t \sin t$ and $t \cos t$, and these terms cannot be $O(1)$ when $t = O(\varepsilon^{-1})$. Therefore

$$2a_0' - b_0 = 0 \quad \text{and} \quad 2b_0' + a_0 = 0.$$

The solutions of these equations that satisfy the initial conditions (6.56b) are

$$a_0(\eta) = \cos \tfrac{1}{2}\eta, \qquad b_0(\eta) = - \sin \tfrac{1}{2}\eta. \qquad (6.56c)$$

This process is analogous to the elimination of **secular terms** in Section 5.4. Finally, from (6.56c) and (6.56a) we obtain

$$X_0(t, \eta) = \cos \tfrac{1}{2}\eta \cos t - \sin \tfrac{1}{2}\eta \sin t$$

$$= \cos(t + \tfrac{1}{2}\eta). \qquad (6.56d)$$

We now have (6.54b) in the form

$$X_1^{(t,t)} + X_1 = 0.$$

The general solution is

$$X_1 = a_1(\eta) \cos t + b_1(\eta) \sin t. \qquad (6.57a)$$

The initial conditions (6.55b) require

$$a_1(0) = 0, \qquad b_1(0) = 0. \qquad (6.57b)$$

Looking ahead, eqn (6.54c) for X_2 must not generate secular terms, so the right-hand side must be zero. By substituting (6.57a) and (6.56a) into the right-hand side we obtain

$$2X_1^{(t,\eta)} + X_1 - X_0^{(\eta,\eta)} = (-2a_1' + b_1) \sin t + (2b_1' + a_1) \cos t$$

$$- \tfrac{1}{4}\cos \tfrac{1}{2}\eta \cos t + \tfrac{1}{4}\sin \tfrac{1}{2}\eta \sin t.$$

This must be zero for all t, so

$$-2a_1' + b_1 = -\tfrac{1}{4}\sin\tfrac{1}{2}\eta, \qquad 2b_1' + a_1 = \tfrac{1}{4}\cos\tfrac{1}{2}\eta.$$

The solution of these equations which satisfies the initial conditions (6.57b) is

$$a_1 = \tfrac{1}{8}\eta \sin \tfrac{1}{2}\eta, \qquad b_1 = \tfrac{1}{8}\eta \cos \tfrac{1}{2}\eta. \tag{6.57c}$$

Therefore, from (6.57a),

$$X_1(t, \eta) = \tfrac{1}{8}\eta \sin \tfrac{1}{2} + \tfrac{1}{8}\eta \cos \tfrac{1}{2}\eta$$

$$= \tfrac{1}{8}\eta \sin(t + \tfrac{1}{2}\eta). \tag{6.57d}$$

From (6.56d) and (6.57d) we obtain the approximation

$$x(t, \varepsilon) = X(t, \eta, \varepsilon) \approx X_0(t, \eta) + \varepsilon X_1(t, \varepsilon)$$

$$= \cos(t + \tfrac{1}{2}\varepsilon t) + \tfrac{1}{8}\varepsilon^2 t \sin(t + \varepsilon t). \tag{6.58}$$

This matches the first two terms of the enhanced Taylor expansion of the exact solution, eqn (6.46). By looking forward, without calculation, to the conditions governing X_2 and X_3, it can be seen that the error in (6.58) is $O(\varepsilon^3)$ for $t = (\varepsilon^{-1})$.

In the following example the differential equation is nonlinear. A variant on the previous procedure is adopted, in which a general solution approximation is found first and the initial conditions are applied to this subsequently. Also, the algebra involved is simplified by expressing the solutions of (6.54a) in the form

$$A_0(\eta)e^{it} + \bar{A}_0(\eta)e^{-it},$$

where A_0 is a complex function.

Example 6.3 *Obtain the general solution of* **Rayleigh's equation**

$$\ddot{x} + \varepsilon(\tfrac{1}{3}\dot{x}^3 - \dot{x}) + x = 0 \tag{i}$$

with error $O(\varepsilon)$ so long as $t = O(\varepsilon^{-1})$. Deduce the particular solution for which $x(0, \varepsilon) = a, \dot{x}(0, \varepsilon) = 0$. Find the zero order approximation to the limit cycle.

Suppose that $x(t, \varepsilon)$ has an expansion of the form

$$x(t, \varepsilon) = X(t, \eta, \varepsilon) = X_0(t, \eta) + \varepsilon X_1(t, \eta) + O(\varepsilon^2) \tag{ii}$$

as $\varepsilon \to 0$, when $t = O(\varepsilon^{-1})$. By substituting (6.51b) and (6.51c) into eqn (i), retaining terms up to order ε after substituting the expansion (ii), and matching powers of ε, we obtain

$$\frac{\partial^2 (X_0)}{\partial t^2} + X_0 = 0, \tag{iii}$$

$$\frac{\partial^2 X_1}{\partial t^2} + X_1 = \frac{\partial X_0}{\partial t} - \frac{1}{3}\left(\frac{\partial X_0}{\partial t}\right)^3 - 2\frac{\partial^2 X_0}{\partial t \partial \eta}. \tag{iv}$$

Equation (iii) has the general solution

$$X_0(t, \eta) = A_0(\eta)e^{it} + \bar{A}_0(\eta)e^{-it}, \tag{v}$$

where A_0 is an arbitrary complex function of η. Equation (iv) becomes

$$\frac{\partial^2 X_1}{\partial t^2} + X_1 = \mathrm{i}\mathrm{e}^{\mathrm{i}t}(A_0 - A_0^2 \bar{A}_0 - 2A_0') + \tfrac{1}{3}\mathrm{i}A_0^3 \mathrm{e}^{3\mathrm{i}t} + \text{complex conjugate.} \tag{vi}$$

(For calculating this result, note that $(z + \bar{z})^3 = (z^3 + 3z^2\bar{z}) + \text{complex conjugate.}$)

Secular terms would arise only from the presence of the terms in (vi) containing $\mathrm{e}^{\mathrm{i}t}$ and $\mathrm{e}^{-\mathrm{i}t}$. They are eliminated by requiring that

$$A_0' - \tfrac{1}{2}A_0 + \tfrac{1}{2}A_0^2 \bar{A}_0 = 0 \tag{vii}$$

(the conjugate equation is then automatically satisfied).

To solve (vii), write

$$A_0(\eta) = \rho(\eta)\mathrm{e}^{\mathrm{i}\alpha(\eta)}, \tag{viii}$$

where ρ and α are real. From (vii) we obtain, by equating real and imaginary parts to zero,

$$d\rho/d\eta = \tfrac{1}{2}\rho - \tfrac{1}{2}\rho^3 \quad \text{and} \quad d\alpha/d\eta = 0.$$

The solutions are

$$\rho(\eta) = (1 + a_0 \mathrm{e}^{-\eta})^{-1/2}, \quad \alpha(\eta) = \alpha_0,$$

where a_0 and α_0 are real arbitrary constants.

Equations (v) and (viii) give the zero-order approximation to $X(t, \eta, \varepsilon)$: replacing η by εt we have

$$x_0(t, \varepsilon) \equiv X_0(t, \varepsilon t) = 2(1 + a_0 \mathrm{e}^{-\varepsilon t})^{-1/2} \cos(t + \alpha_0), \tag{ix}$$

where a_0 and α_0 are arbitrary. This is the approximation to the general solution, valid for $t = O(\varepsilon^{-1})$.

From the initial conditions $x(0, \varepsilon) = a, \dot{x}(0, \varepsilon) = 0$. we obtain

$$a = 2(1 + a_0)^{-1/2} \cos \alpha_0, \qquad 0 = -2(1 + a_0)^{-1/2} \sin \alpha_0.$$

Therefore

$$\alpha_0 = 0, \qquad a_0 = -1 + 4/a^2,$$

so

$$x(t, \varepsilon) = 2\{1 - (1 - 4/a^2)\mathrm{e}^{-\varepsilon t}\}^{-1/2} \cos t + O(\varepsilon),$$

when $t = O(\varepsilon^{-1})$. The limit cycle corresponds to $a = 2$.   ●

Exercise 6.3

Using the slow time $\eta = \varepsilon t$, obtain the general solution of

$$\ddot{x} + \varepsilon(x^2 + \dot{x}^2 - 1)\dot{x} + x = 0$$

with error $O(\varepsilon)$ so long as $t = O(\varepsilon^{-1})$. Assume the initial conditions $x(0, \varepsilon) = a > 0$, $\dot{x}(0, \varepsilon) = 0$ to find the zero-order approximation to the limit cycle. (Refer back to Example 1.9.)

6.5 The multiple-scale technique applied to saddle points and nodes

As a model for the procedure we consider the linear equation

$$\ddot{x} + \varepsilon \dot{x} - x = 0, \quad 0 \le \varepsilon \ll 1. \tag{6.59}$$

The origin in the phase plane with $\dot{x} = y$ is a saddle point. The problem discussed is to approximate to the solutions that correspond to the two separatrices which enter the origin.

The incoming separatrices are given exactly by

$$y = mx,$$

(on either side of the origin) where

$$m = -\tfrac{1}{2}\{\varepsilon + (4 + \varepsilon^2)^{1/2}\}. \tag{6.60}$$

These two phase paths correspond to the family of time solutions

$$x(t, \varepsilon) = Ce^{mt}, \tag{6.61}$$

where C is an arbitrary constant.

We shall aim at an approximation which is valid when $t \ge 0$ and

$$t = O(\varepsilon^{-1}) \quad \text{as } \varepsilon \to 0+, \tag{6.62}$$

that is, where $0 \le t \le k\varepsilon^{-1}$ for some constant value of k. Expand the expression (6.60) for m as a Taylor series; we obtain

$$m = -1 - \tfrac{1}{2}\varepsilon - \tfrac{1}{8}\varepsilon^2 + O(\varepsilon^4).$$

The solutions (6.61) become

$$x(t, \varepsilon) = Ce^{-t - \frac{1}{2}\varepsilon t - \frac{1}{8}\varepsilon^2 t} + O(\varepsilon^4 t). \tag{6.63}$$

Now introduce *two* levels of 'slow time' defined by

$$\eta_1 = \varepsilon t, \quad \eta_2 = \varepsilon^2 t. \tag{6.64}$$

So long as $t = O(\varepsilon^{-1})$, $\eta_1 = O(1)$ and $\eta_2 = O(\varepsilon)$ as $\varepsilon \to 0$. The solutions (6.63) may be written

$$x(t, \varepsilon) = Ce^{-t - \frac{1}{2}\eta_1 - \frac{1}{8}\eta_2} + O(\varepsilon^3 \eta_1) = Ce^{-t - \frac{1}{2}\eta_1 - \frac{1}{8}\eta_2} + O(\varepsilon^3), \tag{6.65}$$

valid for $t = O(\varepsilon^{-1})$.

With (6.65) as a point of reference we solve (6.59) by a procedure similar to that of Section 6.4, but involving the two levels of 'slow time', η_1 and η_2. We seek solutions of the form $x(t, \varepsilon) = X(t, \eta_1, \eta_2, \varepsilon)$, where

$$X(t, \eta_1, \eta_2, \varepsilon) = X_0(t, \eta_1, \eta_2) + \varepsilon X_1(t, \eta_1, \eta_2) + \varepsilon^2 X_2(t, \eta_1, \eta_2) + \cdots, \tag{6.66a}$$

and

$$X_0, X_1, X_2, \ldots = O(1) \quad \text{when } t = O(\varepsilon^{-1}). \tag{6.66b}$$

The function $X(t, \eta_1, \eta_2, \varepsilon)$ in which t, η_1, η_2, are treated as *independent* variables satisfies a certain partial differential equation, which is obtained as follows.

By putting $\mathrm{d}x(t, \varepsilon)/\mathrm{d}t = \mathrm{d}X(t, \varepsilon t, \varepsilon^2 t, \varepsilon)/\mathrm{d}t$, we obtain the differential operators

$$\frac{\mathrm{d}}{\mathrm{d}t} x(t, \varepsilon) = \left(\frac{\partial}{\partial t} + \varepsilon \frac{\partial}{\partial \eta_1} + \varepsilon^2 \frac{\partial}{\partial \eta_2} \right) X(t, \eta_1, \eta_2, \varepsilon),$$

$$\frac{\mathrm{d}^2}{\mathrm{d}t^2} x(t, \varepsilon) = \left(\frac{\partial}{\partial t} + \varepsilon \frac{\partial}{\partial \eta_1} + \varepsilon^2 \frac{\partial}{\partial \eta_2} \right)^2 X(t, \eta_1, \eta_2, \varepsilon),$$

where the index $(\cdots)^2$ indicates that the operation is repeated: the 'square' follows ordinary algebraical rules. Up to order ε^2 we obtain

$$\dot{x} = X^{(t)} + \varepsilon X^{(\eta_1)} + \varepsilon^2 X^{(\eta_2)} + O(\varepsilon^3), \tag{6.67a}$$

$$\ddot{x} = X^{(t,t)} + 2\varepsilon X^{(t,\eta_1)} + \varepsilon^2 (2X^{(t,\eta_2)} + X^{(\eta_1,\eta_2)}) + O(\varepsilon^3). \tag{6.67b}$$

(These formulae are used repeatedly in this section and the next.)

Substitute the expansion (6.66a) for X into the original differential equation $\ddot{x} + \varepsilon \dot{x} + x = 0$, retaining terms upto $O(\varepsilon^2)$. Since the resulting equation must hold good for *all* $0 \leq \varepsilon \ll 1$, we may equate the coefficients of $\varepsilon^0, \varepsilon^1, \ldots$ to zero. After some rearrangement the following system is obtained:

$$X_0^{(t,t)} - X_0 = 0, \tag{6.68a}$$

$$X_1^{(t,t)} - X_1 = -(2X_0^{(t,\eta_1)} + X_0^{(t)}), \tag{6.68b}$$

$$X_2^{(t,t)} - X_1 = -(2X_1^{(t,\eta_1)} + X_1^{(t)}) - (2X_0^{(t,\eta_2)} + X_0^{(\eta_1,\eta_1)} + X_0^{(\eta_1)}). \tag{6.68c}$$

(It is of interest to see how the pattern develops. The next term would be

$$X_3^{(t,t)} - X_3 = -(2X_2^{(t,\eta_1)} + X_2) - (2X_1^{(t,\eta_2)} + X_1^{(\eta_1,\eta_1)} + X^{(\eta_1)})$$
$$- (2X_0^{(t,\eta_3)} + 2X_0^{(\eta_1,\eta_2)} + X_0^{(\eta_2)})$$

in which $\eta_3 = \varepsilon^3 t$. If we do not want η_3, the term $X_0^{(t,\eta_3)}$ is absent.)

These equations are to be solved successively, and at each stage the terms on the right-hand side put to zero. If this is not done then extra factors, which involve powers of t, will appear in the expansion (6.66a): this would invalidate (6.66b), changing the orders of magnitude of the terms in (6.66a) when t is large.

The solutions of (6.68a) which satisfy the boundedness condition (6.66b) are

$$X_0 = a_0(\eta_1, \eta_2) \mathrm{e}^{-t}, \tag{6.69a}$$

where the arbitrary function $a_0(\eta_1, \eta_2)$ is settled when we eliminate the secular terms on the right of (6.68b) and (6.68c) we require

$$2X_0^{(t,\eta_1)} + X_0^{(t)} = 0.$$

From (6.69a) we obtain, equivalently,

$$2a_0^{(\eta_1)} + a_0 = 0,$$

so that

$$a_0(\eta_1, \eta_2) = b_0(\eta_2)e^{-\frac{1}{2}\eta_1}.$$

Therefore

$$X_0 = b_0(\eta_2)e^{-t-\frac{1}{2}\eta_1}, \tag{6.69b}$$

where b_0 will be further restricted at the next stage.

The solutions of (6.68b), with zero on the right, take the form

$$X_1 = a_1(\eta_1, \eta_2)e^{-t}. \tag{6.70a}$$

We require the right-hand side of (6.68c) to be zero identically. From (6.70a) and (6.69b) we obtain

$$-(2a_1^{(\eta_1)} + a_1) - \left(2b_0^{(\eta_2)} + \tfrac{1}{4}b_0\right)e^{-\frac{1}{2}\eta_1} = 0.$$

By using the fact that $a_1 = a_1(\eta_1, \eta_2)$ and $b_0 = b_0(\eta_2)$ it can be shown that the bracketed terms must be zero separately if secular terms are to be absent. Therefore

$$2a_1^{(\eta_1)} + a_1 = 0, \qquad 2b_0^{(\eta_2)} + b_0 = 0,$$

so that

$$a_1 = b_1(\eta_2)e^{-\frac{1}{8}\eta_1}, \qquad b_0 = C_0e^{-\frac{1}{8}\eta_2},$$

where C_0 is a constant. From (6.69b) and (6.70a) we obtain

$$X_0 = C_0e^{-t-\frac{1}{2}\eta_1-\frac{1}{8}\eta_2}, \tag{6.71}$$

and

$$X_1 = b_1(\eta_2)e^{-t-\frac{1}{2}\eta_1}. \tag{6.72}$$

The right-hand side of (6.68c) must be zero:

$$2X_1^{(t,\eta_1)} + X_1^{(t)} = 0.$$

Therefore, from (6.72a),

$$2b_1^{(\eta_2)} - \tfrac{1}{4}b_1 = 0, \quad \text{so } b_1 = C_1e^{-\frac{1}{8}\eta_2},$$

where C_1 is a constant. Finally, we have obtained

$$X(t, \eta_1, \eta_2, \varepsilon) = X_0(t, \eta_1, \eta_2) + \varepsilon X_1(t, \eta_1, \eta_2) + O(\varepsilon^2)$$

$$= Ce^{-t - \frac{1}{2}\eta_1 - \frac{1}{8}\eta_2} + O(\varepsilon^2), \tag{6.73}$$

where $C = C_0 + \varepsilon C_1$. The right-hand sides of the sequence (6.68a) to (6.68c) and beyond imply no further restriction on C_0 and C_1, so C is an arbitrary constant. Equation (6.73) may be written as

$$x(t, \varepsilon) = Ce^{-t - \frac{1}{2}\varepsilon t - \frac{1}{8}\varepsilon^2 t} + O(\varepsilon^2), \tag{6.74a}$$

so long as

$$t = O(\varepsilon^{-1}). \tag{6.74b}$$

We have therefore recovered the expression (6.65), which was obtained from the exact solution of (6.59).

The method may be extended to take in further terms in the series (6.66a) for X, with $t = O(\varepsilon^{-1})$ as before. By introducing further levels of slow time, $\eta_3 = \varepsilon^3 t$ etc, the range of t may be extended to $O(\varepsilon^{-2})$ etc.

The following example shows the same technique applied to a nonlinear differential equation with initial conditions. The problem is motivated by considering how to obtain a fairly accurate starting point for a numerical calculation or a plot of a separatrix which enters the origin in the phase plane. We are free to select the value of $x(0, \varepsilon)$, but need also to find the value of $\dot{x}(0, \varepsilon)$ for this particular curve.

Example 6.4   *The equation*

$$\ddot{x} + \varepsilon(1 - x^2)\dot{x} - x = 0, \tag{i}$$

with $0 \leq \varepsilon \ll 1$ has a saddle point at the origin. (a) Obtain the general time solution for the two phase paths that approach the origin, with error $O(\varepsilon^2)$ when $t = O(\varepsilon^{-1})$. (b) At $t = 0$ a point with abscissa $x(0, \varepsilon) = x_0$ (given), and ordinate $\dot{x}(0, \varepsilon) = y(0, \varepsilon) = y_0 + \varepsilon y_1$ (unknown), lies on one of the paths in (a). Obtain y_0, y_1 and the corresponding time solution, in terms of x_0.

(a) Introduce the scaled times

$$\eta_1 = \varepsilon t, \qquad \eta_2 = \varepsilon^2 t, \tag{ii}$$

and let

$$x(t, \varepsilon) \equiv X(t, \eta_1, \eta_2, \varepsilon) = X_0(t, \eta_1, \eta_2) + \varepsilon X_1(t, \eta_1, \eta_2) + O(\varepsilon^2), \tag{iii}$$

where it is supposed that

$$t = O(\varepsilon^{-1}) \quad \text{as } \varepsilon \to 0, \tag{iv}$$

and that the coefficients in (iii) are $O(1)$. Use eqns (6.67a) and (6.67b) to express $\dot{x}$ and $\ddot{x}$ in terms of X_0 and X_1, and substitute into the given differential equation (i). By assembling the coefficients of $\varepsilon^0, \varepsilon^1, \varepsilon^2$ and rearranging we obtain the sequence

$$X_0^{(t,t)} - X_0 = 0, \tag{v}$$

$$X_1^{(t,t)} - X_1 = -\left(2X_0^{(t,\eta_1)} + X_0^{(t)}\right) + X_0^{(t)}X_0^2, \tag{vi}$$

$$X_2^{(t,t)} - X_2 = -\left(2X_0^{(t,\eta_2)} + X_0^{(\eta_1)}\right) - \left(2X_1^{(t,\eta_1)} + X_1^{(t)} + X_1^{(\eta_1,\eta_1)}\right)$$
$$+ \left\{X_0^{(t)}(X_0^2 + 2X_0X_1) + X_0^{(\eta_1)}X_0^2\right\}. \tag{vii}$$

The solutions of (v) which tend to zero have the form

$$X_0 = a_0(\eta_1, \eta_2)e^{-t}. \tag{viii}$$

The term $X_0^{(t)}X_0^2$ on the right of (vi) becomes

$$X_0^{(t)}X_0^2 = -a_0^3(\eta_1, \eta_2)e^{-3t} \tag{ix}$$

and therefore does not affect the order of magnitude of X_1, when $t = O(\varepsilon^{-1})$. The bracketed terms must be eliminated, so $2X_0^{(t,\eta_1)} + X_0^{(t)} = 0$. Therefore, by (viii),

$$2a_0^{(\eta_1)} + a_0 = 0,$$

so

$$a_0 = b_0(\eta_2)e^{-\frac{1}{2}\eta_1}.$$

Therefore,

$$X_0 = b_0(\eta_2)e^{-t-\frac{1}{2}\eta_1}, \tag{x}$$

and the equation for X_1 becomes (using (ix))

$$X_1^{(t,t)} - X_1 = -b_0^3 e^{-3t-\frac{3}{2}\eta_1}.$$

The appropriate solutions are given by

$$X_1 = a_1(\eta_1, \eta_2)e^{-t} - \frac{1}{8}b_0(\eta_2)e^{-3t-\frac{3}{2}\eta_1}. \tag{xi}$$

By following a similar argument in connection with the right-hand side of (vii), we obtain

$$b_0(\eta_2) = C_0 e^{-\frac{1}{8}\eta_2}, \qquad a_1(\eta_1, \eta_2) = b_1(\eta_2)e^{-\frac{1}{2}\eta_1}. \tag{xii}$$

where C_0 is a constant. Finally we have

$$X = X_0 + \varepsilon X_1 - O(\varepsilon^2)$$
$$= C_0 e^{-t-\frac{1}{2}\eta_1-\frac{1}{8}\eta_2} + \varepsilon\left\{b_1(\eta_2)e^{-t-\frac{1}{2}\eta_1} - \frac{1}{8}C_0^3 e^{-3t-\frac{3}{2}\eta_1}\right\} + O(\varepsilon^2) \tag{xiii}$$

when $t = O(\varepsilon^{-1})$; and in terms of t and ε:

$$x(t, \varepsilon) \approx C_0 e^{-t-\frac{1}{2}\varepsilon t-\frac{1}{8}\varepsilon^2 t} + \varepsilon\left\{b_1(\varepsilon^2 t)e^{-t-\frac{1}{2}\varepsilon t} - \frac{1}{8}C_0^3 e^{-3t-\frac{3}{2}\varepsilon t}\right\}, \tag{xiv}$$

with an error $O(\varepsilon^2)$ when $t = O(\varepsilon^{-1})$.

(b) To apply the initial conditions we require $x(0, \varepsilon)$ and $\dot{x}(0, \varepsilon)$, maintaining an error of $O(\varepsilon^2)$. Note that $db_1/dt = O(\varepsilon^2)$; also write

$$b_1(0) = C_1.$$

Then from (xiv) we obtain

$$x(0, \varepsilon) = C_0 + \varepsilon(C_1 - \tfrac{1}{8}C_0^3) = x_0,$$

$$y(0, \varepsilon) = -C_0 + \varepsilon(-C_1 - \tfrac{1}{2}C_0 + \tfrac{3}{8}C_0^3) = y_0 + \varepsilon y_1,$$

in which C_0 and C_1 are disposable constants, and the initial values are x_0 and $y_0 + \varepsilon y_1$. From the first equation we obtain

$$C_0 = x_0, \qquad C_1 = \tfrac{1}{8}C_0^3 = \tfrac{1}{8}x_0^3. \tag{xv}$$

and from the second,

$$y_0 = -C_0 = -x_0, \qquad y_1 = -\tfrac{1}{2}x_0 + \tfrac{1}{4}x_0^3. \tag{xvi}$$

The constants in (xv) produce the time solution from (xiii); and (xvi) defines the required ordinate to $O(\varepsilon^2)$. The graph of

$$y = y_0 + \varepsilon y_1 = -x_0 + \varepsilon(-\tfrac{1}{2}x_0 + \tfrac{1}{4}x_0^3)$$

is shown in Fig. 6.4, where comparison is made with its computed separatrix. ●

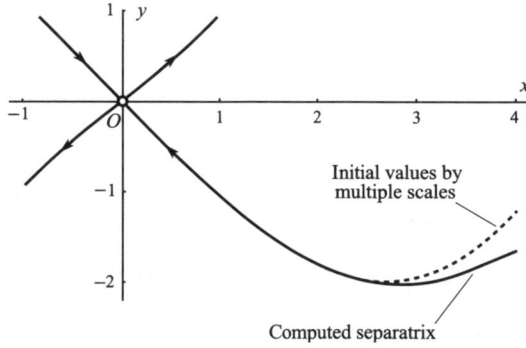

Figure 6.4   Stable separatrix in $x > 0, y < 0$ computed numerically, and the track of initial values given by eqn (xv): $y = y_0 + \varepsilon y_1 = -x_0 + \varepsilon(-\tfrac{1}{2}x_0^2 + \tfrac{1}{4}x_0^3)$ for $\varepsilon = 0.2$.

In the following example the differential equation is not of the type $x + \varepsilon h(x, \dot{x}) \pm x = 0$ considered up to this point. The linear equation $\ddot{x} + \dot{x} + \varepsilon x = 0$, where ε is small and positive (nonzero), is taken to illustrate the principles.

Example 6.5   *(a) By working from the exact solution of the initial value problem*
$$\ddot{x} + \dot{x} + \varepsilon x = 0, \quad x(0, \varepsilon) = x_0, \quad \dot{x}(0, \varepsilon) = 0, \quad 0 < \varepsilon \ll 1, \tag{i}$$

show that $x(t, \varepsilon) = x_0\{e^{-\varepsilon t - \varepsilon^2 t} + \varepsilon(e^{-\varepsilon t} - e^{-t + \varepsilon t})\} + O(\varepsilon^2)$ *as* $\varepsilon \to 0$, *when* $t = O(\varepsilon^{-1})$. *(b) Obtain the same result by using the two time scales* $\eta_1 = \varepsilon t$, $\eta_2 = \varepsilon^2 t$.

(a) The characteristic equation of (i) is $m^2 + m + \varepsilon = 0$, with roots m_1, m_2, where for small ε

$$m_1 = \tfrac{1}{2}\{-1 + (1 - 4\varepsilon)^{1/2}\} = -\varepsilon + \varepsilon^2 + O(\varepsilon^3),$$

$$m_2 = \tfrac{1}{2}\{-1 - (1 - 4\varepsilon)^{1/2}\} = -1 + \varepsilon - \varepsilon^2 + O(\varepsilon^3).$$

Since these are real and negative, the origin in the phase plane is a stable node. The exact solution of the initial value problem is

$$x(t, \varepsilon) = x_0 \{ m_2 e^{m_1 t} - m_1 e^{m_2 t} \} / (m_2 - m_1).$$

Substitute the approximations for m_1, m_2 (noting that when $t = O(\varepsilon^{-1})$, then $\varepsilon^2 t = O(\varepsilon)$ so that $e^{-\varepsilon^2 t} = 1 + O(\varepsilon)$); then

$$x(t, \varepsilon) = x_0 \{ e^{-\varepsilon t - \varepsilon^2 t} + \varepsilon (e^{-\varepsilon t} - e^{-t + \varepsilon t}) \} + O(\varepsilon^2) \tag{ii}$$

is the required approximation, valid when $t = O(\varepsilon^{-1})$.
(b) Proceed as usual, putting $x(t, \varepsilon) = X(t, \eta_1, \eta_2, \varepsilon)$ and

$$X(t, \eta_1, \eta_2, \varepsilon) = X_0(t, \eta_1, \eta_2) + \varepsilon X_1(t, \eta_1, \eta_2) + \varepsilon^2 X_2(t, \eta_1, \eta_2) + O(\varepsilon^3),$$

where $X_0, X_1, X_2, \ldots$ are $O(1)$ when $t = O(\varepsilon^{-1})$. We obtain

$$X_0^{(t,t)} + X_0^{(t)} = 0, \tag{iii}$$

$$X_1^{(t,t)} + X_1^{(t)} = -(2X_0^{(t,\eta_1)} + X_0^{(\eta_1)} + X_0), \tag{iv}$$

$$X_2^{(t,t)} + X_2^{(t)} = -(2X_1^{(t,\eta_1)} + X_1^{(\eta_1)} + X_1) - (2X_0^{(t,\eta_2)} + X_0^{(\eta_1,\eta_1)} + X_0^{(\eta_2)}). \tag{v}$$

The solutions of (iii) take the form

$$X_0 = p_0(\eta_1, \eta_2) + q_0(\eta_1, \eta_2) e^{-t}. \tag{vi}$$

The right-hand side of (iv) then becomes

$$-(p_0^{(\eta_1)} + p_0) + (q_0^{(\eta_1)} - q_0) e^{-t}.$$

Without further action eqn (iv) will therefore have solutions involving t and te^{-t}. Since we need $t = O(\varepsilon^{-1})$ these terms must not occur, so we require

$$p_0^{(\eta_1)} + p_0 = 0, \qquad q_0^{(\eta_1)} - q_0 = 0.$$

Therefore

$$p_0 = r_0(\eta_2) e^{-\eta_1}, \qquad q_0 = s_0(\eta_2) e^{\eta_1},$$

and so from (vi)

$$X_0 = r_0(\eta_2) e^{-\eta_1} + s_0(\eta_2) e^{-t + \eta_1}. \tag{vii}$$

The solutions of (iv), with zero now on the right-hand side, are

$$X_1 = p_1(\eta_1, \eta_2) + q_1(\eta_1, \eta_2) e^{-t}. \tag{viii}$$

By substituting (vii) and (viii), the right-hand side of (v) reduces to

$$-(p_1^{(\eta_1)} + p_1) + (q_1^{(\eta_1)} - q_1) e^{-t} - (r_0^{(\eta_2)} + r_0) + (s_0^{(\eta_2)} + s_0) e^{-t + \eta_1}.$$

Solutions of (v) containing terms of the form t and te^{-t} are eliminated by equating each of the four brackets to zero. We obtain

$$r_0 = C_0 e^{-\eta_2}, \qquad s_0 = D_0 e^{-\eta_2}, \tag{ix}$$

where C_0 and D_0 are constants, and

$$p_1 = u_1(\eta_2)e^{-\eta_1}, \qquad q_1 = v_1(\eta_2)e^{\eta_1}. \tag{x}$$

Substituting these results in (vii) and (viii) we have

$$X_0 = C_0 e^{-\eta_1-\eta_2} + D_0 e^{-t+\eta_1-\eta_2}$$

and

$$X_1 = u_1(\eta_2)e^{-\eta_1} + v_1(\eta_2)e^{-t+\eta_1}.$$

Therefore

$$X = C_0 e^{-\eta_1-\eta_2} + D_0 e^{-t+\eta_1-\eta_2} + \varepsilon\{u_1(\eta_2)e^{-\eta_1} + v_1(\eta_2)e^{-t+\eta_1}\},$$

or

$$x(t,\varepsilon) = C_0 e^{-\varepsilon t-\varepsilon^2 t} + D_0 e^{-t+\varepsilon t-\varepsilon^2 t} + \varepsilon\{u_1(\varepsilon^2 t)e^{-\varepsilon t} + v_1(\varepsilon^2 t)e^{-t+\varepsilon t}\}, \tag{xi}$$

this being valid to an error $O(\varepsilon^2)$ when $t = O(\varepsilon^{-1})$.

Direct application of the initial conditions $x(0,\varepsilon) = x_0$, $\dot{x}(0,\varepsilon) = 0$ (noting that $\mathrm{d}u/\mathrm{d}t$ and $\mathrm{d}v/\mathrm{d}t$ are $O(\varepsilon^2)$) gives the equations

$$x_0 = C_0 + D_0 + \varepsilon(u_1(0) + v_1(0)),$$
$$0 = -D_0 + \varepsilon(-C_0 + D_0 + u_1(0) - v_1(0)),$$

from which we obtain, by matching coefficients,

$$C_0 = x_0, \quad D_0 = 0, \quad u(0) = x_0, \quad v(0) = -x_0. \tag{xii}$$

Since also $u_1(\varepsilon^2 t) = u_1(0) + O(\varepsilon^2 t) = u_1(0) + O(\varepsilon)$ for $t = O(\varepsilon^{-1})$, and similarly for $v_1(\varepsilon^2 t)$, we obtain from (xi) the solution

$$x(t,\varepsilon) = x_0\{e^{-\varepsilon t-\varepsilon^2 t} + \varepsilon(e^{-\varepsilon t} - e^{-t+\varepsilon t})\} + O(\varepsilon^2) \tag{xiii}$$

as $\varepsilon \to 0$ and $t = O(\varepsilon^{-1})$. This agrees with the form (ii) derived from the exact solution.   ●

6.6   Matching approximations on an interval

In this section we shall mainly be concerned with **boundary-value problems**, and we conform with the literature by using y as the dependent and x as the independent variable. For fuller information on the methods of this section the reader is referred to Nayfeh (1973), O'Malley (1974) and Hinch (1991).

As in Section 6.1, we shall illustrate the problem by looking at the approximation to a particular function y:

$$y(\varepsilon, x) = e^{-\frac{1}{2}x} - e^{\frac{1}{2}x}e^{-2x/\varepsilon}, \quad 0 \leq x \leq 1, \ 0 < \varepsilon \ll 1. \tag{6.75}$$

(The particular form chosen appears in the solution to a differential equation later on.) We shall look at the structure of this function as $\varepsilon \to 0$.

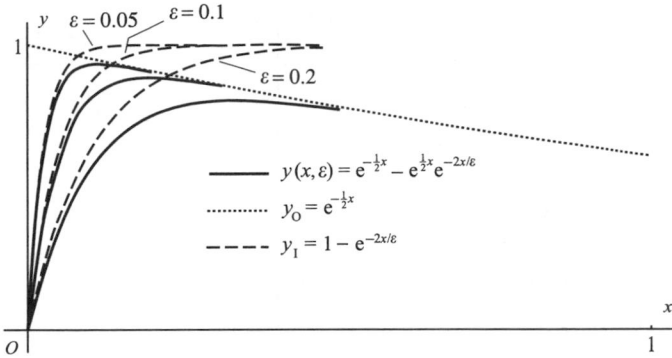

Figure 6.5 Inner approximation y_I and the outer approximation y_O to the function $y(\varepsilon, x)$ in (6.75).

Note that, for $x > 0$,

$$\lim_{\varepsilon \to 0+} (e^{-2x/\varepsilon}/\varepsilon^n) = 0$$

for every positive n: the function tends to zero very rapidly as $\varepsilon \to 0+$. Therefore, for every *fixed* $x > 0$

$$y(\varepsilon, x) \approx e^{-\frac{1}{2}x} = y_O, \quad \text{say,} \tag{6.76}$$

with error $O(\varepsilon^n)$ for every $n > 0$. But by looking back at (6.75) it can be seen that as x takes smaller and smaller values, smaller and smaller values of ε are required before (6.76) becomes an acceptable approximation. It fails altogether at $x = 0$, where (6.75) gives zero and (6.76) gives 1. Therefore (6.76) is not a uniform approximation on $0 \le x \le 1$, and another form is needed near $x = 0$. Figure 6.5 shows the nature of the approximation for some particular values of ε; as ε decreases the interval of good fit becomes extended, but it is always poor near $x = 0$. The approximation begins to fail when x becomes comparable with ε in magnitude. The region of failure is called a **boundary layer** (from a hydrodynamical analogy), having in this case **thickness of order** ε. The function y_O defined in (6.76) is called the **outer approximation** to y.

To get an approximation near $x = 0$ it is no use trying to work with x fixed: this is covered by (6.76). We therefore consider x tending to zero with ε by putting, say,

$$x(\varepsilon) = \xi\varepsilon, \tag{6.77}$$

where ξ may take any value. ξ is called a **stretched variable**: from one point of view we are magnifying the boundary layer to thickness $O(1)$. Then

$$y(\varepsilon, x) = e^{-\frac{1}{2}\xi\varepsilon} - e^{\frac{1}{2}\xi\varepsilon}e^{-2\xi} \approx 1 - e^{-2\xi}, \tag{6.78}$$

where $e^{\pm\frac{1}{2}\xi\varepsilon} = 1 + O(\varepsilon)$. The error is $O(\varepsilon)$ for every fixed ξ, though naturally it works better when ξ is not too large. We can express this idea alternatively by saying

$$y(\varepsilon, x) \approx 1 - e^{-2x/\varepsilon} = y_I, \quad \text{say,} \tag{6.79}$$

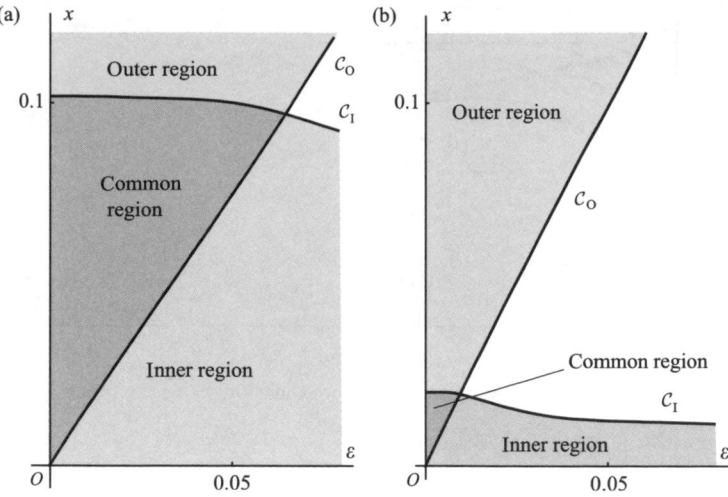

Figure 6.6 Regions of validity for the approximations (6.76) (outer approximation) and (6.79) (inner approximation) with error E: (a) $E = 0.05$; (b) $E = 0.01$.

with error $O(\varepsilon)$ so long as $x = O(\varepsilon)$ (which includes the case $x = o(\varepsilon)$ too). This is rather like the situation described for $t > 0$ at the beginning of Section 6.1. The approximation (6.79) is shown in Fig. 6.5. The function y_I defined in (6.79) is called the **inner approximation** to y.

If our information about y considered only of the approximations (6.76) and (6.79), together with the associated error estimates, it would be pure guesswork, given a value of x and a value of ε, to decide which to use, since an error of low *order* in ε is not necessarily small for any *given* small value of ε . We can see what the regions of validity look like in a general way by experimenting and displaying the results in the plane of ε, x. The regions in which (6.76) and (6.79) give errors of less than 0.05 and 0.01 are shown in Fig. 6.6.

The error boundary of the **outer region** is given by

$$|y(x, \varepsilon) - y_O| = e^{\frac{1}{2}x} e^{-2x/\varepsilon} = E,$$

where $E > 0$ is a specified error. Thus the boundary $\mathcal{C}_O$ is given by

$$x = \frac{2\varepsilon \ln E}{\varepsilon - 4}.$$

The error boundary of the **inner region** is given by

$$|y(x, \varepsilon) - y_I| = |e^{-\frac{1}{2}x} - 1 + e^{-2x/\varepsilon}(1 - e^{\frac{1}{2}x})| = E.$$

This equation can be solved for ε (but not explicitly for x) to give the inner error boundary $\mathcal{C}_I$:

$$\varepsilon = -2x / \ln\left(\frac{1 - E - e^{-\frac{1}{2}x}}{1 - e^{\frac{1}{2}x}}\right).$$

The two boundaries are shown in Fig. 6.6 for the errors $E = 0.05$ and $E = 0.01$.

The figures show that there is a region on the ε, x plane where (6.76) and (6.79) both give a small error. We therefore expect that there may be cases 'between' the cases $x = $ constant and $x = O(\varepsilon)$, in which both approximations have a small remainder. For example, if

$$x = \eta\sqrt{\varepsilon}, \quad \eta \text{ constant}, \tag{6.80}$$

$y(\varepsilon, x)$ becomes

$$e^{-\frac{1}{2}\eta\sqrt{\varepsilon}} - e^{-\frac{1}{2}\eta\sqrt{\varepsilon}}e^{-2\eta/\sqrt{\varepsilon}} = 1 + O(\sqrt{\varepsilon}); \tag{6.81}$$

also (6.76) becomes

$$y_O = e^{-\frac{1}{2}\eta\sqrt{\varepsilon}} = 1 + O(\sqrt{\varepsilon}) \tag{6.82}$$

and (6.79) becomes

$$y_I = 1 - e^{-2\eta/\sqrt{\varepsilon}} = 1 + O(\sqrt{\varepsilon}) \tag{6.83}$$

(in fact, $1 + O(\sqrt{\varepsilon})$). Thus the original function and both approximations have an error tending to zero with ε when $x = \eta\sqrt{\varepsilon}, \eta$ fixed. We say that the functions 'match' to $O(1)$ in the 'overlap region'.

Figure 6.7 indicates the progress of a point $(\varepsilon, \eta\sqrt{\varepsilon}) = (\varepsilon, x)$ as it moves into regions where the two approximations (6.76) and (6.79) have in common an error diminishing to zero with ε. It is desirable to show that there are no 'gaps' in the postulated 'common region'. Instead of (6.80), therefore, consider the more general case

$$x(\varepsilon) = \zeta\psi(\varepsilon),$$

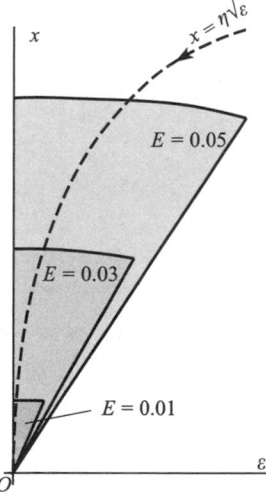

Figure 6.7 As $\varepsilon \to 0$, the point $(\varepsilon, \eta\sqrt{\varepsilon})$ (η constant) always lies ultimately in the 'overlap region' for the approximations (6.76) and (6.79). Here the path is shown for $\eta \approx 0.4$ and errors $E = 0.05, 0.03, 0.01$.

where ζ is any constant and ψ tends to zero, but more slowly than ε, so that

$$\lim_{\varepsilon \to 0} \varepsilon / \psi(\varepsilon) = 0.$$

Then y, from (6.75), becomes

$$y(\varepsilon, x) = e^{-\frac{1}{2}\zeta\psi(\varepsilon)} - e^{\frac{1}{2}\zeta\psi(\varepsilon)} e^{-2\zeta\psi(\varepsilon)/\varepsilon} = 1 + o(1).$$

The outer approximation (6.76) gives

$$e^{-\frac{1}{2}x} = e^{-\frac{1}{2}\zeta\psi(\varepsilon)} = 1 + o(1),$$

and the inner approximation (6.79) gives

$$1 - e^{-2x/\varepsilon} = 1 - e^{-2\zeta\psi(\varepsilon)/\varepsilon} = 1 + o(1).$$

These are in agreement to $o(1)$.

The following example shows how the assumption of a 'common region' is used in connection with differential equations to establish an unknown constant. In this case we have put

$$\psi(\varepsilon) = \varepsilon^{1-\delta}, \quad 0 < \delta < 1,$$

to give a certain generality.

Example 6.6 *A function y has the two approximations on $0 \le x \le 1$: an inner approximation*

$$y(\varepsilon, x) \approx A + (1 - A)e^{-x/\varepsilon} = y_I, \quad \text{say,} \tag{6.84}$$

with error $O(\varepsilon)$ when $x = O(\varepsilon)$, and an outer approximation

$$y(\varepsilon, x) \approx e^{1-x} = y_O, \quad \text{say,} \tag{6.85}$$

with error $O(\varepsilon)$ for x constant. Find the value of A.
Assuming that both approximations are valid simultaneously (though possibly with a larger error) for

$$x = \eta\varepsilon^{1-\delta}, \quad 0 < \delta < 1, \quad \eta \text{ constant}, \quad 0 \le x \le 1,$$

we must have at least

$$\lim_{\substack{\varepsilon \to 0 \\ (\eta \text{ constant})}} [A + (1 - A)e^{-\eta/\varepsilon^\delta}] = \lim_{\substack{\varepsilon \to 0 \\ (\eta \text{ constant})}} e^{1-\eta\varepsilon^{1-\delta}}.$$

Therefore $A = e^1 = e$.   ●

Example 6.7 *From (6.84) and (6.85) make up a single approximate expression applying uniformly on $0 \le x \le 1$.*
Write, (with $A = e$),

$$y_I + y_O = e + (1 - e)e^{-x/\varepsilon} + e^{1-x}.$$

When x is constant and $\varepsilon \to 0$, we obtain

$$e + e^{1-x}$$

instead of e^{1-x} required by (6.85); and when $x = \xi\varepsilon$ and $\varepsilon \to 0$ (ξ constant), we obtain

$$2e + (1 + e)e^{-\xi}$$

instead of $e + (1 - e)e^{-\xi}$ required by (6.84). Therefore the required uniform expansion is given by

$$y_I + y_O - e = (1 - e)e^{-x/\varepsilon} + e^{1-x}. \qquad \bullet$$

Exercise 6.4
Find the outer and inner approximations to

$$y = 1 - e^{-x/\varepsilon} \cos x, \quad 0 \le x \le \tfrac{1}{2}\pi, \quad 0 < \varepsilon \ll 1.$$

Sketch a graph showing the approximations and y.

6.7   A matching technique for differential equations

In the following problems the solution has different approximations in different (x, ε) regions as described in Section 6.5. Characteristic of these is *the presence of ε as the coefficient of the highest derivative appearing in the equation*. The term in the highest derivative is, then, negligible except where the derivative itself is correspondingly large. Thus, over most of the range the equation is effectively of lower order, and can satisfy fewer boundary conditions. However, in certain intervals (boundary layers) the derivative may be large enough for the higher order term to be significant. In such intervals, y will change very rapidly, and may be chosen to satisfy another boundary condition.

Example 6.8 (*An initial-value problem*)   *Find a first-order approximation to the solution of*

$$\varepsilon\frac{dy}{dx} + y = x, \quad x > 0, \quad 0 < \varepsilon \ll 1, \tag{6.86}$$

subject to

$$y(\varepsilon, 0) = 1. \tag{6.87}$$

The obvious first step is to put $\varepsilon = 0$ into (6.86): this gives the 'outer approximation'

$$y \approx y_O = x. \tag{6.88}$$

The error is $O(\varepsilon)$, as can be seen by viewing it as the first step in a perturbation process. It is not clear where this holds, but assume we have numerical or other indications that it is right except near $x = 0$, where there is a boundary layer. We do not know how thick this is, so we allow some freedom in choosing a new, stretched variable, ξ, by writing

$$x = \xi\phi(\varepsilon), \tag{6.89}$$

where

$$\lim_{\varepsilon \to 0} \phi(\varepsilon) = 0.$$

Equation (6.86) becomes

$$\frac{\varepsilon}{\phi(\varepsilon)} \frac{dy}{d\xi} + y = \xi\phi(\varepsilon). \tag{6.90}$$

The choice of $\phi(\varepsilon)$ which retains $dy/d\xi$ as a leading term in order of magnitude, and which gives an equation which can describe a boundary layer is

$$\phi(\varepsilon) = \varepsilon; \tag{6.91}$$

and (6.90) becomes

$$\frac{dy}{d\xi} + y = \varepsilon\xi. \tag{6.92}$$

The first approximation ($\varepsilon = 0$) to the solution fitting the initial condition (6.87) (the inner approximation), is given by

$$y \approx y_I = e^{-\xi} \tag{6.93}$$

with error $O(\varepsilon)$ for ξ constant. This may be interpreted as

$$y_I = e^{-x/\varepsilon}, \quad x = O(\varepsilon). \tag{6.94}$$

There is no scope for 'matching' y_O and y_I here: either they are approximations to the same solution or they are not; there are no arbitrary constants left to be settled. We note, however, that in an 'intermediate region' such as

$$x = \eta\sqrt{\varepsilon}$$

for η constant, both approximations do agree to order $\sqrt{\varepsilon}$.

To construct a uniform approximation, start with the form $y_I + y_O$ as in Example 6.7

$$y_I + y_O = e^{-x/\varepsilon} + x. \tag{6.95}$$

This agrees to order ε with eqns (6.88) and (6.93) for $x = $ constant and $x = \xi\varepsilon$ respectively, so that eqn (6.95) is already a uniform approximation. The reader should compare the exact solution

$$y(\varepsilon, x) = x - \varepsilon + e^{-x/\varepsilon} + \varepsilon e^{-x/\varepsilon}. \qquad \bullet$$

Example 6.9 (*A boundary-value problem*)   *Find a first approximation to the equation*

$$\varepsilon\frac{d^2y}{dx^2} + 2\frac{dy}{dx} + y = 0, \quad 0 < x < 1 \tag{6.96}$$

subject to

$$y(0) = 0, \quad y(1) = 1. \tag{6.97}$$

Putting $\varepsilon = 0$ in (6.96) (or finding the first term in an ordinary perturbation process), gives the differential equation for the outer approximation $y_O(\varepsilon, x)$:

$$2\frac{dy_O}{dx} + y_O = 0. \tag{6.98}$$

Since this is first order, only one boundary condition can be satisfied. We shall assume (if necessary by showing that the contrary assumption leads to failure of the method) that (6.98) holds approximately at $x = 1$. We require $y_O(\varepsilon, 1) = 1$, so

$$y \approx y_O(x, \varepsilon) = e^{\frac{1}{2}}e^{-\frac{1}{2}x} \tag{6.99}$$

with error $O(\varepsilon)$. The non-uniformity is not self-evident, but certainly the boundary condition at $x = 0$ is not satisfied. Assuming that the condition is attained by a sudden change in the nature of the solution near $x = 0$, introduce a new, stretched, variable ξ, where

$$x = \xi \phi(\varepsilon), \tag{6.100}$$

ξ fixed, where

$$\lim_{\varepsilon \to 0} \phi(\varepsilon) = 0. \tag{6.101}$$

Equation (6.96) becomes

$$\frac{\varepsilon}{\phi^2(\varepsilon)} \frac{d^2 y}{d\xi^2} + \frac{2}{\phi(\varepsilon)} \frac{dy}{d\xi} + y = 0. \tag{6.102}$$

The choice $\phi(\varepsilon) = \varepsilon$, which makes the first two terms in (6.101) have the same order in ε, yields the equation

$$\frac{d^2 y}{d\xi^2} + 2\frac{dy}{d\xi} + \varepsilon y = 0. \tag{6.103}$$

This simplifies to the equation for the inner approximation y_I:

$$\frac{d^2 y_I}{d\xi^2} + 2\frac{dy_I}{d\xi} = 0 \tag{6.104}$$

with error order ε. The boundary condition is

$$y_I(\xi, 0) = 0. \tag{6.105}$$

Therefore

$$y \approx y_I(\varepsilon, x) = A(1 - e^{-2\xi}). \tag{6.106}$$

When ξ is constant the error is $O(\varepsilon)$.

The value of A must be determined by the condition that (6.99) and (6.106) should both approximate to the same function (the solution): if there is an 'overlapping region' of approximation to some order, the approximations themselves must agree to this order, and there is only one choice of A making this agreement possible.

Allow the behaviour of x to be given by

$$x = \eta \psi(\varepsilon), \tag{6.107}$$

where η is any constant, and

$$\lim_{\varepsilon \to 0} \psi(\varepsilon) = 0 \tag{6.108}$$

but where ψ does not tend to zero so fast as ϕ, so that

$$\lim_{\varepsilon \to 0} \phi(\varepsilon)/\psi(\varepsilon) = \lim_{\varepsilon \to 0} \varepsilon/\psi(\varepsilon) = 0. \tag{6.109}$$

Choose $\psi(\varepsilon) = \sqrt{\varepsilon}$, which satisfies these conditions. Then

$$y_O(\varepsilon, x) = e^{\frac{1}{2}} e^{-\frac{1}{2}\eta\psi(\varepsilon)} = e^{\frac{1}{2}} + O(\psi) \tag{6.110}$$

by (6.108), and

$$y_I(\varepsilon, x) = A(1 - e^{-2\eta\psi/\varepsilon}) = A + o(1) \tag{6.111}$$

by (6.109). Agreement is only possible if $A = e^{\frac{1}{2}}$, so finally

$$y \approx y_O(\varepsilon, x) = e^{\frac{1}{2}}e^{-\frac{1}{2}x} \quad \text{for } x = O(1) \tag{6.112}$$

and

$$y \approx y_I(\varepsilon, x) = e^{\frac{1}{2}}(1 - e^{-2x/\varepsilon}) \quad \text{for } x = O(\varepsilon). \tag{6.113}$$

To find an approximation uniform on $0 \le x \le 1$, form

$$y_O + y_I - e^{\frac{1}{2}}e^{-\frac{1}{2}x} + e^{\frac{1}{2}}(1 - e^{-2x/\varepsilon}).$$

Then for $x = O(1)$ this becomes $e^{\frac{1}{2}}e^{-\frac{1}{2}x} + e^{\frac{1}{2}}$ as $\varepsilon \to 0$; that is it has the unwanted term $e^{\frac{1}{2}}$ in it. Similarly, when $x = \xi\varepsilon$ it becomes $e^{\frac{1}{2}} + e^{\frac{1}{2}}(1 - e^{-2\xi})$ as $\varepsilon \to 0$, which again contains an unwanted $e^{\frac{1}{2}}$. Therefore the function y_C given by

$$y_C \approx y_O + y_I - e^{\frac{1}{2}} = e^{\frac{1}{2}}(e^{-\frac{1}{2}x} - e^{-2x/\varepsilon}) \tag{6.114}$$

is a uniform approximation on $0 \le x \le 1$, known as the **composite solution**.

The reader should compare the exact solution

$$y(\varepsilon, x) = (e^{\lambda_1 x} - e^{\lambda_2 x})/(e^{\lambda_1} - e^{\lambda_2}), \tag{6.115}$$

where λ_1, λ_2 are the roots of $\varepsilon\lambda^2 + 2\lambda + 1 = 0$: their expansions are

$$\lambda_1 = -\tfrac{1}{2} - \tfrac{1}{8}\varepsilon - \cdots,$$
$$\lambda_2 = -\tfrac{2}{\varepsilon} + \tfrac{1}{2} + \tfrac{1}{8}\varepsilon + \cdots$$

to order ε. Figure 6.8 shows the exact solution and the approximations for $\varepsilon = 0.1$.   ●

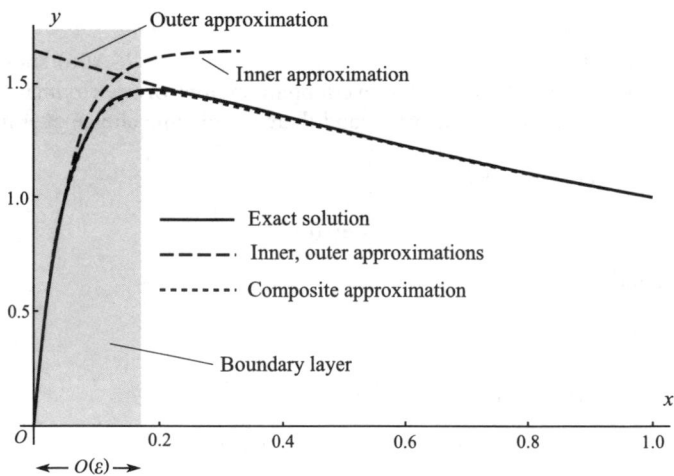

Figure 6.8 Exact solution of (6.96) and its inner, outer and composite approximations are shown for $\varepsilon = 0.1$. The exact solution and the composite approximation are almost indistinguishable.

The expansion and matching process may be extended to take in terms beyond the first-order terms we have considered here. Details are given by Nayfeh (1973).

Example 6.10 *Find the equilibrium points of*

$$\varepsilon \ddot{x} + k\dot{x} - x + x^2 = 0, \quad \dot{x} = y, \quad k > 2\sqrt{\varepsilon}, \tag{6.116}$$

and classify their linear approximations. Consider the initial-value problem with $x(0) = 0$, $\dot{x}(0) = \alpha/\varepsilon$ where $0 < \alpha \leq \frac{1}{2}$. Find matched inner and outer expansions for $0 < \varepsilon \ll 1$.

There are two equilibrium points, at $(0, 0)$ and $(1, 0)$. The former is a saddle point, and the latter a stable node.

To obtain the outer solution use the regular expansion

$$x = x_O = f_0(t) + \varepsilon f_1(t) + \cdots,$$

so that to the lowest order in eqn (6.116), f_0 satisfies

$$k\dot{f}_0 - f_0 + f_0^2 = 0.$$

Hence

$$\int \frac{k\,\mathrm{d}f_0}{f_0(1 - f_0)} = \int \mathrm{d}t = t + C,$$

or

$$\ln\left|\frac{f_0}{1 - f_0}\right| = \frac{1}{k}(t + C),$$

where C is a constant. Then,

$$x = x_O \approx f_0(t) = \frac{1}{1 + e^{-(t+C)/k}}$$

where x_O is the leading order term of the outer solution. Note that $x_O \to 1$ as $t \to \infty$.

For the inner solution let $t = \varepsilon\tau$ so that

$$x'' + kx' - \varepsilon(x - x^2) = 0 \quad (' = \mathrm{d}/\mathrm{d}\tau). \tag{6.117}$$

Now use the expansion

$$x = x_I = g_0(\tau) + \varepsilon g_1(\tau) + \cdots$$

so that, after substitution in (6.117), g_0 satisfies

$$g_0'' + kg_0' = 0.$$

Hence

$$g_0(\tau) = A + Be^{-k\tau}.$$

The initial conditions imply $g_0(0) = 0, g_0'(0) = \alpha$. Therefore

$$B = -A, \quad -Bk = \alpha,$$

which leads to the inner solution x_I:

$$x_I = g_0(\tau) \approx \frac{\alpha}{k}(1 - e^{-k\tau}).$$

The constant C in x_O above is to be determined by matching x_O and x_I. Introduce the time-scaling $t = \eta \varepsilon^\beta$ with $0 < \beta < 1$. Thus $\tau = \eta \varepsilon^{\beta-1}$. In terms of η,

$$x_O = \frac{1}{1 + e^{-(t+C)/k}} = \frac{1}{1 + e^{-C/k}e^{-\varepsilon^\beta \eta/k}}$$

$$= \frac{1}{1 + e^{-C/k}} + O(\varepsilon^\beta)$$

for fixed η. Also

$$x_I = \frac{\alpha}{k}(1 - e^{-k\varepsilon^{\beta-1}}\eta) = \frac{\alpha}{k} + o(1).$$

as $\varepsilon \to 0$. To lowest order these are equal if

$$\frac{1}{1 + e^{-C/k}} = \frac{\alpha}{k} \quad \text{or} \quad e^{-C/k} = \frac{k - \alpha}{\alpha}.$$

Hence, the matched inner and outer expansions to leading orders are

$$x_O = \frac{\alpha}{\alpha + (k - \alpha)e^{-t/k}},$$

$$x_I = \frac{\alpha}{k}(1 - e^{-kt/\varepsilon}).$$

A composite approximation can be obtained from

$$x_C = x_I + x_O - \lim_{\varepsilon \to 0} g_0(t/\varepsilon) \quad (\text{or } \lim_{\varepsilon \to 0} f_0(\varepsilon \tau))$$

$$= \frac{\alpha}{k}(1 - e^{kt/\varepsilon}) + \frac{\alpha}{\alpha + (k - \alpha)e^{-t/k}} - \lim_{\varepsilon \to 0} \frac{\alpha}{k}(1 - e^{-t/\varepsilon})$$

$$= \frac{\alpha}{k}(1 - e^{kt/\varepsilon}) + \frac{\alpha}{\alpha + (k - \alpha)e^{-t/k}} - \frac{\alpha}{k}.$$

Simplifying this expression, we obtain a composite solution valid over $t \geq 0$:

$$x_C = \frac{\alpha}{k}\left[1 - e^{-kt/\varepsilon} + \frac{(k - \alpha)(1 - e^{-t/k})}{k + \alpha(1 - e^{-t/k})}\right].$$

Figure 6.9 shows the exact computed solution and the expansions x_I, x_O and x_C for $\varepsilon = 0.1$, $\alpha = 0.5$ and $k = 0.8$. ●

The previous examples contain versions of **van Dyke's matching rule** which is a useful working method for determining constants, with the advantage of avoiding intermediate variables. We give a two-term version of the method. Suppose that the outer solution to a problem is

$$y_O = f_0(x) + \varepsilon f_1(x),$$

and that the inner approximation is

$$y_I = g_0(\eta) + \varepsilon g_1(\eta), \quad \eta = x/\varepsilon.$$

The rule goes as follows. Put the inner variable η into y_O, and expand in powers of ε to two terms, and then change η back to x:

$$y_O = f_0(\varepsilon\eta) + \varepsilon f_1(\varepsilon\eta) \approx f_0(0) + x f_0'(0) + \varepsilon f_1(0). \tag{6.118}$$

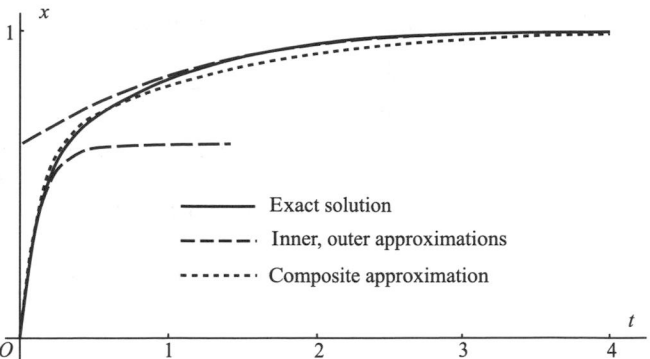

Figure 6.9   Numerical solution of (6.116), and the inner, outer and composite approximations for $\varepsilon = 0.1$, $\alpha = 0.5$, and $k = 0.8$.

In the inner expansion substitute the outer variable x and expand in powers of ε:

$$y_I = g_0(x/\varepsilon) + \varepsilon g_1(x/\varepsilon) \approx p_0(x) + \varepsilon p_1(x). \tag{6.119}$$

Now compare the expansions (6.118) and (6.119). If they can be made *consistent* in terms of x by appropriate choices for any constants then matching is possible, and the inner and outer expansions can be found. It may be that the expansions are not consistent, in which case an alternative form for η might be considered. Further details of Van Dyke's matching rule can be found in the book by Hinch (1991).

Exercise 6.5
Find and classify the equilibrium points of

$$\varepsilon \ddot{x} + k\dot{x} - \sin x = 0, \quad x(0) = 0, \quad \dot{x}(0) = \alpha/\varepsilon$$

assuming $0 < \varepsilon \ll 1$. Find the leading order outer and inner approximations.

Problems

6.1 Work through the details of Example 6.1 to obtain an approximate solution of $\ddot{x} + x = \varepsilon x^3$, with $x(\varepsilon, 0) = 1$, $\dot{x}(\varepsilon, 0) = 0$, with error $O(\varepsilon^3)$ uniformly on $t \geq 0$.

6.2 How does the period obtained by the method of Problem 6.1 compare with that derived in Problem 1.34?

6.3 Apply the method of Problem 6.1 to the equation

$$\ddot{x} + x = \varepsilon x^3 + \varepsilon^2 \alpha x^5$$

with $x(\varepsilon, 0) = 1$, $\dot{x}(\varepsilon, 0) = 0$. Obtain the period to order ε^3, and confirm that the period is correct for the pendulum when the right-hand side is the first two terms in the expansion of $x - \sin x$. (Compare the result of Problem 1.33. To obtain the required equations simply add the appropriate term to the right-hand side of the equation for x_2 in Example 6.1.)

6.4 Use a substitution to show that the case considered in Problem 6.1 covers all boundary conditions $x(\varepsilon, 0) = a$, $\dot{x}(\varepsilon, 0) = 0$.

6.5 The equation for the relativistic perturbation of a planetary orbit is

$$\frac{d^2u}{d\theta^2} + u = k(1 + \varepsilon u^2)$$

(see Problem 5.21). Apply the coordinate perturbation technique to eliminate the secular term in $u_1(\theta)$ in the expansion $u(\varepsilon, \theta) = u_0(\theta) + \varepsilon u_1(\theta) + \cdots$, with $\theta = \phi + \varepsilon T_1(\phi) + \cdots$. Assume the initial conditions $u(0) = k(1+e)$, $du(0)/d\theta = 0$. Confirm that the perihelion of the orbit advances by approximately $2\pi \varepsilon k^2$ in each planetary year.

6.6 Apply the multiple-scale method to van der Pol's equation $\ddot{x} + \varepsilon(x^2 - 1)\dot{x} + x = 0$. Show that, if $x(0) = a$, and $\dot{x}(0) = 0$, then for $t = O(\varepsilon^{-1})$,

$$x = 2\{1 + ((4/a^2) - 1)e^{-\varepsilon t}\}^{-1/2} \cos t.$$

6.7 Apply the multiple-scale method to the equation $\ddot{x} + x - \varepsilon x^3 = 0$, with initial conditions $x(0) = a$, $\dot{x}(0) = 0$. Show that, for $t = O(\varepsilon^{-1})$,

$$x(t) = a\cos\{t(1 - \tfrac{3}{8}\varepsilon a^2)\}.$$

6.8 Obtain the exact solution of Example 6.9, and show that it has the first approximation equal to that obtained by the matching method.

6.9 Consider the problem $\varepsilon y'' + y' + y = 0$, $y(\varepsilon, 0) = 0$, $y(\varepsilon, 1) = 1$, on $0 \leq x \leq 1$, where ε is small and positive.
(a) Obtain the outer approximation

$$y(\varepsilon, x) \approx y_O = e^{1-x}, \quad x \text{ fixed}, \quad \varepsilon \to 0+;$$

and the inner approximation

$$y(\varepsilon, x) \approx y_I - C(1 - e^{-x/\varepsilon}), \quad x = O(\varepsilon), \quad \varepsilon \to 0+,$$

where C is a constant.
(b) Obtain the value of C by matching y_O and y_I in the intermediate region.
(c) Construct from y_O and y_I a first approximation to the solution which is uniform on $0 \leq x \leq 1$.

6.10 Repeat the procedure of Problem 6.9 for the problem

$$\varepsilon y'' + y' + xy = 0, \quad y(0) = 0, \quad y(1) = 1$$

on $0 \leq x \leq 1$.

6.11 Find the outer and inner approximations of

$$\varepsilon y'' + y' + y\sin x = 0, \quad y(0) = 0, \quad y(\tfrac{1}{2}\tau) = 1.$$

6.12 By using the method of multiple scales, with variables x and $\xi = x/\varepsilon$, obtain a first approximation uniformly valid on $0 \leq x \leq 1$ to the solution of

$$\varepsilon y'' + y' + xy = 0, \quad y(\varepsilon, 0) = 0, \quad y(\varepsilon, 1) = 1,$$

on $0 \leq x \leq 1$, with $\varepsilon > 0$. Show that the result agrees to order ε with that of Problem 6.10.

6.13 The steady flow of a conducting liquid between insulated parallel plates at $x = \pm 1$ under the influence of a transverse magnetic field satisfies

$$w'' + Mh' = -1, \quad h'' + Mw' = 0, \quad w(\pm 1) = h(\pm 1) = 0,$$

where, in dimensionless form, w is the fluid velocity, h the induced magnetic field, and M is the Hartmann number. By putting $p = w+h$ and $q = w-h$, find the exact solution. Plot w and h against x for $M = 10$.

This diagram indicates boundary layer adjacent to $x = \pm 1$. From the exact solutions find the outer and inner approximations.

6.14 Obtain an approximation, to order ε and for $t = O(\varepsilon^{-1})$, to the solutions of $\ddot{x} + 2\varepsilon\dot{x} + x = 0$, by using the method of multiple scales with the variables t and $\eta = \varepsilon t$.

6.15 Use the method of multiple scales to obtain a uniform approximation to the solutions of the equation $\ddot{x} + \omega^2 x + \varepsilon x^3 = 0$, in the form

$$x(\varepsilon, t) \approx a_0 \cos[\{\omega_0 + (3\varepsilon a_0^2/8\omega_0)\}t + \alpha],$$

where α is a constant. Explain why the approximation is uniform, and not merely valid for $t = O(\varepsilon^{-1})$.

6.16 Use the coordinate perturbation technique to obtain the first approximation

$$x = \tau^{-1}, \quad t = \tau + \tfrac{1}{2}\varepsilon\tau(1 - \tau^{-2})$$

to the solution of

$$(t + \varepsilon x)\dot{x} + x = 0, \quad x(\varepsilon, 1) = 1, \ 0 \leq x \leq 1.$$

Confirm that the approximation is, in fact, the exact solution, and that an alternative approximation

$$x = \tau^{-1} + \tfrac{1}{2}\varepsilon\tau^{-1}, \quad t = \tau - \tfrac{1}{2}\varepsilon\tau^{-1}$$

is correct to order ε, for fixed τ. Construct a graph showing the exact solution and the approximation for $\varepsilon = 0.2$.

6.17 Apply the method of multiple scales, with variables t and $\eta = \varepsilon t$, to van der Pol's equation $\ddot{x} + \varepsilon(x^2 - 1)\dot{x} + x = 0$. Show that, for $t = O(\varepsilon^{-1})$,

$$x(\varepsilon, t) = \frac{2a_0^{\frac{1}{2}} e^{\frac{1}{2}\varepsilon t}}{\sqrt{(1 + a_0 e^{\varepsilon t})}} \cos(t + \alpha_0) + O(\varepsilon),$$

where a_0 and α_0 are constants.

6.18 Use the method of matched approximations to obtain a uniform approximation to the solution of

$$\varepsilon\left(y'' + \frac{2}{x}y'\right) - y = 0, \quad y(\varepsilon, 0) = 0, \quad y'(\varepsilon, 1) = 1,$$

$(\varepsilon > 0)$ on $0 \leq x \leq 1$. Show that there is a boundary layer of thickness $O(\varepsilon^{1/2})$ near $x = 1$, by putting $1 - x = \xi\phi(\varepsilon)$.

6.19 Use the method of matched approximations to obtain a uniform approximation to the solution of the problem

$$\varepsilon(y'' + y') - y = 0, \quad y(\varepsilon, 0) = 1, \quad y(\varepsilon, 1) = 1,$$

$(\varepsilon > 0)$ given that there are boundary layers at $x = 0$ and $x = 1$. Show that both boundary layers have thickness $O(\varepsilon^{1/2})$. Compare with the exact solution.

6.20 Obtain a first approximation, uniformly valid on $0 \leq x \leq 1$, to the solution of

$$\varepsilon y'' + \frac{1}{1+x}y' + \varepsilon y = 0, \quad \text{with } y(\varepsilon, 0) = 0, \ y(\varepsilon, 1) = 1.$$

6.21 Apply the Lighthill technique to obtain a uniform approximation to the solution of

$$(t + \varepsilon x)\dot{x} + x = 0, \quad x(\varepsilon, 1) = 1, \quad 0 \leq x \leq 1.$$

(Compare Problem 6.16.)

6.22 Obtain a first approximation, uniform on $0 \leq x \leq 1$, to the solution of

$$\varepsilon y' + y = x, \quad y(\varepsilon, 0) = 1,$$

using inner and outer approximations. Compare the exact solution and explain geometrically why the outer approximation is independent of the boundary condition.

6.23 Use the method of multiple scales with variables t and $\eta = \varepsilon t$ to show that, to a first approximation, the response of the van der Pol equation to a 'soft' forcing term described by

$$\ddot{x} + \varepsilon(x^2 - 1)\dot{x} + x = \varepsilon \gamma \cos \omega t$$

is the same as the unforced response assuming that $|\omega|$ is not near 1.

6.24 Repeat Problem 6.23 for $\ddot{x} + \varepsilon(x^2 - 1)\dot{x} + x = \Gamma \cos \omega t$, $\varepsilon > 0$, where $\Gamma = O(1)$ and $|\omega|$ is not near 1. Show that

$$x(\varepsilon, t) = \frac{\Gamma}{1 - \omega^2} \cos \omega t + O(\varepsilon), \quad \Gamma^2 \geq 2(1 - \omega^2)^2;$$

and that for $\Gamma^2 < 2(1 - \omega^2)^2$

$$x(\varepsilon, t) = 2\left\{1 - \frac{\Gamma^2}{2(1 - \omega^2)^2}\right\}^{1/2} \cos t + \frac{\Gamma}{1 - \omega^2} \cos \omega t + O(\varepsilon).$$

6.25 Apply the matching technique to the damped pendulum equation

$$\varepsilon \ddot{x} + \dot{x} + \sin x = 0, \quad x(\varepsilon, 0) = 1, \quad \dot{x}(\varepsilon, 0) = 0.$$

for ε small and positive. Show that the inner and outer approximations given by

$$x_{\mathrm{I}} = 1, \quad x_{\mathrm{O}} = 2 \tan^{-1}\{e^{-t} \tan \tfrac{1}{2}\}.$$

(The pendulum has strong damping and strong restoring action, but the damping dominates.)

6.26 The equation for a tidal bore on a shallow stream is

$$\varepsilon \frac{d^2 \eta}{d\xi^2} - \frac{d\eta}{d\xi} - \eta + \eta^2 = 0;$$

where (in appropriate dimensions) η is the height of the free surface, and $\xi = x - ct$, where c is the wave speed. For $0 < \varepsilon \ll 1$, find the equilibrium points for the equation and classify them according to their linear approximations. Apply the coordinate perturbation method to the equation for the phase paths,

$$\varepsilon \frac{dw}{d\eta} = \frac{w + \eta - \eta^2}{w}, \quad \text{where } w = \frac{d\eta}{d\xi}$$

and show that

$$w = -\zeta + \zeta^2 + O(\varepsilon^2), \quad \eta = \zeta - \varepsilon(-\zeta + \zeta^2) + O(\varepsilon^2).$$

Confirm that, to this degree of approximation, a separatrix from the origin reaches the other equilibrium point. Interpret the result in terms of the shape of the bore.

6.27 The function $x(\varepsilon, t)$ satisfies the differential equation $\varepsilon \ddot{x} + x\dot{x} - x = 0$ $(t \geq 0)$ subject to the initial conditions $x(0) = 0$, $\dot{x}(0) = 1/\varepsilon$. To leading order, obtain inner and outer approximations to the solution for small ε. Show that the composite solution is

$$x_{\mathrm{C}} = t + \sqrt{2} \tanh(t/(\varepsilon\sqrt{2})).$$

6.28 Consider the initial-value problem

$$\varepsilon \ddot{x} + \dot{x} = e^{-t}, \quad x(0) = 0, \quad \dot{x}(0) = 1/\varepsilon \quad (0 < \varepsilon \ll 1).$$

Find inner and outer expansions for x, and confirm that the outer expansion to two terms is

$$x_O = 2 - e^{-t} - \varepsilon e^{-t}.$$

Compare computed graphs of the composite expansion and the *exact* solution of the differential equation for $\varepsilon = 0.1$ and for $\varepsilon = 0.25$.

6.29 Investigate the solution of the initial/boundary-value problem

$$\varepsilon^3 \dddot{x} + \varepsilon \ddot{x} + \dot{x} + x = 0, \quad 0 < \varepsilon \ll 1,$$

with $x(1) = 1$, $x(0) = 0$, $\dot{x}(0) = 1/\varepsilon^2$ using matched approximations. Start by finding, with a regular expansion, the outer solution x_O and an inner solution x_I using $t = \varepsilon\tau$. Confirm that x_I cannot satisfy the conditions at $t = 0$. The boundary-layer thickness $O(\varepsilon)$ at $t = 0$ is insufficient for this problem. Hence we create an *additional* boundary layer of thickness $O(\varepsilon^2)$, and a further time scale η where $t = \varepsilon^2 \eta$. Show that the leading order equation for the **inner–inner approximation** x_{II} is

$$x_{II}''' + x_{II}'' = 0.$$

and confirm that the solution can satisfy the conditions at $t = 0$. Finally match the expansions x_{II} and x_I, and the expansions x_I and x_O. Show that the approximations are

$$x_O = e^{1-t}, \quad x_I = e + (1 - e)e^{-t/\varepsilon}, \quad x_{II} = 1 - e^{-t/\varepsilon^2},$$

to leading order.

Explain why the composite solution is

$$x_C = e^{1-t} + (1 - e)e^{-t/\varepsilon} - e^{-t/\varepsilon^2}.$$

Comparison between the numerical solution of the differential equation and the composite solution is shown in Fig. 6.10. The composite approximation could be improved by taking all approximations to include $O(\varepsilon)$ terms.

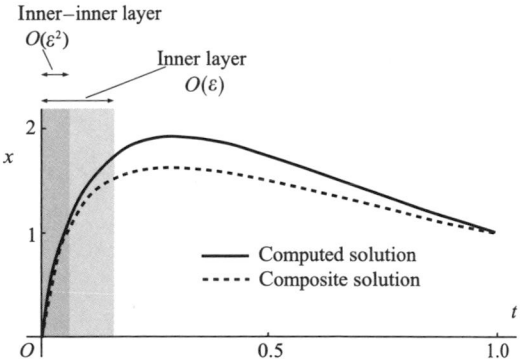

Figure 6.10 Computed solution and the leading composite approximation for $\varepsilon = 0.2$ showing an inner–inner boundary layer for Problem 6.29.

6.30 Let $y(x, \varepsilon)$ satisfy

$$\varepsilon y'' + y' = x,$$

where $y(0, \varepsilon) = 0$, $y(1, \varepsilon) = 1$. Find the outer and inner expansions to order ε using the inner variable $\eta = x/\varepsilon$. Apply the van Dyke matching rule to show that the inner expansion is

$$y_I \approx (\tfrac{1}{2} + \varepsilon)(1 - e^{x/\varepsilon}).$$

6.31 In Example 6.9, a composite solution of

$$\varepsilon \frac{d^2 y}{dx^2} + 2\frac{dy}{dx} + y = 0, \quad y(0) = 0, \quad y(1) = 1,$$

valid over the interval $0 \le x \le 1$, was found to be (eqn (6.114))

$$y_C = e^{\frac{1}{2}}(e^{-\frac{1}{2}x} - e^{-2x/\varepsilon})$$

using matched inner and outer approximations. What *linear* constant coefficient second-order differential equation and boundary conditions does y_C satisfy *exactly*?

7 Forced oscillations: harmonic and subharmonic response, stability, and entrainment

We consider second-order differential equations of the general form $\ddot{x} + f(x, \dot{x}) = F \cos \omega t$, which have a periodic forcing term. When the equation is linear the structure of its solutions is very simple. There are two parts combined additively. One part, the 'free oscillation' is a linear combination of the solutions of the homogeneous equation and involves the initial conditions. The second part, the 'forced solution', is proportional to F and is independent of initial conditions. When damping is present the free oscillation dies away in time. This independence of the free and the forced components, and the ultimate independence of initial conditions when there is damping, allow a very restricted range of phenomena: typically the only things to look at are the amplitude and phase of the ultimate response, the rate of decay of the free oscillation, and the possibility of resonance. When the equation is nonlinear, on the other hand, there is no such simple separation possible, and the resulting interaction between the free and forced terms, especially when a self-excited motion is possible, and the enduring importance of the initial conditions in some cases, generates a range of entirely new phenomena. The present chapter is concerned with using the method of harmonic balance (Chapter 4) and the perturbation method (Chapter 5) to obtain approximate solutions which clearly show these new features.

7.1 General forced periodic solutions

Consider, for definiteness, Duffing's equation

$$\ddot{x} + k\dot{x} + \alpha x + \beta x^3 = \Gamma \cos \omega t, \tag{7.1}$$

of which various special cases have appeared in earlier chapters. Suppose that $x(t)$ is a periodic solution with period $2\pi/\lambda$. Then $x(t)$ can be represented by a Fourier series for all t:

$$x(t) = a_0 + a_1 \cos \lambda t + b_1 \sin \lambda t + a_2 \cos 2\lambda t + \cdots. \tag{7.2}$$

If this series is substituted into (7.1) the nonlinear term x^3 is periodic, and so generates another Fourier series. When the contributions are assembled eqn (7.1) takes the form

$$A_0 + A_1 \cos \lambda t + B_1 \sin \lambda t + A_2 \cos 2\lambda t + \cdots = \Gamma \cos \omega t \tag{7.3}$$

for all t, the coefficients being functions of $a_0, a_1, b_1, a_2, \ldots$. Matching the two sides gives, in principle, an infinite set of equations for $a_0, a_1, a_2, \ldots$ and enables λ to be determined. For example, the obvious matching

$$\lambda = \omega \quad \text{and} \quad A_1 = \Gamma, \quad A_0 = B_1 = A_2 = \cdots = 0$$

resembles the harmonic balance method of Section 4.4; though in the harmonic balance method a drastically truncated series (7.2), consisting of only the single term $a_1 \cos \lambda t$, is employed. The above matching leads to a harmonic (or **autoperiodic**) response of angular frequency ω.

A less obvious matching can sometimes be achieved when

$$\lambda = \omega/n \quad (n \text{ an integer});$$

$$A_n = \Gamma;$$

$$A_i = 0 \, (i \neq n); \quad B_i = 0 \text{ (all } i).$$

If solutions of this set exist, the system will deliver responses of period $2\pi n/\omega$ and angular frequency ω/n. These are called **subharmonics** of order $\frac{1}{2}, \frac{1}{3}, \ldots$. Not all of these actually occur (see Section 7.7 and Problems 7.16 and 7.17).

It is important to realize where the terms in (7.3) come from: there is **throwback** from terms of high order in (7.2) to contribute to terms of low order in (7.3). For example, in the expansion of x^3:

$$x^3(t) = (a_0 + \cdots + a_{11} \cos 11\omega t + \cdots + a_{21} \cos 21\omega t + \cdots)^3,$$

one of the terms is

$$3a_{11}^2 a_{21} \cos^2 11\omega t \cos 21\omega t = 3a_{11}^2 a_{21} (\tfrac{1}{2} \cos 21\omega t + \tfrac{1}{4} \cos 43\omega t + \tfrac{1}{4} \cos \omega t),$$

which includes a term in $\cos \omega t$. Normally we assume that terms in (7.2) above a certain small order are negligible, that is, that the coefficients are small. Hopefully, then, the combined throwback effect, as measured by the modification of coefficients by high-order terms, will be small.

An alternative way of looking at the effect of a nonlinear term is as a **feedback**. Consider the undamped form of (7.1) (with $k = 0$) written as

$$\mathcal{L}(x) \equiv \left(\frac{d^2}{dt^2} + \alpha \right) x = -\beta x^3 + \Gamma \cos \omega t.$$

We can represent this equation by the block diagram shown in Fig. 7.1. Regard $\Gamma \cos \omega t$ as the **input** to the system and $x(t)$ as the **output**. The box A represents the inverse operator $\mathcal{L}^{-1}$ which solves the linear equation $\ddot{x} + \alpha x = f(t)$ for a given input f and for assigned initial conditions. Here, the input to A is equal to the current value of $-\beta x^3 + \Gamma \cos \omega t$. Its output is $x(t)$. Suppose the output from A is assumed to be simple, containing only a few harmonic components. Then B generates a shower of harmonics of higher and possibly of lower orders which are fed back into A. The higher harmonics are the most attenuated on passing through A (roughly like n^{-2} where n is the order). It is therefore to be expected that a satisfactory consistency between the inputs might be obtained by a representation of $x(t)$ in terms only of

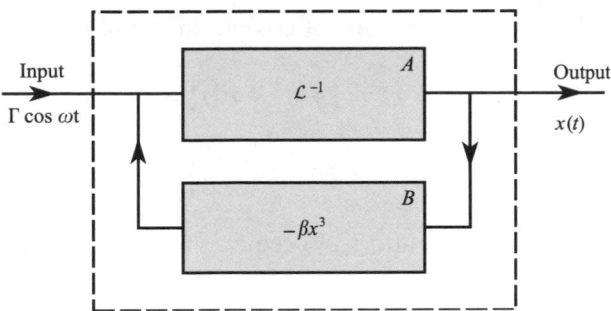

Figure 7.1   Feedback diagram for $\ddot{x} + \alpha x + \beta x^3 = \Gamma \cos \omega t$.

the lowest harmonics present. The low-order approximation should be most adequate when the lowest harmonic present is **amplified** by $\mathcal{L}^{-1}$; that is, when $\omega^2 \approx n^2 \alpha$, being a condition of near resonance.

7.2   Harmonic solutions, transients, and stability for Duffing's equation

As a direct illustration of the method described in Section 7.1, consider the undamped Duffing equation (7.1) in the form

$$\ddot{x} + x + \beta x^3 = \Gamma \cos \omega t, \tag{7.4}$$

where $\Gamma > 0$. As an approximation to the solution we use the truncated Fourier series

$$x(t) = a \cos \omega t + b \sin \omega t. \tag{7.5}$$

This form allows for a possible phase difference between the forcing term and the solution. The omission of the constant term involves foreknowledge that the only relevant solutions of reasonably small amplitude will have zero mean value (see Section 4.3). Then

$$\ddot{x}(t) = -a\omega^2 \cos \omega t - b\omega^2 \sin \omega t \tag{7.6}$$

and

$$
\begin{aligned}
x^3(t) &= a^3 \cos^3 \omega t + 3a^2 b \cos^2 \omega t \sin \omega t + 3ab^2 \cos \omega t \sin^2 \omega t + b^3 \sin^3 \omega t \\
&= \tfrac{3}{4} a(a^2 + b^2) \cos \omega t + \tfrac{3}{4} b(a^2 + b^2) \sin \omega t \\
&\quad + \tfrac{1}{4} a(a^2 - 3b^2) \cos 3\omega t + \tfrac{1}{4} b(3a^2 - b^2) \sin 3\omega t.
\end{aligned} \tag{7.7}
$$

(See Appendix E for trigonometric identities.) We disregard the terms in $\cos 3\omega t$, $\sin 3\omega t$ on the grounds that, regarded as feedback input to the differential equation, the attenuation will be large compared with that of terms having the fundamental frequency. When (7.5), (7.6), (7.7) are substituted into (7.4) and the coefficients of $\cos \omega t, \sin \omega t$ are matched we obtain

$$b\left\{(w^2 - 1) - \tfrac{3}{4}\beta(a^2 + b^2)\right\} = 0, \qquad (7.8)$$

$$a\left\{(w^2 - 1) - \tfrac{3}{4}\beta(a^2 + b^2)\right\} = -\Gamma. \qquad (7.9)$$

The only admissible solution of (7.8) and (7.9) requires $b = 0$, and the corresponding values of a are the solutions of (7.9):

$$\tfrac{3}{4}\beta a^3 - (\omega^2 - 1)a - \Gamma = 0. \qquad (7.10)$$

The roots are given by the intersections, in the plane of a and z, of the curves

$$z = -\Gamma, \quad z = -\tfrac{3}{4}\beta a^3 + (\omega^2 - 1)a.$$

From Fig. 7.2 it can be seen that, for $\beta < 0$, there are three solutions for Γ small and one for Γ larger when $\omega^2 < 1$, and that when $\omega^2 > 1$ there is one solution only. This reproduces the results of Sections 5.4 and 5.5. The oscillations are in phase with the forcing term when the critical value of a is positive and out of phase by a half cycle when a is negative.

We do not yet know the *stability* of these various oscillations and it is necessary to decide this matter since an unstable oscillation will not occur in practice. To investigate stability we shall look at the 'transient' states, which lead to or away from periodic states, by supposing the coefficients a and b to be slowly varying functions of time, at any rate near to periodic states. Assume that

$$x(t) = a(t) \cos \omega t + b(t) \sin \omega t, \qquad (7.11)$$

where a and b are **slowly varying amplitudes** (compared with $\cos \omega t$ and $\sin \omega t$).

Then

$$\dot{x}(t) = (\dot{a} + \omega b) \cos \omega t + (-\omega a + \dot{b}) \sin \omega t, \qquad (7.12)$$

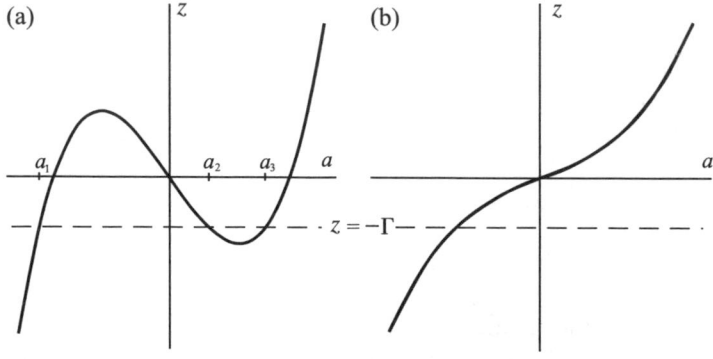

Figure 7.2   Graph of $z = -\tfrac{3}{4}\beta a^3 + (\omega^2 - 1)a$ for $\beta < 0$ and (a) $\omega^2 < 1$; (b) $\omega^2 > 1$.

and (neglecting $\ddot{a}$, $\ddot{b}$)

$$\ddot{x}(t) \approx (-\omega^2 a + 2\omega\dot{b})\cos\omega t + (-2\omega\dot{a} - \omega^2 b)\sin\omega t. \tag{7.13}$$

Also, as in (7.7),

$$x^3(t) = \tfrac{3}{4}a(a^2 + b^2)\cos\omega t + \tfrac{3}{4}b(a^2 + b^2)\sin\omega t$$

$$+ \text{ harmonics in } \cos 3\omega t, \sin 3\omega t. \tag{7.14}$$

As before we shall ignore the terms in $\cos 3\omega t$, $\sin 3\omega t$.

When (7.12), (7.13), (7.14) are substituted into the differential equation (7.4), and the terms are rearranged, we have

$$[2\omega\dot{b} - a\{(\omega^2 - 1) - \tfrac{3}{4}\beta(a^2 + b^2)\}]\cos\omega t$$

$$+ [-2\omega\dot{a} - b\{(\omega^2 - 1) - \tfrac{3}{4}\beta(a^2 + b^2)\}]\sin\omega t = \Gamma\cos\omega t. \tag{7.15}$$

Appealing again to the supposition that a and b are slowly varying we may approximately match the coefficients of $\cos\omega t$ and $\sin\omega t$, giving the **autonomous system**

$$\dot{a} = -\frac{1}{2\omega}b\{(\omega^2 - 1) - \tfrac{3}{4}\beta(a^2 + b^2)\} \equiv A(a,b), \quad \text{say}; \tag{7.16}$$

$$\dot{b} = \frac{1}{2\omega}a\{(\omega^2 - 1) - \tfrac{3}{4}\beta(a^2 + b^2)\} + \frac{\Gamma}{2\omega} \equiv B(a,b), \quad \text{say}. \tag{7.17}$$

Initial conditions are given in terms of those for the original equation, (7.4), by (assuming that $\dot{a}(0)$ is small)

$$a(0) = x(0), \quad b(0) = \dot{x}(0)/\omega. \tag{7.18}$$

The phase plane for a, b in the system above is called the **van der Pol plane**. The equilibrium points, given by $A(a,b) = B(a,b) = 0$, represent the steady periodic solutions already obtained. The other paths correspond to solutions of (7.4) which are non-periodic in general. The phase diagram computed for a particular case is shown in Fig. 7.3. The point $(a_3, 0)$ is a saddle and represents an unstable oscillation, and $(a_1, 0)$, $(a_2, 0)$ are centres.

The equilibrium points may be analysed algebraically as follows. Consider a case when there are three equilibrium points

$$\omega^2 < 1, \quad \beta < 0 \tag{7.19}$$

and let a_0 represent any one of the values $a = a_1, a_2$, or a_3 of Fig. 7.2(a). Putting

$$a = a_0 + \xi, \tag{7.20}$$

the local linear approximation to (7.16), (7.17) is

$$\dot{\xi} = A_1(a_0, 0)\xi + A_2(a_0, 0)b,$$
$$\dot{b} = B_1(a_0, 0)\xi + B_2(a_0, 0)b, \tag{7.21}$$

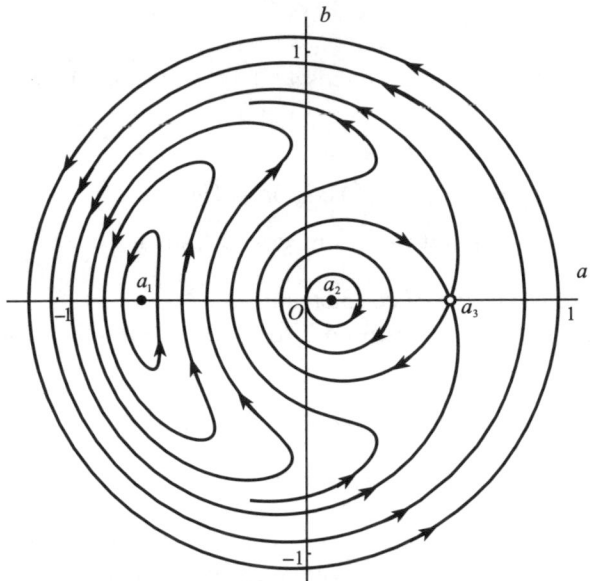

Figure 7.3 Paths in the van der Pol phase plane for the undamped pendulum (see eqns (7.16), (7.17)) with parameter values $\Gamma = 0.005$, $\omega = 0.975$, $\beta = -0.167$. The equilibrium points are located on $b = 0$ at $a_1 = -0.673$, $a_2 = 0.104$, $a_3 = 0.569$.

where $A_1(a, b) = \partial A(a, b)/\partial a$, and so on. It is easy to confirm that

$$A_1(a_0, 0) = B_2(a_0, 0) = 0,$$

$$A_2(a_0, 0) = -\frac{\omega^2 - 1}{2\omega} + \frac{3\beta a_0^2}{8\omega} = -\frac{z_0}{2\omega a_0},$$

(where, in Fig. 7.2(a), z_0 is the ordinate at a_0), and that

$$B_1(a_0, 0) = s_0/(2\omega),$$

where s_0 is the slope of the curve in Fig. 7.2(a) at a_0. Therefore (7.21) can be written

$$\dot{\xi} = -\frac{z_0}{2\omega a_0}b, \quad \dot{b} = \frac{s_0}{2\omega}\xi.$$

By considering the signs of a_0, z_0, s_0 in Fig. 7.2(a) it can be seen that $(a_1, 0), (a_2, 0)$ are centres for the linear system (7.21), and that $(a_3, 0)$ is a saddle. The saddle will never be observed exactly, though if a state can be set up near enough to this point, $\dot{a}$ and $\dot{b}$ will be very small and a nearly periodic motion at the forcing frequency may linger long enough to be observed.

We may introduce a damping term into (7.4) so that it becomes

$$\ddot{x} + k\dot{x} + x + \beta x^3 = \Gamma \cos \omega t, \quad k > 0. \tag{7.22}$$

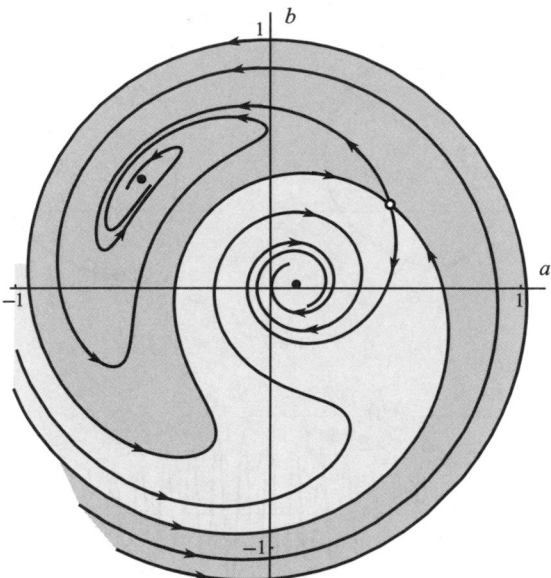

Figure 7.4 Paths in the van der Pol phase plane for the damped pendulum (7.22) with $\Gamma = 0.005$, $\omega = 0.975$, $k = 0.005$, $\beta = -0.167$. The equilibrium points are located at $(0.479, 0.329)$, $(0.103, 0.010)$, $(-0.506, 0.429)$. The light and dark shading indicate the domains of initial points for the two spirals.

The equilibrium points (Problem 7.10) are given by

$$b\{\omega^2 - 1 - \tfrac{3}{4}\beta(a^2 + b^2)\} + k\omega a = 0,$$

$$a\{\omega^2 - 1 - \tfrac{3}{4}\beta(a^2 + b^2)\} - k\omega b = -\Gamma;$$

so that, after squaring and adding these equations,

$$r^2\left\{k^2\omega^2 + \left(\omega^2 - 1 - \tfrac{3}{4}\beta r^2\right)^2\right\} = \Gamma^2, \quad r = \sqrt{(a^2 + b^2)}. \tag{7.23}$$

and the closed paths of Fig. 7.3 become spirals. The present theory predicts that one of two stable periodic states is approached from any initial state (we shall later show, however, that from some initial states we arrive at sub-harmonics). A calculated example is shown in Fig. 7.4. Typical response curves are shown in Fig. 5.3; the governing equations for the equilibrium points, and the conclusions about stability, being the same as in Chapter 5.

Referring back to Fig. 7.3, notice that the existence of closed curves, which implies periodicity of $a(t)$ and $b(t)$, does not in general imply periodicity of $x(t)$, for since $x(t) = a(t)\cos\omega t + b(t)\sin\omega t$ the motion is periodic only if the period of $a(t)$ and $b(t)$ is a rational multiple of $2\pi/\omega$. Though this may occur infinitely often in a family of closed paths, the theory is too rough to decide which periodic motions, possibly of very long period, are actually present. Normally the closed paths represent special cases of 'almost periodic motion'.

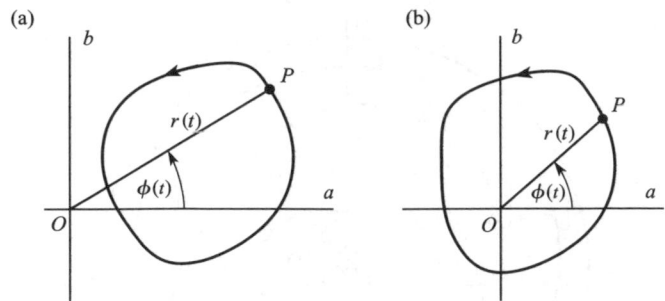

Figure 7.5

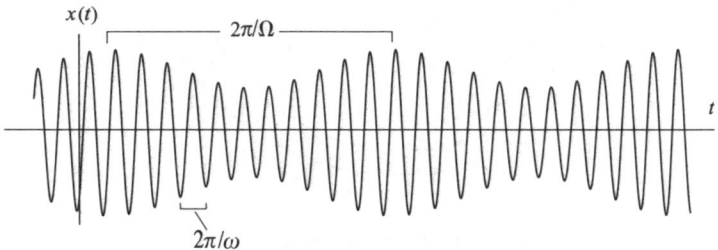

Figure 7.6   A time solution showing beats.

If P is a representative point on a given closed path with polar representation $r(t)$, $\phi(t)$, then (7.5) can be written as

$$x(t) = r(t)\cos(\omega t - \phi(t)).$$

Let $2\pi/\Omega$ be the period of $a(t)$, $b(t)$, and hence of $r(t)$, $\phi(t)$, on the path. Then since a and b are slowly varying, $\Omega \ll \omega$. When the path does not encircle the origin (Fig. 7.5(a)) the phase ϕ is restored after each complete circuit. The effect is of an oscillation with frequency ω modulated by a slowly varying periodic amplitude. The phenomenon of 'beats' (Fig. 7.6) appears.

If the path encircles the orgin as in Fig. 7.5(b) the phase increases by 2π in every circuit, which effectively modulates the approximate 'frequency' of the oscillation from ω to $\omega + \Omega$.

Exercise 7.1

In the resonant case $\omega = 1$, the amplitude equations (7.16) and (7.17) in the (a, b) van der Pol plane are

$$\dot{a} = \tfrac{3}{8}\beta b(a^2 + b^2), \quad \dot{b} = -\tfrac{3}{8}\beta a(a^2 + b^2) + \tfrac{1}{2}\Gamma.$$

Using polar variables r and θ, show that phase paths in the (a, b) plane are given by the equation

$$r(r^3 - \mu \cos \theta) = \text{constant},$$

when $\mu = 16\Gamma/(3\beta)$. Sketch or compute the phase diagram for $\mu = -1$.

7.3 The jump phenomenon

Equation (7.22) for the damped Duffing oscillator has periodic solutions which are approximately of the form $a \cos \omega t + b \sin \omega t$, where the amplitude $r = \sqrt{(a^2 + b^2)}$ satisfies (see eqn (7.23))

$$r^2 \left\{ k^2 \omega^2 + (\omega^2 - 1 - \tfrac{3}{4}\beta r^2)^2 \right\} = \Gamma^2 \quad (k > 0, \, \omega > 0, \, \Gamma > 0).$$

Suppose that $\beta < 0$, as in the case of the pendulum, and we put $\rho = -\beta r^2$, $\gamma = \Gamma \sqrt{(-\beta)}$. The amplitude equation can then be expressed as

$$\gamma^2 = G(\rho) = \rho \left\{ k^2 \omega^2 + (\omega^2 - 1 + \tfrac{3}{4}\rho)^2 \right\}. \tag{7.24}$$

We could represent this amplitude equation as a surface in a four-dimensional parameter space with variables ρ, k, ω, γ, but we shall simplify the representation by assuming that the damping parameter k is specified (interesting phenomena are generally associated with variations in the forcing frequency ω and the forcing amplitude Γ parameters).

The amplitude function $G(\rho)$ is a cubic in ρ, which can be rewritten as

$$G(\rho) = \tfrac{9}{16}\rho^3 + \tfrac{3}{2}(\omega^2 - 1)\rho^2 + \{k^2\omega^2 + (\omega^2 - 1)^2\}\rho.$$

Since $G(0) = 0$ and $G(\rho) \to \infty$ as $\rho \to \infty$, $G(\rho) = \gamma^2$ must have at least one positive root (we are only interested in $\rho \geq 0$). There will be three real roots for *some* parameter values if the equation $G'(\rho) = 0$ has two distinct solutions for $\rho \geq 0$. Thus

$$G'(\rho) = \tfrac{27}{16}\rho^2 + 3(\omega^2 - 1)\rho + k^2\omega^2 + (\omega^2 - 1)^2 = 0 \tag{7.25}$$

has two real roots

$$\rho_1, \rho_2 = \tfrac{8}{9}(1 - \omega^2) \pm \tfrac{4}{9}\sqrt{[(1 - \omega^2)^2 - 3k^2\omega^2]}, \tag{7.26}$$

provided that $|1 - \omega^2| > \sqrt{3}k\omega$. From (7.25) and (7.26)

$$\rho_1 \rho_2 = \tfrac{16}{27}\{(1 - \omega^2)^2 + k^2\omega^2\} > 0,$$

$$\rho_1 + \rho_2 = \tfrac{16}{9}(1 - \omega^2).$$

We are only interested in $\rho_1, \, \rho_2 \geq 0$, and two real positive roots are only possible if

$$0 < \omega < 1 \quad \text{and} \quad \omega^2 + k\omega\sqrt{3} - 1 > 0,$$

which are both satisfied if

$$0 < \omega < \tfrac{1}{2}\{\sqrt{(3k^2 + 4)} - k\sqrt{3}\}.$$

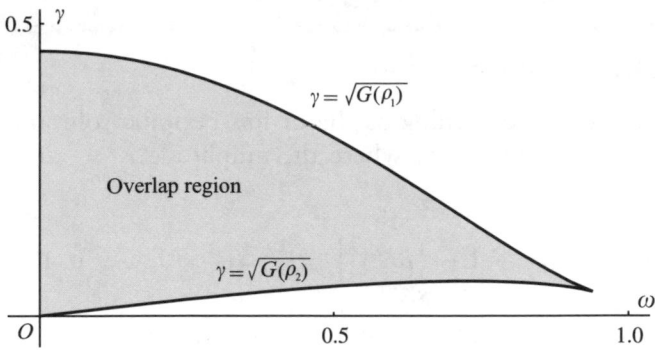

Figure 7.7   Overlap region for the surface $G(\rho) = \gamma^2$ with $k = 0.1$.

Thus $G(\rho) = \gamma^2$ has three distinct real roots if

$$G(\rho_2) < \gamma^2 < G(\rho_1) \quad \text{and} \quad \omega < \tfrac{1}{2}\{\sqrt{(3k^2 + 4)} - k\sqrt{3}\}.$$

The boundary of the overlap on the surface $G(\rho) = \gamma^2$ is shown in Fig. 7.7 projected on to the (γ, ω) plane. The region shaded in Fig. 7.7 has a **cusp** at

$$\omega = \tfrac{1}{2}\{\sqrt{(3k^2 + 4)} - k\sqrt{3}\}.$$

Sections through $\omega = 0.85$ and $\gamma = 0.05$ are shown in Fig. 7.8(a) and (b). The surface $\gamma = \sqrt{G(\rho)}$ drawn in the (γ, ω, ρ) space (essentially showing response amplitude in terms of forcing amplitude and frequency) is shown in Fig. 7.9 in the neighbourhood of the cusp. The graphs in Fig. 7.8(a) and (b) are vertical sections of this surface at $\omega = 0.85$ and at $\gamma = 0.05$, respectively.

The theory of Section 7.2 predicts that for values of γ and ω where there exists only a single response then this response is stable; and that where there are three responses, the oscillations with greatest and least amplitudes are stable and the remaining intermediate one is unstable (Fig. 7.8(a) and (b)). In Fig. 7.9, the surface under the fold corresponds to unstable oscillations which will not be attainable. The surface exhibits what is known as a **fold catastrophe** which leads to the so-called **jump phenomenon** (see Poston and Stewart (1978) for an extensive account of catastrophe theory).

To illustrate the consequences of this fold consider what happens in an experiment in which γ is held constant at a value which intersects the fold as in Fig. 7.8(b), and the applied frequency is varied in such a way that it crosses the edges of the fold.

Begin the experiment at applied frequency $\omega = \omega_1$, steadily increase to ω_2, the bring it back again to ω_1. Starting at A_1, the response moves to A' at frequency ω'. As ω increases beyong ω', the oscillation will, after some irregular motion, settle down to the codition represented by B'; that is to say, the amplitude 'jumps' at the critical frequency ω'. After this it follows the smooth curve from B' to A_2. On the way back the response point moves along the upper curve as far as A''. Here it must drop to B'' on the lower curve and then back to A_1.

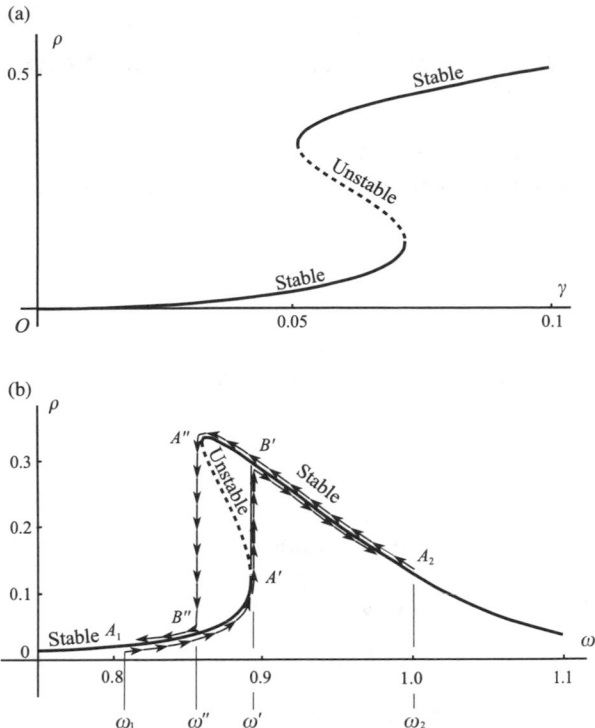

Figure 7.8 (a) Amplitude ρ versus γ for $k = 0.1$ and $\omega = 0.85$; (b) amplitude ρ versus ω for $k = 0.1$ and $\gamma = 0.05$.

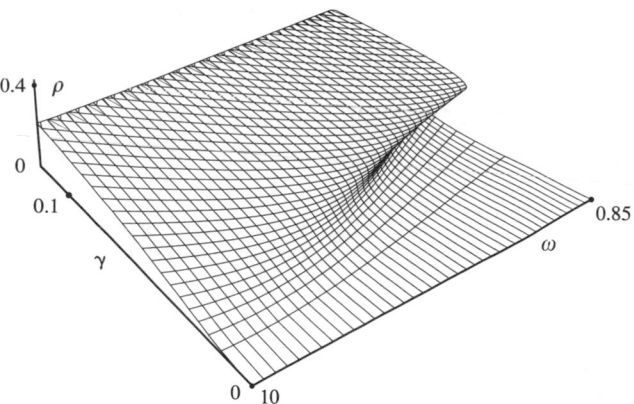

Figure 7.9 The amplitude surface $\gamma = \sqrt{[G(\rho)]}$ showing ρ in terms of γ and ω, and the fold.

Exercise 7.2
Find the forcing amplitude γ in terms of k of the Duffing equation at the cusp where the frequency is $\omega = \frac{1}{2}\{\sqrt{(3k^2 + 4)} - k\sqrt{3}\}$. Compute how the amplitude γ varies with k.

7.4   Harmonic oscillations, stability, and transients for the forced van der Pol equation

In this section we look at the effects of harmonic forcing on a nonlinear system which also has an autonomous **self-excited oscillation**. The equation considered is the **forced van der Pol oscillator**

$$\ddot{x} + \varepsilon(x^2 - 1)\dot{x} + x = \Gamma \cos \omega t, \tag{7.27}$$

where $\varepsilon > 0$. In the absence of a forcing term there is single, self-excited oscillation which is approached from all non-zero initial conditions. We look for responses approximately of the form

$$x(t) = a(t) \cos \omega t + b(t) \sin \omega t, \tag{7.28}$$

where a, b are slowly varying functions. Then, neglecting $\ddot{a}$, $\ddot{b}$,

$$\dot{x}(t) = (\dot{a} + \omega b) \cos \omega t + (-\omega a + \dot{b}) \sin \omega t, \tag{7.29a}$$

$$\ddot{x}(t) = (-\omega^2 a + 2\omega \dot{b}) \cos \omega t + (-2\omega \dot{a} - \omega^2 b) \sin \omega t. \tag{7.29b}$$

After some algebra (see Appendix E for trigonometric reduction formulas),

$$(x^2 - 1)\dot{x} = \{(\tfrac{3}{4}a^2 + \tfrac{1}{4}b^2 - 1)\dot{a} + \tfrac{1}{2}ab\dot{b} - \omega b(1 - \tfrac{1}{4}a^2 - \tfrac{1}{4}b^2)\} \cos \omega t$$
$$+ \{\tfrac{1}{2}ab\dot{a} + (\tfrac{1}{4}a^2 + \tfrac{3}{4}b^2 - 1)\dot{b} + \omega a(1 - \tfrac{1}{4}a^2 - \tfrac{1}{4}b^2)\} \sin \omega t$$
$$+ \text{higher harmonics.} \tag{7.30}$$

Finally substitute (7.28) to (7.30) into (7.27), ignoring the higher harmonics. As in Section 7.2, the equation is satisfied approximately if the coefficients of $\cos \omega t$, $\sin \omega t$ are equated to zero. We obtain

$$(2\omega - \tfrac{1}{2}\varepsilon ab)\dot{a} + \varepsilon(1 - \tfrac{1}{4}a^2 - \tfrac{3}{4}b^2)\dot{b} = \varepsilon \omega a(1 - \tfrac{1}{4}r^2) - (\omega^2 - 1)b, \tag{7.31a}$$

$$-\varepsilon(1 - \tfrac{3}{4}a^2 - \tfrac{1}{4}b^2)\dot{a} + (2\omega + \tfrac{1}{2}\varepsilon ab)\dot{b} = (\omega^2 - 1)a + \varepsilon \omega b(1 - \tfrac{1}{4}r^2) + \Gamma, \tag{7.31b}$$

where

$$r = \sqrt{(a^2 + b^2)}.$$

The equilibrium points occur when $\dot{a} = \dot{b} = 0$, that is, when the right-hand sides are zero. We can reduce the parameters $(\omega, \Gamma, \varepsilon)$ to two by multiplying through by $1/\varepsilon\omega$ and putting

$$\nu = (\omega^2 - 1)/\varepsilon\omega, \quad \gamma = \Gamma/\varepsilon\omega; \tag{7.32}$$

(the quantity v is a measure of the '**detuning**'). The equation for the equilibrium points become

$$a(1 - \tfrac{1}{4}r^2) - vb = 0, \tag{7.33a}$$

$$va + b(1 - \tfrac{1}{4}r^2) = -\gamma. \tag{7.33b}$$

By squaring and adding these equations we obtain

$$r^2 \left\{ v^2 + \left(1 - \tfrac{1}{4}r^2\right)^2 \right\} = \gamma^2, \tag{7.34}$$

and when this is solved the values of a, b can be recovered from (7.33a,b). Equation (7.34) may have either 1 or 3 real roots (since $r > 0$) depending on the parameter values v and γ (or ω, Γ, ε). This dependence can be conveniently represented on a single figure, the 'response diagram', Fig. 7.10(a).

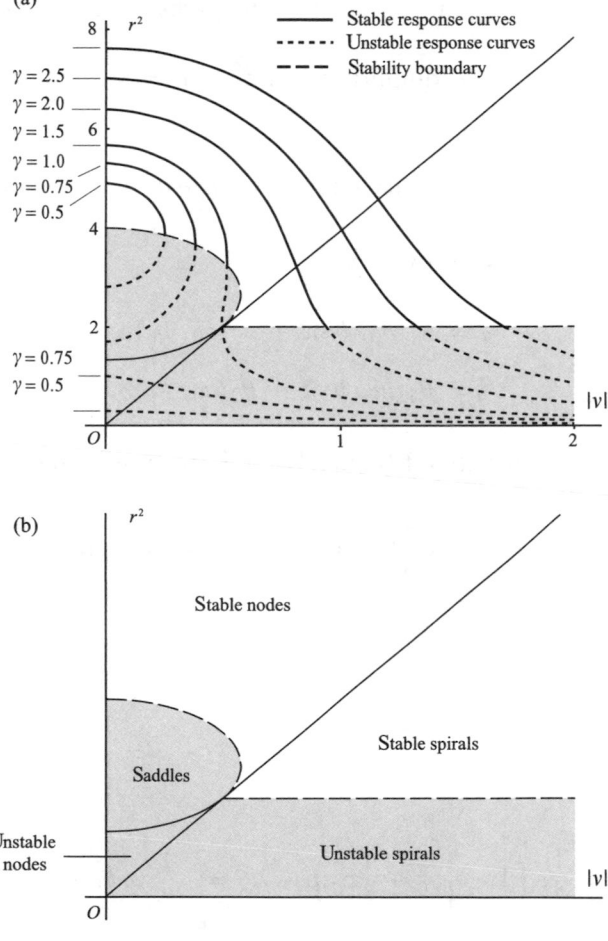

Figure 7.10 (a) response curves and stability region for the forced van der Pol equation; (b) classification of equilibrium points in the van der Pol plane.

There are two questions which can be settled by a study of (7.31): that of the stability of the periodic solutions just found and, connected with this, the changes in the general behaviour of the system when one of the parameters ω or Γ varies. We shall simplify equations (7.31) by supposing that ε is small enough for $\varepsilon \dot{a}$, $\varepsilon \dot{b}$ to be of negligible effect. Then (7.31) becomes

$$
\dot{a} = \frac{1}{2}\varepsilon \left(1 - \frac{1}{4}r^2\right)a - \frac{\omega^2 - 1}{2\omega}b = \frac{1}{2}\varepsilon \left[\left(1 - \frac{1}{4}r^2\right)a - vb\right]
$$

$$
\equiv A(a,b), \quad \text{say}, \tag{7.35a}
$$

$$
\dot{b} = \frac{\omega^2 - 1}{2\omega}a + \frac{1}{2}\varepsilon\left(1 - \frac{1}{4}r^2\right)b + \frac{\Gamma}{2\omega} = \frac{1}{2}\varepsilon\left[va + \left(1 - \frac{1}{4}r^2\right)b + \gamma\right]
$$

$$
\equiv B(a,b), \quad \text{say}. \tag{7.35b}
$$

Effectively we can eliminate ε by rescaling time using a transformation $\tau = \varepsilon t$. The precise value of ε will not affect the general character of the paths in the van der Pol plane, although it does appear indirectly in v and γ.

Now consider the stability of an equilibrium point at a $a = a_0$, $b = b_0$. In the neighbourhood of the point put

$$
a = a_0 + \xi, \quad b = b_0 + \eta. \tag{7.36}
$$

The corresponding linearized equations are

$$
\dot{\xi} = A_1(a_0, \ b_0)\xi + A_2(a_0, \ b_0)\eta,
$$

$$
\dot{\eta} = B_1(a_0, \ b_0)\xi + B_2(a_0, \ b_0)\eta.
$$

The point is a stable equilibrium point (Fig. 2.10) if, in the notation of (2.61),

$$
q = \det \begin{bmatrix} A_1 & A_2 \\ B_1 & B_2 \end{bmatrix} = A_1 B_2 - A_2 B_1 > 0, \tag{7.37}
$$

and

$$
p = A_1 + B_2 \le 0. \tag{7.38}
$$

From (7.35) and (7.36),

$$
A_1 = \frac{1}{2}\varepsilon\left(1 - \frac{3}{4}a_0^2 - \frac{1}{4}b_0^2\right), \quad A_2 = -\frac{1}{4}\varepsilon a_0 b_0 - \frac{\omega^2 - 1}{2\omega},
$$

$$
B_1 = -\frac{1}{4}\varepsilon a_0 b_0 + \frac{\omega^2 - 1}{2\omega}, \quad B_2 = \frac{1}{2}\varepsilon\left(1 - \frac{1}{4}a_0^2 - \frac{3}{4}b_0^2\right).
$$

Therefore, after simplification, we have

$$q = A_1 B_2 - A_2 B_1 = \tfrac{1}{4}\varepsilon^2(\tfrac{3}{16}r_0^4 - r_0^2 + 1) + \tfrac{1}{4}\varepsilon^2 v^2, \tag{7.39}$$

$$p = A_1 + B_2 = \tfrac{1}{2}\varepsilon(2 - r_0^2), \tag{7.40}$$

where $r_0^2 = \sqrt{(a_0^2 + b_0^2)}$ and $v = (\omega^2 - 1)/(\varepsilon\omega)$. In terms of the parameter v, (7.32), the conditions for stability, (7.37) and (7.38), become

$$\tfrac{3}{16}r_0^4 - r_0^2 + 1 + v^2 > 0, \tag{7.41}$$

$$2 - r_0^2 \leq 0. \tag{7.42}$$

The response curves derived from (7.34), and the stability regions given by (7.41) and (7.42), are exhibited in Fig. 7.10, together with some further detail about the nature of the equilibrium points in various regions.

For example, the boundary between nodes and spirals is given by

$$\Delta = p^2 - 4q = \tfrac{1}{4}\varepsilon^2(2 - r_0^2)^2 - \varepsilon^2(\tfrac{3}{16}r_0^4 - r_0^2 + 1) - \varepsilon^2 v^2 = \varepsilon^2(\tfrac{1}{16}r_0^4 - v^2) = 0$$

on the line $r_0^2 = 4|v|$ as shown in Fig. 7.10(b).

It can be seen that the stable and unstable regions are correctly given by the argument in Section 5.5, (vi). Note that the *responses are stable when* $\mathrm{d}\Gamma/\mathrm{d}r > 0$ *and unstable when* $\mathrm{d}\Gamma/\mathrm{d}r < 0$. In the stability diagram (Fig. 7.10), r^2 is plotted against $|v|$. If v is given then by (7.32) there will be two corresponding forcing frequencies

$$\omega = \tfrac{1}{2}[\varepsilon v \pm \sqrt{(\varepsilon^2\omega^2 + 4)}].$$

If we attempt to force an oscillation corresponding to an equilibrium point lying in the unstable region of Fig. 7.10(a) the system will drift away from this into another state: into a stable equilibrium point if one is available, or into a limit cycle, which corresponds to an almost-periodic motion. A guide to the transition states is provided by the phase paths of (7.35). Although we are not directly concerned with the detail of the phase plane it is interesting to see the variety of patterns that can arise according to the parameter values concerned, and three typical cases are shown in Figs 7.12, 7.13, and 7.14.

We can show numerically how the settling-down takes place from a given initial condition to a periodic oscillation. For

$$\dot{x} = y, \quad \dot{y} = -\varepsilon(x^2 - 1)y - x + \Gamma\cos\omega t, \tag{7.43}$$

Fig. 7.11(a) shows a solution with initial conditions $x(0) = -2.6$, $\dot{x}(0) = 0.3$ with the parameter values $\varepsilon = 1$, $\Gamma = 5$ and $\omega = 2$. The amplitude of the oscillation settles down to a value of about 1.63. In the van der Pol plane, the phase path for a and b starts from $a(0) = -2.6$, $b(0) = \dot{x}(0)/\omega = 0.15$, and spirals into an equilibrium point at approximately $(-1.57, -0.27)$. In Fig. 7.10(b), $\gamma = 2.5$ and $v = 1.5$, which confirms that the equilibrium point in the van der Pol plane is a stable spiral.

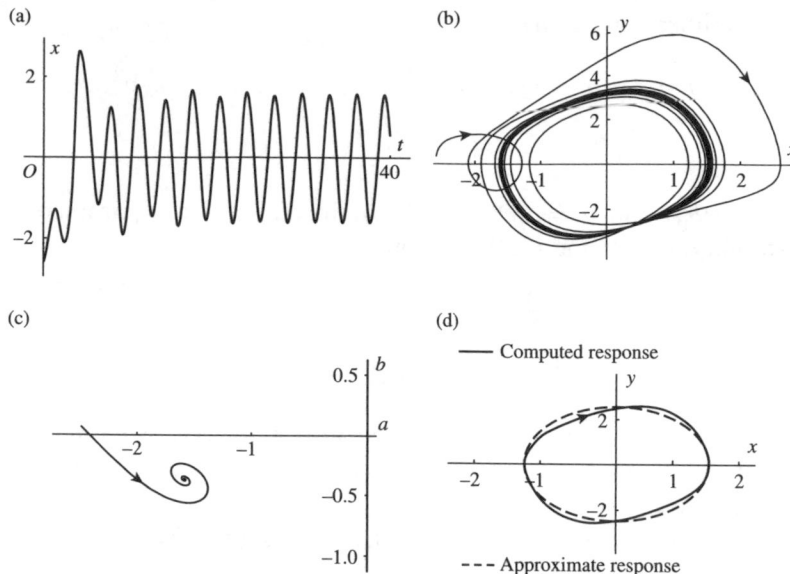

Figure 7.11   Different aspects of a solution of the forced van der Pol oscillator for $\varepsilon = 1$, $\gamma = 5$, and $\omega = 2$ (that is, $\gamma = 2.5$ and $\nu = 1.5$), with the initial conditions $x(0) = -2.6$, $\dot{x}(0) = 0.3$. (a) (x, t) time solution; (b) phase diagram showing the path approaching a stable limit cycle; (c) amplitude curve in the van der Pol phase plane for the approximation given by (7.28) with (7.31a,b) and $a(0) = -2.6$, $b(0) = 0.15$; (d) the computed limit cycle showing the approximation given by (7.28) with a and b derived from (7.33a,b).

Exercise 7.3

Using (7.39) and (7.40) show that the equation of the boundary curve between saddles and nodes in Fig. 7.10(b) is the ellipse

$$\tfrac{3}{16}(r_0^2 - \tfrac{8}{3})^2 + v^2 = \tfrac{1}{3}$$

in the $(|\nu|, r_0^2)$ plane.

Exercise 7.4

Show that the system

$$\ddot{x} + k\dot{x} + (x^2 + \dot{x}^2)x = \Gamma \cos t$$

has the exact solution $x = a \cos t + b \sin t$ where $a(r^2 - 1) + kb = \Gamma$, $-ka + b(r^2 - 1) = 0$. Show that

$$r^2(1 - r^2)^2 + k^2 r^2 = \Gamma^2,$$

where $r^2 = a^2 + b^2$, and that $d\Gamma/dr = 0$ where $r^2 = \tfrac{1}{3}[2 \pm \sqrt{(1 - 3k^2)}](k < 1/\sqrt{3})$. Sketch the (Γ, r) curves for selected values of k.

7.5   Frequency entrainment for the van der Pol equation

The description offered by the treatment of Sections 7.3 and 7.4 is incomplete because of the approximate representation used. For example, no sub-harmonics (of period $2\pi n/\omega$, $n > 1$) can be revealed by the representation (7.28) if a and b are slowly varying. Since subharmonics do exist they must be arrived at from some region of initial conditions $x(0)$, $\dot{x}(0)$; (see Section 7.8), but such regions are not identifiable on the van der Pol plane for a, b. *References to 'all paths' and 'any initial condition' in the remarks below must therefore be taken as merely a broad descriptive wording.*

Referring to Fig. 7.10, the phenomena to be expected for different ranges of v, γ (or ω, Γ, ε, eqn (7.32)) fall into three main types.

(I) When $\gamma < 1.08$ approximately, and v is appropriately small enough, there are three equilibrium points. The value $\gamma = 1.08$ can be found using the equation of the ellipse in Exercise 7.3. The right-hand limit of the ellipse occurs where $r_0^2 = \frac{8}{3}$ and $v = 1/\sqrt{3}$. Put $r_0^2 = \frac{8}{3}$ and $v^2 = \frac{1}{3}$ into (7.34), and compute γ: thus

$$\gamma^2 = r_0^2\left\{v^2 + (1 - \tfrac{1}{4}r_0^2)^2\right\} = \tfrac{8}{3}\left\{\tfrac{1}{3} + (1 - \tfrac{2}{3})^2\right\} = \tfrac{32}{27},$$

so that $\gamma = 1.08$ which separates the cases of three equilibrium points from that of a single one. One of the equilibrium points is a stable node, and the other two are unstable, namely a saddle point and an unstable spiral. Starting from any initial condition, the corresponding path will in practice run into the stable node. Figure 7.12(a) and (b) illustrate this case.

(II) For any $\gamma > 1.08$, and v large enough, there is a single stable point, either a node or spiral, which all paths approach (Fig. 7.13).

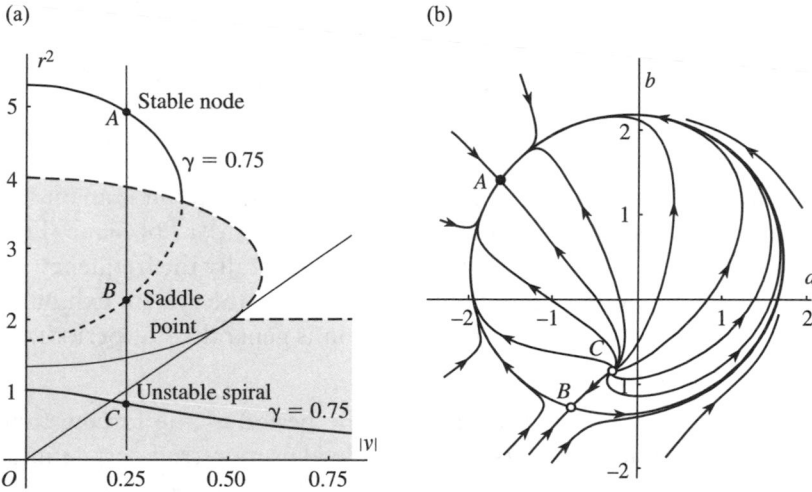

Figure 7.12   (a) response region for case I with three equilibrium points; (b) phase paths in the van der Pol plane ($\gamma = 0.75$, $v = 0.25$).

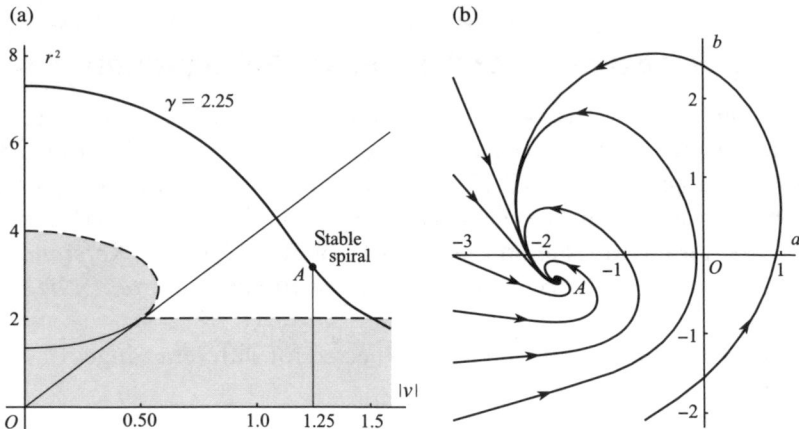

Figure 7.13   (a) response region for case II (stable spiral); (b) phase paths in the van der Pol plane ($\gamma = 2.25$, $\nu = 1.25$).

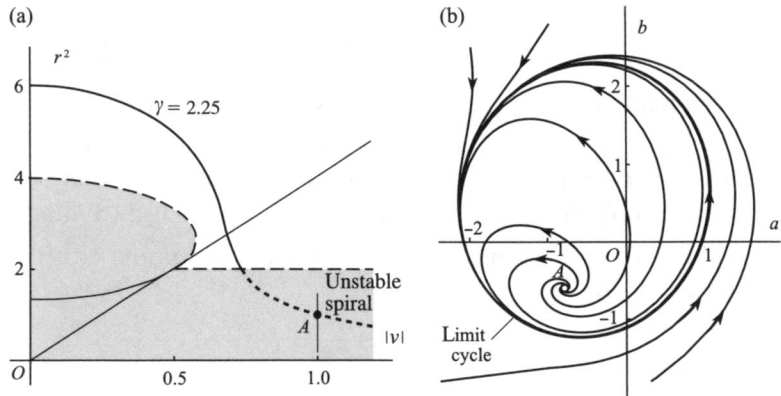

Figure 7.14   (a) response region for case III (unstable spiral); (b) phase paths in the van der Pol plane ($\gamma = 1.25$, $\nu = 1$).

(III) For any $\gamma > 1.08$, and ν small enough, the only equilibrium point is an unstable one. It can be shown that all paths approach a limit cycle in the van der Pol plane (Fig. 7.14). There are no stable harmonic solutions in this case. Since generally the frequency associated with the limit cycle is not related to the forcing frequency ω, the system exhibits an oscillation with two frequencies so that the limiting solution is generally not periodic.

In cases I and II the final state is periodic with the period of the forcing function and the natural oscillation of the system appears to be completely suppressed despite its self-sustaining nature. This condition is arrived at from arbitrary initial states (at any rate, so far as this approximate theory tells us). The system is said to be **entrained** at the frequency of the forcing function. This phenomenon *does not depend critically on exact parameter values*: if any of the

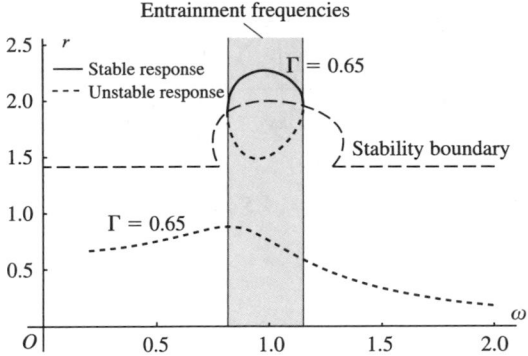

Figure 7.15   Harmonic entrainment region in the (r, ω) plane for the forced van der Pol equation: in the case shown, $\varepsilon = 1$, $\Gamma = 0.65$.

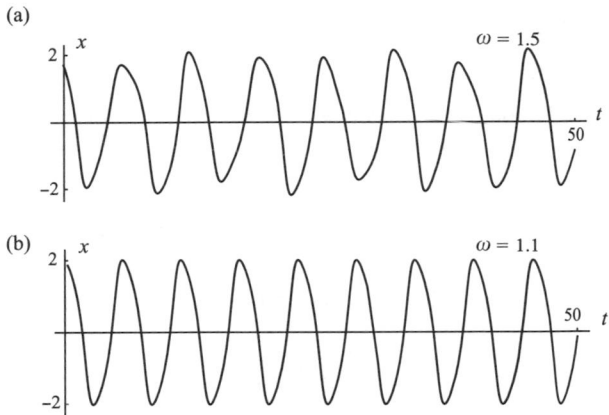

Figure 7.16   Forced outputs from the van der Pol equation for $\varepsilon = 1$, $\Gamma = 0.65$. In (a) at $\omega = 1.5$ the output is not periodic (Fig. 7.15). In (b) $\omega = 1.1$ which is within the stable region (Fig. 7.15), and entrainment occurs. The frequency of the limit cycle of the unforced van der Pol equation is approximately 0.95.

parameters ω, Γ, ε of the system fluctuate a little, the system will continue to be entrained at the prevailing frequency.

Figure 7.15 is a rendering of Fig. 7.10 in terms of ω and Γ instead of ν and γ. The sample values $\Gamma = 0.65$ and $\varepsilon = 1$ are used in the figure. Figure 7.16 shows two outputs for $\omega = 1.5$ and $\omega = 1.1$ after any transient behaviour has subsided. For $\omega = 1.5$ and $\Gamma = 0.65$, there is no stable amplitude and the system responds in 7.16(a) with a bounded nonperiodic output. For $\omega = 1.1$, there is a stable portion with $\Gamma = 0.65$ within the entrainment region and the system responds with a periodic output with the forced frequency ω.

The phenomenon of **entrainment** is related to that of **synchronization** (Minorsky 1962), in which two coupled systems having slightly different natural frequencies may fall into a common frequency of oscillation.

Results obtained by this method should always be treated with caution. Generally the forcing frequency ω should be not drastically different from the natural frequency (namely 1 in this application) so that higher harmonics do not become significant. Also the forcing amplitude

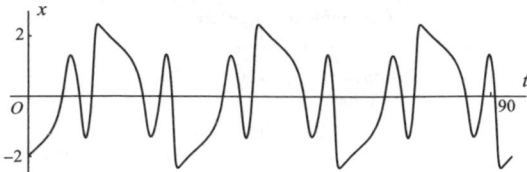

Figure 7.17   Entrainment output for $\omega = 1$, $\Gamma = 1.5$ and $\varepsilon = 0.2$ in the forced van der Pol equation.

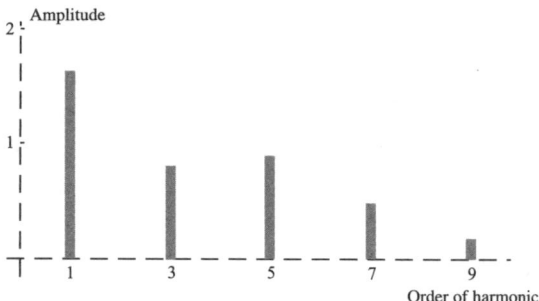

Figure 7.18   Amplitudes of the Fourier coefficients in the Fourier expansion of the entrained output in Fig. 7.17. Only odd harmonics have significant amplitudes.

should not be too large. Figure 7.17 shows the steady entrained output for $\varepsilon = 1$, $\Gamma = 1.5$ and $\omega = 0.2$. Three periods are shown and the oscillations indicate the presence of significant higher harmonics. This is confirmed by the amplitudes shown in Fig. 7.18.

> **Exercise 7.5**
> It was stated at the beginning of this section that the curve which separates the existence of one equilibrium point from three equilibrium points in the van der Pol plane is
>
> $$\gamma^2 = r_0^2\{\nu^2 + (1 - \tfrac{1}{4}r_0^2)^2\}$$
>
> where $\gamma^2 = \frac{32}{27}$, $\nu^2 = \frac{1}{3}$, $r_0^2 = \frac{8}{3}$. Prove the result by investigating the zeros of $\mathrm{d}(\nu^2)/\mathrm{d}(r_0^2)$.

7.6   Subharmonics of Duffing's equation by perturbation

In considering periodic solutions of differential equations with a forcing term we have so far looked only for solutions having the period of the forcing term. But even a linear equation may have a periodic solution with a different principal period. For example, all solutions of

$$\ddot{x} + \frac{1}{n^2}x = \Gamma \cos t$$

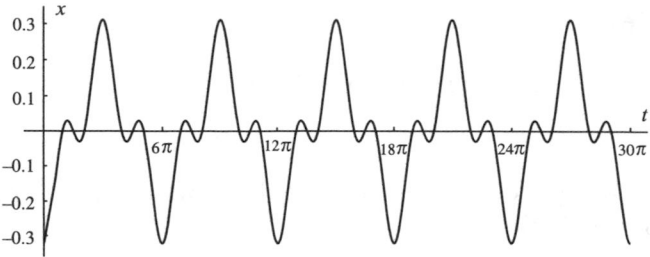

Figure 7.19   A subharmonic of order 1/3 with $a = -0.2$, $b = 0$, $\Gamma = 0.1$.

are of the form

$$x(t) = a \cos \frac{1}{n}t + b \sin \frac{1}{n}t - \frac{\Gamma}{1 - n^{-2}} \cos t.$$

If $n(>1)$ is an integer the period is $2\pi n$ instead of 2π. The response is said to be a **subharmonic of order** $1/n$. Figure 7.19 shows a case when $n = 3$.

For the linear equation this periodic motion appears to be merely an anomalous case of the usual almost-periodic motion, depending on a precise relation between the forcing and natural frequencies. Also, any damping will cause its disappearance. When the equation is nonlinear, however, the generation of alien harmonics by the nonlinear terms may cause a stable sub-harmonic to appear for a *range of the parameters*, and in particular for a range of applied frequencies. Also, the forcing amplitude plays a part in generating and sustaining the subharmonic even in the presence of damping. Thus there will exist the tolerance of slightly varying conditions necessary for the consistent appearance of a subharmonic and its use in physical systems. The possibility of such a response was pointed out in Section 7.1.

Assuming that a subharmonic exists, its form can be established by using the perturbation method. Consider Duffing's equation without damping

$$\ddot{x} + \alpha x + \beta x^3 = \Gamma \cos \omega t, \tag{7.44}$$

$\alpha, \Gamma, \omega > 0$. Take β to be the 'small parameter' of the perturbation method, so that we regard (7.44) as a member of a family of equations for which β lies in a small interval including $\beta = 0$. The values of ω for which the subharmonics occur are unknown. To simplify the subsequent algebra we put

$$\omega t = \tau \tag{7.45}$$

and look for periodic solutions of the transformed equation

$$\omega^2 x'' + \alpha x + \beta x^3 = \Gamma \cos \tau, \tag{7.46}$$

where the derivatives are with respect to τ. There are no subharmonics of order 1/2 (this is also true for the damped case as in Problem 7.16), and we shall look for those of order 1/3. The solution of (7.44) will then have period $6\pi/\omega$ and the solution of (7.46) will have fixed period 6π.

Now write, in (7.46),

$$x(\tau) = x_0(\tau) + \beta x_1(\tau) + \cdots, \tag{7.47}$$

$$\omega = \omega_0 + \beta \omega_1 + \cdots. \tag{7.48}$$

The condition that x has period 6π for all β implies that $x_0, x_1, \ldots$ have period 6π. After substitution in (7.46) we obtain

$$\omega_0^2 x_0'' + \alpha x_0 = \Gamma \cos \tau, \tag{7.49a}$$

$$\omega_0^2 x_1'' + \alpha x_1 = -2\omega_0 \omega_1 x_0'' - x_0^3. \tag{7.49b}$$

The periodicity condition applied to the general solution of (7.49a) gives $\alpha/\omega_0^2 = 1/9$, or

$$\omega_0 = 3\sqrt{\alpha}. \tag{7.50}$$

The 1/3 subharmonic is therefore stimulated by applied frequencies in the neighbourhood of three times the natural frequency of the linearized equation ($\beta = 0$). Then $x_0(\tau)$ takes the form

$$x_0(\tau) = a_{1/3} \cos \tfrac{1}{3}\tau + b_{1/3} \sin \tfrac{1}{3}\tau - \frac{\Gamma}{8\alpha} \cos \tau, \tag{7.51}$$

where $a_{1/3}$, $b_{1/3}$ are constants, to be settled at the next stage. The solution therefore bifurcates from a solution of the linearized version of (7.46). For (7.49b), we have

$$x_0''(\tau) = -\frac{1}{9}a_{1/3} \cos \tfrac{1}{3}\tau - \frac{1}{9}b_{1/3} \sin \tfrac{1}{3}\tau + \frac{\Gamma}{8\alpha} \cos \tau, \tag{7.52}$$

and

$$x_0^3(\tau) = \frac{3}{4} \left\{ a_{1/3} \left(a_{1/3}^2 + b_{1/3}^2 + 2 \left(\frac{\Gamma}{8\alpha} \right)^2 \right) - \frac{\Gamma}{8\alpha}(a_{1/3}^2 - b_{1/3}^2) \right\} \cos \tfrac{1}{3}\tau$$

$$+ \frac{3}{4} \left\{ b_{1/3} \left(a_{1/3}^2 + b_{1/3}^2 + 2 \left(\frac{\Gamma}{8\alpha} \right)^2 \right) + \frac{\Gamma}{4\alpha}a_{1/3}b_{1/3} \right\} \sin \tfrac{1}{3}\tau$$

$$+ \text{terms in } \cos \tau, \sin \tau \text{ and higher harmonics.} \tag{7.53}$$

In order that $x_1(\tau)$ should be periodic it is necessary that resonance should not occur in (7.49b). The coefficients of $\cos \tfrac{1}{3}\tau$, $\sin \tfrac{1}{3}\tau$ on the right must therefore be zero. Clearly this condition is insufficient to determine all three of $a_{1/3}$, $b_{1/3}$, and ω_1. Further stages will still give no condition to determine ω_1 so we shall consider it to be arbitrary, and determine $a_{1/3}$ and $b_{1/3}$ in terms of ω_0, ω_1 and so (to order β^2) in terms of ω. It is convenient to write

$$2\omega_0\omega_1 \approx (\omega^2 - 9\alpha)/\beta \tag{7.54}$$

this being correct with an error of order β, so that the arbitrary status of ω is clearly displayed. (Remember, however, that ω must remain near ω_0, or $3\sqrt{\alpha}$, by (7.50).) Then the conditions that the coefficients of $\cos \frac{1}{3}\tau$, $\sin \frac{1}{3}\tau$ on the right of (7.49b) should be zero are

$$a_{1/3}\left(a_{1/3}^2 + b_{1/3}^2 + \frac{\Gamma^2}{32\alpha^2} - \frac{4}{27}\frac{(\omega^2 - 9\alpha)}{\beta}\right) - \frac{\Gamma}{8\alpha}(a_{1/3}^2 - b_{1/3}^2) = 0, \qquad (7.55a)$$

$$b_{1/3}\left(a_{1/3}^2 + b_{1/3}^2 + \frac{\Gamma^2}{32\alpha^2} - \frac{4}{27}\frac{(\omega^2 - 9\alpha)}{\beta}\right) + \frac{\Gamma}{4\alpha}a_{1/3}b_{1/3} = 0. \qquad (7.55b)$$

The process of solution can be simplified in the following way. One solution of (7.55b) is given by

$$b_{1/3} = 0, \qquad (7.56)$$

and then (7.55a) becomes

$$a_{1/3}^2 - \frac{\Gamma}{8\alpha}a_{1/3} + \left(\frac{\Gamma^2}{32\alpha^2} - \frac{4}{27}\frac{(\omega^2 - 9\alpha)}{\beta}\right) = 0 \qquad (7.57)$$

(rejecting the trivial case $a_{1/3} = 0$). Thus there are either two real solutions of (7.57), or none according as, respectively,

$$\Gamma^2 > \text{or} < \frac{1024}{189}\frac{\alpha^2}{\beta}(\omega^2 - 9\alpha).$$

We shall assume that there are two real solutions. The roots are of different signs if additionally

$$\Gamma^2 < \frac{128}{27}\frac{\alpha^2}{\beta}(\omega^2 - 9\alpha),$$

and of the same sign (positive) if

$$\frac{128}{27}\frac{\alpha^2}{\beta}(\omega^2 - 9\alpha) < \Gamma^2 < \frac{1024}{189}\frac{\alpha^2}{\beta}(\omega^2 - 9\alpha).$$

All the inequalities assume that $(\omega^2 - 9\alpha)/\beta > 0$.

Now consider the case when $b_{1/3} \neq 0$. Solve (7.55b) for $b_{1/3}$ and substitute for $b_{1/3}^2$ in (7.55a). We obtain

$$b_{1/3} = \pm\sqrt{3}a_{1/3} \qquad (7.58)$$

and (compare (7.57))

$$(-2a_{1/3})^2 - \frac{\Gamma}{8\alpha}(-2a_{1/3}) + \left(\frac{\Gamma^2}{32\alpha^2} - \frac{4}{27}\frac{(\omega^2 - 9\alpha)}{\beta}\right) = 0. \qquad (7.59)$$

If the roots are displayed on the $a_{1/3}$, $b_{1/3}$ plane, those arising from (7.58) and (7.59) are obtained from the first pair, (7.56) and (7.57), by rotations through an angle $\pm\frac{2}{3}\pi$ (Fig. 7.20). Another interpretation is obtained by noting that the differential equation (7.46) is invariant

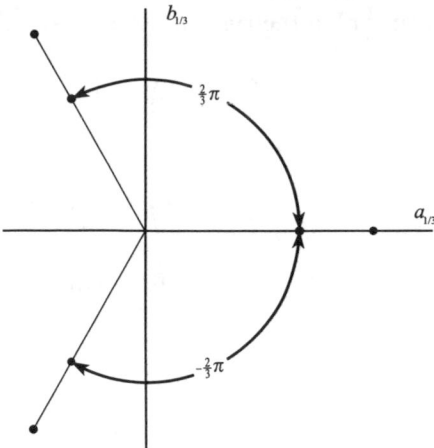

Figure 7.20   This illustration assumes that the roots of (7.57) are both positive.

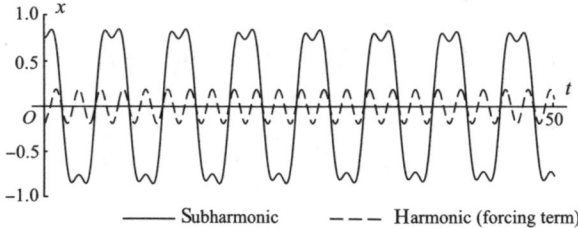

Figure 7.21   Subharmonic of order 1/3 and harmonic solutions plotted from (7.51) for the equation $\ddot{x} + x - \frac{1}{6}x^3 =$ $1.5\cos(2.85t)$. The cases $a_{1/3} = 0.9414$, $b_{1/3} = 0$ and $a_{1/3} = 0$, $b_{1/3} = 0$ are shown. Both approximations are very close to the numerical solutions of the differential equation, although, since the equation is undamped, it is difficult to eliminate natural oscillations completely.

under the change of variable $\tau \to \tau \pm 2\pi$. If $b_{1/3} = 0$ and $a_{1/3}$ represent one solution of (7.56) and (7.57), and $x^*(\tau)$ is the corresponding solution of (7.46), then

$$x^*(\tau \pm 2\pi) = a_{1/3}\cos\left(\tfrac{1}{3}\tau \pm \tfrac{2}{3}\pi\right) - \frac{\Gamma}{8\alpha}\cos\tau$$

$$= -\tfrac{1}{2}a_{1/3}\cos\tfrac{1}{3}\tau \pm \tfrac{\sqrt{3}}{2}a_{1/3}\sin\tfrac{1}{3}\tau - \frac{\Gamma}{8\alpha}\cos\tau$$

are also solutions. The new coefficients are clearly the solutions of (7.58), (7.59). The stability properties of corresponding pairs of equilibrium points are the same.

Figure 7.21 shows the subharmonic for a special case. Note that the sub-harmonic state will only be entered when the initial conditions are suitable.

We should expect that the perturbation method would be satisfactory only if $\alpha = O(1)$, $|\beta| \ll 1$ and $(\omega^2 - 9\alpha)/\beta \approx 2\omega_0\omega_1$ is of order 1. However, put

$$y = x/C, \tag{7.60}$$

where C is constant, into (7.46). This becomes

$$y'' + c_1 y + c_3 y^3 = \gamma \cos \tau, \tag{7.61}$$

where

$$c_1 = \alpha/\omega^2, \quad c_3 = \beta C^2/\omega^2, \quad \gamma = \Gamma/(C\omega^2). \tag{7.62}$$

If C has the same physical dimensions as x, then (7.61) is dimensionless. Now choose

$$C^2 = (\omega^2 - 9\alpha)/\beta \tag{7.63}$$

so that

$$9c_1 + c_3 = 1. \tag{7.64}$$

The perturbation method applied to (7.61) in the same way as to (7.46) (and obviously leading to the same result), requires at most

$$|c_3| \ll 1, \quad (1 - 9c_1)/c_3 = O(1). \tag{7.65}$$

But from (7.64)

$$(1 - 9c_1)/c_3 = 1. \tag{7.66}$$

So the second condition in (7.65) is satisfied, and $|c_3| \ll 1$ requires only

$$|c_3| = \left| \beta \left(\frac{\omega^2 - 9\alpha}{\beta} \right) \frac{1}{\omega^2} \right| \ll 1,$$

or

$$\alpha \simeq \tfrac{1}{9}\omega^2. \tag{7.67}$$

In other words, so long as ω is near enough to $3\sqrt{\alpha}$ (a near-resonant state for the subharmonic), the calculations are valid without the restriction on β. The usefulness of the approximate form (7.51) is even wider than this, and the reader is referred to Hayashi (1964) for numerical comparisons.

7.7   Stability and transients for subharmonics of Duffing's equation

Some of the subharmonics obtained in Section 7.7 might not be stable, and so will not appear in practice. When there are stable subharmonics, the question arises of how to stimulate them. We know that there are stable solutions having the period of the forcing term; there may also be subharmonics of order other than 1/3 (so far as we know at the moment). Which state of oscillation is ultimately adopted by a system depends on the *initial conditions*. The following method, due to Mandelstam and Papalexi (see Cesari 1971), is similar to the use of the van der

Pol plane in Section 7.4. It enables the question of stability to be settled and gives an idea of the domain of initial conditions leading ultimately to a subharmonic of order 1/3.

The method involves accepting the form (7.51) as a sufficiently good approximation to the subharmonic, and assuming that the solutions for the 'transient' states of

$$\ddot{x} + \alpha x + \beta x^3 = \Gamma \cos \omega t \tag{7.68}$$

are approximately of the form

$$x(t) = a_{1/3}(t) \cos \tfrac{1}{3}\omega t + b_{1/3}(t) \sin \tfrac{1}{3}\omega t - \frac{\Gamma}{8\alpha} \cos \omega t, \tag{7.69}$$

where $a_{1/3}$ and $b_{1/3}$ are slowly varying functions in the sense of Section 7.4. Further justification for the use of the form (7.69) will be found in Hayashi (1964) and McLachlan (1956).

From (7.69)

$$\dot{x}(t) = (\dot{a}_{1/3} + \tfrac{1}{3}\omega b_{1/3}) \cos \tfrac{1}{3}\omega t + (-\tfrac{1}{3}\omega a_{1/3} + \dot{b}_{1/3}) \sin \tfrac{1}{3}\omega t$$
$$+ (\omega\Gamma/8\alpha) \sin \omega t,$$

and

$$\ddot{x}(t) = (-\tfrac{1}{9}\omega^2 a_{1/3} + \tfrac{2}{3}\omega \dot{b}_{1/3}) \cos \tfrac{1}{3}\omega t + (-\tfrac{2}{3}\omega \dot{a}_{1/3} - \tfrac{1}{9}\omega^2 b_{1/3}) \sin \tfrac{1}{3}\omega t$$
$$- (\omega^2 \Gamma/8\alpha) \cos \omega t, \tag{7.70}$$

where $\ddot{a}_{1/3}, \ddot{b}_{1/3}$ have been neglected. Equations (7.69) and (7.70) are substituted into (7.68). When the terms are assembled, neglect all harmonics of higher order than $\tfrac{1}{3}$, and balance the coefficients of $\cos \tfrac{1}{3}\omega t$, $\sin \tfrac{1}{3}\omega t$. This leads to

$$\dot{a}_{1/3} = \frac{9\beta}{8\omega} \left\{ b_{1/3} \left(a_{1/3}^2 + b_{1/3}^2 + \frac{\Gamma^2}{32\alpha^2} - \frac{4}{27} \frac{(\omega^2 - 9\alpha)}{\beta} \right) + \frac{\Gamma}{4\alpha} a_{1/3} b_{1/3} \right\}$$

$$\equiv A(a_{1/3}, b_{1/3}), \tag{7.71a}$$

$$\dot{b}_{1/3} = \frac{9\beta}{8\omega} \left\{ -a_{1/3} \left(a_{1/3}^2 + b_{1/3}^2 + \frac{\Gamma^2}{32\alpha^2} - \frac{4}{27} \frac{(\omega^2 - 9\alpha)}{\beta} \right) + \frac{\Gamma}{8\alpha} (a_{1/3}^2 - b_{1/3}^2) \right\}$$

$$\equiv B(a_{1/3}, b_{1/3}). \tag{7.71b}$$

The phase paths of these autonomous equations, representing 'transients' for $x(t)$, can be displayed on a van der Pol plane of $a_{1/3}$, $b_{1/3}$ (e.g., see Fig. 7.22). Stable solutions for the form (7.69) correspond to stable equilibrium points of the system (7.71), and unstable solutions to unstable equilibrium points. By comparing with (7.55) we see that the equilibrium points are the same as those found earlier by the perturbation method. The point

$$a_{1/3} = b_{1/3} = 0$$

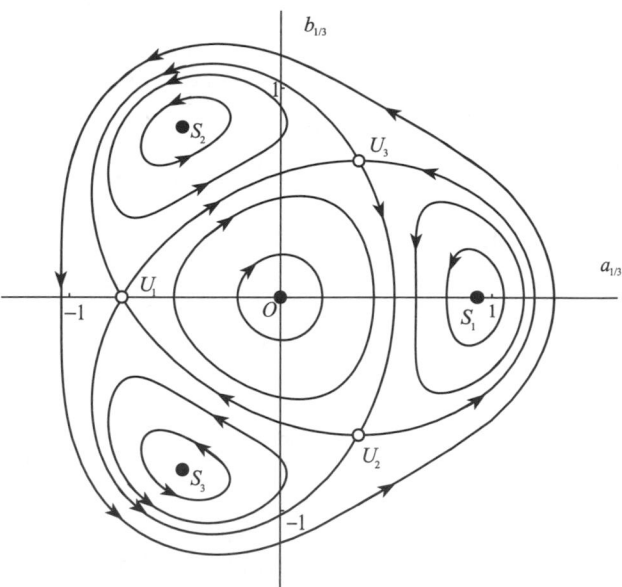

Figure 7.22 Van der Pol phase plane for subharmonics of order $\frac{1}{3}$ for $\ddot{x} + x - \frac{1}{6}x^3 = 1.5\cos(2.85t)$ (see also Fig. 7.21). The phase paths have been computed from eqns (7.71a, b). The equilibrium points U_1, U_2, and U_3, are (unstable) saddle points, and S_1, S_2, and S_3 centres.

is also an equilibrium point and this may be taken to acknowledge the possibility that from some range of initial conditions the harmonic, rather than a subharmonic, oscillation is approached, namely, the solution

$$x(t) = -\frac{\Gamma}{8\alpha}\cos\omega t.$$

This, as expected, is approximately equal to the forced solution of the linearized equation, for the exact solution of $\ddot{x} + \alpha x = \Gamma\cos\omega t$ is

$$\frac{\Gamma}{\alpha - \omega^2}\cos\omega t \simeq -\frac{\Gamma}{8\alpha}\cos\omega t$$

so long as $\omega^2 \approx 9\alpha$ (eqn (7.67)).

We may determine the stability of the equilibrium points as follows. It is only necessary to consider the pair for which $b_{1/3} = 0$, by the argument following eqn (7.59). Let $a_{1/3}^* = 0$, $b_{1/3}^* = 0$, be such a point, and consider the neighbourhood of this point by writing

$$a_{1/3} = a_{1/3}^* + \xi, \quad b_{1/3} = b_{1/3}^* + \eta = \eta$$

with ξ, η small. Then the system (7.71) becomes

$$\left.\begin{aligned}
\dot{\xi} &= A_1(a_{1/3}^*, 0)\xi + A_2(a_{1/3}^*, 0)\eta, \\
\dot{\eta} &= B_1(a_{1/3}^*, 0)\xi + B_2(a_{1/3}^*, 0)\eta.
\end{aligned}\right\}
\tag{7.72}$$

Writing

$$\gamma = \frac{\Gamma}{8\alpha}, \quad \nu = \frac{\Gamma^2}{32\alpha^2} - \frac{4}{27}\frac{(\omega^2 - 9\alpha)}{\beta}, \tag{7.73}$$

we find that the coefficients in (7.72), with $b^*_{1/3} = 0$, are given by

$$A_1(a^*_{1/3}, 0) = 0,$$

$$A_2(a^*_{1/3}, 0) = \frac{9\beta}{8\omega}(a^{*2}_{1/3} + 2\gamma a^*_{1/3} + \nu) = \frac{9\beta}{8\omega}3\gamma a^*_{1/3} \tag{7.74}$$

(from (7.57)),

$$B_1(a^*_{1/3}, 0) = \frac{9\beta}{8\omega}(-3a^{*2}_{1/3} + 2\gamma a^*_{1/3} - \nu) = \frac{9\beta}{8\omega}(-\gamma a^*_{1/3} + 2\nu),$$

$$B_2(a^*_{1/3}, 0) = 0.$$

The equilibrium points are therefore either centres (in the linear approximation) or saddles. The origin can similarly be shown to be a centre. A typical layout for the $a_{1/3}$, $b_{1/3}$ plane, using the data of Fig. 7.21, is shown in Fig. 7.22.

When **damping** is taken into account the closed paths become spirals. Figure 7.23(a) shows a typical pattern (see also Hayashi 1964) obtained by adding terms $\frac{3}{2}k(\dot{b}_{1/3} - \frac{1}{3}\omega a_{1/3})$ and $-\frac{3}{2}k(\dot{a}_{1/3} + \frac{1}{3}\omega b_{1/3})$ respectively, to the right-hand sides of (7.71a) and (7.71b). The additional

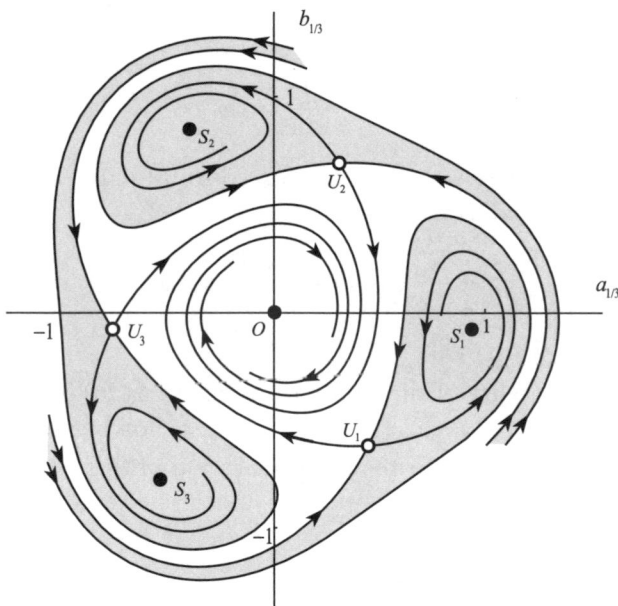

Figure 7.23   Van der Pol plane: subharmonics of order 1/3 for $\ddot{x} + 0.002\dot{x} + x - \frac{1}{6}x^3 = 1.5\cos(2.85t)$ The equilibrium points S_1, S_2, and S_3 are stable spirals, and U_1, U_2, and U_3 saddles. The amplitudes of the subharmonics are about the same but their phases differ.

terms arise from the damping term in the Duffing equation

$$\ddot{x} + k\dot{x} + \alpha x + \beta x^3 = \Gamma \cos \omega t.$$

It is assumed that $0 < k \ll 1$ and that $0 < k \ll \Gamma$, so that the damping term does not significantly affect the amplitude and phase of the harmonic $-(\Gamma/8\alpha)\cos \omega t$. The damping causes the centres in Fig. 7.22 to be transformed into stable spirals in Fig. 7.23. The shaded areas are **domains of attraction** for the equilibrium points to which they relate. It can be seen that in this case the subharmonics occur over restricted ranges of initial conditions.

It may be inferred that when the parameters of the system (such as the applied amplitude or frequency) are varied slowly, the system will suddenly become entrained at a subharmonic if the domain of attraction of a non-zero equilibrium point, appropriate to the instantaneous value of the changing parameter, is encountered. Figures such as Fig. 7.22 have to be treated with caution so far as the details are concerned. The nature and positions of the equilibrium points are likely to be nearly correct, but the phase paths are not necessarily very close to the correct ones.

Problems

7.1 Show that eqns (7.16) and (7.17), for the undamped Duffing equation in the van der Pol plane, have the exact solution

$$r^2\{(\omega^2 - 1) - \tfrac{3}{8}\beta r^2\} + 2\Gamma a = \text{constant}, \quad r = \sqrt{(a^2 + b^2)}.$$

Show that these approximate to circles when r is large. Estimate the period on such a path of $a(t), b(t)$.

7.2 Express eqns (7.16) and (7.17) in polar coordinates. Deduce the approximate period of $a(t)$ and $b(t)$ for large r. Find the approximate equations for these distant paths. Show how frequency modulation occurs, by deriving an expression for $x(t)$.

7.3 Consider the equation $\ddot{x} + \text{sgn}(x) = \Gamma \cos \omega t$. Assume solutions of the form $x = a \cos \omega + b \sin \omega t$. Show that solutions of period $2\pi/\omega$ exist when $|\Gamma| \le 4/\pi$. Show also that

$$a(4 - \pi\omega^2|a|) = \pi\Gamma|a|, \quad b = 0.$$

$$\left(\text{Hint: } \text{sgn}\{x(t)\} = \frac{4a}{\pi\sqrt{(a^2 + b^2)}} \cos \omega t + \frac{4b}{\pi\sqrt{(a^2 + b^2)}} \sin \omega t + \text{higher harmonics.}\right)$$

7.4 Show that solutions, period 2π, of the equation $\ddot{x} + x^3 = \Gamma \cos t$ are given approximately by $x = a \cos t$, where a is a solution of $3a^3 - 4a = 4\Gamma$.

7.5 Show that solutions, period 2π, of $\ddot{x} + k\dot{x} + x + x^3 = \Gamma \cos t$ are given approximately by $x = a \cos t + b \sin t$, where

$$ka - \tfrac{3}{4}br^2 = 0, \quad kb + \tfrac{3}{4}ar^2 = \Gamma, \quad r = \sqrt{(a^2 + b^2)}.$$

Deduce that the response curves are given by $r^2(k^2 + \tfrac{9}{16}r^4) = \Gamma^2$.

7.6 Obtain approximate solutions, period $2\pi/\omega$, of $\ddot{x} + \alpha x + \beta x^2 = \Gamma \cos \omega t$, by assuming the form $x = c + a \cos \omega t$, and deducing equations for c and a.

Show that if β is small, $\Gamma = O(\beta)$, and $\omega^2 - \alpha = O(\beta)$, then there is a solution with $c \approx -\beta a^2/(2\alpha)$, and $a \approx \Gamma/(\alpha - \omega^2)$.

7.7 Consider the equation $\ddot{x} + x^3 = \Gamma \cos t$. Substitute $x = a \cos t + b \sin t$, and obtain the solution $x = a \cos t$, where $\tfrac{3}{4}a^3 - a = \Gamma$ (see Problem 7.4).

Now fit x^3, by a least squares procedure, to a straight line of the form px, where p is constant, on $-A \leq x \leq A$, so that

$$\int_{-A}^{A} (x^3 - px)^2 \, dx$$

is a minimum with respect to p. Deduce that this linear approximation to the restoring force is compatible with an oscillation, period 2π, of amplitude A, provided $\frac{3}{5}A^3 - A = \Gamma$.

Compare a with A when $\Gamma = 0.1$.

7.8 Consider the equation

$$\ddot{x} + 0.16x^2 = 1 + 0.2 \cos t.$$

By linearizing the restoring force about the equilibrium points of the unforced system (without $\cos t$), show that there are two modes of oscillation, period 2π, given by

$$x \approx 2.5 - \cos t, \quad x \approx -2.5 - 0.11 \cos t.$$

Find to what extent the predicted modes differ when a substitution of the form $x = c + a \cos t + b \sin t$ is used instead.

7.9 By examining the nonperiodic solutions of the linearized equations obtained in the first part of Problem 7.8, show that the two solutions, period 2π, obtained are respectively stable and unstable.

7.10 Show that the equations giving the equilibrium points in the van der Pol plane for solutions period $2\pi/\omega$ for the forced, damped pendulum equation

$$\ddot{x} + k\dot{x} + x - \tfrac{1}{6}x^3 = \Gamma \cos \omega t, \quad k > 0$$

are

$$k\omega a + b\{\omega^2 - 1 + \tfrac{1}{8}(a^2 + b^2)\} = 0,$$

$$-k\omega b + a\{\omega^2 - 1 + \tfrac{1}{8}(a^2 + b^2)\} = -\Gamma.$$

Deduce that

$$r^2(\omega^2 - 1 + \tfrac{1}{8}r^2)^2 + \omega^2 k^2 r^2 = \Gamma^2, \quad \omega k r^2 = \Gamma b,$$

where $r = \sqrt{(a^2 + b^2)}$.

7.11 For the equation $\ddot{x} + x - \tfrac{1}{6}x^3 = \Gamma \cos \omega t$, find the frequency-amplitude equations in the van der Pol plane. Show that there are three equilibrium points in the van der Pol plane if $\omega^2 < 1$ and $|\Gamma| > \tfrac{2}{3}\sqrt{(\tfrac{8}{3})}(1-\omega^2)^{3/2}$, and one otherwise. Investigate their stability.

7.12 For the equation $\ddot{x} + \alpha x + \beta x^2 = \Gamma \cos t$, substitute $x = c(t) + a(t) \cos t + b(t) \sin t$, and show that, neglecting $\ddot{a}$ and $\ddot{b}$,

$$\dot{a} = \tfrac{1}{2}b(\alpha - 1 + 2\beta c),$$

$$\dot{b} = -\tfrac{1}{2}a(\alpha - 1 + 2\beta c) + \Gamma,$$

$$\ddot{c} = -\alpha c - \beta\{c^2 + \tfrac{1}{2}(a^2 + b^2)\}.$$

Deduce that if $|\Gamma|$ is large there are no solutions of period 2π, and that if $\alpha < 1$ and Γ is sufficiently small there are two solutions of period 2π.

7.13 Substitute $x = c(t) + a(t) \cos t + b(t) \sin t$ into the equation $\ddot{x} + \alpha x^2 = 1 + \Gamma \cos t$ (compare Problem 7.8), and show that if $\ddot{a}$ and $\ddot{b}$ are neglected, then

$$2\dot{a} = b(2\alpha c - 1), \quad 2\dot{b} = a(1 - 2\alpha c) + \Gamma, \quad \ddot{c} + \alpha(c^2 + \tfrac{1}{2}a^2 + \tfrac{1}{2}b^2) = 1.$$

Use a graphical argument to show that there are two equilibrium points when $\alpha < \tfrac{1}{4}$ and $\Gamma < \sqrt{(2/\alpha)}$.

7.14 In the forced Duffing equation $\ddot{x} + k\dot{x} + x - \tfrac{1}{6}x^3 = \Gamma \cos \omega t$, $(k > 0)$ substitute $x = a(t) \cos \omega t + b(t) \sin \omega t$ to investigate the solutions of period $2\pi/\omega$. Assume that a and b are slowly varying and that $k\dot{a}$, $k\dot{b}$ can be neglected. Show that the paths in the van der Pol plane are given by

$$\dot{a} = -\frac{b}{2\omega}\left\{\omega^2 - 1 + \frac{1}{8}(a^2 + b^2)\right\} - \frac{1}{2}ka,$$

$$\dot{b} = \frac{a}{2\omega}\left\{\omega^2 - 1 + \frac{1}{8}(a^2 + b^2)\right\} - \frac{1}{2}kb + \frac{\Gamma}{2\omega}.$$

Show that there is one equilibrium point if $\omega^2 > 1$.

Find the linear approximation in the neighbourhood of the equilibrium point when $\omega^2 > 1$, and show that it is a stable node or spiral when $k > 0$.

7.15 For the equation $\ddot{x} + \alpha x + \beta x^3 = \Gamma \cos \omega t$, show that the restoring force $\alpha x + \beta x^3$ is represented in the linear least-squares approximation on $-A \leq x \leq A$ by $(\alpha + \tfrac{3}{5}\beta A^2)x$. Obtain the general solution of the approximating equation corresponding to a solution of amplitude A. Deduce that there may be a subharmonic of order $\tfrac{1}{3}$ if $\alpha + \tfrac{3}{5}\beta A^2 = \tfrac{1}{9}\omega^2$ has a real solution A. Compare eqn (7.57) for the case when $\Gamma/(8\alpha)$ is small. Deduce that when $\alpha \approx \tfrac{1}{9}\omega^2$ (close to subharmonic resonance), the subharmonic has the approximate form

$$A \cos\left(\frac{1}{3}\omega t + \phi\right) - \frac{\Gamma}{8\alpha}\cos \omega t,$$

where ϕ is a constant.

(The interpretation is that when $\Gamma/(8\alpha)$ is small enough for the oscillation to lie in $[-A, A]$, A can be adjusted so that the slope of the straight-line fit on $[-A, A]$ is appropriate to the generation of a natural oscillation which is a subharmonic. The phase cannot be determined by this method.)

Show that the amplitude predicted for the equation $\ddot{x} + 0.15x - 0.1x^3 = 0.1 \cos t$ is $A = 0.805$.

7.16 Use the perturbation method to show that

$$\ddot{x} + k\dot{x} + \alpha x + \beta x^3 = \Gamma \cos \omega t$$

has no subharmonic of order $\tfrac{1}{2}$ when β is small and $k = O(\beta)$. (Assume that

$$(a \cos \tfrac{1}{2}\tau + b \sin \tfrac{1}{2}\tau + c \cos \tau)^3$$

$$= \tfrac{3}{4}c(a^2 - b^2) + \tfrac{3}{4}a(a^2 + b^2 + 2c^2) \cos \tfrac{1}{2}\tau$$

$$+ \tfrac{3}{4}b(a^2 + b^2 + 2c^2) \sin \tfrac{1}{2}\tau + \text{ higher harmonics.)}$$

7.17 Use the perturbation method to show that

$$\ddot{x} + k\dot{x} + \alpha x + \beta x^3 = \Gamma \cos \omega t$$

has no subharmonic of order other than $\tfrac{1}{3}$ when β is small and $k = O(\beta)$. (Use the identity

$$(a \cos \tfrac{1}{n}\tau + b \sin \tfrac{1}{n}\tau + c \cos \tau)^n$$

$$= \tfrac{3}{4}a(a^2 + b^2 + 2c^2) \cos \tau + \tfrac{3}{4}b(a^2 + b^2 + 2c^2) \sin \tau$$

$$+ \text{ (higher harmonics)} \quad \text{for } n \neq 3.)$$

7.18 Look for subharmonics of order $\frac{1}{2}$ for the equation

$$\ddot{x} + \varepsilon(x^2 - 1)\dot{x} + x = \Gamma \cos \omega t$$

using the perturbation method with $\tau = \omega t$. If $\omega = \omega_0 + \varepsilon\omega_1 + \cdots$, show that this subharmonic is only possible if $\omega_1 = 0$ and $\Gamma^2 < 18$. (Hint: let $x_0 = a \cos \frac{1}{2}\tau + b \sin \frac{1}{2}\tau - \frac{1}{3}\Gamma \cos \tau$, and use the expansion)

$$(x_0^2 - 1)x_0' = \frac{1}{72}[-36 + 9(a^2 + b^2) + 2\Gamma^2](b \cos \frac{1}{2}\tau - a \sin \frac{1}{2}\tau)$$

$$+ \text{ (higher harmonics)}.$$

7.19 Extend the analysis of the equation

$$\ddot{x} + \varepsilon(x^2 - 1)\dot{x} + x = \Gamma \cos \omega t$$

in Problem 7.18 by assuming that

$$x = a(t) \cos \tfrac{1}{2}\omega t + b(t) \sin \tfrac{1}{2}\omega t - \tfrac{1}{3}\Gamma \cos \omega t,$$

where a and b are slowly varying. Show that when $\ddot{a}, \ddot{b}, \varepsilon\dot{a}, \varepsilon\dot{b}$, are neglected,

$$\tfrac{1}{2}\omega\dot{a} = (1 - \tfrac{1}{4}\omega^2)b - \tfrac{1}{8}\varepsilon\omega a(a^2 + b^2 + \tfrac{2}{9}\Gamma^2 - 4),$$

$$\tfrac{1}{2}\omega\dot{b} = -(1 - \tfrac{1}{4}\omega^2)a - \tfrac{1}{8}\varepsilon\omega b(a^2 + b^2 + \tfrac{2}{9}\Gamma^2 - 4),$$

on the van der Pol plane for the subharmonic.

By using $\rho = a^2 + b^2$ and ϕ the polar angle on the plane show that

$$\dot{\rho} = -\tfrac{1}{4}\varepsilon\rho(\rho + K), \quad \dot{\phi} = -(1 - \tfrac{1}{4}\omega^2)/(2\omega), \quad K = \tfrac{2}{9}\Gamma^2 - 4.$$

Deduce that

(i) When $\omega \neq 2$ and $K \geq 0$, all paths spiral into the origin, which is the only equilibrium point (so no subharmonic exists).

(ii) When $\omega = 2$ and $K \geq 0$, all paths are radial straight lines entering the origin (so there is no subharmonic).

(iii) When $\omega \neq 2$ and $K < 0$, all paths spiral on to a limit cycle, which is a circle, radius $-K$ and centre the origin (so x is not periodic).

(iv) When $\omega = 2$ and $K < 0$, the circle centre the origin and radius $-K$ consists entirely of equilibrium points, and all paths are radial straight lines approaching these points (each such point represents a subharmonic).

(Since subharmonics are expected only in case (iv), and for a critical value of ω, entrainment cannot occur. For practical purposes, even if the theory were exact we could never expect to observe the subharmonic, though solutions near to it may occur.)

7.20 Given eqns (7.34), (7.41), and (7.42) for the response curves and the stability boundaries for van der Pol's equation (Fig. 7.10), eliminate r^2 to show that the boundary of the entrainment region in the γ, ν-plane is given by

$$\gamma^2 = 8\{1 + 9\nu^2 - (1 - 3\nu^2)^{3/2}\}/27$$

for $\nu^2 < \frac{1}{3}$. Show that, for small ν, $\gamma \approx \pm 2\nu$, or $\gamma \approx \pm\frac{2}{3\sqrt{3}}(1 - \frac{9}{8}\nu^2)$.

7.21 Consider the equation

$$\ddot{x} + \varepsilon(x^2 + \dot{x}^2 - 1)\dot{x} + x = \Gamma \cos \omega t.$$

To obtain solutions of period $2\pi/\omega$, substitute

$$x = a(t)\cos\omega t + b(t)\sin\omega t$$

and deduce that, if $\ddot{a}, \ddot{b}, \varepsilon\dot{a}, \varepsilon\dot{b}$ can be neglected, then

$$\dot{a} = \tfrac{1}{2}\varepsilon\{a - vb - \tfrac{1}{4}\mu a(a^2 + b^2)\},$$
$$\dot{b} = \tfrac{1}{2}\varepsilon\{va + b - \tfrac{1}{4}\mu b(a^2 + b^2)\} + \tfrac{1}{2}\varepsilon\gamma,$$

where

$$\mu = 1 + 3\omega^2, \quad v = (\omega^2 - 1)/(\varepsilon\omega), \quad \text{and} \quad \gamma = \Gamma/(\varepsilon\omega).$$

Show that the stability boundaries are given by

$$1 + v^2 - \mu r^2 + \tfrac{3}{16}\mu^2 r^4 = 0, \quad 2 - \mu r^2 = 0.$$

7.22 Show that the equation

$$\ddot{x}(1 - x\dot{x}) + (\dot{x}^2 - 1)\dot{x} + x = 0$$

has an *exact* periodic solution $x = \cos t$. Show that the corresponding forced equation:

$$\ddot{x}(1 - x\dot{x}) + (\dot{x}^2 - 1)\dot{x} + x = \Gamma\cos\omega t$$

has an *exact* solution of the form $a\cos\omega t + b\sin\omega t$, where

$$a(1 - \omega^2) - \omega b + \omega^3 b(a^2 + b^2) = \Gamma,$$
$$b(1 - \omega^2) + \omega a - \omega^3 a(a^2 + b^2) = 0.$$

Deduce that the amplitude $r = \sqrt{(a^2 + b^2)}$ satisfies

$$r^2\{(1 - \omega^2)^2 + \omega^2(1 - r^2\omega^2)^2\} = \Gamma^2.$$

7.23 The frequency–amplitude relation for the damped forced pendulum is (eqn (7.23), with $\beta - 1/6$)

$$r^2\{k^2\omega^2 + (\omega^2 - 1 + \tfrac{1}{8}r^2)^2\} = \Gamma^2.$$

Show that the vertex of the cusp bounding the fold in Fig. 7.7 occurs where

$$\omega = \tfrac{1}{2}\left\{\sqrt{(3k^2 + 4)} - k\sqrt{3}\right\}.$$

Find the corresponding value for Γ^2.

7.24 **(Combination tones)** Consider the equation

$$\ddot{x} + \alpha x + \beta x^2 = \Gamma_1\cos\omega_1 t + \Gamma_2\cos\omega_2 t, \quad \alpha > 0, \ |\beta| \ll 1,$$

where the forcing term contains two distinct frequencies ω_1 and ω_2. To find an approximation to the response, construct an iterative process leading to the sequence of approximations $x^{(0)}, x^{(1)}, \ldots$, and starting with

$$\ddot{x}^{(0)} + \alpha x^{(0)} = \Gamma_1\cos\omega_1 t + \Gamma_2\cos\omega_2 t,$$
$$\ddot{x}^{(1)} + \alpha x^{(1)} = \Gamma_1\cos\omega_1 t + \Gamma_2\cos\omega_2 t - \beta(x^{(0)})^2,$$

show that a particular solution is given approximately by

$$x(t) = -\frac{\beta}{2\alpha}(a^2 + b^2) + a\cos\omega_1 t + b\cos\omega_2 t + \frac{\beta a^2}{2(4\omega_1^2 - \alpha)}\cos 2\omega_1 t$$

$$+ \frac{\beta b^2}{2(4\omega_2^2 - \alpha)}\cos 2\omega_2 t + \frac{\beta ab}{(\omega_1 + \omega_2)^2 - \alpha}\cos(\omega_1 + \omega_2)t$$

$$+ \frac{\beta ab}{(\omega_1 - \omega_2)^2 - \alpha}\cos(\omega_1 - \omega_2)t,$$

where

$$a \approx \Gamma_1/(\alpha - \omega_1^2), \quad b \approx \Gamma_2/(\alpha - \omega_2^2).$$

(The presence of 'sum and difference tones' with frequencies $\omega_1 \pm \omega_2$ can be detected in sound resonators having suitable nonlinear characteristics, or as an auditory illusion attributed to the nonlinear detection mechanism in the ear (McLachlan 1956). The iterative method of solution can be adapted to simpler forced oscillation problems involving a single input frequency.)

7.25 Apply the method of Problem 7.24 to the Duffing equation

$$\ddot{x} + \alpha x + \beta x^3 = \Gamma_1 \cos\omega_1 t + \Gamma_2 \cos\omega_2 t.$$

7.26 Investigate the resonant solutions of Duffing's equation in the form

$$\ddot{x} + x + \varepsilon^3 x^3 = \cos t, \quad |\varepsilon| \ll 1,$$

by the method of multiple scales (Section 6.4) using a slow time $\eta = \varepsilon t$ and a solution of the form

$$x(\varepsilon, t) = \frac{1}{\varepsilon}X(\varepsilon, t, \eta) = \frac{1}{\varepsilon}\sum_{n=0}^{\infty}\varepsilon^n X_n(t, \eta).$$

Show that $X_0 = a_0(\eta)\cos t + b_0(\eta)\sin t$, where

$$8a_0' - 3b_0(a_0^2 + b_0^2) = 0, \quad 8b_0' + 3a_0(a_0^2 + b_0^2) = 4.$$

(This example illustrates that even a small nonlinear term may inhibit the growth of resonant solutions.)

7.27 Repeat the multiple scale procedure of the previous exercise for the equation

$$\ddot{x} + x + \varepsilon^3 x^2 = \cos t, \quad |\varepsilon| \ll 1,$$

which has an unsymmetrical, quadratic departure from linearity. Use a slow time $\eta = \varepsilon^2 t$ and an expansion $x(\varepsilon, t) = \varepsilon^{-2}\sum_{n=0}^{\infty}\varepsilon^n X_n(t, \eta)$.

7.28 Let $\ddot{x} - x + bx^3 = c\cos t$. Show that this system has an exact subharmonic $k\cos\frac{1}{3}t$ if b, c, k satisfy

$$k = \frac{27}{10}c, \quad b = \frac{4c}{k^3}.$$

7.29 Noting that $y = 0$ is a solution of the second equation in the forced system

$$\dot{x} = -x(1 + y) + \gamma\cos t, \quad \dot{y} = -y(x + 1),$$

obtain the forced periodic solution of the system.

7.30 Show that, if

$$\dot{x} = \alpha y \sin t - (x^2 + y^2 - 1)x, \quad \dot{y} = -\alpha x \sin t - (x^2 + y^2 - 1)y,$$

where $0 < \alpha < \pi$, then $2\dot{r} = (r^2 - 1)r$. Find r as a function of t, and show that $r \to 1$ as $t \to \infty$. Discuss the periodic oscillations which occur on the circle $r = 1$.

7.31 Show that

$$\ddot{x} + kx + x^2 = \Gamma \cos t$$

has an exact subharmonic of the form $x = A + B\cos(\frac{1}{2}t)$ provided $16k^2 > 1$. Find A and B.

7.32 Computed solutions of the particular two-parameter Duffing equation

$$\ddot{x} + k\dot{x} + x^3 = \Gamma \cos t$$

have been investigated in considerable detail by Ueda (1980). Using $x = a(t)\cos t + b(t)\sin t$, and assuming that $a(t)$ and $b(t)$ are slowly varying amplitudes, obtain the equations for $\dot{a}(t)$ and $\dot{b}(t)$ as in Section 7.2. Show that the response amplitude, r, and the forcing amplitude, Γ, satisfy

$$r^2\{k^2 + (1 - \tfrac{3}{4}r^2)^2\} = \Gamma^2$$

for 2π-periodic solutions. By investigating the zeros of $\mathrm{d}(\Gamma^2)/\mathrm{d}r^2$, show that there are three response amplitudes if $0 < k < 1/\sqrt{3}$. Sketch this region in the (Γ, k) plane.

7.33 Show that there exists a Hamiltonian

$$H(x, y, t) = \tfrac{1}{2}(x^2 + y^2) + \tfrac{1}{4}\beta x^4 - \Gamma x \cos \omega t$$

for the undamped Duffing equation

$$\ddot{x} + x + \beta x^3 = \Gamma \cos \omega t, \quad \dot{x} = y$$

(see eqn (7.4)).

   Show also that the autonomous system for the slowly varying amplitudes a and b in the van der Pol plane (eqns (7.16) and (7.17)) is also Hamiltonian (see Section 2.8). What are the implications for the types of equilibrium points in the van der Pol plane?

7.34 Show that the exact solution of the equation $\ddot{x} + x = \Gamma \cos \omega t$ ($\omega \neq 1$) is

$$x(t) = A\cos t + B\sin t + \frac{\Gamma}{1 - \omega^2}\cos \omega t,$$

where A and B are arbitrary constants.

   Introduce the van der Pol variables $a(t)$ and $b(t)$ through

$$x(t) = a(t)\cos \omega t + b(t)\sin \omega t,$$

and show that $x(t)$ satisfies the differential equation if $a(t)$ and $b(t)$ satisfy

$$\ddot{a} + 2\omega\dot{b} + (1 - \omega^2)a = \Gamma, \quad \ddot{b} - 2\omega\dot{a} + (1 - \omega^2)b = 0.$$

Solve these equations for a and b by combining them into an equation in $z = a + ib$. Solve this equation, and confirm that, although the equations for a and b contain four constants, these constants combine in such a way that the solution for x still contains just two arbitrary constants.

7.35 Show that the system

$$\ddot{x} + (k - x^2 - \dot{x}^2)\dot{x} + \beta x = \Gamma \cos t \quad (k, \Gamma > 0, \beta \neq 1),$$

has exact harmonic solutions of the form $x(t) = a \cos t + b \sin t$, if the amplitude $r = \sqrt{(a^2 + b^2)}$ satisfies

$$r^2[(\beta - 1)^2 + (k - r^2)^2] = \Gamma^2.$$

By investigating the solution of $d(\Gamma^2)/d(r^2) = 0$, show that there are three harmonic solutions for an interval of values of Γ if $k^2 > 3(\beta - 1)^2$. Find this interval if $k = \beta = 2$. Draw the amplitude diagram r against Γ in this case.

7.36 Show that the equation $\ddot{x} + k\dot{x} - x + \omega^2 x^2 + \dot{x}^2 = \Gamma \cos \omega t$ has exact solutions of the form $x = c + a \cos \omega t + b \sin \omega t$, where the translation c and the amplitude $r = \sqrt{(a^2 + b^2)}$ satisfy

$$\Gamma^2 = r^2[(1 + \omega^2(1 - 2c))^2 + k^2\omega^2],$$

and

$$\omega^2 r^2 = c(1 - c\omega^2).$$

Sketch a graph showing response amplitude (r) against the forcing amplitude (Γ).

8 Stability

The word 'stability' has been used freely as a descriptive term in earlier chapters. In Chapter 2, **equilibrium points** are classified as stable or unstable according to their appearance in the phase diagram. Roughly speaking, if every initial state close enough to equilibrium leads to states which continue permanently to be close, then the equilibrium is stable. Some equilibrium states may be thought of as being more stable than others; for example, all initial states near a stable node lead ultimately into the node, but in the case of a centre there always remains a residual oscillation after disturbance of the equilibrium. **Limit cycles** are classified as stable or unstable according to whether or not nearby paths spiral into the limit cycle. In Chapter 7 we considered the stability of **forced periodic oscillations**, reducing the question to that of the stability of the corresponding equilibrium points on a van der Pol plane.

These cases are formally different, as well as being imprecisely defined. Moreover, they are not exhaustive: the definition of stability chosen must be that appropriate to the types of phenomena we want to distinguish between. The question to be answered is usually of this kind. If a system is in some way disturbed, will its subsequent behaviour differ from its undisturbed behaviour by an acceptably small amount? In practice, physical and other systems are always subject to small, unpredictable influences: to variable switch-on conditions, to maladjustment, to variation in physical properties, and the like. If such variations produce large changes in the operating conditions the system is probably unusable, and its normal operating condition would be described as unstable. Even if the effect of small disturbances does not grow, the system may not be 'stable enough' for its intended purpose: for proper working it might be necessary for the system to re-approach the normal operating condition, rather than to maintain a permanent superimposed deviation, however small. We might even require it to approach its normal state without ultimately suffering any time delay.

Some of these possibilities are treated in the following chapter. We are concerned with **regular systems** throughout (see Appendix A). The treatment is not restricted merely to second-order systems.

The general *autonomous* system in n dimensions can be written as

$$\dot{x} = X(x),$$

where $x(t) = [x_1(t), x_2(t), \ldots, x_n(t)]^{\mathrm{T}}$ is a vector with n components $x_1(t), x_2(t), \ldots, x_n(t)$. Here, the index T stands for the **transpose**: $x(t)$ is a column vector. A **time solution** $x = x(t)$ defines a phase path in the $(x_1, x_2, \ldots, x_n)$ phase space, and a set of phase paths defines a phase diagram for the system. A point x_0 is an equilibrium point if the constant vector $x(t) = x_0$ is a solution for all t. Thus equilibrium points are given by solutions of $X(x) = 0$. As described in Chapter 2 for the case $n = 2$ the system can be linearized in the neighbourhood of x_0. The local behaviour can then be interpreted from the eigenvalues of the linearized system. Obviously for $n \geq 3$ the

classification of linear equations in terms of their eigenvalues becomes extensive. The general
nth-order *non-autonomous* system can be represented by

$$\dot{x} = X(x, t).$$

8.1 Poincaré stability (stability of paths)

This is a relatively undemanding criterion, which applies to *autonomous systems*. We shall treat
it intuitively, in the *plane* case only. It agrees with the tentative classification of equilibrium
points as stable or unstable in Chapter 2, and of limit cycles in Section 3.4.

We identify a phase path (or an equilibrium point) of the system $\dot{x} = X(x)$ representing a
solution $x^*(t)$ whose stability is in question, called the **standard path**. We are usually interested
in this path only from a particular point a^* onwards, that is, for all $t \geq t_0$: then we have a
positive half-path $\mathcal{H}^*$ with **initial point** a^* (Fig. 8.1). The solution which $\mathcal{H}^*$ represents is

$$x^*(t), \quad t \geq t_0,$$

where

$$x^*(t_0) = a^*,$$

and since the system is autonomous, all choices of t_0 lead to the same $\mathcal{H}^*$. Some types of
half-paths are shown in Fig. 8.1.

The solution $x^*(t), t \geq t_0$ is **Poincaré stable** (or **orbitally stable**), if all sufficiently small disturbances of the initial value a^* lead to half-paths remaining for all later time at a small distance
from $\mathcal{H}^*$. To form an analytical definition it is necessary to reverse this statement. First, choose
$\varepsilon > 0$ arbitrarily and construct a strip of whose edges are at a distance ε from $\mathcal{H}^*$ (Fig. 8.2).
This represents the permitted deviation from $\mathcal{H}^*$. Then for stability it must be shown, for every
ε, that we can find a δ such that *all* paths starting within a distance δ of a a^* (where $\delta \leq \varepsilon$
necessarily), remain *permanently* within the strip. Such a condition must hold for *every* ε: in
general, the smaller ε the smaller δ must be.

The formal definition is as follows.

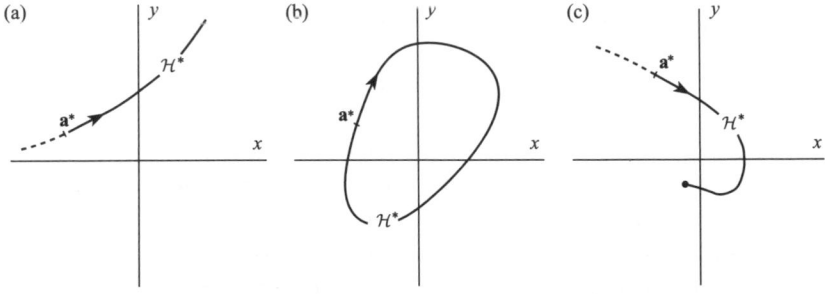

Figure 8.1 Typical half-paths $\mathcal{H}^*$; in (b) the half-path is a closed curve repeated indefinitely; in (c) the half-path
approaches an equilibrium point.

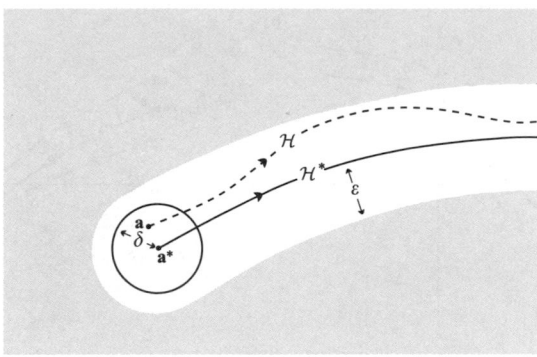

Figure 8.2

Definition 8.1 (Poincaré, or orbital stability)   *Let $\mathcal{H}^*$ be the half-path for the solution $\boldsymbol{x}^*(t)$ of $\dot{\boldsymbol{x}} = \boldsymbol{X}(\boldsymbol{x})$ which starts when $\boldsymbol{a}^*$ when $t = t_0$. Suppose that for every $\varepsilon > 0$ there exists $\delta(\varepsilon) > 0$ such that if $\mathcal{H}$ is the half-path starting at $\boldsymbol{a}$,*

$$|\boldsymbol{a} - \boldsymbol{a}^*| < \delta \Rightarrow \max_{\boldsymbol{x} \in \mathcal{H}} \mathrm{dist}(\boldsymbol{x}, \mathcal{H}^*) < \varepsilon. \tag{8.1}$$

Then $\mathcal{H}^*$ (or the corresponding time-solution) is said to be Poincaré stable. Otherwise $\mathcal{H}^*$ is unstable.

Here, the distance from a point $\boldsymbol{x}$ to a curve $\mathcal{C}$ is defined by

$$\mathrm{dist}(\boldsymbol{x}, \mathcal{C}) = \min_{\boldsymbol{y} \in \mathcal{C}} |\boldsymbol{x} - \boldsymbol{y}| = \min_{\boldsymbol{y} \in \mathcal{C}} [(x_1 - y_1)^2 + (x_2 - y_2)^2]^{1/2},$$

in the plane case. The maximum permitted separation of $\mathcal{H}$ from $\mathcal{H}^*$ is limited, in two dimensions, by the construction illustrated in Fig. 8.2: two discs, each of diameter ε, are rolled along either side of $\mathcal{H}^*$ to sweep out a strip; for stability, there must exist a $\delta > 0$ such that every $\mathcal{H}$ with starting point $\mathbf{a}$, $|\mathbf{a} - \mathbf{a}^*| < \delta$, lies permanently in the strip.

Figure 8.3 illustrates schematically some cases of stability and instability. (a), (b) and (c) show stable cases. Case (d) represents the system

$$\dot{x} = x, \qquad \dot{y} = 2y$$

with solutions

$$x = Ae^t, \qquad y = Be^{2t}$$

where A and B are constants. In the case illustrated the half-path $\mathcal{H}^*$, corresponding to the solution with $A = B = 0$, starts at $\boldsymbol{a}^* = (0, x_0)$ with $x_0 > 0$. Choose $\varepsilon > 0$ and sketch the tolerance region as shown in Fig. 8.3(d). Since the paths are given by the family of parabolas $y = Cx^2$, every half-path $\mathcal{H}$, with $C \neq 0$ escapes from the permitted region at some time, no matter how close to a $\boldsymbol{a}^*$ it started off. The half-path $\mathcal{H}^*$ is therefore unstable. In fact, by a similar argument all the half-paths are unstable.

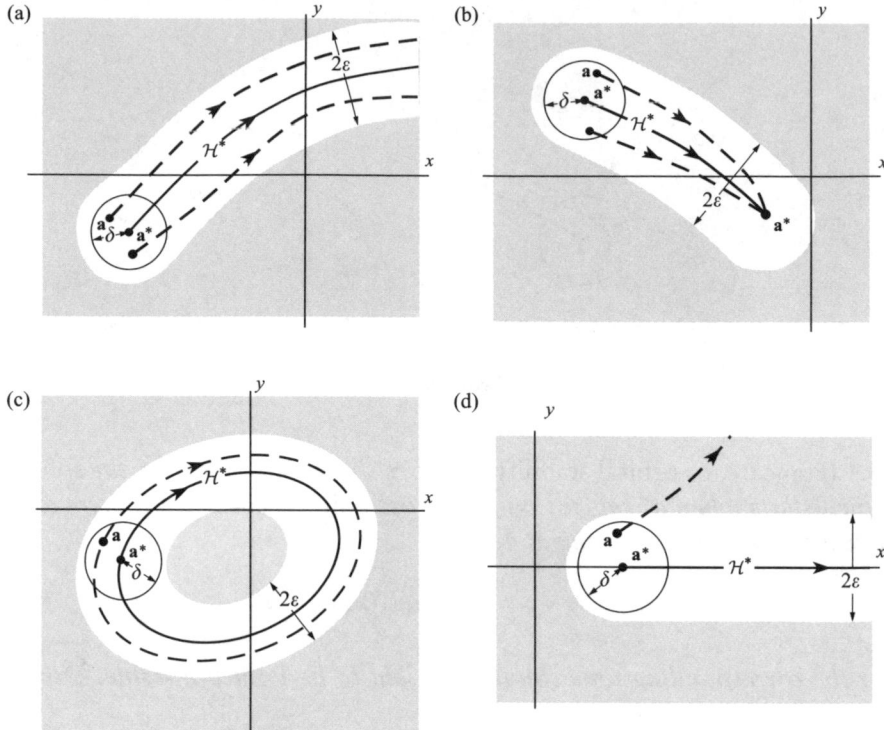

Figure 8.3 — half-path $\mathcal{H}^*$, ----- neighbouring paths for some typical cases. (a) general half-path; (b) half-path approaching an equilibrium point; (c) periodic half = path; (d) an unstable case.

A simple example of a centre is given by the system

$$\dot{x} = y, \qquad \dot{y} = -x,$$

whose phase diagram consists of circles, centre the origin. The positive half-paths consist of these circles, traversed endlessly. It is clear from Fig. 8.4 that every circular path is Poincaré stable; for if the path considered is on the circle of radius $(x_0^2 + y_0^2)^{1/2}$, starting at (x_0, y_0), the strip of arbitrary given width 2ε is bounded by two more circles, and every half-path starting within a distance $\delta = \varepsilon$ of (x_0, y_0) lies permanently in the strip.

By the same argument we can show that *the closed paths which make up any centre are all Poincaré stable paths*, because they are sandwiched between two other closed paths. The stability or instability of **limit cycles** (which are *isolated* closed paths) depends on the asymptotic behaviour of neighbouring paths. *If all the neighbouring paths approach the limit cycle then it is Poincaré stable.* If not, it is unstable. In Fig. 8.5, the first limit cycle is stable, but the other two are unstable.

Consider the system

$$\dot{x} = y, \qquad \dot{y} = x,$$

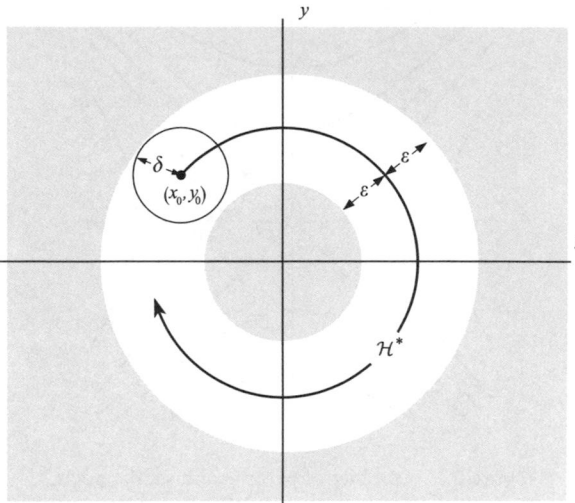

Figure 8.4 Poincaré stability of the system $\dot{x} = y, \quad \dot{y} = -x$.

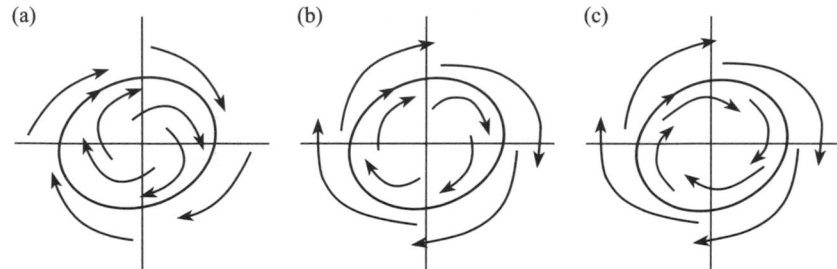

Figure 8.5 Limit cycles: (a) is Poincaré stable; (b) and (c) are not.

the paths being given by $x^2 - y^2 = C$ where C is the parameter of the family. The phase diagram shows a saddle point (Fig. 8.6).

On the basis of the previous discussion it can be seen that the half-paths starting typically at A and C, and all the curved half-paths, are Poincaré stable. Those starting typically at B and D are unstable despite the fact that they approach the origin, since every nearby path eventually goes to infinity.

Consider now the special half-paths which are **equilibrium points**, representing constant solutions. A saddle (for example, Fig. 8.6) is unstable, since there is no circle with centre at the equilibrium point, of however small radius, such that *every* half-path starting in it remains within an arbitrary preassigned distance ε of the equilibrium point.

Those types of equilibrium point which we earlier designated as being stable—the centre, the stable node, and the stable spiral—are indeed Poincaré stable according to the formal definition (8.1), though a rigorous proof is difficult. In the case of a centre, however, a simple geometrical argument is convincing. Suppose the centre to be the origin (Fig. 8.7). Then Definition 8.1 requires that if, given any $\varepsilon > 0$, a circle is drawn, $\mathcal{C}_\varepsilon$, centre the origin

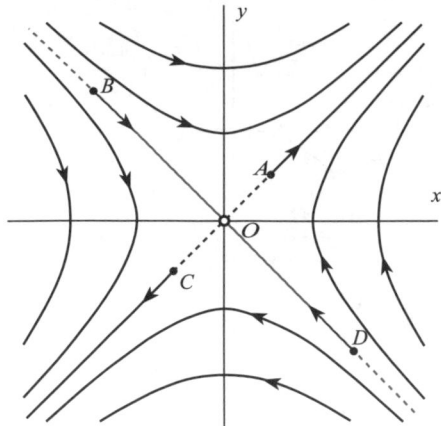

Figure 8.6   Stability of paths near a saddle point.

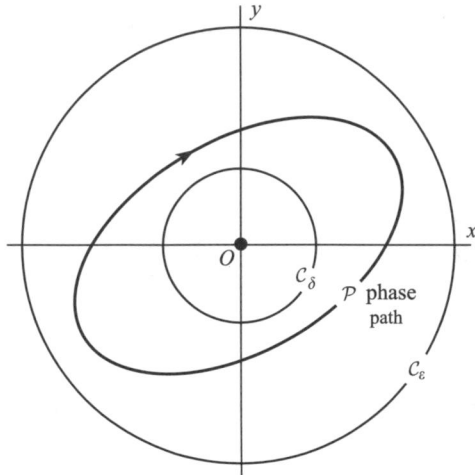

Figure 8.7   Stability of a centre at O.

and radius ε, then a circle $\mathcal{C}_\delta$ of radius δ can be found such that no half-path starting in $\mathcal{C}_\delta$ will ever reach $\mathcal{C}_\varepsilon$. The construction is clear from the figure: first a path $\mathcal{P}$ is chosen wholly inside $\mathcal{C}_\varepsilon$, then $\mathcal{C}_\delta$ is drawn inside $\mathcal{P}$. No path starting in $\mathcal{C}_\delta$ can cross $\mathcal{P}$, so it can never get outside $\mathcal{C}_\varepsilon$.

Exercise 8.1
Find the equation of the phase paths of $\dot{x} = 1 + x^2$, $\dot{y} = -2xy$. It is obvious from the phase diagram that $y = 0$ in Poincaré stable. Show that for the path $y = 0$, all paths which start in $(x + 1)^2 + y^2 = \delta^2$ subsequently remain in a circle of radius $\delta[1 + (1 + \delta)^2]$ centred on $y = 0$.

8.2 Paths and solution curves for general systems

Poincaré stability is concerned with autonomous systems; another concept is required for non-autonomous systems. Even when a system is autonomous, however, a more sensitive criterion may be needed which cannot be decided from mere inspection of the phase diagram.

Consider, for example, the system

$$\dot{x} = y, \qquad \dot{y} = -\sin x,$$

representing the motion of a pendulum. Near the origin, which is a centre, the phase paths have the form shown in Fig. 8.8. Consider a time solution $x^*(t)$, $t \geq t_0$, with $x^*(t_0) = a^*$, its half-path $\mathcal{H}^*$ being the heavy line starting at P; and a second time solution $x(t)$, $t \geq t_0$, with $x(t_0) = a$, its half-path $\mathcal{H}$ starting at Q.

The half-path $\mathcal{H}^*$ is Poincaré stable: if Q is close to P the corresponding paths remain close. However, *the representative points P', Q' do not remain close.* It is known (Problem 33, Chapter 1) that the period of $x(t)$ in Fig. 8.8 is greater than that of $x^*(t)$. Therefore the representative point Q' lags increasingly behind P'. No matter how close Q and P are at time t_0, after a sufficient time Q' will be as much as half a cycle behind P' and the difference between the *time solutions* at this time will therefore become large. This condition recurs at regular intervals. The representative points P' and Q' are like two planets in stable orbits; they are sometimes close, sometimes at opposite ends of their orbits. If we choose Q closer to P at the start it will take longer for Q' and P' to reach the maximum separation, but this will occur eventually. If the pendulum (or other nonlinear oscillating system) were simply used to take a system through a sequence of states at more or less predetermined intervals this might not be important, but for timekeeping, for example, the accumulating error arising from a

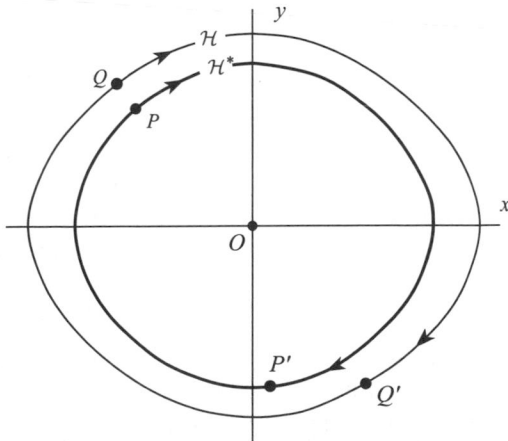

Figure 8.8 Paths for the pendulum equation are Poincaré stable, but representative points do not stay in step. P is the point a^*, Q the point a, both at time t_0: P' and Q' are their subsequent positions.

(a)   (b)

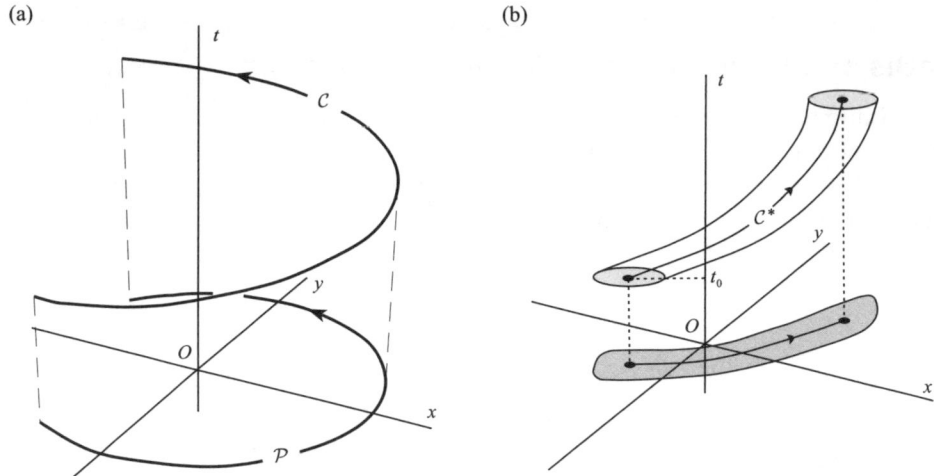

Figure 8.9   (a) A solution $\mathcal{C}$ in (x, y, t) space and its projection $\mathcal{P}$ on to the (x, y) plane which is the corresponding phase path; (b) a tube of solutions surrounding a particular solution $\mathcal{C}^*$ and its projection in the phase plane.

small change in amplitude would make an uncontrolled pendulum useless. We require a more stringent form of stability criterion to take this into account.

Time-dependent behaviour may be illustrated by displaying the time solutions $x(t), y(t)$ in a three-dimensional graph, rather than as phase paths in a two-dimensional graph as in Chapter 1, Section 1.8. This depiction is also applicable to nonautonomous systems, which have the form $\dot{x} = X(x, y, t), \ \dot{y} = Y(x, y, t)$.

In Fig. 8.9(a), the axes are x, y, t, and a solution $x(t), \ y(t)$ appears as a curve $\mathcal{C}$, which advances steadily along the t axis. The corresponding phase path $\mathcal{P}$ is the projection of $\mathcal{C}$ on to the (x, y) plane. According to the uniqueness theorem (see Appendix A), different solution curves do not intersect.

Figure 8.9(b) represents a narrow tube or bundle of time-solutions, all starting at the same time $t = t_0$ and surrounding a particular solution curve $\mathcal{C}^*$, whose stability we shall consider. If the tube continues to be narrow for $t \to \infty$, as suggested by the figure, then so does its projection on the (x, y) plane; this indicates that the half-path corresponding to $\mathcal{C}^*$ in the phase plane is Poincaré stable. We have seen (Fig. 8.8) that a phase path may be Poincaré stable even when the representative points become widely separated, but in the case of Fig. 8.9(b) this is not to; solutions which start by being close remain close for all time. This represents a situation which is in a sense more stable than is required by the Poincaré criterion.

In contrast Fig. 8.10 illustrates the case of a centre in the phase plane displayed in the x, y, t frame, where the period of the oscillations varies with the amplitude (compare Fig. 8.8). The solution curves corresponding to neighbouring phase paths are helices having different pitches. However close the initial points may be, the representative points become widely separated in these axes at certain times, although the phase paths imply Poincaré stability.

We shall formulate a more demanding criterion for stability, which distinguishes such cases, in the next section.

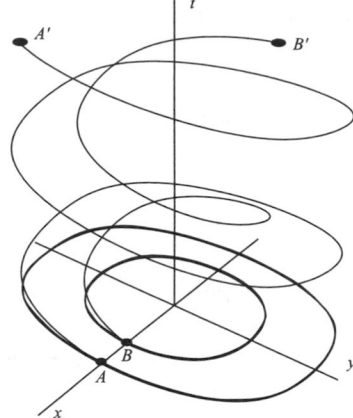

Figure 8.10 Solution curves corresponding to paths about a centre. They start close initially at A and B but become widely separated at A' and B' at a later time.

8.3 Stability of time solutions: Liapunov stability

Consider a regular dynamical system (see Appendix A), not necessarily autonomous, in n dimensions, written in vector form as

$$\dot{\boldsymbol{x}} = \boldsymbol{X}(\boldsymbol{x}, t), \qquad (8.2)$$

or in component form as the simultaneous system

$$\dot{x}_1 = X_1(x_1, x_2, \ldots, x_n, t),$$
$$\dot{x}_2 = X_2(x_1, x_2, \ldots, x_n, t),$$
$$\ldots$$
$$\dot{x}_n = X_n(x_1, x_2, \ldots, x_n, t).$$

We need a measure of the separation between pairs of solutions at particular *times*. Here we shall use the extension to n dimensions of the ordinary idea of the separation between two points with given position vectors, as in vector geometry. However, we shall also need to consider complex solutions at a later stage, and this must also be taken into account. It is sometimes convenient to use terms such as $\mathrm{Re}(\mathrm{e}^{it})$ rather than $\cos t$.

Suppose that $\boldsymbol{x}^*(t)$ and $\boldsymbol{x}(t)$ are two real or complex solution vectors of (8.2) with components

$$\boldsymbol{x}^*(t) = [x_1^*(t), x_2^*(t), \ldots, x_n^*(t)]^{\mathrm{T}},$$
$$\boldsymbol{x}(t) = [x_1(t), x_2(t), \ldots, x_n(t)]^{\mathrm{T}}.$$

(where T stands for transpose: these are column vectors). The separation between them at any time t will be denoted by the symbol $\|\boldsymbol{x}(t) - \boldsymbol{x}^*(t)\|$, defined by

$$\|\boldsymbol{x}(t) - \boldsymbol{x}^*(t)\| = \left(\sum_{i=1}^{n} |x_i(t) - x_i^*(t)|^2\right)^{1/2}, \tag{8.3}$$

where $|\ldots|$ denotes the modulus in the complex number sense, and the ordinary magnitude when x_i and x_i^* are real solutions.

There are many different measures of separation between two vectors which may be used. All are usually denoted by $\|\ldots\|$; such a measure is called a **metric** or **distance function** on the space. Most of the theory here would be unaffected by the choice of metric since they all have similar properties; for example they satisfy the triangle inequality:

$$\|\boldsymbol{u} + \boldsymbol{v}\| \leq \|\boldsymbol{u}\| + \|\boldsymbol{v}\|,$$

where $\boldsymbol{u}$ and $\boldsymbol{v}$ are any complex n-dimensional vectors. Corresponding to the distance function (8.3), the **norm** of vector $\boldsymbol{u}$ is defined by

$$\|\boldsymbol{u}\| = \left(\sum_{i=1}^{n} |u_i|^2\right)^{1/2}. \tag{8.4}$$

The norm (8.4) measures the magnitude of the vector concerned; hence the term 'triangle inequality'. For further information see Appendix C.

The enhanced type of stability described at the end of the previous section is called **Liapunov stability**, or **solution stability**. The formal definition is as follows:

Definition 8.2 (Liapunov stability)   *Let $\boldsymbol{x}^*(t)$ be a given real or complex solution vector of the n-dimensional system $\dot{\boldsymbol{x}} = X(\boldsymbol{x}, t)$. Then*

  (i) *$\boldsymbol{x}^*(t)$ is Liapunov stable for $t \geq t_0$ if, and only if, to each value of $\varepsilon > 0$, however small, there corresponds a value of $\delta > 0$ (where δ may depend only on ε and t_0) such that*

$$\|\boldsymbol{x}(t_0) - \boldsymbol{x}^*(t_0)\| < \delta \implies \|\boldsymbol{x}(t) - \boldsymbol{x}^*(t)\| < \varepsilon \tag{8.5}$$

  for all $t \geq t_0$, where $\boldsymbol{x}(t)$ represents any other neighbouring solution.
 (ii) *If the given system is autonomous, the reference to t_0 in (i) may be disregarded; the solution $\boldsymbol{x}^*(t)$ is either Liapunov stable, or not, for all t_0.*
(iii) *Otherwise the solution $\boldsymbol{x}^*(t)$ is unstable in the sense of Liapunov.*

In other words, (8.5) requires that no matter how small is the permitted deviation, measured by ε, there always exists a nonzero tolerance, δ, in the initial conditions when the system is activated, allowing it to run satisfactorily.

It can be also be shown (Cesari 1971) that if (i) is satisfied for initial conditions at time t_0, then a similar condition is satisfied when any $t_1 > t_0$ is substituted for t_0: that is, if $\boldsymbol{x}^*(t)$ is stable for $t \geq t_0$, it is stable for $t \geq t_1 \geq t_0$.

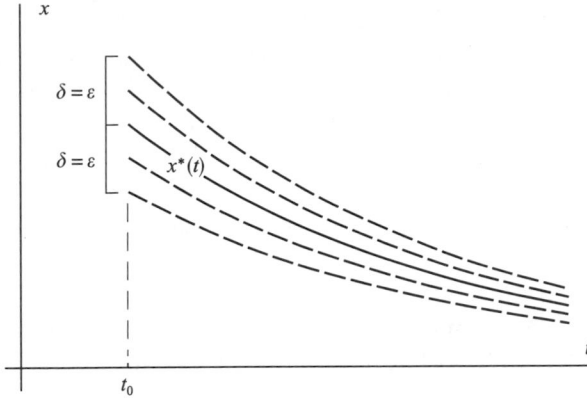

Figure 8.11   Liapunov stability of the solution $x^*(t) = x(t_0)\mathrm{e}^{-(t-t_0)}$.

It can be proved that if the system equations are *autonomous*, then *Liapunov stability* implies Poincaré stability (Cesari 1971).

Example 8.1   *Show that all solutions of $\dot{x} = -x$ (a one-dimensional system) are stable in the Liapunov sense.*
We will consider real solutions only, with initial condition at t_0. The general solution can be written in the form

$$x(t) = x(t_0)\mathrm{e}^{-(t-t_0)},$$

where $x(t_0)$ is the initial value. Consider the stability of $x^*(t)$ starting at $(t_0, x^*(t_0))$ (Fig. 8.11). Choose $\varepsilon > 0$ arbitrarily and consider the strip $x^*(t) - \varepsilon < x < x^*(t) + \varepsilon$. For stability, we must show that all solutions starting at t_0 and sufficiently close to $x^*(t_0)$ lie inside this strip. Since all solutions 'close up' on $x^*(t)$ as t increases it is clear that the conditions of the definition are satisfied by the choice $\delta = \varepsilon$. This is true for *all* initial values $x^*(t_0)$.   ●

The following example shows that the corresponding n-dimensional problem leads to the same conclusion.

Example 8.2   *Show that all solutions of the n-dimensional system $\dot{\boldsymbol{x}} = -\boldsymbol{x}$ are stable for $t \geq 0$ in the Liapunov sense.*
The general solution is given by

$$\boldsymbol{x}(t) = \boldsymbol{x}(0)\mathrm{e}^{-t}.$$

Consider the stability of $x^*(t)$ for $t > 0$, where

$$\boldsymbol{x}^*(t) = \boldsymbol{x}^*(0)\mathrm{e}^{-t}.$$

For any $x^*(0)$ we have

$$\|\boldsymbol{x}(t) - \boldsymbol{x}^*(t)\| \leq \|\boldsymbol{x}(0) - \boldsymbol{x}^*(0)\|\mathrm{e}^{-t} \leq \|\boldsymbol{x}(0) - \boldsymbol{x}^*(0)\|, \quad t \geq 0.$$

Therefore, given any $\varepsilon > 0$,

$$\|\boldsymbol{x}(0) - \boldsymbol{x}^*(0)\| < \varepsilon \Rightarrow \|\boldsymbol{x}(t) - \boldsymbol{x}^*(t)\| < \varepsilon, \quad t > 0.$$

Thus $\delta = \varepsilon$ in Definition 8.2. (The process can be modified to hold for any t_0. However, the system is autonomous, so this is automatic.)   ●

The implication of (ii) in Definition 8.2, which also applies to non-autonomous systems, is as follows. (i) taken alone implies that arbitrarily small enough changes in the initial condition at $t = t_0$ lead to uniformly small departures from the tested solution on $t_0 \leq t < \infty$. Such behaviour could be described as stability with respect only to variations in initial conditions at t_0. Armed with (ii), however, it can be said that the solution is stable with respect to variation in initial conditions at all times t_1 where $t_1 \geq t_0$, and that the stability is therefore a property of the solution as a whole. (ii) holds by virtue of (i) under very general smoothness conditions that we do not discuss here.

If (i) is satisfied, and therefore (ii), we may still want to be assured that a particular solution does not become 'less stable' as time goes on. It is possible that a system's sensitivity to disturbance might increase indefinitely with time although it remains technically stable, the symptom being that $\delta(\varepsilon, t_0)$ decreases to zero as t_0 increases.

To see that this is possible, consider the family of curves with parameter c:

$$x = f(c, t) = ce^{(c^2-1)t}/t, \quad t > 0. \tag{8.6}$$

These are certainly solutions of *some* first-order equation. The equation is of no particular interest: the Definition 8.2 refers only to the family of solutions $x(t)$, and does not require us to refer to the differential equation.

In Fig. 8.12, the curves separating those which tend to infinity $(c^2 > 1)$ from those tending to zero $(c^2 < 1)$ are the curves $f(1, t) = 1/t$ and $f(-1, t) = -1/t$, which also tend to zero. Now consider the stability of the 'solution' having $c = 0 : f(0, t) \equiv 0$. It is apparent from the diagram that this function is stable for all $t_0 > 0$. However, the system can afford a disturbance from the initial state $x(t_0) = 0$ at a time t_0 no greater than the height of the ordinate $A_0 B_0$; otherwise curves which approach infinity will be included. As t_0 increases, the greatest permitted

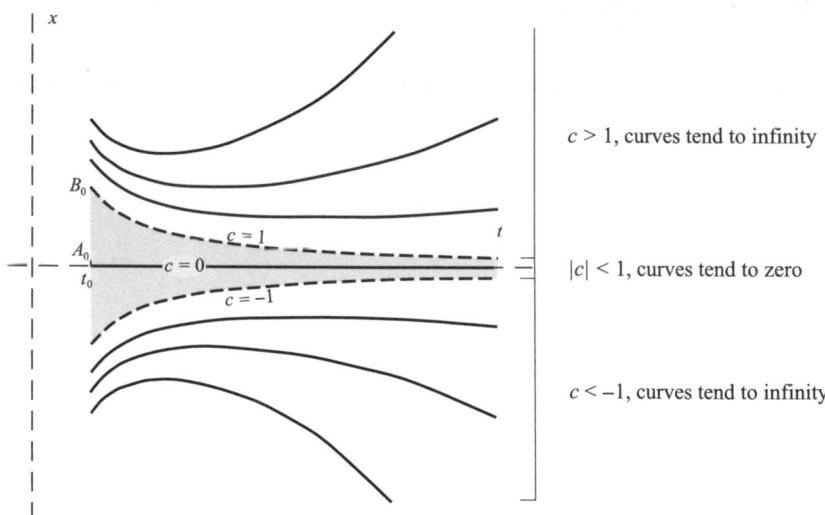

Figure 8.12   Family of curves $x(t) = ce^{(c^2-1)t}/t$.

disturbance decreases to zero. Therefore the system becomes progressively more sensitive as t_0 increases, and approaches zero tolerance of disturbance as $t_0 \to \infty$.

The stability is said to be **non-uniform** in this case. The definition of **uniform stability** is as follows.

Definition 8.3 (Uniform stability)   *If a solution is stable for $t \geq t_0$, and the δ of Definition 8.2 is independent of t_0, the solution is uniformly stable on $t \geq t_0$.*

It is clear that any stable solutions of an **autonomous** system are uniformly stable, since the system is invariant with respect to time translation.

A third desirable property is **asymptotic stability**. The system $\dot{x} = 0$ has the general solution

$$x(t) = x(t_0) = C,$$

where C is an arbitrary constant vector, and t_0 is any value of t.

These are all stable on $t \geq t_0$ for any t_0 (and also uniformly stable); however, a disturbed solution shows no tendency to return to the original solution: it remains a constant distance away. On the other hand, solutions of the type examined in Figs 8.11 and 8.12 do re-approach the undisturbed solution after being disturbed; thus the system tends to return to its original operating curve. Such solutions are said to be **asymptotically stable**, uniformly and nonuniformly respectively in the cases of Figs 8.11 and 8.12.

Definition 8.4 (Asymptotic stability)   *Let x^* be a stable (or uniformly stable) solution for $t \geq t_0$. If additionally there exists $\eta(t_0) > 0$ such that*

$$\| x(t_0) - x^*(t_0) \| < \eta \Rightarrow \lim_{t \to \infty} \| x(t) - x^*(t) \| = 0, \tag{8.7}$$

then the solution is said to be asymptotically stable (or uniformly and asymptotically stable).

In the rest of this chapter we prove theorems enabling general statements to be made about the stability of solutions of certain classes of equations.

Exercise 8.2
Find the time solutions of

$$x\ddot{x} - \dot{x}^2 = 0, \quad x > 1$$

which satisfy $x(0) = 1$. Decide which solutions are asymptotically stable.

8.4 Liapunov stability of plane autonomous linear systems

We consider first the stability in the Liapunov sense of the **constant solutions** (equilibrium points) of the two-dimensional, constant-coefficient systems classified in Chapter 2. Without loss of generality we can place the equilibrium point at the origin, since this only involves modifying the solutions by additive constants. We have then

$$\dot{x}_1 = ax_1 + bx_2, \qquad \dot{x}_2 = cx_1 + dx_2, \tag{8.8}$$

where a, b, c, d are constants.

The stability properties of the constant solutions $x^*(t) = [x_1^*(t), x_2^*(t)]^T = 0$, $t \geq t_0$ can sometimes be read off from the phase diagram: for example a saddle is obviously unstable.

Consider also the case of a *centre*. Figure 8.13(a) shows the phase plane and 8.13(b) the solution space: the solution $x^*(t) = 0$ whose stability is being considered lies along the t axis.

Choose an arbitrary value of $\varepsilon > 0$. In the phase diagram, Fig. 8.13(a), the region in which the perturbed paths are required to remain is the interior of the outer circle C_ε:

$$\| x \| < \varepsilon.$$

The initial time t_0 is immaterial since the equations (8.8) are autonomous.

This region contains closed phase paths: choose any one of them, say $\mathcal{P}$ in Fig. 8.13(a). Construct a region of initial values about the origin (which corresponds to the zero solution $x^* = 0$) bounded by the circle C_δ:

$$\| x(t_0) - x^*(t_0) \| = \| x(t_0) \| < \delta,$$

by choosing δ small enough for this region to lie entirely within $\mathcal{P}$. Then any half-path originating in the interior of C_δ remains permanently inside the closed path $\mathcal{P}$, and therefore also satisfies (8.9) for all $t \geq t_0$. Since ε is arbitrary the condition (8.5) for Liapunov stability of the zero solution $x^*(t)$ is satisfied.

Since δ is independent of t_0, the zero solution is uniformly stable; this is also a consequence of the autonomous property of the system (8.8). It is not asymptotically stable since the perturbed solutions do not approach zero as $t \to \infty$.

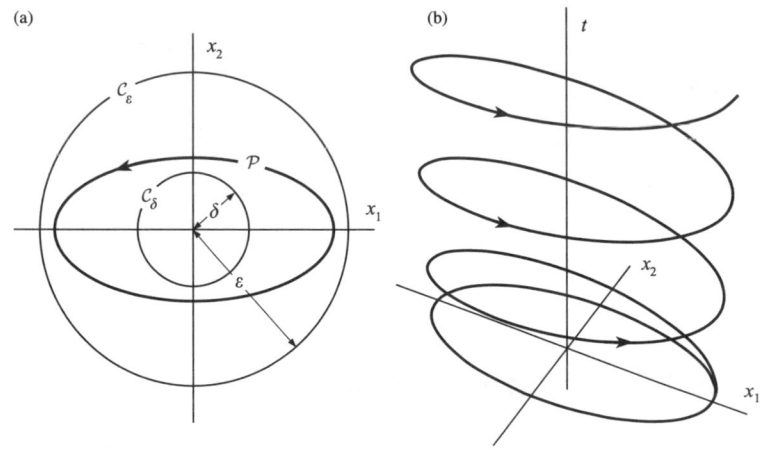

Figure 8.13   Phase path and solution space near a centre.

The other types of equilibrium point (zero solutions of (8.8)) are more difficult, and we merely summarize the results:

Stability properties of the zero solutions of $\dot{x} = ax + by, \quad \dot{y} = cx + dy$

Phase plane feature	Liapunov stability property
Centre: Poincaré stable	Uniformly stable
Stable spiral: Poincaré stable; paths approach zero	Uniformly, asymptotically stable
Stable node: Poincaré stable; paths approach zero	Uniformly, asymptotically stable
Saddle: Poincaré unstable	Unstable
Unstable spiral and node	Unstable

(8.9)

We shall show that the classification (8.9) applies to all solutions, and not merely to zero solutions

The most general linear system is the nonautonomous and nonhomogeneous equation in n variables given by

$$\dot{x} = A(t)x + f(t), \tag{8.10}$$

where $A(t)$ is an $n \times n$ matrix. We wish to investigate the stability of a solution $x^*(t)$. Let $x(t)$ represent any other solution, and define $\xi(t)$ by

$$\xi(t) = x(t) - x^*(t). \tag{8.11}$$

Then $\xi(t)$ tracks the difference between the 'test' solution and a solution having a different initial value at time t_0. The initial condition for ξ is

$$\xi(t_0) = x(t_0) - x^*(t_0). \tag{8.12}$$

Also, ξ satisfies the homogeneous equation derived from (8.10):

$$\dot{\xi} = A(t)\xi. \tag{8.13}$$

By comparison of (8.11), (8.12), and (8.13) with Definition (8.2), it can be seen that the Liapunov stability property of $x^*(t)$ is the same as the stability of the zero solution of (8.13). $\xi(t)$ is called a **perturbation** of the solution $x^*(t)$.

Since this new formulation of the problem is independent of the solution of (8.10) initially chosen, we can make the following statement:

Theorem 8.1 *All solutions of the regular linear system $\dot{x} = A(t)x + f(t)$ have the same Liapunov stability property (unstable, stable, uniformly stable, asymptotically stable, uniformly and asymptotically stable). This is the same as that of the zero (or any other) solution of the homogeneous equation $\dot{\xi} = A(t)\xi$.* ■

Notice that the stability of time solutions of linear systems does not depend at all on the forcing term $f(t)$.

Example 8.3 *All the solutions of the system $\dot{x}_1 = x_2$, $\dot{x}_2 = -\omega^2 x_1 + f(t)$ are uniformly stable, but not asymptotically stable (forced linear oscillator).*

Equation (8.13) becomes $\dot{\xi}_1 = \xi_2$, $\dot{\xi}_2 = -\omega^2 \xi_1$. The zero solution is a centre, which is uniformly stable (see (8.9)). By Theorem 8.1, all solutions of the non-homogeneous equation are also uniformly stable. ●

Example 8.4 *All the solutions of the equations $\dot{x}_1 = x_2$, $\dot{x}_2 = -kx_2 - \omega^2 x_1 + f(t)$ are uniformly and asymptotically stable (damped, forced linear oscillator).*

Equation (8.13) becomes

$$\dot{\xi}_1 = \xi_2, \qquad \dot{\xi}_2 = -\omega^2 \xi_1 - k\xi_2.$$

The zero solution corresponds to the equilibrium point $\xi_1 = \xi_2 = 0$ which (Chapter 2) is a 'stable' node or spiral (as named in that chapter). By (8.9), these are uniformly and asymptotically stable. This result corresponds to the damping out of free oscillations set up by the initial conditions: the amplitude and phase of the forced oscillation are independent of these. ●

Exercise 8.3
Solve the forced linear system

$$\begin{bmatrix} \dot{x} \\ \dot{y} \end{bmatrix} = \begin{bmatrix} -1 & 1 \\ 0 & -2 \end{bmatrix} \begin{bmatrix} x \\ y \end{bmatrix} + \begin{bmatrix} 0 \\ e^t \end{bmatrix}.$$

Explain why all solutions are asymptotically stable confirming Theorem 8.1.

8.5   Structure of the solutions of n-dimensional linear systems

The need to consider the general properties of **linear systems** arises in connection with nonlinear systems which are in some sense close to linear. In such cases certain characteristics of their solutions, notably stability properties, may follow those of the approximating linear systems.

The general homogeneous, first-order linear system of n dimensions is

$$\dot{x} = A(t)x, \tag{8.14}$$

where $A(t)$ is an $n \times n$ matrix whose elements $a_{ij}(t)$ are functions of time, and $x(t)$ is a column vector of the n dependent variables. The explicit appearance of t indicates that the system may be nonautonomous. In component form,

$$\begin{bmatrix} \dot{x}_1 \\ \dot{x}_2 \\ \vdots \\ \dot{x}_n \end{bmatrix} = \begin{bmatrix} a_{11} & a_{12} & \cdots & a_{1n} \\ a_{21} & a_{22} & \cdots & a_{2n} \\ \vdots & \vdots & \vdots & \vdots \\ a_{n1} & a_{n2} & \cdots & a_{nn} \end{bmatrix} \begin{bmatrix} x_1 \\ x_2 \\ \vdots \\ x_n \end{bmatrix},$$

or, when expanded,

$$\dot{x}_i = \sum_{j=1}^{n} a_{ij}(t) x_j; \quad i = 1, 2, \dots, n.$$

We shall assume that each $a_{ij}(t)$ is continuous on $-\infty < t < \infty$, so that the system is regular (Appendix A). In that case, given any initial conditions $x(t_0) = x_0$, there is a unique solution satisfying this condition. Moreover, it exists for $-\infty < t < \infty$ (the same is not true in general of nonlinear equations).

If $x_1(t), x_2(t), \dots, x_m(t)$ are real or complex vector solutions of (8.14), then so is $\alpha_1 x_1(t) + \alpha_2 x_2(t) + \cdots + \alpha_m x_m(t)$ where $\alpha_1, \alpha_2, \dots, \alpha_m$ are any constants, real or complex. A central question to be settled is the dimension of the solution space: the minimum number of solutions whose linear combinations generate every solution of the homogeneous system (8.14).

Definition 8.5 (linearly dependent vector functions)   *Let $\psi_1(t), \psi_2(t), \dots, \psi_m(t)$ be vector functions (real or complex), continuous on $-\infty < t < \infty$, none being identically zero. If there exist constants (real or complex) $\alpha_1, \alpha_2, \dots, \alpha_m$, not all zero, such that*

$$\alpha_1 \psi_1(t) + \alpha_2 \psi_2(t) + \dots + \alpha_m \psi_m(t) = 0$$

for $-\infty < t < \infty$, the functions are **linearly dependent**. *Otherwise they are* **linearly independent**.

(As a consequence of the definition, note that the vector functions $[1, 1]^{\mathrm{T}}, [t, t]^{\mathrm{T}}$ are *linearly independent*, although the constant vectors $[1, 1]^{\mathrm{T}}, [t_0, t_0]^{\mathrm{T}}$ are *linearly dependent* for every t_0.)

Example 8.5 $\cos t$ *and* $\sin t$ *are linearly independent on* $-\infty < t < \infty$.
In amplitude/phase form

$$\alpha_1 \cos t + \alpha_2 \sin t = \sqrt{(\alpha_1^2 + \alpha_2^2)} \sin(t + \beta)$$

where β is given by

$$\alpha_1 = \sqrt{(\alpha_1^2 + \alpha_2^2)} \sin \beta, \qquad \alpha_2 = \sqrt{(\alpha_1^2 + \alpha_2^2)} \cos \beta.$$

There is no choice of α_1, α_2, except for $\alpha_1 = \alpha_2 = 0$, which makes this function zero for all t.   ●

Example 8.6   *The functions* $\cos t$, $\sin t$, $2 \sin t$ *are linearly dependent.*
Since, for example,

$$0(\cos t) + 2(\sin t) - 1(2 \sin t) = 0$$

for all t, the result follows. Here $\alpha_1 = 0$, $\alpha_2 = 2$ and $\alpha_3 = -1$.   ●

If functions are linearly dependent, at least one of them can be expressed as a linear combination of the others. From now on we shall be concerned with the linear dependence of solutions of the homogeneous linear system (8.14). Given any set of n solution (column) vectors $\boldsymbol{\phi}_1, \boldsymbol{\phi}_2, \ldots, \boldsymbol{\phi}_n$, real or complex, where $\boldsymbol{\phi}_j$ has elements $\phi_{1j}(t), \phi_{2j}(t), \ldots, \phi_{nj}(t)$, we shall use the notation

$$\boldsymbol{\Phi}(t) = [\boldsymbol{\phi}_1, \boldsymbol{\phi}_2, \ldots, \boldsymbol{\phi}_n] = \begin{bmatrix} \phi_{11} & \phi_{12} & \cdots & \phi_{1n} \\ \phi_{21} & \phi_{22} & \cdots & \phi_{2n} \\ \vdots & \vdots & & \vdots \\ \phi_{n1} & \phi_{n2} & \cdots & \phi_{nn} \end{bmatrix}$$

for the matrix of these solutions. Then from (8.14), since $\dot{\boldsymbol{\phi}}_j(t) = A(t)\boldsymbol{\phi}_j(t)$,

$$\dot{\boldsymbol{\Phi}}(t) = [\dot{\boldsymbol{\phi}}_1, \dot{\boldsymbol{\phi}}_2, \ldots, \dot{\boldsymbol{\phi}}_n] = A(t)[\boldsymbol{\phi}_1, \boldsymbol{\phi}_2, \ldots, \boldsymbol{\phi}_n] = A(t)\boldsymbol{\Phi}(t), \tag{8.15}$$

where $\dot{\boldsymbol{\Phi}}(t)$ represents the matrix $\boldsymbol{\Phi}(t)$ with all its elements differentiated.

Remember that the constituent solutions may be real or complex; for example it is convenient to be able to choose a matrix of solutions for

$$\dot{x}_1 = x_2, \qquad \dot{x}_2 = -x_1$$

of the form

$$\boldsymbol{\Phi}(t) = \begin{bmatrix} e^{it} & e^{-it} \\ ie^{it} & -ie^{-it} \end{bmatrix}.$$

Correspondingly, the constants occurring in linear dependence, and the eigenvalues and eigenvectors, may all be complex. This possibility will not always be specially pointed out in the rest of this chapter.

Theorem 8.2   *Any $n+1$ nonzero solutions of the homogeneous system $\dot{x} = A(t)x$ are linearly dependent.*

Proof   Let the solutions be $\boldsymbol{\phi}_1(t), \boldsymbol{\phi}_2(t), \ldots, \boldsymbol{\phi}_{n+1}(t)$. Let t_0 be any value of t. Then the $n+1$ *constant* vectors $\boldsymbol{\phi}_1(t_0), \boldsymbol{\phi}_2(t_0), \ldots, \boldsymbol{\phi}_{n+1}(t_0)$ are linearly dependent (as are any set of $n+1$ constant vectors of dimension n); that is, there exist constants $\alpha_1, \alpha_2, \ldots, \alpha_{n+1}$, not all zero, such that

$$\sum_{j=1}^{n+1} \alpha_j \boldsymbol{\phi}_j(t_0) = 0.$$

Let

$$x(t) = \sum_{j=1}^{n+1} \alpha_j \boldsymbol{\phi}_j(t).$$

Then $x(t_0) = 0$, and $x(t)$ is a solution of (1). Therefore, by the Uniqueness Theorem (Appendix A), $x(t) = 0$ for all t: that is to say, the solutions $\boldsymbol{\phi}_j(t), j = 1, 2, \ldots, n+1$ are linearly dependent. ∎

Theorem 8.3   *There exists a set of n linearly independent solutions of $\dot{x} = A(t)x$.*

Proof   By the Existence Theorem (Appendix A), there is a set of *n* solutions $\psi_1(t), \psi_2(t), \ldots, \psi_n(t)$ whose matrix $\Psi(t)$ satisfies

$$\Psi(0) = I,$$

where *I* is the identity matrix. Since the columns of $\Psi(0)$; $\psi_1(0), \ldots, \psi_n(0)$; are linearly independent, so are the solution vectors $\psi_1(t), \psi_2(t), \ldots, \psi_n(t)$. For if not, there exist numbers $\alpha_1, \alpha_2, \ldots, \alpha_n$, not all zero, such that $\sum_{j=1}^{n} \alpha_j \psi_j(t) = 0$ for all *t* (including $t = 0$), which is a contradiction. ∎

These two theorems settle the dimension of the solution space: every solution is a linear combination of the solutions ψ_j, $j = 1, 2, \ldots, n$ of Theorem 8.3; but since these solutions are themselves linearly independent, we cannot do without any of them. Instead of the special solutions $\psi_j(t)$ we may take any set of *n* linearly independent solutions as the basis, as in shown in Theorem 8.4.

Theorem 8.4   *Let $\phi_1(t), \phi_2(t), \ldots \phi_n(t)$ be any set of linearly independent vector solutions (real or complex) of the homogeneous system $\dot{x} = A(t)x$. Then every solution is a linear combination of these solutions.*

Proof   Let $\phi(t)$ be any non-trivial solution of (8.14). Then by Theorem 8.2, $\phi, \phi_1, \phi_2, \ldots, \phi_n$ are linearly dependent. The coefficient of ϕ in any relation of linear dependence is non-zero, or we should violate the requirement that the ϕ_j are linearly independent. Therefore ϕ is a linear combination of $\phi_1, \ldots, \phi_n$. ∎

Definition 8.6   *Let $\phi_1(t), \phi_2(t), \ldots, \phi_n(t)$ be n linearly independent solutions of the homogeneous system $\dot{x} = A(t)x$. Then the matrix*

$$\Phi(t) = [\phi_1, \phi_2, \ldots, \phi_n] = \begin{bmatrix} \phi_{11} & \phi_{12} & \cdots & \phi_{1n} \\ \phi_{21} & \phi_{22} & \cdots & \phi_{2n} \\ \vdots & \vdots & & \vdots \\ \phi_{n1} & \phi_{n2} & \cdots & \phi_{nn} \end{bmatrix}$$

*is called a **fundamental matrix** of the homogeneous system.*

Note that any two fundamental matrices Φ_1 and Φ_2 are related by

$$\Phi_2(t) = \Phi_1(t)C, \tag{8.16}$$

where *C* is a nonsingular $n \times n$ matrix (since by Theorem 8.4, each column of Φ_2 is a linear combination of the columns of Φ_1, and vice versa).

Theorem 8.5   *Given any $n \times n$ solution matrix $\Phi(t) = [\phi_1(t), \ldots, \phi_n(t)]$ of the homogeneous system $\dot{x} = A(t)x$, then either (i) for all t, $\det\{\Phi(t)\} = 0$, or (ii) for all t, $\det\{\Phi(t)\} \neq 0$. Case (i) occurs if and only if the solutions are linearly dependent, and case (ii) implies that $\Phi(t)$ is a fundamental matrix.*

Proof   Suppose $\det\{\mathbf{\Phi}(t_0)\} = 0$ for some t_0. Then the columns of $\mathbf{\Phi}(t_0)$ are linearly dependent: there exist constants $\alpha_1, \ldots, \alpha_n$, not all zero, such that $\alpha_1\boldsymbol{\phi}_1(t_0) + \ldots + \alpha_n\boldsymbol{\phi}_n(t_0) = 0$. Define $\boldsymbol{\phi}(t)$ by

$$\boldsymbol{\phi}(t) = \alpha_1\boldsymbol{\phi}_1(t) + \ldots + \alpha_n\boldsymbol{\phi}_n(t).$$

Then $\boldsymbol{\phi}(t)$ is a solution satisfying the initial condition $\boldsymbol{\phi}(t_0) = 0$. By the Uniqueness Theorem (Appendix A), $\boldsymbol{\phi}(t) = 0$ for all t. Therefore

$$\alpha_1\boldsymbol{\phi}_1(t) + \alpha_2\boldsymbol{\phi}_2(t) + \ldots + \alpha_n\boldsymbol{\phi}_n(t) = 0$$

for all t, and $\det\{\mathbf{\Phi}(t)\}$ is therefore either zero everywhere, or non-zero everywhere.   ∎

Example 8.7   *Verify that*

$$\begin{bmatrix} e^t \\ e^{2t} \end{bmatrix}, \quad \begin{bmatrix} e^{2t} \\ e^{4t} \end{bmatrix}$$

cannot be a pair of solutions of a second-order linear homogeneous system.
For, writing

$$\mathbf{\Phi}(t) = \begin{bmatrix} e^t & e^{2t} \\ e^{2t} & e^{4t} \end{bmatrix},$$

$\det\{\mathbf{\Phi}(t)\} = e^{5t} - e^{4t}$, which is zero at $t = 0$ but nowhere else. By Theorem 8.5(ii), $\mathbf{\Phi}(t)$ cannot be fundamental matrix.   ●

Theorem 8.6   *The solution of the homogeneous system $\dot{x} = A(t)x$ with the initial conditions $x(t_0) = x_0$ is given by $x(t) = \mathbf{\Phi}(t)\mathbf{\Phi}^{-1}(t_0)x_0$, where $\mathbf{\Phi}$ is any fundamental matrix of the system.*

Proof   The solution must be of the form

$$x(t) = \mathbf{\Phi}(t)a \tag{8.17}$$

where a is a constant vector, by Theorem 8.4. The initial conditions give $x_0 = \mathbf{\Phi}(t_0)a$. The columns of $\mathbf{\Phi}(t_0)$ (regarded as columns of constants) are linearly independent by Theorem 8.5, so $\mathbf{\Phi}(t_0)$ has an inverse $\mathbf{\Phi}^{-1}(t_0)$. Therefore $a = \mathbf{\Phi}^{-1}(t_0)x_0$ and the result follows from (8.17).   ∎

Example 8.8   *Verify that*

$$\begin{bmatrix} 2 \\ e^t \end{bmatrix}, \quad \begin{bmatrix} e^{-t} \\ 1 \end{bmatrix}$$

are solutions of

$$\begin{bmatrix} \dot{x}_1 \\ \dot{x}_2 \end{bmatrix} = \begin{bmatrix} 1 & -2e^{-t} \\ e^t & -1 \end{bmatrix} \begin{bmatrix} x_1 \\ x_2 \end{bmatrix},$$

and find the solution $x(t)$ *such that*

$$x(0) = \begin{bmatrix} 3 \\ 1 \end{bmatrix}.$$

Direct substitution confirms that the given functions are solutions. They are linearly independent, so

$$\Phi(t) = \begin{bmatrix} 2 & e^{-1} \\ e^t & 1 \end{bmatrix}$$

is a fundamental matrix. We have

$$\Phi(0) = \begin{bmatrix} 2 & 1 \\ 1 & 1 \end{bmatrix}, \qquad \Phi^{-1}(0) = \begin{bmatrix} 1 & -1 \\ -1 & 2 \end{bmatrix}.$$

Hence, by Theorem 8.6,

$$x(t) = \Phi(t)\Phi^{-1}(0)x(0) = \begin{bmatrix} 2 & e^{-t} \\ e^t & 1 \end{bmatrix}\begin{bmatrix} 1 & -1 \\ -1 & 2 \end{bmatrix}\begin{bmatrix} 3 \\ 1 \end{bmatrix} = \begin{bmatrix} 4 - e^{-t} \\ 2e^t - 1 \end{bmatrix}.$$

●

Exercise 8.4
By eliminating y solve the system

$$\dot{x} = \begin{bmatrix} \dot{x} \\ \dot{y} \end{bmatrix} = \begin{bmatrix} 1 & 1 \\ (-6 + 4t - t^2)/t^2 & (4 - t)/t \end{bmatrix}\begin{bmatrix} x \\ y \end{bmatrix}$$

for $t > 0$, and obtain a fundamental matrix solution for x.

8.6 Structure of n-dimensional inhomogeneous linear systems

The general **inhomogeneous linear system** is

$$\dot{x} = A(t)x + f(t), \tag{8.18}$$

where $f(t)$ is a column vector. The **associated homogeneous system** is

$$\dot{\phi} = A(t)\phi. \tag{8.19}$$

The following properties are readily verified.

(I) Let $x = x_{\mathrm{p}}(t)$ be any solution of (8.18) (called a **particular solution** of the given system) and $\phi = \phi_{\mathrm{c}}(t)$ any solution of (8.19) (called a **complementary function** for the given system). Then $x_{\mathrm{p}}(t) + \phi_{\mathrm{c}}(t)$ is a solution of (8.18).

(II) Let $x_{\mathrm{p}_1}(t)$ and $x_{\mathrm{p}_2}(t)$ be any solutions of (8.18). Then $x_{\mathrm{p}_1}(t) - x_{\mathrm{p}_2}(t)$ is a solution of (8.19); that is, it is a complementary function.

Theorem 8.7 follows immediately:

Theorem 8.7   *Let $x_{\mathrm{p}}(t)$ be any one particular solution of $\dot{x} = A(t)x + f(t)$. Then every solution of this equation is of the form $x(t) = x_{\mathrm{p}}(t) + \phi_{\mathrm{c}}(t)$, where ϕ_{c} is a complementary function, and conversely.*   ∎

The strategy for finding all solutions of (8.18) is therefore to obtain, somehow, any *one* solution of (8.18), then to find *all* the solutions of the associated homogeneous system (8.19).

Example 8.9 *Find all solutions of the system $\dot{x}_1 = x_2$, $\dot{x}_2 = -x_1 + t$.*
In the notation of (8.18),

$$\begin{bmatrix} \dot{x}_1 \\ \dot{x}_2 \end{bmatrix} = \begin{bmatrix} 0 & 1 \\ -1 & 0 \end{bmatrix} \begin{bmatrix} x_1 \\ x_2 \end{bmatrix} + \begin{bmatrix} 0 \\ t \end{bmatrix},$$

where

$$A = \begin{bmatrix} 0 & 1 \\ -1 & 0 \end{bmatrix}, \quad f(t) = \begin{bmatrix} 0 \\ t \end{bmatrix}.$$

The corresponding homogeneous system is $\dot{\phi}_1 = \phi_2$, $\dot{\phi}_2 = -\phi_1$, which is equivalent to $\ddot{\phi}_1 + \phi_1 = 0$. The linearly independent solutions $\phi_1 = \cos t, \sin t$ correspond respectively to $\phi_2 = -\sin t, \cos t$. Therefore all solutions of the corresponding homogeneous system are the linear combinations of

$$\begin{bmatrix} \cos t \\ -\sin t \end{bmatrix}, \quad \begin{bmatrix} \sin t \\ \cos t \end{bmatrix};$$

which are given in matrix form by

$$\phi(t) = \begin{bmatrix} \cos t & \sin t \\ -\sin t & \cos t \end{bmatrix} \begin{bmatrix} a_1 \\ a_2 \end{bmatrix}$$

where a_1, a_2 are arbitrary.
   It can be confirmed that $x_1 = t, x_2 = 1$ is a particular solution of the original system. Therefore all solutions are given by

$$x(t) = \begin{bmatrix} t \\ 1 \end{bmatrix} + \begin{bmatrix} \cos t & \sin t \\ -\sin t & \cos t \end{bmatrix} \begin{bmatrix} a_1 \\ a_2 \end{bmatrix},$$

where a_1, a_2 are arbitrary.   ●

Theorem 8.8 *The solution of the system $\dot{x} = A(t)x + f(t)$ with initial conditions $x(t_0) = x_0$ is given by*

$$x(t) = \Phi(t)\Phi^{-1}(t_0)x_0 + \Phi(t) \int_{t_0}^{t} \Phi^{-1}(s)f(s)ds,$$

where $\Phi(t)$ is any fundamental solution matrix of the corresponding homogeneous system $\dot{\Phi} = A(t)\phi$.

Proof   Let $x(t)$ be the required solution, for which the following form is postulated

$$x(t) = \Phi(t)\Phi^{-1}(t_0)\{x_0 + \phi(t)\}. \tag{8.20}$$

The inverses of $\Phi(t)$ and $\Phi^{-1}(t_0)$ exist since, by Theorem 8.5, they are non-singular. Then by the initial condition $x(t_0) = x_0$, or $x_0 + \phi(t_0)$ by (8.20), and so

$$\phi(t_0) = 0.$$

To find the equation satisfied by $\boldsymbol{\phi}(t)$, substitute (8.20) into the equation, which becomes

$$\dot{\boldsymbol{\Phi}}(t)\boldsymbol{\Phi}^{-1}(t_0)\{x_0 + \boldsymbol{\phi}(t)\} + \boldsymbol{\Phi}(t)\boldsymbol{\Phi}^{-1}(t_0)\dot{\boldsymbol{\phi}}(t)$$
$$= A(t)\boldsymbol{\Phi}(t)\boldsymbol{\Phi}^{-1}(t_0)\{x_0 + \boldsymbol{\phi}(t)\} + \boldsymbol{f}(t).$$

Since $\boldsymbol{\Phi}(t)$ is a solution matrix of the homogeneous equation, $\dot{\boldsymbol{\Phi}}(t) = A(t)\boldsymbol{\Phi}t$, and the previous equation then becomes

$$\boldsymbol{\Phi}(t)\boldsymbol{\Phi}^{-1}(t_0)\dot{\boldsymbol{\phi}}(t) = \boldsymbol{f}(t).$$

Therefore,

$$\dot{\boldsymbol{\phi}}(t) = \boldsymbol{\Phi}(t_0)\boldsymbol{\Phi}^{-1}(t)\boldsymbol{f}(t),$$

whose solution satisfying the initial condition is

$$\boldsymbol{\phi}(t) = \boldsymbol{\Phi}(t_0)\int_{t_0}^{t}\boldsymbol{\Phi}^{-1}(s)\boldsymbol{f}(s)\mathrm{d}s.$$

Therefore, by (8.20),

$$x(t) = \boldsymbol{\Phi}(t)\boldsymbol{\Phi}^{-1}(t_0)x_0 + \boldsymbol{\Phi}(t)\int_{t_0}^{t}\boldsymbol{\Phi}^{-1}(s)\boldsymbol{f}(s)\mathrm{d}s.\qquad\blacksquare$$

For an alternative form of solution when A is a constant matrix, see Theorem 8.13 in Section 8.10.

Example 8.10   *Find the solution of*

$$\dot{x} = A(t)x + \boldsymbol{f}(t),$$

where

$$x = \begin{bmatrix} x_1 \\ x_2 \\ x_3 \end{bmatrix}, \quad A(t) = \begin{bmatrix} 0 & 1 & 0 \\ 1 & 0 & 0 \\ te^{-t} & te^{-t} & 1 \end{bmatrix}, \quad \boldsymbol{f}(t) = \begin{bmatrix} e^t \\ 0 \\ 1 \end{bmatrix},$$

which satisfies the initial conditions $x(0) = [0, 1, -1]^{\mathrm{T}}$.

The solution is given by the formula in Theorem 8.8. We first require a fundamental solution matrix of the associated homogeneous system $\dot{\boldsymbol{\phi}} = A(t)\boldsymbol{\phi}$. In component form, this equation separates into

$$\dot{\phi}_1 = \phi_2, \quad \dot{\phi}_2 = \phi_1, \quad \dot{\phi}_3 - \phi_3 = te^{-t}(\phi_1 + \phi_2).$$

From the first two equations

$$\phi_1 = Ae^t + Be^{-t}, \quad \phi_2 = Ae^t - Be^{-t}.$$

The third equation now becomes

$$\dot{\phi}_3 - \phi_3 = 2At.$$

which has the general solution

$$\phi_3 = -2A(1+t) + Ce^t.$$

Hence a fundamental solution matrix is

$$\Phi(t) = \begin{bmatrix} e^t & e^{-t} & 0 \\ e^t & -e^{-t} & 0 \\ -2(1+t) & 0 & e^t \end{bmatrix}.$$

Since $\det[\Phi(t)] = -2e^t$,

$$\Phi^{-1}(t) = \frac{1}{2} \begin{bmatrix} e^{-t} & e^{-t} & 0 \\ e^t & -e^t & 0 \\ 2(1+t)e^{-2t} & 2(1+t)e^{-2t} & 2e^{-t} \end{bmatrix},$$

and

$$\Phi^{-1}(0) = \frac{1}{2} \begin{bmatrix} 1 & 1 & 0 \\ 1 & -1 & 0 \\ 2 & 2 & 2 \end{bmatrix}.$$

Thus the required solution is

$$x(t) = \begin{bmatrix} e^t & e^{-t} & 0 \\ e^t & -e^{-t} & 0 \\ -2(1+t) & 0 & e^t \end{bmatrix} \left\{ \begin{bmatrix} \frac{1}{2} & \frac{1}{2} & 0 \\ \frac{1}{2} & -\frac{1}{2} & 0 \\ 1 & 1 & 1 \end{bmatrix} \begin{bmatrix} 0 \\ 1 \\ -1 \end{bmatrix} \right.$$

$$\left. + \frac{1}{2} \int_0^t \begin{bmatrix} e^{-s} & e^{-s} & 0 \\ e^s & -e^s & 0 \\ 2(1+s)e^{-2s} & 2(1+s)e^{-2s} & 2e^{-s} \end{bmatrix} \begin{bmatrix} e^s \\ 0 \\ 1 \end{bmatrix} ds \right\}$$

$$= \begin{bmatrix} e^t & e^{-t} & 0 \\ e^t & -e^{-t} & 0 \\ -2(1+t) & 0 & e^t \end{bmatrix} \left\{ \begin{bmatrix} \frac{1}{2} \\ -\frac{1}{2} \\ 0 \end{bmatrix} + \frac{1}{2} \int_0^t \begin{bmatrix} 1 \\ e^{2s} \\ (4+2s)e^{-s} \end{bmatrix} ds \right\}$$

$$= \begin{bmatrix} e^t & e^{-t} & 0 \\ e^t & -e^{-t} & 0 \\ -2(1+t) & 0 & e^t \end{bmatrix} \left\{ \begin{bmatrix} \frac{1}{2} \\ -\frac{1}{2} \\ 0 \end{bmatrix} + \frac{1}{2} \begin{bmatrix} t \\ \frac{1}{2}e^{2t} - \frac{1}{2} \\ 6 - 2e^{-t}(3+t) \end{bmatrix} \right\}.$$

Hence the solution is

$$x_1(t) = (\tfrac{3}{4} + \tfrac{1}{2}t)e^t - \tfrac{3}{4}e^{-t},$$

$$x_2(t) = (\tfrac{1}{4} + \tfrac{1}{2}t)e^t + \tfrac{3}{4}e^{-t},$$

$$x_3(t) = 3e^t - t^2 - 3t - 4.$$

Exercise 8.5
Using Theorem 8.8 find the solution of $\dot{x} = A(t)x + f(t)$, where

$$x = \begin{bmatrix} x_1 \\ x_2 \\ x_3 \end{bmatrix}, \quad A(t) = \begin{bmatrix} 0 & 0 & 1 \\ te^{-t} & 2 & te^{-t} \\ 1 & 0 & 0 \end{bmatrix}, \quad f(t) = \begin{bmatrix} 0 \\ e^{2t} \\ 0 \end{bmatrix},$$

and $x(0) = [0, 1, 1]^T$.

8.7 Stability and boundedness for linear systems

The following theorem requires a suitable **norm** (see Appendix D) for a matrix. For any matrix A, with elements $a_{ij}, i, j = 1, \ldots, n$, which are real or complex, we define

$$\|A\| = \left[\sum_{i,j} |a_{ij}|^2 \right]^{\frac{1}{2}}. \tag{8.21}$$

The norm serves as a measure of magnitude for A. It combines with the norm (8.1) for a vector to produce the inequality

$$\|Aa\| \leq \|A\| \|a\| \tag{8.22}$$

if a is a vector of dimension n.

Theorem 8.9 *For the regular linear system $\dot{x} = A(t)x$ the zero solution, and hence, by Theorem 8.1, all solutions, are Liapunov stable on $t \geq t_0$, t_0 arbitrary, if and only if every solution is bounded as $t \to \infty$. If A is constant and every solution is bounded, the solutions are uniformly stable.*

Proof By Theorem 8.1 we need consider only the stability of the zero solution.
First, suppose that the zero solution, $x^*(t) \equiv 0$ is stable. Choose any $\varepsilon > 0$. Then there exists a corresponding δ for Definition 8.2. Let

$$\Psi(t) = [\psi_1(t), \psi_2(t), \ldots, \psi_n(t)]$$

be the fundamental matrix satisfying the initial condition

$$\Psi(t_0) = \tfrac{1}{2}\delta I$$

where I is the unit matrix. (This is a diagonal matrix with elements $\frac{1}{2}\delta$ on the principal diagonal.) By Definition 8.2

$$\|\psi_i(t_0)\| = \tfrac{1}{2}\delta < \delta \Rightarrow \|\psi_i(t)\| < \varepsilon, \quad t \geq t_0.$$

Therefore every solution is bounded since any other solution is a linear combination of the $\psi_i(t)$.

Suppose, conversely, that every solution is bounded. Let $\Phi(t)$ be any fundamental matrix; then there exists, by hypothesis, $M > 0$ such that $\|\Phi(t)\| < M$, $t \geq t_0$. Given any $\varepsilon > 0$ let

$$\delta = \frac{\varepsilon}{M \|\Phi^{-1}(t_0)\|}$$

Let $x(t)$ be any solution; we will test the stability of the zero solution. We have $x(t) = \Phi(t)\Phi^{-1}(t_0)x(t_0)$, (Theorem 8.6), and if

$$\|x(t_0)\| < \delta,$$

then

$$\|x(t)\| \le \|\Phi(t)\| \, \|\Phi^{-1}(t_0)\| \, \|x(t_0)\| < M \frac{\varepsilon}{M\delta}\delta = \varepsilon. \tag{8.23}$$

Thus Definition 8.2 of stability for the zero solution is satisfied.

When A is a constant matrix, the autonomous nature of the system ensures that stability is uniform. ∎

Linearity is essential for this theorem: it predicts that the system $\dot{x}_1 = x_2$, $\dot{x}_2 = -x_1$ has only stable solutions; but the solutions of $\dot{x}_1 = x_2$, $\dot{x}_2 = -\sin x_1$ are not all stable. Note that the stability of *forced* solutions (Theorem 8.1) is not determined by whether they are bounded or not; it is the boundedness of the *unforced* solutions which determines this.

> **Exercise 8.6**
> Find the norm of
>
> $$A(t) = \begin{bmatrix} e^{-t} & -1 \\ 1/(t^2+1) & \sin t \end{bmatrix},$$
>
> and show that $\|A(t)\| \le 2$.

8.8   Stability of linear systems with constant coefficients

When the coefficients $a_{ij}(t)$ in (8.14) are functions of t, it will usually be impossible to construct an explicit fundamental matrix for the system. When the coefficients are all constants, however, the solutions are of comparatively elementary form, and we shall obtain a simple statement (Theorem 8.12) summarizing their stability properties. The lines of the proof follow those of Section 2.4, where the system of dimension 2 is solved.

Consider the system

$$\dot{x} = Ax, \tag{8.24}$$

where A is a constant $n \times n$ matrix with real elements. We look for solutions of the form

$$x = re^{\lambda t}, \tag{8.25}$$

where λ is a constant, and r is a constant column vector. In order to satisfy (8.24) we must have

$$Are^{\lambda t} - \lambda re^{\lambda t} = (A - \lambda I)re^{\lambda t} = 0$$

for all t, or

$$(A - \lambda I)r = 0. \tag{8.26}$$

Given a value for λ, this is equivalent to n linear equations for the components of r, and has non-trivial solutions if and only if

$$\det(A - \lambda I) = 0. \tag{8.27a}$$

In component form (8.27a) becomes

$$\begin{vmatrix} a_{11} - \lambda & a_{12} & \cdots & a_{1n} \\ a_{21} & a_{22} - \lambda & \cdots & a_{2n} \\ \cdots & \cdots & \cdots & \cdots \\ a_{n1} & a_{n2} & \cdots & a_{nm} - \lambda \end{vmatrix} = 0. \tag{8.27b}$$

Equation (8.27) is a polynomial equation of degree n for λ, called the **characteristic equation**. It therefore has n roots, real or complex, some of which may be repeated roots. If λ is a complex solution, then so is $\bar{\lambda}$, since A is a real matrix. The values of λ given by (8.27) are the **eigenvalues** of A. Equation (8.24) therefore has a solution of the form (8.25) if and only if λ is an eigenvalue of A.

Now suppose that *the eigenvalues are all different*, so that there are exactly n distinct eigenvalues, real, complex or zero, $\lambda_1, \lambda_2, \ldots, \lambda_n$. For each λ_i there exist nonzero solutions $r = r_i$ of equation (8.26). These are **eigenvectors** corresponding to λ_i. It is known that in the present case of n distinct eigenvalues, all the eigenvectors of a particular eigenvalue are simply multiples of each other: therefore we have essentially n solutions of (8.24):

$$[r_1 e^{\lambda_1 t}, r_2 e^{\lambda_2 t}, \ldots, r_n e^{\lambda_n t}],$$

where r_i is any one of the eigenvectors of λ_i. The solutions are linearly independent. Therefore:

Theorem 8.10   *For the system $\dot{x} = Ax$, with A a real, constant matrix whose eigenvalues $\lambda_1, \lambda_2, \ldots, \lambda_n$ are all different,*

$$\Phi(t) = [r_1 e^{\lambda_1 t}, r_2 e^{\lambda_2 t}, \ldots, r_n e^{\lambda_n t}] \tag{8.28}$$

is a fundamental matrix (complex in general), where r_i is any eigenvector corresponding to λ_i. ∎

Example 8.11   *Find a fundamental matrix for the system*

$$\dot{x}_1 = x_2 - x_3, \quad \dot{x}_2 = x_3, \quad \dot{x}_3 = x_2.$$

We have $\dot{x} = Ax$ with

$$A = \begin{bmatrix} 0 & 1 & -1 \\ 0 & 0 & 1 \\ 0 & 1 & 0 \end{bmatrix}.$$

To find the eigenvalues, we require

$$\det(A - \lambda I) = \begin{vmatrix} -\lambda & 1 & -1 \\ 0 & -\lambda & 1 \\ 0 & 1 & -\lambda \end{vmatrix} = -\lambda(\lambda - 1)(\lambda + 1),$$

so the eigenvalues are $\lambda_1 = 0$, $\lambda_2 = 1$, $\lambda_3 = -1$. The equations for the eigenvectors, r_1, r_2, r_3 are as follows.

For $\lambda_1 = 0, r_1 = [\alpha_1, \beta_1, \gamma_1]^T$ where

$$\beta_1 - \gamma_1 = 0, \quad \beta_1 = 0, \quad \gamma_1 = 0.$$

A solution is $r_1 = [1, 0, 0]^T$.
For $\lambda_2 = 0, r_2 = [\alpha_2, \beta_2, \gamma_2]^T$ where

$$-\alpha_2 + \beta_2 - \gamma_2 = 0, \quad -\beta_2 + \gamma_2 = 0, \quad \beta_2 - \gamma_2 = 0.$$

A solution is $r_2 = [0, 1, 1]^T$.
For $\lambda_3 = -1, r_3 = [\alpha_3, \beta_3, \gamma_3]^T$ where

$$\alpha_3 + \beta_3 - \gamma_3 = 0, \quad \beta_3 + \gamma_3 = 0, \quad \beta_3 + \gamma_3 = 0.$$

A solution is $r = [2, -1, 1]^T$.
A fundamental matrix is therefore given by (8.28), namely

$$\Phi(t) = \begin{bmatrix} 1 & 0 & 2e^{-t} \\ 0 & e^t & -e^{-t} \\ 0 & e^t & e^{-t} \end{bmatrix}. \qquad \bullet$$

When the eigenvalues are not all distinct the formal situation is more complicated, and for the theory the reader is referred, for example, to Wilson (1971) or Boyce and DiPrima (1996). We illustrate some possibilities in the following examples.

Example 8.12   *Find a fundamental matrix for the system*

$$\dot{x}_1 = x_1, \quad \dot{x}_2 = x_2, \quad \dot{x}_3 = x_3.$$

The coefficient matrix for the system is

$$A = \begin{bmatrix} 1 & 0 & 0 \\ 0 & 1 & 0 \\ 0 & 0 & 1 \end{bmatrix}.$$

The characteristic equation is $\det(A - \lambda I) = 0$, which becomes $(1 - \lambda)^3 = 0$. Thus $\lambda = 1$ is a threefold repeated root. The equation for the eigenvectors, $(A - \lambda I)r = 0$, gives no restriction on r, so the linearly independent vectors $[1, 0, 0]^T, [0, 1, 0]^T, [0, 0, 1]^T$ can be chosen as eigenvectors. This leads to the fundamental matrix Φ given by

$$\Phi(t) = \begin{bmatrix} e^t & 0 & 0 \\ 0 & e^t & 0 \\ 0 & 0 & e^t \end{bmatrix}. \qquad \bullet$$

Example 8.13   *Find a fundamental matrix for the system*

$$\dot{x}_1 = x_1 + x_2, \quad \dot{x}_2 = x_2, \quad \dot{x}_3 = x_3.$$

We have

$$A = \begin{bmatrix} 1 & 1 & 0 \\ 0 & 1 & 0 \\ 0 & 0 & 1 \end{bmatrix}.$$

The characteristic equation becomes $(1 - \lambda)^3 = 0$, so $\lambda = 1$ is a threefold root. The equation (8.26) for the eigenvectors gives $r_2 = 0$ with r_1 and r_3 unrestricted. We choose the two simplest linearly independent eigenvectors, $[1, 0, 0]^T$ and $[0, 0, 1]^T$, say, satisfying this condition. The corresponding solutions are

$$[e^t, 0, 0]^T, \quad [0, 0, e^t]^T.$$

To find a third solution let

$$x = rte^t + se^t.$$

Then

$$\dot{x} - Ax = -[(A - I)s - rI]e^t - (A - I)rte^t = 0$$

if r and s satisfy

$$(A - I)r = 0, \quad (A - I)s = r.$$

Choose $r = [1, 0, 0]^T$. Then $s = [0, 1, 0]^T$, so that a third solution is $[te^t, e^{-t}, 0]^T$. A fundamental matrix Φ is therefore given by

$$\Phi(t) = \begin{bmatrix} e^t & 0 & te^t \\ 0 & 0 & e^t \\ 0 & e^t & 0 \end{bmatrix}.$$ ●

Example 8.14 *Find a fundamental matrix for the system*

$$\dot{x}_1 = x_1 + x_2, \quad \dot{x}_2 = x_2 + x_3, \quad \dot{x}_3 = x_3.$$

We have

$$A = \begin{bmatrix} 1 & 1 & 0 \\ 0 & 1 & 1 \\ 0 & 0 & 1 \end{bmatrix}.$$

Once again, $\lambda = 1$ is a threefold root of the characteristic equation. The equation for the eigenvectors gives $r_2 = r_3 = 0$. One solution is $r = [1, 0, 0]^T$ and there is no other which is linearly independent of this one. The corresponding solution is $x = [e^t, 0, 0]^T$. A second solution is given by

$$x = rte^t + se^t,$$

where

$$(A - I)s = r,$$

and a third solution by

$$x = r\frac{t^2}{2!}e^t + ste^t + ue^t,$$

where

$$(A - I)u = s.$$

By finding vectors s and u, it can be confirmed that two more linearly independent solutions are

$$x = [te^t, e^t, 0]^T, \qquad x = [\tfrac{1}{2}t^2e^t, te^t, e^t]^T$$

leading to the fundamental matrix

$$\Phi(t) = \begin{bmatrix} e^t & te^t & \frac{1}{2}t^2e^t \\ 0 & e^t & te^t \\ 0 & 0 & e^t \end{bmatrix}.$$ ●

Inspection of these examples suggests the following theorem on the form of linearly independent solutions associated with multiple eigenvalues. We state the theorem without proof.

Theorem 8.11   *Corresponding to an eigenvalue of* A, $\lambda = \lambda_i$, *of multiplicity* $m \leq n$ *there are* m *linearly independent solutions of the system* $\dot{x} = Ax$, *where* A *is a constant matrix. These are of the form*

$$p_1(t)e^{\lambda_i t}, \ldots, p_m(t)e^{\lambda_i t}, \tag{8.29}$$

where the $p_j(t)$ *are vector polynomials of degree less than* m. ∎

Note that when an eigenvalue is complex, the eigenvectors and the polynomials in (8.29) will be complex, and the arrays consist of complex-valued solutions. Since the elements of A are real, the characteristic equation (8.27) has real coefficients. Therefore, if λ_i is one eigenvalue, $\bar{\lambda}_i$ is another. The corresponding polynomial coefficients in (8.29) are, similarly, complex conjugates. In place, therefore, of the pair of complex solutions $\phi_i(t)$, $\bar{\phi}_i(t)$ corresponding to λ_i and $\bar{\lambda}_i$ respectively, we could take the real solutions, Re$\{\phi_i(t)\}$, Im$\{\phi_i(t)\}$.

According to Theorems 8.9, 8.10 and 8.11 it is possible to make a simple statement about the stability of the solutions of systems with constant coefficients, $\dot{x} = Ax$.

Theorem 8.12   *Let* A *be a constant matrix in the system* $\dot{x} = Ax$, *with eigenvalues* λ_i, $i = 1, 2, \ldots n$.

(i) *If the system is stable, then* Re$\{\lambda_i\} \leq 0, i = 1, 2, \ldots, n$.

(ii) *If either* Re$\{\lambda_i\} < 0, i = 1, 2, \ldots, n$; *or if* Re$\{\lambda_i\} \leq 0, i = 1, 2, \ldots, n$ *and there is no zero repeated eigenvalue; then the system is uniformly stable.*

(iii) *The system is asymptotically stable if and only if* Re$\{\lambda_i\} < 0, i = 1, 2, \ldots, n$ *(and then it is also uniformly stable, by* (ii)).

(iv) *If* Re$\{\lambda_i\} > 0$ *for any* i, *the solution is unstable.* ∎

In connection with (ii), note that if there is a zero repeated eigenvalue the system may be stable or unstable. For example, the system $\dot{x}_1 = 0$, $\dot{x}_2 = 0$ has a fundamental matrix

$$\begin{bmatrix} 1 & 0 \\ 0 & 1 \end{bmatrix}$$

which implies boundedness and therefore stability, but $\dot{x}_1 = x_2$, $\dot{x}_2 = 0$ has a fundamental matrix

$$\begin{bmatrix} 1 & t \\ 0 & 1 \end{bmatrix}$$

showing that the system is unstable.

Exercise 8.7
Find the eigenvalues of

$$A = \begin{bmatrix} -1 & 1 & 1 \\ 0 & 1 & 2 \\ -1 & 0 & 1 \end{bmatrix},$$

and find a fundamental matrix solution of $\dot{x} = Ax$.

Exercise 8.8
Show that

$$A = \begin{bmatrix} 2 & 0 & 1 \\ 1 & 1 & 1 \\ 1 & -1 & 2 \end{bmatrix}$$

has a repeated eigenvalue. Find a fundamental matrix solution of $\dot{x} = Ax$.

8.9   Linear approximation at equilibrium points for first-order systems in n variables

The equilibrium points of the autonomous first-order system in n variables (often described simply as an nth order system)

$$\dot{x} = X(x) \tag{8.30}$$

occur at solutions of the n simultaneous equations given by $X(x) = 0$. For nonlinear systems there are no general methods of solving this equation, and we must rely on *ad hoc* eliminations, or numerically computed solutions. The phase space of this system is the n-dimensional space of the components of x: phase paths are curves drawn in this space.

Example 8.15   *Find all equilibrium points of the third-order system*

$$\dot{x} = \begin{bmatrix} \dot{x}_1 \\ \dot{x}_2 \\ \dot{x}_3 \end{bmatrix} = X(x) = \begin{bmatrix} x_1^2 - x_2 + x_3 \\ x_1 - x_2 \\ 2x_2^2 + x_3 - 2 \end{bmatrix}. \tag{i}$$

We require all simultaneous solutions of $X(x) = 0$, that is,

$$x_1^2 - x_2 + x_3 = 0, \tag{ii}$$

$$x_1 - x_2 = 0, \tag{iii}$$

$$2x_2^2 + x_3 - 2 = 0. \tag{iv}$$

Using (iii), eliminate x_2 in (ii) and (iv):

$$x_1^2 - x_1 + x_3 = 0, \qquad 2x_1^2 + x_3 - 2 = 0.$$

Now eliminate x_3 between these equations, so that

$$x_1^2 + x_1 - 2 = (x_1 + 2)(x_1 - 1) = 0.$$

Thus $x_1 = -2$ or 1, and the coordinates of the two equilibrium points are $(-2, -2, -6)$ and $(1, 1, 0)$. ⬤

As in Chapter 2, we can investigate the nature of an equilibrium point by examining its linear approximation. Suppose that the system (8.30) has an equilibrium point at $x = x_c$. Consider a perturbation $x = x_c + \xi$ about the equilibrium point, where it is assumed that the magnitude of ξ is small. Substitution into (8.30) leads to

$$\dot{\xi} = X(x_c + \xi) = X(x_c) + J\xi + o(\|\xi\|), = J\xi + o(\|\xi\|),$$

where J is the $n \times n$ **Jacobian** matrix of $X = [X_1(x), X_2(x), \ldots, X_n(x)]^{\mathrm{T}}$ evaluated at the equilibrium point x_c, namely the matrix with elements J_{ij} given by

$$J = [J_{ij}] = \left[\frac{\partial X_i(x)}{\partial x_j} \right]_{x = x_c}. \tag{8.31}$$

Then the linear approximation is

$$\dot{\xi} = J\xi. \tag{8.32}$$

Example 8.16 *Find the linear approximations at the equilibrium points of*

$$\dot{x} = \begin{bmatrix} x_1^2 - x_2 + x_3 \\ x_1 - x_2 \\ 2x_2^2 + x_3 - 2 \end{bmatrix}$$

(see Example 8.15).
From Example 8.15 the equilibrium points are located at $(-2, -2, -6)$ and $(1, 1, 0)$. In this case the Jacobian is

$$JX = \begin{bmatrix} 2x_1 & -1 & 1 \\ 1 & -1 & 0 \\ 0 & 4x_2 & 1 \end{bmatrix}.$$

Hence at $(-2, -2, -6)$ the linear approximation is

$$\dot{\xi} = \begin{bmatrix} -4 & -1 & 1 \\ 1 & -1 & 0 \\ 0 & -8 & 1 \end{bmatrix} \xi,$$

and at $(1, 1, 0)$,

$$\dot{\xi} = \begin{bmatrix} 2 & -1 & 1 \\ 1 & -1 & 0 \\ 0 & 4 & 1 \end{bmatrix} \xi. \qquad ⬤$$

As we explained in Theorem 8.12, the stability of the solutions of a linear system with constant coefficients depends on the sign of the real parts of the eigenvalues of the Jacobian matrix evaluated at the equilibrium point. If the eigenvalues are listed as $[\lambda_1, \lambda_2, \ldots, \lambda_n]$ and $\mathrm{Re}\{\lambda_i\} < 0$, $i = 1, 2, \ldots, n$), then the linear approximation will be asymptotically stable. We shall look briefly at the stability of the equilibrium points of some first-order linear systems in three variables,

$$\dot{x} = Ax,$$

where A is a constant matrix. Since the equation for the eigenvalues is a real cubic equation, the eigenvalues will either be all real, or one real and two complex. The eigenvalues satisfy the **characteristic equation**

$$\det(A - \lambda I) = 0,$$

where I_n represents the identity matrix of order n.

Example 8.17   *Find the eigenvalues of*

$$A = \begin{bmatrix} -3 & 0 & 2 \\ -1 & -3 & 5 \\ -1 & 0 & 0 \end{bmatrix},$$

and discuss the phase diagram near $x = 0$ of $\dot{x} = Ax$.
The eigenvalues are given by the characteristic equation

$$|A - \lambda I| = \begin{vmatrix} -3 - \lambda & 0 & 2 \\ -1 & -3 - \lambda & 5 \\ -1 & 0 & -\lambda \end{vmatrix} = -(1 + \lambda)(2 + \lambda)(3 + \lambda) = 0.$$

Denoting the eigenvalues by $\lambda_1 = -1$, $\lambda_2 = -2$, $\lambda_3 = -3$, then the corresponding eigenvectors are

$$r_1 = \begin{bmatrix} 1 \\ 2 \\ 1 \end{bmatrix}, \quad r_2 = \begin{bmatrix} 2 \\ 3 \\ 1 \end{bmatrix}, \quad r_3 = \begin{bmatrix} 0 \\ 1 \\ 0 \end{bmatrix},$$

and the general solution is

$$\begin{bmatrix} x_1 \\ x_2 \\ x_3 \end{bmatrix} = \alpha \begin{bmatrix} 1 \\ 2 \\ 1 \end{bmatrix} e^{-t} + \beta \begin{bmatrix} 2 \\ 3 \\ 1 \end{bmatrix} e^{-2t} + \gamma \begin{bmatrix} 0 \\ 1 \\ 0 \end{bmatrix} e^{-3t},$$

where α, β, and γ are constants. Since all the eigenvalues are negative the origin will be asymptotically stable. This equilibrium point has strong damping with a node-like structure.   ●

Example 8.18   *Find the eigenvalues of*

$$A = \begin{bmatrix} -\frac{9}{4} & -2 & -3 \\ 1 & \frac{3}{4} & 1 \\ 1 & 1 & \frac{3}{4} \end{bmatrix},$$

and sketch some phase paths of $\dot{x} = Ax$ near $x = 0$.

The eigenvalues are given by

$$\begin{vmatrix} -\frac{9}{4} - \lambda & -2 & -3 \\ 1 & \frac{3}{4} - \lambda & 1 \\ 1 & 1 & \frac{3}{4} - \lambda \end{vmatrix} = -\frac{1}{64}(1 + 4\lambda)\left(\lambda + \frac{1}{4} + i\right)\left(\lambda + \frac{1}{4} - i\right) = 0.$$

Denoting them by $\lambda_1 = -\frac{1}{4}, \lambda_2 = -\frac{1}{4} - i, \lambda_3 = -\frac{1}{4} + i$, the corresponding eigenvectors are

$$r_1 = \begin{bmatrix} -1 \\ 1 \\ 0 \end{bmatrix}, \quad r_2 = \begin{bmatrix} -2 - i \\ 1 \\ 1 \end{bmatrix}, \quad r_3 = \begin{bmatrix} -2 + i \\ 1 \\ 1 \end{bmatrix}.$$

Hence the general solution can be expressed in real form as

$$x = \alpha r_1 e^{-\frac{1}{4}t} + (\beta + i\gamma)r_2 e^{-\left(\frac{1}{4} + i\right)t} + (\beta - i\gamma)r_3 e^{-\left(\frac{1}{4} - i\right)t},$$

where α, β, and γ are arbitrary real constants. Since all the eigenvalues have negative real part, the equilibrium point is asymptotically stable. The system has an exponentially damped solution of $x_1 + x_2 = 0$, $x_3 = 0$ if $\beta = \gamma = 0$. All other solutions are damped spirals: an example is shown in Fig. 8.14. ●

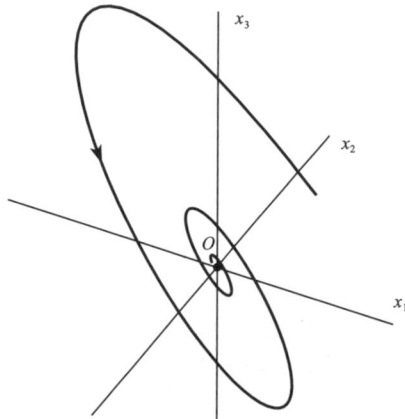

Figure 8.14   A single spiral solution of Example 8.18.

Exercise 8.9
Find the general solution of $\dot{x} = Ax$ where

$$A = \begin{bmatrix} -2 & 0 & -1 \\ -1 & -1 & -1 \\ -1 & 1 & -3 \end{bmatrix}.$$

Show that the equilibrium point at the origin is asymptotically stable, and find the equations of the straight line phase paths which approach the origin.

8.10 Stability of a class of non-autonomous linear systems in n dimensions

The system considered is

$$\dot{x} = \{A + C(t)\}x, \tag{8.33}$$

where A is a constant matrix. Under quite general conditions the stability is determined by the stability of the solutions of the autonomous system

$$\dot{x} = Ax. \tag{8.34}$$

We require the following theorems, the first of which expresses the solutions of the system

$$\dot{x} = Ax + f(t)$$

in a form alternative to that given in Theorem 8.8.

Lemma   *Let $\Phi(t)$ be any fundamental matrix of the system $\dot{\phi} = A\phi$, A constant. Then for any two parameters s, t_0,*

$$\Phi(t)\Phi^{-1}(s) = \Phi(t - s + t_0)\Phi^{-1}(t_0). \tag{8.35}$$

In particular,

$$\Phi(t)\Phi^{-1}(s) = \Phi(t - s)\Phi^{-1}(0). \tag{8.36}$$

Proof   Since $\dot{\Phi} = A\Phi$, if we define $U(t)$ by $U(t) = \Phi(t)\Phi^{-1}(s)$, then $\dot{U}(t) = AU(t)$, and $U(s) = I$.

Now consider $V(t) = \phi(t - s + t_0)\Phi^{-1}(t_0)$. Then $\dot{V}(t) = AV(t)$. (for since A is constant, $\Phi(t)$ and $\Phi(t - s + t_0)$ satisfy the same equation), and $V(s) = I$.

Therefore, the corresponding columns of U and V satisfy the same equation with the same initial conditions, and are therefore identical by the Uniqueness Theorem (Appendix A).   ■

The following theorem is obtained by applying the above Lemma to Theorem 8.8.

Theorem 8.13   *Let A be a constant matrix. The solution of the system $\dot{x} = Ax + f(t)$, with initial conditions $x(t_0) = x_0$, is given by*

$$x(t) = \Phi(t)\Phi^{-1}(t_0)x_0 + \int_{t_0}^{t} \Phi(t - s + t_0)\Phi^{-1}(t_0)f(s)\,\mathrm{d}s, \tag{8.37}$$

where $\Phi(t)$ is any fundamental matrix for the system $\dot{\phi} = A\phi$. In particular, if $\Psi(t)$ is the fundamental matrix satisfying $\Psi(t_0) = I$), then

$$x(t) = \Psi(t)x_0 + \int_{t_0}^{t} \Psi(t - s)f(s)\,\mathrm{d}s. \tag{8.38}$$

■

Example 8.19   *Express the solution of the second-order equation $\ddot{x} - x = h(t)$, with $x(0) = 0$, $\dot{x}(0) = 1$ as an integral.*

An equivalent first-order pair is

$$\dot{x} = y, \qquad \dot{y} = x + h(t),$$

and

$$\begin{bmatrix} \dot{x} \\ \dot{y} \end{bmatrix} = \begin{bmatrix} 0 & 1 \\ 1 & 0 \end{bmatrix} \begin{bmatrix} x \\ y \end{bmatrix} + \begin{bmatrix} 0 \\ h(t) \end{bmatrix} = A \begin{bmatrix} x \\ y \end{bmatrix} + f(t).$$

Since the eigenvalues of A are $\lambda_1 = 1$, $\lambda_2 = -1$, and the corresponding eigenvectors are $r_1 = [1, 1]^T$ and $r_2 = [1, -1]^T$, a fundamental matrix for the homogeneous system is

$$\Phi(t) = \begin{bmatrix} e^t & e^{-t} \\ e^t & -e^{-t} \end{bmatrix}.$$

Then, following Theorem 8.13

$$\Phi(0) = \begin{bmatrix} 1 & 1 \\ 1 & -1 \end{bmatrix}, \quad \Phi^{-1}(0) = \begin{bmatrix} \frac{1}{2} & \frac{1}{2} \\ \frac{1}{2} & -\frac{1}{2} \end{bmatrix},$$

and

$$\begin{bmatrix} x \\ y \end{bmatrix} = \begin{bmatrix} e^t & e^{-t} \\ e^t & -e^{-t} \end{bmatrix} \begin{bmatrix} \frac{1}{2} & \frac{1}{2} \\ \frac{1}{2} & -\frac{1}{2} \end{bmatrix} \begin{bmatrix} 0 \\ 1 \end{bmatrix} + \int_0^t \begin{bmatrix} e^{t-s} & e^{-t+s} \\ e^{t-s} & -e^{-t+s} \end{bmatrix} \begin{bmatrix} \frac{1}{2} & \frac{1}{2} \\ \frac{1}{2} & -\frac{1}{2} \end{bmatrix} \begin{bmatrix} 0 \\ h(s) \end{bmatrix} ds$$

$$= \begin{bmatrix} \sinh & t \\ \cosh & -t \end{bmatrix} + \int_0^t \begin{bmatrix} h(s) \sinh(t-s) \\ h(s) \cosh(t-s) \end{bmatrix} ds.$$

●

Theorem 8.14 (Gronwall's lemma)   *If, for $t \geq t_0$,*

(i) *$u(t)$ and $v(t)$ are continuous and $u(t) \geq 0$, $v(t) \geq 0$;*

(ii) *$u(t) \leq K + \int_{t_0}^t u(s)v(s)ds$, $K > 0$;*   (8.39)

   then

$$u(t) \leq K + \exp\left(\int_{t_0}^t v(s)\,ds\right), \qquad t \geq t_0.$$ (8.40)

Proof   The right-hand side of (8.39) is positive, since $K > 0$ and $u(t)$, $v(t) \geq 0$. Therefore (8.39) gives

$$\frac{u(t)v(t)}{K + \int_{t_0}^t u(s)v(s)ds} \leq v(t).$$

Integrate both sides from t_0 to t:

$$\log\left\{K + \int_{t_0}^t u(s)v(s)ds\right\} - \log K \leq \int_{t_0}^t v(s)\,ds.$$

Therefore

$$K + \int_{t_0}^{t} u(s)v(s)\, ds \leq K \exp\left(\int_{t_0}^{t} v(s)\, ds\right).$$

The application of (8.39), again gives the required result. ∎

Gronwall's lemma implies that if

$$u(t) - \int_{t_0}^{t} u(s)v(s)\, ds$$

is bounded, then

$$u(t)/\exp\left(\int_{t_0}^{t} v(s)\, ds\right)$$

is also bounded by the same constant. Equality occurs if $\dot{u} = uv$. Given v, the optimum u can be found from this equation. For example, if $v(t) = t$ and $t_0 = 0$, then we could choose $u(t) = e^{\frac{1}{2}t^2}$ (or any multiple), in which case $K = 1$.

Theorem 8.15   *Suppose that*

 (i) A *is a constant $n \times n$ matrix whose eigenvalues have negative real parts;*

(ii) *For $t_0 \leq t < \infty$, $C(t)$, is continuous and*

$$\int_{t_0}^{t} \|C(t)\|\, dt \quad \text{is bounded.} \tag{8.41}$$

Then all solutions of the linear, homogeneous system $\dot{x} = \{A + C(t)\}x$ are asymptotically stable.

Proof   Write the system in the form

$$\dot{x} = Ax + C(t)x. \tag{8.42}$$

If $x(t)$ is a solution, then $C(t)x(t)$ is a function of t which may play the part of $f(t)$ in Theorem 8.13. Therefore (8.42) implies

$$x(t) = \Phi(t)\Phi^{-1}(t_0)x_0 + \int_{t_0}^{t} \Phi(t - s + t_0)\Phi^{-1}(t_0)C(s)x(s)\, ds,$$

where Φ is any fundamental matrix for the system $\dot{x} = Ax$, and $x(t_0) = x_0$. Using the properties of norms (Appendix C), which parallel the properties of vector magnitudes, we obtain

$$\|x(t)\| \leq \|\Phi(t)\|\,\|\Phi^{-1}(t_0)\|\,\|x_0\|$$

$$+ \|\Phi^{-1}(t_0)\| \int_{t_0}^{t} \|\Phi(t - s + t_0)\|\,\|C(s)\|\,\|x(s)\|\, ds. \tag{8.43}$$

Since A has eigenvalues with negative real part, Theorem 8.12 shows that for some positive M and m,

$$\|\Phi(t)\| \le Me^{-mt}, \quad t \ge t_0. \tag{8.44}$$

Therefore, putting

$$\|\Phi^{-1}(t_0)\| = \beta,$$

(8.43) implies, after some regrouping, that for $t \ge t_0$

$$\|x(t)\|e^{mt} \le M\beta\|x_0\| + \int_{t_0}^{t} \{\|x(s)\|e^{ms}\}\{C(s)M\beta e^{-mt_0}\}\,ds. \tag{8.45}$$

In Theorem 8.2, let

$$u(t) = \|x(t)\|e^{mt}, \quad v(t) = \|C(t)\|\beta Me^{-mt_0},$$

and

$$K = M\beta\|x_0\|.$$

Then from (8.45) and Theorem 8.13,

$$\|x(t)\|e^{mt} \le M\beta\|x_0\| \exp\left(\beta Me^{-mt_0}\int_{t_0}^{t}\|C(s)\|\,ds\right)$$

or

$$\|x(t)\| \le M\beta\|x_0\| \exp\left(\beta Me^{-mt_0}\int_{t_0}^{t}\|C(s)\|ds - mt\right). \tag{8.46}$$

Therefore, by (8.41) every solution of (8.42) is bounded for $t \ge t_0$ and is therefore stable by Theorem 8.9. Also every solution tends to zero as $t \to \infty$ and is therefore asymptotically stable. ∎

Corollary 8.15 *If $C(t)$ satisfies the conditions of the theorem but all solutions of $\dot{x} = Ax$ are merely bounded, then all solutions of $\dot{x} = \{A + C(t)\}x$ are bounded and therefore stable.*

Proof This follows from (8.46) by writing $m = 0$ in (8.45). Note that $\text{Re}\{\lambda_i\} \le 0$ for all i is not in itself sufficient to establish the boundedness of all solutions of $\dot{x} = Ax$. ∎

The stability of nth order *differential equations* can be discussed in the same terms as that of suitable n-dimensional *systems*. If we replace the equation

$$x_1^{(n)} + a_1(t)x_1^{(n-1)} + \cdots + a_n(t)x_1 = f(t)$$

by the equivalent system

$$\dot{x}_1 = x_2, \quad \dot{x}_2 = x_3, \ldots, \dot{x}_{n-1} = x_n, \quad \dot{x}_n = a_1(t)x_n - \cdots - a_n(t)\,x_1 + f(t), \tag{8.47}$$

the set of initial conditions for the *system*, $(x_1(t_0), \ldots, x_n(t_0))$, correspond with the usual initial conditions appropriate to the equation: the set

$$[x_1(t_0), x_1^{(1)}(t_0), \ldots, x_1^{(n-1)}(t_0)].$$

Therefore, to discuss the stability of the solutions of equations in terms of the definitions for systems we use the presentation (8.47) rather than any other. It can, in fact, be shown (Problem 8.7) that a system obtained by a general transformation of the variables need not retain the stability properties of the original system.

Example 8.20   *Show that when $a > 0$ and $b > 0$ all solutions of*

$$\ddot{x} + a\dot{x} + (b + ce^{-t}\cos t)x = 0$$

are asymptotically stable for $t \geq t_0$, for any t_0.
The appropriate equivalent system (with $\dot{x} = y$) is

$$\begin{bmatrix} \dot{x} \\ \dot{y} \end{bmatrix} = \begin{bmatrix} 0 & 1 \\ -b & -a \end{bmatrix}\begin{bmatrix} x \\ y \end{bmatrix} + \begin{bmatrix} 0 & 0 \\ -ce^{-t}\cos t & 0 \end{bmatrix}\begin{bmatrix} x \\ y \end{bmatrix}.$$

In the notation of the above theorem

$$A = \begin{bmatrix} 0 & 1 \\ -b & -a \end{bmatrix}, \quad C(t) = \begin{bmatrix} 0 & 0 \\ -ce^{-t}\cos t & 0 \end{bmatrix}.$$

The eigenvalues of A are negative if $a > 0$ and $b > 0$. Also

$$\int_{t_0}^{\infty} \|C(t)\|dt = |c|\int_{t_0}^{\infty} e^{-t}|\cos t|dt < \infty.$$

The conditions of the theorem are satisfied, so all solutions are asymptotically stable.   ●

Example 8.21   *Show that all solutions of $\ddot{x} + \{a + c(1 + t^2)^{-1}\}x = f(t)$ are stable if $a > 0$.*
Let $\dot{x} = y$. The equation is equivalent to

$$\begin{bmatrix} \dot{x} \\ \dot{y} \end{bmatrix} = \begin{bmatrix} 0 & 1 \\ -a & 0 \end{bmatrix}\begin{bmatrix} x \\ y \end{bmatrix} + \begin{bmatrix} 0 & 0 \\ -c(1 + t^2)^{-1} & 0 \end{bmatrix}\begin{bmatrix} x \\ y \end{bmatrix} + \begin{bmatrix} 0 \\ f(t) \end{bmatrix}.$$

By Theorem 8.1, all solutions of the given system have the same stability property as the zero solution (or any other) of the corresponding homogeneous system $\dot{\xi} = \{A + C(t)\}\xi$, where

$$A = \begin{bmatrix} 0 & 1 \\ -a & 0 \end{bmatrix}, \quad C(t) = \begin{bmatrix} 0 & 0 \\ -c(1 + t^2)^{-1} & 0 \end{bmatrix}.$$

The solutions of $\dot{\xi} = A\xi$ are bounded when $a > 0$ (the zero solution is a centre on the phase plane). Also

$$\int_{t_0}^{\infty} \|C(t)\|dt = |c|\int_{t_0}^{\infty} \frac{dt}{1 + t^2} < \infty.$$

By Corollary 8.15 all solutions of $\dot{\xi} = \{A + C(t)\}\xi$ are bounded and are therefore stable. (Note that the **inhomogeneous** equation may, depending on $f(t)$, have **unbounded** solutions which are stable.)   ●

Exercise 8.10
Show that all solutions of

$$
\begin{bmatrix} \dot{x}_1 \\ \dot{x}_2 \\ \dot{x}_3 \end{bmatrix} = \begin{bmatrix} -2 + (1+t^2)^{-1} & 0 & -1 \\ -1 & -1 + (1+t^2)^{-1} & -1 \\ -1 & 1 & -3 + (1+t^2)^{-1} \end{bmatrix} \begin{bmatrix} x_1 \\ x_2 \\ x_3 \end{bmatrix}
$$

are asymptotically stable.

8.11   Stability of the zero solutions of nearly linear systems

Certain nonlinear systems can be regarded as perturbed linear systems in respect of their stability properties, the stability (or instability) of the linearized system being preserved. The following theorem refers to the stability of the **zero** solutions.

Theorem 8.16   *If $h(0,t) = 0$, A is constant, and*

(i) *the solutions of $\dot{x} = Ax$ are asymptotically stable;*

(ii) *$\lim_{\|x\|\to 0}\{\|h(x,t)\|/\|x\|\} = 0$ uniformly in t, $0 \le t < \infty$;*

then the zero solution, $x(t) = 0$ for $t \ge 0$, is an asymptotically stable solution of the regular system

$$
\dot{x} = Ax + h(x,t). \tag{8.48}
$$

Proof   Regularity implies that h is such that the conditions of the existence theorem (Appendix A) hold for all x and for $t \ge 0$. However, it does not of itself imply that any particular solution actually persists for all t: for a proof see Cesari (1971, p. 92).

Let $\Psi(t)$ be the fundamental matrix for the system $\dot{x} = Ax$ for which $\Psi(0) = I$. Then by Theorem 8.13 with h in place of f, every solution of (8.48) with initial values at $t = 0$ satisfies

$$
x(t) = \Psi(t)x(0) + \int_0^t \Psi(t-s)h(x(s),s)ds. \tag{8.49}
$$

By (i) and (8.44) there exist $M > 0$, $m > 0$ such that

$$
\|\Psi(t)\| \le Me^{-mt}, \quad t \ge 0. \tag{8.50}
$$

Also, by (ii), there exists δ_0 such that

$$
\|x\| < \delta_0 \Rightarrow \|h(x,t)\| < \frac{m}{2M}\|x\|, \quad t \ge 0. \tag{8.51}
$$

Now let δ be chosen arbitrarily subject to

$$
0 < \delta < \delta_0, \tag{8.52}
$$

and consider solutions for which

$$\|\boldsymbol{x}(0)\| < \delta. \tag{8.53}$$

From (8.51) to (8.53), (8.49) gives

$$\|\boldsymbol{x}(t)\| \le M\mathrm{e}^{-mt}\|\boldsymbol{x}(0)\| + \int_0^t M\mathrm{e}^{-mt}\mathrm{e}^{ms}\frac{m}{2M}\|\boldsymbol{x}(s)\}\mathrm{d}s, \quad 0 \le t < \infty,$$

or

$$\|\boldsymbol{x}(t)\|\mathrm{e}^{mt} \le M\|\boldsymbol{x}(0)\| + \int_0^t \frac{1}{2}m\|\boldsymbol{x}(s)\|\mathrm{e}^{ms}\mathrm{d}s, \quad 0 \le t \le \infty. \tag{8.54}$$

By Gronwall's Lemma, Theorem 8.13 with $u(t) = \|\boldsymbol{x}(t)\|\mathrm{e}^{mt}$, $v(t) = \frac{1}{2}m$, $K = M\|\boldsymbol{x}(0)\|$, applied to this inequality we obtain

$$\|\boldsymbol{x}(t)\| \le M\|\boldsymbol{x}(0)\|\mathrm{e}^{-\frac{1}{2}mt} \le M\delta\mathrm{e}^{-\frac{1}{2}mt}, \quad 0 \le t < \infty. \tag{8.55}$$

by (8.53).

Now δ is arbitrarily small, so we have proved that

$$\|\boldsymbol{x}(0)\| < \delta \Rightarrow \|\boldsymbol{x}(t)\| < M\delta\mathrm{e}^{-\frac{1}{2}mt}, \quad t \ge 0, \tag{8.56}$$

which implies asymptotic stability of the zero solution.

The asymptotic stability of the linearized system ensures that $\|\boldsymbol{x}(t)\|$ tends to decrease, reducing the relative effect of the nonlinear term. This process is self-reinforcing. ∎

The example in which

$$\boldsymbol{x} = \begin{bmatrix} x_1 \\ x_2 \end{bmatrix}, \quad \boldsymbol{A} = \begin{bmatrix} -1 & 0 \\ 1 & -2 \end{bmatrix}, \quad \boldsymbol{h} = \begin{bmatrix} x_1^2 \\ 0 \end{bmatrix}$$

satisfies the conditions of the theorem since $\boldsymbol{A}$ has eigenvalues -1 and -2. In this case

$$\boldsymbol{\Psi}(t) = \begin{bmatrix} \mathrm{e}^{-t} & 0 \\ -\mathrm{e}^{-t} + \mathrm{e}^{-2t} & \mathrm{e}^{-2t} \end{bmatrix}.$$

In the bound $\|\boldsymbol{\Psi}\| \le M\mathrm{e}^{-mt}$ we can choose $M = \sqrt{6}$ and $m = 1$.

Example 8.22 *Show that van der Pol's equation $\ddot{x} + v(x^2 - 1)\dot{x} + x = 0$ has an asymptotically stable zero solution when $v < 0$.*

Replace the equation by the system

$$\begin{pmatrix} \dot{x} \\ \dot{y} \end{pmatrix} = \begin{pmatrix} 0 & 1 \\ -1 & v \end{pmatrix}\begin{pmatrix} x \\ y \end{pmatrix} + \begin{pmatrix} 0 \\ -vx^2y \end{pmatrix} = \boldsymbol{A}\boldsymbol{x} + \boldsymbol{h}(x, y).l$$

The eigenvalues of $\boldsymbol{A}$ are negative when $v < 0$; therefore, by Theorem 8.13, all the solutions of $\dot{\boldsymbol{x}} = \boldsymbol{A}\boldsymbol{x}$ are asymptotically stable. Condition (ii) of Theorem 8.13 is satisfied since

$$\|\boldsymbol{h}(x, y)\| = |v|x^2|y| \le |v|(|x| + |y|)^2(|x| + |y|) = |v|\|\boldsymbol{x}\|^3. \quad \bullet$$

Example 8.23 *Show that the zero solution of the equation $\ddot{x} + k\dot{x} + \sin x = 0$ is asymptotically stable for $k > 0$.*

The equivalent system is

$$\begin{pmatrix} \dot{x} \\ \dot{y} \end{pmatrix} = \begin{pmatrix} 0 & 1 \\ -1 & -k \end{pmatrix} \begin{pmatrix} x \\ y \end{pmatrix} + \begin{pmatrix} 0 \\ x - \sin x \end{pmatrix} = Ax + h(x), \quad x = [x, y]^{\mathrm{T}},$$

which satisfies the conditions of Theorem 8.15, since, using the Taylor series for $\sin x$,

$$\|h(x)\| = |x - \sin x| = \left| \frac{x^3}{3!} - \frac{x^5}{5!} + \cdots \right|$$

$$\leq |x^3| \left[\frac{1}{3!} + \frac{x^2}{5!} + \cdots \right]$$

$$\leq \|x\| \|x\|^2 \left[\frac{1}{3!} + \frac{\|x\|^2}{5!} + \cdots \right].$$

Hence

$$\frac{\|h(x)\|}{\|x\|} \rightarrow 0 \quad \text{as } \|x\| \rightarrow 0. \qquad \bullet$$

Exercise 8.11
Show that the solution $x = 0$, $y = 0$ of

$$\begin{bmatrix} \dot{x} \\ \dot{y} \end{bmatrix} = \begin{bmatrix} -3 & 2 \\ 2 & -3 \end{bmatrix} \begin{bmatrix} x \\ y \end{bmatrix} + \begin{bmatrix} xe^{-x}\cos t \\ 0 \end{bmatrix}$$

is asymptotically stable.

Problems

8.1 Use the phase diagram for the pendulum equation, $\ddot{x} + \sin x = 0$, to say which paths are not Poincaré stable (see Fig. 1.2).

8.2 Show that all the paths of $\dot{x} = x$, $\dot{y} = y$ are Poincaré unstable.

8.3 Find the limit cycles of the system

$$\dot{x} = -y + x\sin r, \quad \dot{y} = x + y\sin r, \quad r = \sqrt{(x^2 + y^2)}.$$

Which cycles are Poincaré stable?

8.4 Find the phase paths for $\dot{x} = x$, $\dot{y} = y\log y$, in the half-plane $y > 0$. Which paths are Poincaré stable?

8.5 Show that every nonzero solution of $\dot{x} = x$ is unbounded and Liapunov unstable, but that every solution of $\dot{x} = 1$ is unbounded and stable.

8.6 Show that the solutions of the system $\dot{x} = 1$, $\dot{y} = 0$, are Poincaré and Liapunov stable, but that the system $\dot{x} = y$, $\dot{y} = 0$, is Poincaré but not Liapunov stable.

8.7 Solve the equations $\dot{x} = -y(x^2 + y^2)$, $\dot{y} = x(x^2 + y^2)$, and show that the zero solution is Liapunov stable and that all other solutions are unstable.

Replace the coordinates x, y by r, ϕ where $x = r\cos(r^2 t + \phi)$, $y = r\sin(r^2 t + \phi)$. and deduce that $\dot{r} = 0$, $\dot{\phi} = 0$. Show that in this coordinate system the solutions are stable. (Change of coordinates can affect the stability of a system: see Cesari (1971, p. 12).

8.8 Prove that Liapunov stability of a solution implies Poincaré stability for plane autonomous systems but not conversely: see Problem 8.6.

8.9 Determine the stability of the solutions of
(i) $\dot{x}_1 = x_2 \sin t$, $\dot{x}_2 = 0$.
(ii) $\ddot{x}_1 = 0$, $\dot{x}_2 = x_1 + x_2$.

8.10 Determine the stability of the solutions of
(i) $\begin{bmatrix} \dot{x}_1 \\ \dot{x}_2 \end{bmatrix} = \begin{bmatrix} -2 & 1 \\ 1 & -2 \end{bmatrix} \begin{bmatrix} x_1 \\ x_2 \end{bmatrix} + \begin{bmatrix} 1 \\ -2 \end{bmatrix} e^t$;
(ii) $\ddot{x} + e^{-t}\dot{x} + x = e^t$.

8.11 Show that every solution of the system

$$\dot{x} = -t^2 x, \quad \dot{y} = -ty$$

is asymptotically stable.

8.12 The motion of a heavy particle on a smooth surface of revolution with vertical axis z and shape $z = f(r)$ in cylindrical polar coordinates is

$$\frac{1}{r^4}\{1 + f'^{2(r)}\}\frac{d^2 r}{d\theta^2} + \left[\frac{1}{r^4}f'(r)f''(r) - \frac{2}{r^5}\{1 + f'^{2(r)}\}\right]\left(\frac{dr}{d\theta}\right)^2 - \frac{1}{r^3} = -\frac{g}{h^2}f'(r),$$

where h is the constant angular momentum ($h = r^2\dot{\theta}$). Show that plane, horizontal motion $r = a$, $z = f(a)$, is stable for perturbations leaving h unaltered provided $3 + [af''(a)/f'(a)] > 0$.

8.13 Determine the linear dependence or independence of the following:
(i) $(1, 1, -1)$, $(2, 1, 1)$, $(0, 1, -3)$.
(ii) $(t, 2t)$, $(3t, 4t)$, $(5t, 6t)$.
(iii) (e^t, e^{-t}), (e^{-t}, e^t). Could these both be solutions of a 2×2 homogeneous linear system?

8.14 Construct a fundamental matrix $\boldsymbol{\Phi}$ for the system $\dot{x} = y$, $\dot{y} = -x - 2y$. Deduce a fundamental matrix $\boldsymbol{\Psi}$ satisfying $\boldsymbol{\Psi}(0) = I$.

8.15 Construct a fundamental matrix for the system $\dot{x}_1 = -x_1$, $\dot{x}_2 = x_1 + x_2 + x_3$, $\dot{x}_3 = -x_2$.

8.16 Construct a fundamental matrix for the system $\dot{x}_1 = x_2$, $\dot{x}_2 = x_1$, and deduce the solution satisfying $x_1 = 1$, $x_2 = 0$, at $t = 0$.

8.17 Construct a fundamental matrix for the system $\dot{x}_1 = x_2$, $\dot{x}_2 = x_3$, $\dot{x}_3 = -2x_1 + x_2 + 2x_3$, and deduce the solution of $\dot{x}_1 = x_2 + e^t$, $\dot{x}_2 = x_3$, $\dot{x}_3 = -2x_1 + x_2 + 2x_3$, with $x(0) = (1, 0, 0)^{\mathrm{T}}$.

8.18 Show that the differential equation $x^{(n)} + a_1 x^{(n-1)} + \cdots + a_n x = 0$ is equivalent to the system

$$\dot{x} = x_2, \quad \dot{x}_2 = x_3, \ldots, \quad \dot{x}_{n-1} = x_n, \quad \dot{x}_n = -a_n x_1 - \cdots - a_1 x_n,$$

with $x = x_1$. Show that the equation for the eigenvalues is

$$\lambda^n + a_1\lambda^{n-1} + \cdots + a_n = 0.$$

8.19 A bird population, $p(t)$, is governed by the differential equation $\dot{p} = \mu(t)p - kp$, where k is the death rate and $\mu(t)$ represents a variable periodic birth rate with period 1 year. Derive a condition which ensures that the mean annual population remains constant. Assuming that this condition is fulfilled, does it seem likely that, in practice, the average population will remain constant? (This is asking a question about a particular kind of stability.)

8.20 Are the periodic solutions of

$$\ddot{x} + \mathrm{sgn}(x) = 0.$$

(i) Poincaré stable? (ii) Liapunov stable?

8.21 Give a descriptive argument to show that if the index of an equilibrium point in a plane autonomous system is not unity, then the equilibrium point is not stable.

8.22 Show that the system

$$\dot{x} = x + y - x(x^2 + y^2), \quad \dot{y} = -x + y - y(x^2 + y^2), \quad \dot{z} = -z,$$

has a limit cycle $x^2 + y^2 = 1$, $z = 0$. Find the linear approximation at the origin and so confirm that the origin is unstable. Use cylindrical polar coordinates $r = \sqrt{(x^2 + y^2)}$, z to show that the limit cycle is stable. Sketch the phase diagram in x, y, z space.

8.23 Show that the nth-order nonautonomous system $\dot{x} = X(x, t)$ can be reduced to an $(n + 1)$th-order autonomous system by introducing a new variable, $x_{n+1} = t$. (The $(n + 1)$-dimensional phase diagram for the modified system is then of the type suggested by Fig. 8.9. The system has no equilibrium points.)

8.24 Show that all phase paths of

$$\ddot{x} = x - x^3$$

are Poincaré stable except the homoclinic paths (see Section 3.6).

8.25 Investigate the equilibrium points of

$$\dot{x} = y, \quad \dot{y} = z - y - x^3, \quad \dot{z} = y + x - x^3.$$

Confirm that the origin has homoclinic paths given by

$$x = \pm\sqrt{2}\,\mathrm{sech}\,t, \quad y = \mp\sqrt{2}\,\mathrm{sech}^2 t \sinh t, \quad z = \pm\sqrt{2}\,\mathrm{sech}\,t \mp \sqrt{2}\sec\mathrm{h}^2 t\,\mathrm{sech}^2 t \sinh t.$$

In which directions do the solutions approach the origin as $t \to \pm\infty$?

8.26 By using linear approximations investigate the equilibrium points of the **Lorenz equations**

$$\dot{x} = a(y - x), \quad \dot{y} = bx - y - xz, \quad \dot{z} = xy - cz.$$

where $a, b, c > 0$ are constants. Show that if $b \le 1$, then the origin is the only equilibrium point, and that there are three equilibrium points if $b > 1$. Discuss the stability of the zero solution.

8.27 Test the stability of the linear system

$$\dot{x}_1 = t^{-2}x_1 - 4x_2 - 2x_3 + t^2,$$

$$\dot{x}_2 = -x_1 + t^{-2}x_2 + x_3 + t,$$

$$\dot{x}_3 = t^{-2}x_1 - 9x_2 - 4x_3 + 1.$$

8.28 Test the stability of the solutions of the linear system

$$\dot{x}_1 = 2x_1 + e^{-t}x_2 - 3x_3 + e',$$

$$\dot{x}_2 = -2x_1 + e^{-t}x_2 + x_3 + 1,$$

$$\dot{x}_3 = (4 + e^{-t})x_1 - x_2 - 4x_3 + e^t.$$

8.29 Test the stability of the zero solution of the system

$$\dot{x} = y + xy/(1 + t^2), \quad \dot{y} = -x - y + y^2/(1 + t^2).$$

8.30 Test the stability of the zero solution of the system

$$\dot{x}_1 = e^{-x_1-x_2} - 1, \quad \dot{x}_2 = e^{-x_2-x_3} - 1, \quad \dot{x}_3 = -x_3.$$

8.31 Test the stability of the zero solution of the equation

$$\ddot{x} + [\{1 + (t-1)|\dot{x}|\}/\{1 + t|\dot{x}|\}]\dot{x} + \tfrac{1}{4}x = 0.$$

8.32 Consider the restricted three-body problem in planetary dynamics in which one body (possibly a satellite) has negligible mass in comparison with the other two. Suppose that the two massive bodies (gravitational masses μ_1 and μ_2) remain at a fixed distance a apart, so that the line joining them must rotate with spin $\omega = \sqrt{[(\mu_1 + \mu_2)/a^3]}$. It can be shown (see Hill 1964) that the equations of motion of the third body are given by

$$\ddot{\xi} - 2\omega\dot{\eta} = \partial U/\partial\xi, \quad \ddot{\eta} + 2\omega\dot{\xi} = \partial U/\partial\eta,$$

where the gravitational field

$$U(\xi, \eta) = \frac{1}{2}\omega^2(\xi^2 + \eta^2) + \frac{\mu_1}{d_1} + \frac{\mu_2}{d_2},$$

and

$$d_1 = \sqrt{\left[\left(\xi + \frac{\mu_1 a}{\mu_1 + \mu_2}\right)^2 + \eta^2\right]}, \quad d_2 = \sqrt{\left[\left(\xi - \frac{\mu_2 a}{\mu_1 + \mu_2}\right)^2 + \eta^2\right]}.$$

The origin of the rotating (ξ, η) plane is at the mass centre with ξ axis along the common radius of the two massive bodies in the direction of μ_2.

Consider the special case in which $\mu_1 = \mu_2 = \mu$. Show that there are three equilibrium points along the ξ axis (use a computed graph to establish this), and two equilibrium points at the triangulation points of μ_1 and μ_2.

8.33 Express the equations

$$\dot{x} = x[1 - \sqrt{(x^2 + y^2)}] - \tfrac{1}{2}y[\sqrt{(x^2 + y^2)} - x],$$

$$\dot{y} = y[1 - \sqrt{(x^2 + y^2)}] + \tfrac{1}{2}x[\sqrt{(x^2 + y^2)} - x],$$

in polar form in terms of r and θ. Show that the system has two equilibrium points at $(0, 0)$ and $(1, 0)$. Solve the equations for the phase paths in terms of r and θ, and confirm that all paths which start at any point other than the origin approach $(1, 0)$ as $t \to \infty$. Sketch the phase diagram for the system.

Consider the half-path which starts at $(0, 1)$. Is this path stable in the Poincaré sense? Is the equilibrium point at $(1, 0)$ stable?

8.34 Consider the system

$$\dot{x} = -y, \quad \dot{y} = x + \lambda(1 - y^2 - z^2)y, \quad \dot{z} = -y + \mu(1 - x^2 - y^2)z.$$

Classify the linear approximation of the equilibrium point at the origin in terms of the parameters $\lambda \neq 0$ and $\mu \neq 0$. Verify that the system has a periodic solution

$$x = \cos(t - t_0), \quad y = \sin(t - t_0), \quad z = \cos(t - t_0),$$

for any t_0.

9 Stability by solution perturbation: Mathieu's equation

Stability or instability of nonlinear systems can often be tested by an approximate procedure which leads to a *linear* equation describing the growth of the difference between the test solution and its neighbours. By Theorem 8.9 the stability or instability of the original system resolves itself into the question of the boundedness or otherwise of the solutions of the linear equation. This 'variational equation' often turns out to have a periodic coefficient (Mathieu's equation) and the properties of such equations are derived in this chapter. The fact that the solutions to be tested are themselves usually known only approximately can also be assimilated into this theory.

9.1 The stability of forced oscillations by solution perturbation

Consider the general n-dimensional autonomous system

$$\dot{x} = f(x,t). \tag{9.1}$$

The stability of a solution $x^*(t)$ can be reduced to consideration of the zero solution of a related system. Let $x(t)$ be any other solution, and write

$$x(t) = x^*(t) + \xi(t). \tag{9.2}$$

Then $\xi(t)$ represents a **perturbation**, or disturbance, of the original solution: it seems reasonable to see what happens to $\xi(t)$, since the question of stability is whether such (small) disturbances grow or not. Equation (9.1) can be written in the form

$$\dot{x}^* + \dot{\xi} = f(x^*,t) + \{f(x^* + \xi, t) - f(x^*,t)\}.$$

Since x^* satisfies (9.1), this becomes

$$\dot{\xi} = f(x^* + \xi, t) - f(x^*,t) = h(\xi, t), \tag{9.3}$$

say, since $x^*(t)$ is assumed known. By (9.2), the stability properties of $x^*(t)$ are the same as those of the zero solution of (9.3), $\xi(t) \equiv 0$.

The right-hand side of (9.3) may have a linear approximation for small ξ, in which case

$$\dot{\xi} = h(\xi, t) \approx A(t)\dot{\xi}. \tag{9.4}$$

Here, $A(t) = J(0,t)$, where $J(\xi,t)$ is the Jacobian matrix (see also Section 8.9) of first partial derivatives given by

$$J[\xi, t] = \left[\frac{\partial h_i(\xi, t)}{\partial \xi_j} \right] \quad (i = 1, 2, \ldots, n; \; j = 1, 2, \ldots, n).$$

The properties of this linear system may correctly indicate that of the zero solution of the exact system (9.3). The approximation (9.4) is called the **first variational equation**. This process is not rigorous: it is generally necessary to invoke an approximation not only at the stage (9.4), but also in representing $x^*(t)$, which, of course, will not generally be known exactly.

We shall illustrate the procedure in the case of the two-dimensional forced, undamped pendulum-type equation (a form of Duffing's equation)

$$\ddot{x} + x + \varepsilon x^3 = \Gamma \cos \omega t. \tag{9.5}$$

In order to match the notation of the theory of linear systems of Chapter 8 we will express it in the form

$$\dot{x} = \begin{bmatrix} \dot{x} \\ \dot{y} \end{bmatrix} = \begin{bmatrix} y \\ -x - \varepsilon x^3 + \Gamma \cos \omega t \end{bmatrix}. \tag{9.6}$$

To obtain the variational equation define $\xi = (\xi, \eta)^{\mathrm{T}}$ by

$$\xi = x - x^*, \tag{9.7}$$

where

$$x^* = (x^*, y^*)^{\mathrm{T}}$$

and x^* is the solution to be tested. Substitution for x and y from (9.7) into (9.6) gives

$$\dot{\xi} + \dot{x}^* = \eta + y^*,$$

$$\dot{\eta} + \dot{y}^* = -\xi - x^* - \varepsilon(\xi + x^*)^3 + \Gamma \cos \omega t.$$

By neglecting powers of ξ higher than the first, and using the fact that x^*, y^* satisfy (9.6), the system simplifies to

$$\dot{\xi} = \eta, \quad \dot{\eta} = -\xi - 3\varepsilon x^{*2}\xi, \tag{9.8}$$

corresponding to (9.4).

From Section 7.2 we know that there are periodic solutions of (9.5) which are approximately of the form

$$x = a \cos \omega t,$$

where possible real values of the amplitude a are given by the equation

$$\tfrac{3}{4}\varepsilon a^3 - (\omega^2 - 1)a - \Gamma = 0. \tag{9.9}$$

We shall test the stability of one of these solutions by treating it as being sufficiently close to the corresponding exact form of x^* required by (9.8). By eliminating η between eqns (9.8) we obtain

$$\ddot{\xi} + (1 + 3\varepsilon x^{*2})\xi = 0.$$

When x^* is replaced by its appropriate estimate, $x^* = a \cos \omega t$, with a given by (9.9), this equation becomes

$$\ddot{\xi} + (1 + \tfrac{3}{2}\varepsilon a^2 + \tfrac{3}{2}\varepsilon a^2 \cos 2\omega t)\xi = 0,$$

and we expect that the stability property of x^* and ξ will be the same. The previous equation can be reduced to a standard form

$$\xi'' + (\alpha + \beta \cos \tau)\xi = 0 \tag{9.10}$$

by the substitutions

$$\tau = 2\omega t, \quad \xi' = d\xi/d\tau, \quad \alpha = (2 + 3\varepsilon a^2)/8\omega^2, \quad \beta = 3\varepsilon a^2/8\omega^2. \tag{9.11}$$

For general values of α and β equation (9.10) is known as **Mathieu's equation**. By Theorem 8.8 its solutions are stable for values of the parameters, α, β for which all its solutions are bounded. We shall return to the special case under discussion at the end of Section 9.4 after studying the stability of solutions of Mathieu's general equation, to which problems of this kind may often be reduced.

A pendulum suspended from a support vibrating vertically is a simple model which leads to an equation with a periodic coefficient. Assuming that friction is negligible, consider a rigid pendulum of length a with a bob of mass m suspended from a point which is constrained to oscillate vertically with prescribed displacement $\zeta(t)$ as shown in Fig. 9.1.

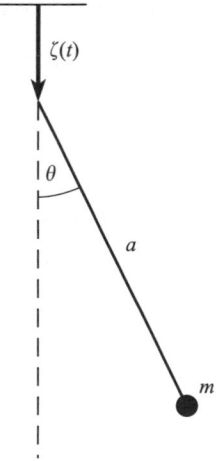

Figure 9.1   Pendulum with vertical forcing: $\zeta(t)$ is the displacement of the support.

The kinetic energy $\mathcal{T}$ and potential energy $\mathcal{V}$ are given by

$$\mathcal{T} = \tfrac{1}{2}m[(\dot{\zeta} - a\sin\theta\dot{\theta})^2 + a^2\cos^2\theta\dot{\theta}^2],$$

$$\mathcal{V} = -mg(\zeta + a\cos\theta).$$

Lagranges's equation of motion

$$\frac{\mathrm{d}}{\mathrm{d}t}\left(\frac{\partial\mathcal{T}}{\partial\dot{\theta}}\right) - \frac{\partial\mathcal{T}}{\partial\theta} = -\frac{\partial\mathcal{V}}{\partial\theta}$$

becomes

$$a\ddot{\theta} + (g - \ddot{\zeta})\sin\theta = 0,$$

which, for oscillations of small amplitude, reduces to

$$a\ddot{\theta} + (g - \ddot{\zeta})\theta = 0.$$

As a standardized form for this equation we may write

$$\ddot{x} + (\alpha + p(t))x = 0.$$

When $p(t)$ is periodic this equation is known as **Hill's equation**. For the special case $p(t) = \beta\cos t$,

$$\ddot{x} + (\alpha + \beta\cos t)x = 0$$

which is Mathieu's equation (9.10). This type of forced motion, in which $p(t)$ acts as an energy source, is an instance of **parametric excitation**.

Exercise 9.1
Show that the damped equation

$$\ddot{x} + k\dot{x} + (\gamma + \beta\cos t)x = 0$$

can be transformed into a Mathieu equation by the change of variable $x = ze^{\mu t}$ for a suitable choice for μ.

9.2 Equations with periodic coefficients (Floquet theory)

Equation (9.10) is a particular example of an equation associated with the general n-dimensional first-order system

$$\dot{x} = P(t)x, \tag{9.12}$$

where $P(t)$ is periodic with minimal period T; that is, T is the smallest positive number for which

$$P(t + T) = P(t), \quad -\infty < t < \infty. \tag{9.13}$$

($P(t)$, of course, also has periods $2T$, $3T$, ...) The *solutions* are not necessarily periodic, as can be seen from the one-dimensional example

$$\dot{x} = P(t)x = (1 + \sin t)x;$$

the coefficient $P(t)$ has period 2π, but all solutions are given by

$$x = ce^{t-\cos t},$$

where c is any constant, so only the solution $x = 0$ is periodic. Similarly, the system

$$\begin{bmatrix} \dot{x} \\ \dot{y} \end{bmatrix} = \begin{bmatrix} 1 & \cos t \\ 0 & -1 \end{bmatrix} \begin{bmatrix} x \\ y \end{bmatrix}$$

has no periodic solutions apart from the trivial case $x = y = 0$.

In the following discussions remember that the displayed solution vectors may consist of complex solutions.

Theorem 9.1 (Floquet's theorem)   *The regular system $\dot{x} = P(t)x$, where P is an $n \times n$ matrix function with minimal period T, has at least one non-trivial solution $x = \chi(t)$ such that*

$$\chi(t + T) = \mu\chi(t), \quad -\infty < t < \infty, \tag{9.14}$$

where μ is a constant.

Proof   Let $\Phi(t) = [\phi_{ij}(t)]$ be a fundamental matrix for the system. Then $\dot{\Phi}(t) = P(t)\Phi(t)$. Since $P(t+T) = P(t)$, $\Phi(t+T)$ satisfies the same equation, and by Theorem 8.5, $\det \Phi(t+T) \neq 0$, so $\Phi(t+T)$ is another fundamental matrix. The columns (solutions) in $\Phi(t+T)$ are linear combinations of those in $\Phi(t)$ by Theorem 8.4:

$$\phi_{ij}(t + T) = \sum_{k=1}^{n} \phi_{ik}(t)e_{kj}$$

for some constants e_{kj}, so that

$$\Phi(t + T) = \Phi(t)E, \tag{9.15}$$

where $E = [e_{kj}]$. E is nonsingular, since $\det \Phi(t+T) = \det \Phi(t) \det(E)$, and therefore $\det(E) \neq 0$. The matrix E can be found from $\Phi(t_0 + T) = \Phi(t_0)E$ where t_0 is a convenient value of t. Thus

$$E = \Phi^{-1}(t_0)\Phi(t_0 + T).$$

Let μ be an eigenvalue of E:

$$\det(E - \mu I) = 0, \tag{9.16}$$

and let *s* be an eigenvector corresponding to μ:

$$(E - \mu I)s = 0. \tag{9.17}$$

Consider the solution $x = \Phi(t)s = \chi(t)$ (being a linear combination of the columns of Φ, χ is a solution of (9.12)). Then

$$
\begin{aligned}
\chi(t + T) &= \Phi(t + T)s \\
&= \Phi(t)Es = \Phi(t)\mu s \quad \text{(by (9.17))} \\
&= \mu\chi(t). \quad\quad\quad\quad\quad \text{(by (9.14))}
\end{aligned}
$$

∎

The eigenvalues μ of E are called **characteristic numbers** or **multipliers** of eqn (9.12) (not to be confused with the *eigenvalues* of $P(t)$, which will usually be dependent on t). The importance of this theorem is the possibility of a characteristic number with a special value implying the existence of a periodic solution (though not necessarily of period T).

Example 9.1 *Find a fundamental matrix for the periodic differential equation*

$$
\begin{bmatrix} \dot{x}_1 \\ \dot{x}_2 \end{bmatrix} = P(t)x = \begin{bmatrix} 1 & 1 \\ 0 & h(t) \end{bmatrix} \begin{bmatrix} x_1 \\ x_2 \end{bmatrix}, \tag{9.18}
$$

where $h(t) = (\cos t + \sin t)/(2 + \sin t - \cos t)$, and determine the characteristic numbers.
From (9.18),

$$(2 + \sin t - \cos t)\dot{x}_2 = (\cos t + \sin t)x_2,$$

which has the solution

$$x_2 = b(2 + \sin t - \cos t)$$

where *b* is any constant. Then x_1 satisfies

$$\dot{x}_1 - x_1 = x_2 = b(2 + \sin t - \cos t)$$

and therefore

$$x_1 = ae^t - b(2 + \sin t).$$

where *a* is any constant. A fundamental matrix $\Phi(t)$ can be obtained by putting, say, $a = 0$, $b = 1$, and $a = 1$, $b = 0$:

$$
\Phi(t) = \begin{bmatrix} -2 - \sin t & e^t \\ 2 + \sin t - \cos t & 0 \end{bmatrix}.
$$

The matrix $P(t)$ has minimal period $T = 2\pi$, and E in (9.15) must satisfy $\Phi(t + 2\pi) = \Phi(t)E$ for all t. Therefore $\Phi(2\pi) = \Phi(0)E$ and

$$
E = \Phi^{-1}(0)\Phi(2\pi) = \begin{bmatrix} 1 & 0 \\ 0 & e^{2\pi} \end{bmatrix}.
$$

The eigenvalues μ of E satisfy

$$\begin{vmatrix} 1-\mu & 0 \\ 0 & e^{2\pi}-\mu \end{vmatrix} = 0,$$

so $\mu = 1$ or $e^{2\pi}$. From (9.14), since one eigenvalue is unity there exist solutions such that $\chi(t+2\pi) = \chi(t)$: that is, solutions with period 2π. We have already found these: they correspond to a $a = 0$.   ●

Theorem 9.2   *The constants μ in Theorem 9.1 are independent of the choice of Φ.*

Proof   Let $\Phi(t)$, $\Phi^*(t)$ be two fundamental matrices; then

$$\Phi^*(t) = \Phi(t)C, \tag{9.19}$$

where C is some constant, nonsingular matrix (nonsingular since $\Phi(t)$ and $\Phi^*(t)$ are nonsingular by Theorem 8.5). Let T be the minimal period of $P(t)$. Then

$$\begin{aligned} \Phi^*(t+T) &= \Phi(t+T)C && \text{(by (9.19))} \\ &= \Phi(t)EC && \text{(by (9.15))} \\ &= \Phi^*(t)C^{-1}EC && \text{(by (9.19))} \\ &= \Phi^*(t)D, \end{aligned}$$

say, where $D = C^{-1}EC$ and C is nonsingular. We may write

$$\begin{aligned} \det(D-\mu I) &= \det(C^{-1}EC-\mu I) = \det[C^{-1}(E-\mu I)C] \\ &= \det(C^{-1}C)\det(E-\mu I) = \det(E-\mu I) \end{aligned}$$

(using the product rule for determinants). Since $\det(D-\mu I)$ is zero if and only if $\det(E-\mu I)$ is zero, D and E have the same eigenvalues.   ∎

We can therefore properly refer to 'the characteristic numbers of the system'. Note that when Φ is chosen as real E is real, and the characteristic equation for the numbers μ has real coefficients. Therefore if μ (complex) is a characteristic number, then so is its complex conjugate $\bar{\mu}$.

Definition 9.1   *A solution of (9.12) satisfying (9.14) is called a **normal solution**.*

Definition 9.2 (Characteristic exponent)   *Let μ be a characteristic number, real or complex, of the system (9.12), corresponding to the minimal period T of $P(t)$. Then ρ, defined by*

$$e^{\rho T} = \mu \tag{9.20}$$

*is called a **characteristic exponent** of the system. Note that ρ is defined only to an additive multiple of $2\pi i/T$. It will be fixed by requiring $-\pi < \mathrm{Im}(\rho T) \leq \pi$, or by $\rho = (1/T)\mathrm{Ln}(\mu)$, where the principal value of the logarithm is taken.*

Theorem 9.3   *Suppose that* E *of Theorem 9.1 has n distinct eigenvalues,* μ_i, $i = 1, 2, \ldots, n$. *Then (9.12) has n linearly independent* **normal solutions** *of the form*

$$x_i = p_i(t)e^{\rho_i t} \tag{9.21}$$

(ρ_i are the characteristic exponents corresponding to μ_i), where the $p_i(t)$ are vector functions with period T.

Proof   To each μ_i corresponds a solution $x_i(t)$ satisfying (9.14): $x_i(t + T) = \mu_i x_i(t) = e^{\rho_i T} x_i(t)$. Therefore, for every t,

$$x_i(t + T)e^{-\rho_i(t+T)} = x_i(t)e^{-\rho_i t}. \tag{9.22}$$

Writing

$$p_i(t) = e^{-\rho_i t} x_i(t),$$

(9.22) implies that $p_i(t)$ has period T.

The linear independence of the $x_i(t)$ is implied by their method of construction in Theorem 9.1: from (9.17), they are given by $x_i(t) = \Phi(t)s_i$; s_i are the eigenvectors corresponding to the *different* eigenvalues μ_i, and are therefore linearly independent. Since $\Phi(t)$ is non-singular it follows that the $x_i(t)$ are also linearly independent.   ■

When the eigenvalues of E are not all distinct, the coefficients corresponding to the $p_i(t)$ are more complicated.

Under the conditions of Theorem 9.3, periodic solutions of period T exist when E has an eigenvalue

$$\mu = 1.$$

The corresponding normal solutions have period T, the minimal period of $P(t)$. This can be seen from (9.14) or from the fact that the corresponding ρ is zero.

There are periodic solutions whenever E has an eigenvalue μ which is one of the mth roots of unity:

$$\mu = 1^{1/m}, \quad m \text{ a positive integer}. \tag{9.23a}$$

In this case, from (9.14),

$$\chi(t + mT) = \mu\chi\{t + (m-1)T\} = \cdots = \mu^m \chi(t) = \chi(t), \tag{9.23b}$$

so that $\chi(t)$ has period mT.

Example 9.2   *Identify the periodic vectors $p_i(t)$ (see eqn (9.21)) in the solution of the periodic differential equation in Example 9.1.*

The characteristic numbers were shown to be $\mu_1 = 1$, $\mu_2 = e^{2\pi}$. The corresponding characteristic exponents (Definition 9.2) are $\rho_1 = 0$, $\rho_2 = 1$. From Example 9.1, a fundamental matrix is

$$\Phi(t) = \begin{bmatrix} -2 - \sin t & e^t \\ 2 + \sin t - \cos t & 0 \end{bmatrix}.$$

From the columns we can identify the 2π-periodic vectors

$$p_1(t) = a \begin{bmatrix} -2 - \sin t \\ -2 + \sin t - \cos t \end{bmatrix}, \quad p_2(t) = b \begin{bmatrix} 1 \\ 0 \end{bmatrix},$$

where a and b are any constants. In terms of normal solutions

$$\begin{bmatrix} x_1 \\ x_2 \end{bmatrix} = a \begin{bmatrix} -2 - \sin t \\ 2 + \sin t - \cos t \end{bmatrix} e^0 + b \begin{bmatrix} 1 \\ 0 \end{bmatrix} e^t. \qquad \bullet$$

In the preceding theory, $\det \mathbf{\Phi}(t)$ appeared repeatedly, where $\mathbf{\Phi}$ is a fundamental matrix of the regular system $\dot{x} = A(t)x$. This has a simple representation, as follows.

Definition 9.3 *Let $[\boldsymbol{\phi}_1(t), \boldsymbol{\phi}_2(t), \dots, \boldsymbol{\phi}_n(t)]$ be a matrix whose columns are any solutions of the n-dimensional system $\dot{x} = A(t)x$. Then*

$$W(t) = \det[\boldsymbol{\phi}_1(t), \boldsymbol{\phi}_2(t), \dots, \boldsymbol{\phi}_n(t)] \qquad (9.24)$$

*is called the **Wronskian** of this set of solutions, taken in order.*

Theorem 9.4 *For any t_0, the Wronskian of $\dot{x} = A(t)x$ is*

$$W(t) = W(t_0) \exp \left(\int_{t_0}^{t} \mathrm{tr}\{A(s)\}ds \right), \qquad (9.25)$$

*where $\mathrm{tr}\{A(s)\}$ is the **trace** of $A(s)$ (the sum of the elements of its principal diagonal).*

Proof If the solutions are linearly dependent, $W(t) \equiv 0$ by Theorem 8.5, and the result is true trivially.

If not, let $\mathbf{\Phi}(t)$ be any fundamental matrix of solutions, with $\mathbf{\Phi}(t) = [\phi_{ij}(t)]$. Then dW/dt is equal to the sum of n determinants Δ_k, $k = 1, 2, \dots, n$, where Δ_k is the same as $\det[\phi_{ij}(t)]$, except for having $\dot{\phi}_{kj}(t)$, $j = 1, 2, \dots, n$ in place of $\phi_{kj}(t)$ in its kth row. Consider one of the Δ_k, say Δ_1:

$$\Delta_1 = \begin{vmatrix} \dot{\phi}_{11} & \dot{\phi}_{12} & \cdots & \dot{\phi}_{1n} \\ \phi_{21} & \phi_{22} & \cdots & \phi_{2n} \\ \cdots & \cdots & \cdots & \cdots \\ \phi_{n1} & \phi_{n2} & \cdots & \phi_{nn} \end{vmatrix} = \begin{vmatrix} \sum_{m=1}^{n} a_{1m}\phi_{m1} & \sum_{m=1}^{n} a_{1m}\phi_{m2} & \cdots & \sum_{m=1}^{n} a_{1m}\phi_{mn} \\ \phi_{21} & \phi_{22} & \cdots & \phi_{2n} \\ \cdots & \cdots & \cdots & \cdots \\ \phi_{n1} & \phi_{n2} & \cdots & \phi_{nn} \end{vmatrix}$$

(from eqn (8.14))

$$= \sum_{m=1}^{n} a_{1m} \begin{vmatrix} \phi_{m1} & \phi_{m2} & \cdots & \phi_{mn} \\ \phi_{21} & \phi_{22} & \cdots & \phi_{2n} \\ \cdots & \cdots & \cdots & \cdots \\ \phi_{n1} & \phi_{n2} & \cdots & \phi_{nn} \end{vmatrix} = a_{11} \begin{vmatrix} \phi_{11} & \phi_{12} & \cdots & \phi_{1n} \\ \phi_{21} & \phi_{22} & \cdots & \phi_{2n} \\ \cdots & \cdots & \cdots & \cdots \\ \phi_{n1} & \phi_{n2} & \cdots & \phi_{nn} \end{vmatrix} = a_{11}W(t),$$

since all the other determinants have repeated rows, and therefore vanish. In general $\Delta_k = a_{kk}W(t)$. Therefore

$$\frac{\mathrm{d}W(t)}{\mathrm{d}t} = \mathrm{tr}\{A(t)\}W(t),$$

which is a differential equation for W having solution (9.25). ∎

For periodic systems we have the following result.

Theorem 9.5 *For the system* $\dot{x} = P(t)x$, *where* $P(t)$ *has minimal period* T, *let the characteristic numbers of the system be* $\mu_1, \mu_2, \ldots, \mu_n$. *Then*

$$\mu_1 \mu_2 \ldots \mu_n = \exp\left(\int_0^T \mathrm{tr}\{P(s)\}\mathrm{d}s\right),$$

a repeated characteristic number being counted according to its multiplicity.

Proof Let $\Psi(t)$ be the fundamental matrix of the system for which

$$\Psi(0) = I. \tag{9.26}$$

Then, (eqn (9.15)),

$$\Psi(T) = \Psi(0)E = E, \tag{9.27}$$

in the notation of Theorem 9.1. The characteristic numbers μ_i are the eigenvalues of E, given by

$$\det(E - \mu I) = 0.$$

This is an nth-degree polynomial in μ, and the product of the roots is equal to the constant term: that is, equal to the value taken when $\mu = 0$. Thus, by (9.27),

$$\mu_1 \mu_2 \ldots \mu_n = \det(E) = \det \Psi(T) = W(T),$$

but by Theorem 9.4 with $t_0 = 0$ and $t = T$,

$$W(T) = W(0)\int_0^T \mathrm{tr}\{P(s)\}\mathrm{d}s$$

and $W(0) = 1$ by (9.26). ∎

Example 9.3 *Verify the formula in Theorem 9.5 for the product of the characteristic numbers of Example 9.1.*
In Example 9.1, $T = 2\pi$ and

$$P(t) = \begin{bmatrix} 1 & 1 \\ 0 & (\cos t + \sin t)/(2 + \sin t - \cos t) \end{bmatrix}.$$

Then

$$\int_0^{2\pi} \text{tr}\{P(s)\} ds = \int_0^{2\pi} \left[1 + \frac{\cos s + \sin s}{2 + \sin s - \cos s} \right] ds = \int_0^{2\pi} \left[1 + \frac{d(\sin s - \cos s)/ds}{2 + \sin s - \cos s} \right] ds$$

$$= [s + \log(2 + \sin s - \cos s)]_0^{2\pi} = 2\pi.$$

Therefore

$$\exp \left[\int_0^{2\pi} \text{tr}\{P(s)\} ds \right] = e^{2\pi} = \mu_1 \mu_2,$$

by Example 9.1.

●

Exercise 9.2

Find the matrix E for the system

$$\begin{bmatrix} \dot{x}_1 \\ \dot{x}_2 \end{bmatrix} = \begin{bmatrix} 1 & \cos t - 1 \\ 0 & \cos t \end{bmatrix} \begin{bmatrix} x_1 \\ x_2 \end{bmatrix},$$

and obtain its characteristic numbers. Verify the result in Theorem 9.5.

9.3 Mathieu's equation arising from a Duffing equation

We now return to look in more detail at Mathieu's equation (9.10)

$$\ddot{x} + (\alpha + \beta \cos t)x = 0, \tag{9.28}$$

As a first-order system it can be expressed as

$$\begin{bmatrix} \dot{x} \\ \dot{y} \end{bmatrix} = \begin{bmatrix} 0 & 1 \\ -\alpha - \beta \cos t & 0 \end{bmatrix} \begin{bmatrix} x \\ y \end{bmatrix}, \tag{9.29}$$

In the notation of the previous section,

$$P(t) = \begin{bmatrix} 0 & 1 \\ -\alpha - \beta \cos t & 0 \end{bmatrix}. \tag{9.30}$$

Clearly $P(t)$ is periodic with minimal period 2π. The general structure of the solution is determined by Theorem 9.3, whilst the question of the stability of a solution can be decided, through Theorem 8.9, by the boundedness or otherwise of the solution for given values of the parameters α and β. We are not particularly interested in periodic solutions as such, though we shall need them to settle the stability question.

From eqn (9.30),

$$\text{tr}\{P(t)\} = 0. \tag{9.31}$$

Therefore, by Theorem 9.5,

$$\mu_1 \mu_2 = e^0 = 1, \tag{9.32}$$

where μ_1, μ_2 are the characteristic numbers of $P(t)$. They are solutions of a quadratic characteristic equation (9.16), with real coefficients, which by (9.16) has the form

$$\mu^2 - \phi(\alpha, \beta)\mu + 1 = 0.$$

where the value of ϕ, depending on E (eqn (9.27)) can, in principle, be found in a particular case. The solutions μ are given by

$$\mu_1, \mu_2 = \tfrac{1}{2}[\phi \pm \sqrt{(\phi^2 - 4)}]. \tag{9.33}$$

Although $\phi(\alpha, \beta)$ is not specified explicitly, we can make the following deductions.

(i) $\phi > 2$. The characteristic numbers are real, different, and positive, and by (9.32), one of them, say μ_1, exceeds unity. The corresponding characteristic exponents (9.20) are real and have the form $\rho_1 = \sigma > 0$, $\rho_2 = -\sigma < 0$. The general solution is therefore of the form (Theorem 9.3)

$$x(t) = c_1 e^{\sigma t} p_1(t) + c_2 e^{-\sigma t} p_2(t),$$

where c_1, c_2 are constants and p_1, p_2 have minimal period 2π. The parameter region $\phi(\alpha, \beta) > 2$ therefore contains unbounded solutions, and is called an **unstable parameter region**.

(ii) $\phi = 2$. Then $\mu_1 = \mu_2 = 1$, $\rho_1 = \rho_2 = 0$. By (9.21), there is *one solution of period 2π on the curves $\phi(\alpha, \beta) = 2$*. (The other solution is unbounded.)

(iii) $-2 < \phi < 2$. The characteristic numbers are complex, and $\mu_2 = \bar{\mu}_1$. Since also $|\mu_1| = |\mu_2| = 1$, we must have $\rho_1 = i\nu$, $\rho_2 = -i\nu$, ν real. The general solution is of the form

$$x(t) = c_1 e^{i\nu t} p_1(t) + c_2 e^{-i\nu t} p_2(t) \quad (p_1, p_2 \text{ period } 2\pi).$$

and *all solutions in the parameter region $-2 < \phi(\alpha, \beta) < 2$ are bounded*. This is called the **stable parameter region**. The solutions are oscillatory, but not in general periodic, since the two frequencies ν and 2π are present.

(iv) $\phi = -2$. Then $\mu_1 = \mu_2 = -1 (\rho_1 = \rho_2 = \tfrac{1}{2} i)$, so by Theorem 9.1, eqn (9.14), there is *one solution with period 4π at every point on $\phi(\alpha, \beta) = -2$*. (The other solution is in fact unbounded.)

(v) $\phi < -2$. Then μ_1 and μ_2 are real and negative. Since, also, $\mu_1 \mu_2 = 1$, the general solution is of the form

$$x(t) = c_1 e^{(\sigma + \tfrac{1}{2} i)t} p_1(t) + c_2 e^{\left(-\sigma + \tfrac{1}{2} i\right)t} p_2(t),$$

where $\sigma > 0$ and p_1, p_2 have period 2π. For later purposes it is important to notice that *the solutions have the alternative form*

$$c_1 e^{\sigma t} q_1(t) + c_2 e^{-\sigma t} q_2(t), \tag{9.34}$$

where q_1, q_2 have period 4π.

From (i) to (v) it can be seen that certain curves, of the form

$$\phi(\alpha, \beta) = \pm 2,$$

separate parameter regions where unbounded solutions exist ($|\phi(\alpha, \beta)| > 2$) from regions where all solutions are bounded ($|\phi(\alpha, \beta)| < 2$) (Fig. 9.2). We do not specify the function $\phi(\alpha, \beta)$ explicitly, but we do know that these are also the curves on which periodic solutions, period 2π or 4π, occur. Therefore, if we can establish, by any method, the parameter values for which such periodic solutions can occur, then we have also found the boundaries between the stable and unstable region by Theorem 8.8. These boundaries are called **transition curves**.

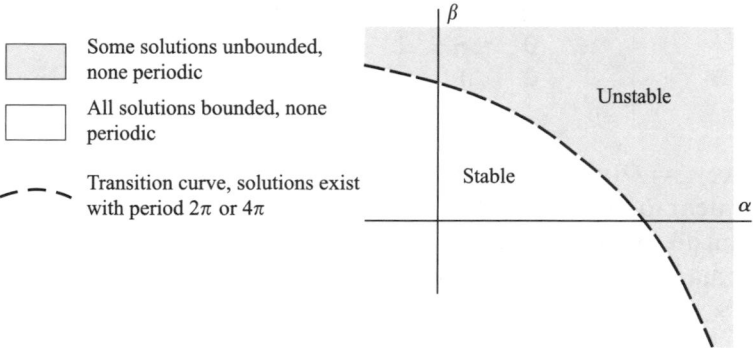

Figure 9.2

First, we find what parameter values α, β give periodic solutions of period 2π. Represent such a solution by the complex Fourier series

$$x(t) = \sum_{n=-\infty}^{\infty} c_n e^{int}.$$

We now adopt the following formal procedure which assumes convergence where necessary. Substitute the series into Mathieu's equation

$$\ddot{x} + (\alpha + \beta \cos t)x = 0$$

replacing $\cos t$ by $\frac{1}{2}(e^{it} + e^{-it})$. The result is

$$-\sum_{n=-\infty}^{\infty} c_n n^2 e^{int} + \left[\alpha + \frac{1}{2}\beta(e^{it} + e^{-it})\right] \sum_{n=-\infty}^{\infty} c_n e^{int} = 0.$$

which becomes after re-ordering the summation,

$$\sum_{n=-\infty}^{\infty} [\beta c_{n+1} + 2(\alpha - n^2)c_n + \beta c_{n-1}]e^{int} = 0.$$

This equation can only be satisfied for all t if the coefficients of e^{int} are all zero, that is if

$$\beta c_{n+1} + 2(\alpha - n^2)c_n + \beta c_{n-1} = 0, \quad n = 0, \pm 1, \pm 2, \ldots.$$

Assume that $\alpha \neq n^2$, and express this equation in the form

$$\gamma_n c_{n+1} + c_n + \gamma_n c_{n-1} = 0, \quad \text{where } \gamma_n = \frac{\beta}{2(\alpha - n^2)}, (n = 0, \pm 1, \pm 2, \ldots), \tag{9.35}$$

but observe that $\gamma_{-n} = \gamma_n$. The infinite set of homogeneous linear equations in (9.35) for the sequence $\{c_n\}$ has nonzero solutions if the **infinite determinant** (Whittaker and Watson, 1962), known as a **Hill determinant**, formed by their coefficients is zero, namely if

$$\begin{vmatrix} \cdots & \cdots & \cdots & \cdots & \cdots & \cdots & \cdots \\ \cdots & \gamma_1 & 1 & \gamma_1 & 0 & 0 & \cdots \\ \cdots & 0 & \gamma_0 & 1 & \gamma_0 & 0 & \cdots \\ \cdots & 0 & 0 & \gamma_1 & 1 & \gamma_1 & \cdots \\ \cdots & \cdots & \cdots & \cdots & \cdots & \cdots & \cdots \end{vmatrix} = 0. \tag{9.36}$$

The condition that $\gamma_n = O(n^{-2})$ (from (9.35)) ensures the convergence of the determinant. This equation is equivalent to $\phi(\alpha, \beta) = 2$ (see Section 9.3(ii)).

The determinant in (9.36) is **tridiagonal** (zero elements everywhere except on the leading diagonal and the diagonals immediately above and below it), and a recurrence relation can be established for $n \times n$ approximations. Let

$$D_{m,n} = \begin{vmatrix} 1 & \gamma_m & 0 & 0 & \cdots & \cdots & \cdots & \cdots & \cdots & \cdots & \cdots \\ \gamma_{m-1} & 1 & \gamma_{m-1} & 0 & \cdots & \cdots & \cdots & \cdots & \cdots & \cdots & \cdots \\ \cdots & \cdots & \cdots & \cdots & \cdots & \cdots & \cdots & \cdots & \cdots & \cdots & \cdots \\ \cdots & \cdots & \cdots & 0 & \gamma_0 & 1 & \gamma_0 & 0 & \cdots & \cdots & \cdots \\ \cdots & \cdots & \cdots & \cdots & \cdots & \cdots & \cdots & \cdots & \cdots & \cdots & \cdots \\ \cdots & \cdots & \cdots & \cdots & \cdots & \cdots & \cdots & 0 & \gamma_{n-1} & 1 & \gamma_{n-1} \\ \cdots & \cdots & \cdots & \cdots & \cdots & \cdots & \cdots & 0 & 0 & \gamma_n & 1 \end{vmatrix}. \tag{9.37}$$

$D_{m,n}$ is a determinant with $m + n + 1$ rows and columns. Expansion by the first row leads to

$$D_{m,n} = D_{m-1,n} - \gamma_m \gamma_{m-1} D_{m-2,n}.$$

Note that $D_{m,n} = D_{n,m}$. Let $E_n = D_{n,n}$, $P_n = D_{n-1,n}$ and $Q_n = D_{n-2,n}$. Put $m = n, n+1, n+2$ successively in (9.37) resulting in

$$E_n = P_n - \gamma_n \gamma_{n-1} Q_n, \tag{9.38}$$

$$P_{n+1} = E_n - \gamma_{n+1} \gamma_n P_n, \tag{9.39}$$

$$Q_{n+2} = P_{n+1} - \gamma_{n+2} \gamma_{n+1} E_n. \tag{9.40}$$

Eliminate Q_n between (9.38) and (9.40), so that

$$E_{n+2} = P_{n+2} - \gamma_{n+2}\,\gamma_{n+1}P_{n+1} + \gamma_{n+2}^2\,\gamma_{n+1}^2 E_n. \tag{9.41}$$

Now eliminate E_n between (9.39) and (9.41), so that

$$2\gamma_{n+1}\,\gamma_{n+2}P_{n+1} = E_{n+1} - E_{n+2} + \gamma_{n+1}^2\,\gamma_{n+2}^2 E_n.$$

Finally substitute this formula for P_n back into (9.39) to obtain the following third-order difference equation

$$E_{n+2} = (1 - \gamma_{n+1}\,\gamma_{n+2})E_{n+1} - \gamma_{n+1}\,\gamma_{n+2}(1 - \gamma_{n+1}\,\gamma_{n+2})E_n + \gamma_n^2\gamma_{n+1}^3\,\gamma_{n+2}E_{n-1},$$

for $n \geq 1$. In order to solve this difference equation we require E_0, E_1 and E_2, which are given by

$$E_0 = 1, \quad E_1 = \begin{vmatrix} 1 & \gamma_1 & 0 \\ \gamma_0 & 1 & \gamma_0 \\ 0 & \gamma_0 & 1 \end{vmatrix} = 1 - 2\gamma_0\gamma_1,$$

$$E_2 = \begin{vmatrix} 1 & \gamma_2 & 0 & 0 & 0 \\ \gamma_1 & 1 & \gamma_1 & 0 & 0 \\ 0 & \gamma_0 & 1 & \gamma_0 & 0 \\ 0 & 0 & \gamma_1 & 1 & \gamma_1 \\ 0 & 0 & 0 & \gamma_2 & 1 \end{vmatrix} = (\gamma_1\gamma_2 - 1)(\gamma_1\gamma_2 - 1 + 2\gamma_0\gamma_1).$$

The sequence of determinants $\{E_n\}$ is said to converge if there exists a number E such that

$$\lim_{n \to \infty} E_n = E.$$

It can be shown (see Whittaker and Watson (1962), Section 2.8) that E_n converges if the sum of the non-diagonal elements converges absolutely. The sum is

$$2\gamma_0 + 4\sum_{i=1}^{\infty} \gamma_i,$$

which is absolutely convergent since $|\gamma_n| = O(n^{-2})$ as $n \to \infty$.

Given β we solve the equations $E_i = 0$ for α for i increasing from 1 until α is obtained to the required accuracy. However there can be convergence problems if α is close to $1, 2^2, 3^2, \ldots$. To avoid this numerical problem rescale the rows in E to eliminate the denominators $\alpha - n^2$. Hence we consider instead the zeros of (we need not consider E_0)

$$H_1(\alpha, \beta) = \begin{vmatrix} 2(\alpha - 1^2) & \beta & 0 \\ \beta & 2\alpha & \beta \\ 0 & \beta & 2(\alpha - 1^2) \end{vmatrix} = 2^3\alpha(\alpha - 1^2)^2 E_1,$$

$$H_2(\alpha, \beta) = \begin{vmatrix} 2(\alpha - 2^2) & \beta & 0 & 0 & 0 \\ \beta & 2(\alpha - 1^2) & \beta & 0 & 0 \\ 0 & \beta & 2\alpha & \beta & 0 \\ 0 & 0 & \beta & 2(\alpha - 1^2) & \beta \\ 0 & 0 & 0 & \beta & 2(\alpha - 2^2) \end{vmatrix}$$

$$= 2^5 \alpha (\alpha - 1^2)^2 (\alpha - 2^2)^2 E_2,$$

and so on. The evaluations of the first three determinants lead to

$$H_1(\alpha, \beta) = 4(\alpha - 1)(-2\alpha + 2\alpha^2 - \beta^2),$$

$$H_2(\alpha, \beta) = 2(16 - 20\alpha + 4\alpha^2 - \beta^2)(16\alpha - 20\alpha^2 + 4\alpha^3 + 8\beta^2 - 3\alpha\beta^2),$$

$$H_3(\alpha, \beta) = 8(-72 + 98\alpha - 28\alpha^2 + 2\alpha^3 + 5\beta^2 - \alpha\beta^2)$$

$$(-288\alpha + 392\alpha^2 - 112\alpha^3 + 8\alpha^4 - 144\beta^2 + 72\alpha\beta^2 - 8\alpha^2\beta^2 + \beta^4),$$

(computer software is needed to expand and factorize these determinants). It can be seen from the determinants $H_i(\alpha, 0)$ that $H_i(\alpha, 0) = 0$ if $\alpha_j = j^2$ for $j \leq i$. These are the critical values on the α axis shown in Fig. 9.3.

The table shows the solutions of the equations $H_i(\alpha, \beta) = 0$ for $i = 1, 2, 3$ for values of $\beta = 0, 0.4, 0.8, 1.2$. As the order of the determinant is increased an increasing number of solutions for α for fixed β appear. The results of a more comprehensive computation are shown

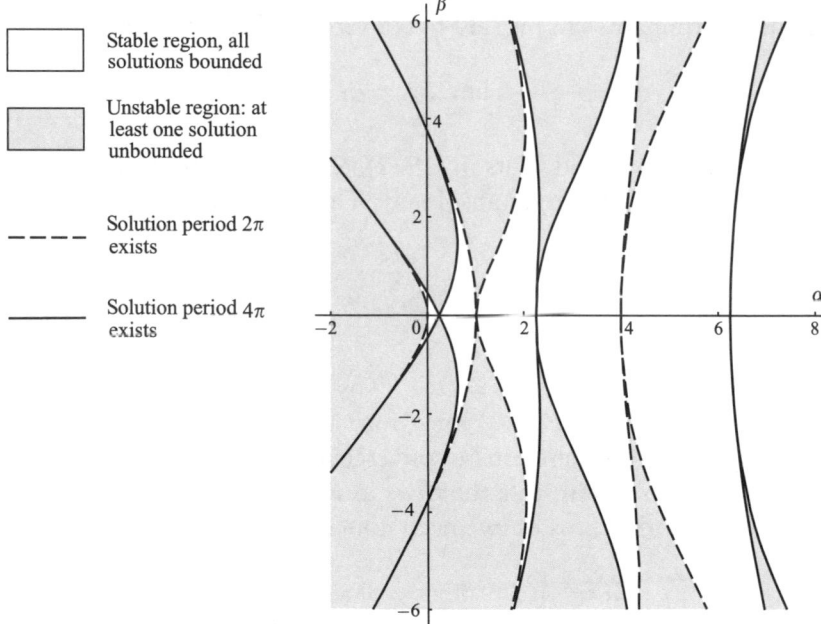

Figure 9.3   Stability diagram for Mathieu's equation $\ddot{x} + (\alpha + \beta \cos t)x = 0$.

in Fig. 9.3. The dashed curves show the parameter values of α and β on which 2π-periodic solutions of the Mathieu equation exist. The curves are symmetric about the α axis, one passes through the origin, and the others have cusps at $\alpha = 1, 4, 9, \ldots$.

	$H_1(\alpha, \beta) = 0$	$H_2(\alpha, \beta) = 0$	$H_3(\alpha, \beta) = 0$
$\beta = 0$	$\alpha = 0$	$\alpha = 0$	$\alpha = 0$
	$\alpha = 1$	$\alpha = 1$	$\alpha = 1$
		$\alpha = 4$	$\alpha = 4$
			$\alpha = 9$
$\beta = 0.4$	$\alpha = -0.074$	$\alpha = -0.075$	$\alpha = -0.075$
	$\alpha = 1.000$	$\alpha = 0.987$	$\alpha = 0.987$
	$\alpha = 1.074$	$\alpha = 1.062$	$\alpha = 1.062$
		$\alpha = 4.013$	$\alpha = 4.005$
		$\alpha = 4.013$	$\alpha = 4.005$
$\beta = 0.8$	$\alpha = -0.255$	$\alpha = -0.261$	$\alpha = -0.261$
	$\alpha = 1.000$	$\alpha = 0.948$	$\alpha = 0.947$
	$\alpha = 1.256$	$\alpha = 1.208$	$\alpha = 1.207$
		$\alpha = 4.052$	$\alpha = 4.021$
		$\alpha = 4.053$	$\alpha = 4.022$
$\beta = 1.2$	$\alpha = -0.485$	$\alpha = -0.505$	$\alpha = -0.505$
	$\alpha = 1.000$	$\alpha = 0.884$	$\alpha = 0.883$
	$\alpha = 1.485$	$\alpha = 1.383$	$\alpha = 1.381$
		$\alpha = 4.116$	$\alpha = 4.046$
		$\alpha = 4.122$	$\alpha = 4.052$

For the 4π periodic solutions, let

$$x(t) = \sum_{n=-\infty}^{\infty} d_n e^{\frac{1}{2}int}.$$

As in the previous case substitute $x(t)$ into eqn (9.28) and equate to zero the coefficients of $e^{\frac{1}{2}int}$ so that

$$\tfrac{1}{2}\beta d_{n+2} + (\alpha - \tfrac{1}{4}n^2)d_n + \tfrac{1}{2}d_{n-1} = 0, \quad (n = 0, 1, 2 \ldots).$$

This set of equations split into two independent sets for $\{d_n\}$. If n is even then the equations reproduce those of (9.35) for the 2π period solutions. Therefore we need only consider solutions for n odd, and can put $d_{2m} = 0$ for all m. For n odd, the set of equations have a nontrivial solutions if, and only if,

$$
\begin{vmatrix}
\cdots & \cdots & \cdots & \cdots & \cdots & \cdots & \cdots & \cdots \\
\cdots & \delta_2 & 1 & \delta_2 & 0 & 0 & 0 & \cdots \\
\cdots & 0 & \delta_1 & 1 & \delta_1 & 0 & 0 & \cdots \\
\cdots & 0 & 0 & \delta_1 & 1 & \delta_1 & 0 & \cdots \\
\cdots & 0 & 0 & 0 & \delta_2 & 1 & \delta_2 & \cdots \\
\cdots & \cdots & \cdots & \cdots & \cdots & \cdots & \cdots & \cdots
\end{vmatrix} = 0,
$$

where

$$
\delta_1 = \frac{\beta}{2(\alpha - \frac{1}{4}1^2)}, \qquad \delta_2 = \frac{\beta}{2(\alpha - \frac{1}{4}3^2)},
$$

$$
\delta_m = \frac{\beta}{2[\alpha - \frac{1}{4}(2m-1)^2]}, \qquad m = 1, 2, \ldots,
$$

provided $\alpha \neq \frac{1}{4}(2m-1)^2$. The numerical relation between α and β can be computed by taking finite approximations to the infinite determinant. The transition curves corresponding to the 4π periodic solutions are shown in Fig. 9.3. The curves pass through the critical points $\beta = 0$, $\alpha = \frac{1}{4}(2m-1)^2$, $(m = 1, 2, 3, \ldots)$.

Exercise 9.3
For the 4π-periodic solutions of Mathieu's equation, let

$$
G_1(\alpha, \beta) = \begin{vmatrix} 2(\alpha - \frac{1}{4}) & \beta \\ \beta & 2(\alpha - \frac{1}{4}) \end{vmatrix},
$$

$$
G_2(\alpha, \beta) = \begin{vmatrix}
2(\alpha - \frac{9}{4}) & \beta & 0 & 0 \\
\beta & 2(\alpha - \frac{1}{4}) & \beta & 0 \\
0 & \beta & 2(\alpha - \frac{1}{4}) & \beta \\
0 & 0 & \beta & 2(\alpha - \frac{9}{4})
\end{vmatrix}.
$$

Obtain the relations between β and α in the first two approximations to the zeros of $G_1(\alpha, \beta) = 0$ and $G_2(\alpha, \beta) = 0$.

9.4   Transition curves for Mathieu's equation by perturbation

For small values of $|\beta|$ a perturbation method can be used to establish the transition curves. In the equation

$$
\ddot{x} + (\alpha + \beta \cos t)x = 0, \tag{9.42}
$$

suppose that the transition curves are given by

$$\alpha = \alpha(\beta) = \alpha_0 + \beta\alpha_1 + \beta^2\alpha_2 \cdots, \tag{9.43}$$

and that the corresponding solutions have the form

$$x(t) = x_0(t) + \beta x_1(t) + \beta^2 x_2(t) + \cdots, \tag{9.44}$$

where $x_0, x_1, \ldots$ all have either minimal period 2π or 4π.

When (9.43) and (9.44) are substituted into (9.42) and the coefficients of powers of β are equated to zero in the usual perturbation way, we have

$$\ddot{x}_0 + \alpha_0 x_0 = 0, \tag{9.45a}$$

$$\ddot{x}_1 + \alpha_0 x_1 = -(\alpha_1 + \cos t)x_0, \tag{9.45b}$$

$$\ddot{x}_2 + \alpha_0 x_2 = -\alpha_2 x_0 - (\alpha_1 + \cos t)x_1, \tag{9.45c}$$

$$\ddot{x}_3 + \alpha_0 x_3 = -\alpha_3 x_0 - \alpha_2 x_1 - (\alpha_1 + \cos t)x_2, \tag{9.45d}$$

and so on.

From the analysis in the Section 9.3, we are searching for solutions with minimum period 2π if $\alpha_0 = n^2$, $n - 0, 1, 2, \ldots$, and for solutions of minimum period 4π if $\alpha_0 = (n + \frac{1}{2})^2$, $n = 0, 1, 2, \ldots$. Both cases can be covered by defining $\alpha_0 = \frac{1}{4}n^2$, $n = 0, 1, 2, \ldots$ We consider the cases $n = 0$ and $n = 1$.

(i) $n = 0$. In this case $\alpha_0 = 0$ so that $\ddot{x}_0 = 0$. The periodic solution of (9.45a) is $x_0 = a_0$, where we assume that a_0 is any nonzero constant. Equation (9.45b) becomes

$$\ddot{x}_1 = -(\alpha_1 + \cos t)a_0,$$

which has periodic solutions only if $\alpha_1 = 0$. We need only choose the *particular* solution

$$x_1 = a_0 \cos t.$$

(inclusion of complementary solutions does not add generality since further arbitrary constants can be amalgamated). Equation (9.45c) becomes

$$\ddot{x}_2 = -a_0\alpha_2 - \tfrac{1}{2}a_0 - a_0 \cos^2 t = -a_0\alpha_2 - \tfrac{1}{2}a_0 - \tfrac{1}{2}a_0 \cos 2t,$$

which generates a periodic solution of 2π (and π) only if $\alpha_2 + \frac{1}{2}a_0 = 0$, that is, if $\alpha_2 = -\frac{1}{2}a_0$. Therefore choose

$$x_2 = \tfrac{1}{8}a_0 \cos 2t.$$

From (9.45d),

$$\ddot{x}_3 = -\alpha_3 x_0 - \alpha_2 a_0 \cos t - \tfrac{1}{8}a_0 \cos t \cos 2t$$

$$= a_0[-\alpha_3 - (\alpha_2 + \tfrac{1}{16})\cos t - \tfrac{1}{16}\cos 3t].$$

Solutions will only be periodic if $\alpha_3 = 0$. Therefore, for small $|\beta|$,

$$\alpha = -\tfrac{1}{2}\beta^2 + O(\beta^4),$$

which is a parabolic approximation to the curve through the origin in the α, β shown in Fig. 9.3. The corresponding 2π periodic solution is

$$x = a_0[1 + \beta \cos t + \tfrac{1}{8}\beta^2 \cos 2t] + O(\beta^3).$$

(ii) $n = 1$. In this case $\alpha_0 = \tfrac{1}{4}$, and $x_0 = a_0 \cos \tfrac{1}{2}t + b_0 \sin \tfrac{1}{2}t$. Equation (9.45b) becomes

$$\ddot{x}_1 + \tfrac{1}{4}x_1 = -(\alpha_1 + \cos t)(a_0 \cos \tfrac{1}{2}t + b_0 \sin \tfrac{1}{2}t)$$
$$= -a_0(\alpha_1 + \tfrac{1}{2}) \cos \tfrac{1}{2}t - b_0(\alpha_1 - \tfrac{1}{2}) \sin \tfrac{1}{2}t - \tfrac{1}{2}a_0 \cos \tfrac{3}{2}t - \tfrac{1}{2}b_0 \sin \tfrac{3}{2}t \qquad (9.46)$$

There are periodic solutions of period 4π only if either $b_0 = 0$, $\alpha_1 = -\tfrac{1}{2}$, or $a_0 = 0$, $\alpha_1 = \tfrac{1}{2}$. Here are two cases to consider.

(a) $b_0 = 0$, $\alpha_1 = -\tfrac{1}{2}$. It follows that the particular solution of (9.46) is $x_1 = \tfrac{1}{4}a_0 \cos \tfrac{3}{2}t$. Equation (9.45c) for x_2 is

$$\ddot{x}_2 + \tfrac{1}{4}x_2 = -(\alpha_2 + \tfrac{1}{8})a_0 \cos \tfrac{1}{2}t + \tfrac{1}{8}a_0 \cos \tfrac{3}{2}t - \tfrac{1}{8}a_0 \cos \tfrac{5}{2}t.$$

Secular terms can be eliminated by putting $\alpha_2 = -\tfrac{1}{8}$. Hence one transition curve through $\alpha = \tfrac{1}{4}$, $\beta = 0$ is

$$\alpha = \tfrac{1}{4} - \tfrac{1}{2}\beta - \tfrac{1}{8}\beta^2 + O(\beta^3). \qquad (9.47)$$

(b) $a_0 = 0$, $\alpha_1 = \tfrac{1}{2}$. From (9.46), $x_1 = \tfrac{1}{4}b_0 \sin \tfrac{3}{2}t$. Equation (9.445c) becomes

$$\ddot{x}_2 + \tfrac{1}{4}x_2 = -(\alpha_2 + \tfrac{1}{8})b_0 \sin \tfrac{1}{2}t - \tfrac{1}{8}b_0 \sin \tfrac{3}{2}t - \tfrac{1}{8}b_0 \sin \tfrac{5}{2}t.$$

Secular terms can be eliminated by putting $\alpha_2 = -\tfrac{1}{8}$. Therefore the other transition curve is given by

$$\alpha = \tfrac{1}{4} + \tfrac{1}{2}\beta - \tfrac{1}{8}\beta^2 + O(\beta^3). \qquad (9.48)$$

The transition curves given by (9.47) and (9.48) approximate to the computed curves through $\alpha = \tfrac{1}{4}$, $\beta = 0$ shown in Fig. 9.3.

The same perturbation method can be applied to approximate to the transition curves through $\alpha = 1, \tfrac{9}{4}, 4, \tfrac{25}{4} \ldots$. A more extensive investigation of perturbation methods applied to Mathieu's equation is given by Nayfeh and Mook (1979, Chapter 5).

9.5 Mathieu's damped equation arising from a Duffing equation

As we saw in Section 9.1, the variational equation for the undamped forced, pendulum is Mathieu's equation (9.10) or (9.28). With dissipation included, the Duffing equation in standardized form is

$$\ddot{x} + k\dot{x} + x + \varepsilon x^3 = \Gamma \cos \omega t. \tag{9.49}$$

In Chapter 7 we also showed that this equation has periodic solutions which are approximately of the form $a \cos \omega t + b \sin \omega t$ where $r = \sqrt{(a^2 + b^2)}$ satisfies

$$\left\{ (\omega^2 - 1 - \tfrac{3}{4}\varepsilon r^2)^2 + \omega^2 k^2 \right\} r^2 = \Gamma^2, \tag{9.50}$$

which reduces to eqn (7.23) if ε replaces β in the earlier notation. Following the notation and procedure of Section 9.1, write (9.49) as the first-order system

$$\dot{\boldsymbol{x}} = \begin{bmatrix} \dot{x} \\ \dot{y} \end{bmatrix} = \begin{bmatrix} y \\ -ky - x - \varepsilon x^3 + \Gamma \cos \omega t \end{bmatrix} \tag{9.51}$$

and put (approximately)

$$\boldsymbol{x}^* = a \cos \omega t + b \sin \omega t, \quad \boldsymbol{y}^* = -a\omega \sin \omega t + b\omega \cos \omega t.$$

The variations $\xi = x - x^*$ and $\eta = y - y^*$ satisfy

$$\dot{\xi} + \dot{x}^* = \eta + y^*,$$
$$\dot{\eta} + \dot{y}^* = -k(\eta + y^*) - (\xi + x^*) - \varepsilon(\xi + x^*)^3 + \Gamma \cos \omega t.$$

By using (9.51) and retaining only the first powers of ξ and η we obtain corresponding linearized equations

$$\dot{\xi} = \eta, \quad \dot{\eta} = -k\eta - \xi - 3\varepsilon x^{*2}\xi.$$

Elimination of η leads to the second-order equation

$$\ddot{\xi} + k\dot{\xi} + (1 + 3\varepsilon x^{*2})\xi = 0.$$

By substituting for x^* its approximate form $a \cos \omega t + b \sin \omega t$ we obtain

$$\ddot{\xi} + k\dot{\xi} + \left\{ 1 + \tfrac{3}{2}\varepsilon r^2 + \tfrac{3}{2}\varepsilon r^2 \cos(2\omega t + 2c) \right\} \xi = 0, \tag{9.52}$$

where r, c are defined by

$$a \cos \omega t + b \sin \omega t = r \cos(\omega t + c).$$

We can reduce the eqn (9.52) to 'standard' form by putting

$$\tau = 2\omega t + 2\gamma, \quad \xi' \equiv \frac{d\xi}{d\tau},$$

$$\kappa = \frac{k}{2\omega} = \kappa_1 \varepsilon, \quad \nu = \frac{2 + 3\varepsilon r^2}{8\omega^2} = \nu_0 + \varepsilon \nu_1, \quad \beta = \frac{3\varepsilon r^2}{8\omega^2} = \beta_1 \varepsilon,$$

say, so that

$$\xi'' + \kappa \xi' + (\nu + \beta \cos \tau)\xi = 0, \tag{9.53}$$

This is known as **Mathieu's equation with damping**.

We assume that $0 < \varepsilon \ll 1$, $k = O(\varepsilon)$ and $\omega \approx 1 = 1 + O(\varepsilon)$ (near resonance). For near resonance $\nu = \frac{1}{4} + O(\varepsilon)$. Let $\eta = \xi'$ in (9.53). Then the corresponding first-order system is

$$\zeta' = \begin{bmatrix} \xi' \\ \eta' \end{bmatrix} = \begin{bmatrix} 0 & 1 \\ -\nu - \beta \cos \tau & -\kappa \end{bmatrix} \begin{bmatrix} \xi \\ \eta \end{bmatrix} = P(\tau)\zeta, \tag{9.54}$$

say. The characteristic numbers of $P(\tau)$ satisfy (see Theorem 9.5)

$$\mu_1 \mu_2 = \exp\left[\int_0^{2\pi} \text{tr}\{P(\tau)\}d\tau\right] = \exp\left[-\int_0^{2\pi} \kappa \, d\tau\right] = e^{-2\pi\kappa}.$$

The numbers μ_1 and μ_2 are solutions of a characteristic equation of the form

$$\mu^2 - \phi(\nu, \beta, \kappa)\mu + e^{-2\pi\kappa} = 0. \tag{9.55}$$

The two solutions are

$$\mu_1, \mu_2 = \tfrac{1}{2}[\phi \pm \sqrt{\{\phi^2 - 4e^{-2\pi\kappa}\}}]. \tag{9.56}$$

For distinct values of μ_1 and μ_2, (9.54) has 2 linearly independent solutions of the form (see Theorem 9.3)

$$\zeta_i = p_i(\tau)e^{\rho_i \tau} \quad (i = 1, 2),$$

where $e^{2\rho_i \pi} = \mu_i$, $(i = 1, 2)$ and p_i are functions of period 2π.

From (9.55), the general solution for ξ, the first component of ζ, is given by

$$\xi = c_1 q_1(\tau)e^{\rho_1 \tau} + c_2 q_2(\tau)e^{\rho_2 \tau}, \tag{9.57}$$

where c_1, c_2 are constants and $q_1(\tau)$, $q_2(\tau)$ have minimum period 2π: η can then be found from $\eta' = \xi$. The stability or otherwise of the periodic solution of (9.53) will be determined by the behaviour of ξ in (9.57). If the solution for ξ is damped then we can infer its stability. The characteristic exponents may be complex, so that the limit $\xi \to 0$ as $\tau \to \infty$ will occur if both

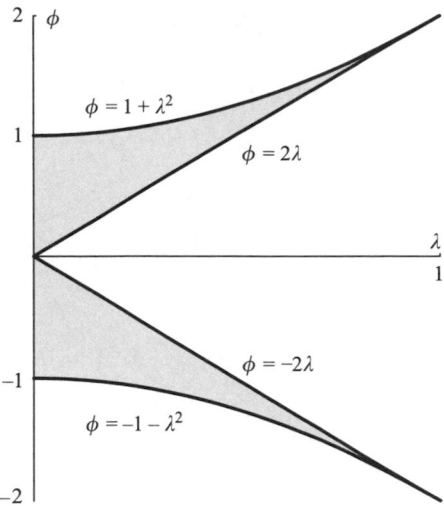

Figure 9.4   The boundaries of the shaded region are $\phi = 2\lambda$, $\phi = 1 + \lambda^2$, and $\phi - 2\lambda, \phi = -1 - \lambda^2$, where $\lambda = e^{-\pi\kappa}$.

$\mathrm{Re}(\rho_1) < 0$ and $\mathrm{Re}(\rho_2) < 0$. This is equivalent to $|\mu_1| < 1$ and $|\mu_2| < 1$. There are three cases to consider.

- $\phi^2 > 4e^{-2\pi\kappa}$. μ_1 and μ_2 are both real and positive, or both real and negative according to the sign of ϕ: in both cases $\mu_2 < \mu_1$. If they are both positive, then the periodic solution is stable if

$$\mu_1 = \tfrac{1}{2}[\phi + \sqrt{(\phi^2 - 4e^{-2\pi\kappa})}] < 1, \quad \text{or,} \quad \phi < 1 + e^{-2\pi\kappa}. \tag{9.58}$$

For $\kappa > 0$, this lower bound is always greater than $2e^{-\pi\kappa}$. The shaded region in $\phi > 0$ in Fig. 9.4 shows the stability domain. Similarly if $\phi < -2e^{-\pi\kappa}$, then the stability boundaries are $\phi = -2e^{-\pi\kappa}$ and $\phi = -1 - e^{-2\pi\kappa}$, which are also shown in Fig. 9.4.
- $\phi^2 = 4e^{-2\pi\kappa}$. In this case $\mu_1 = \mu_2 = \tfrac{1}{2}\phi = \pm e^{-\pi\kappa} = \mu$, say. If $\mu = e^{-\pi\kappa}$, then $\rho = -\tfrac{1}{2}\kappa$, and if $\mu = -e^{-\pi\kappa}$, then $\mathrm{Re}(\rho) = -\tfrac{1}{2}\kappa$. In both cases the solution is stable, also shown shaded in Fig. 9.4.
- $\phi^2 < 4e^{-2\pi\kappa}$. μ_1 and μ_2 are complex conjugates given by $\tfrac{1}{2}(\phi \pm i\theta)$, where $\theta = \sqrt{[4e^{-2\pi\kappa} - \phi^2]}$. The system is therefore stable if $|\phi| < 2$.

As in Section 9.5, we can search for periodic solutions of periods 2π and 4π by using Fourier series. Let

$$\xi(\tau) = \sum_{n=-\infty}^{\infty} c_n e^{in\tau}.$$

Substitute this series into (9.50) so that

$$\sum_{n=-\infty}^{\infty} [\beta c_{n+1} + 2\{\nu - n^2 + i\kappa n\}c_n + \beta c_{n-1}]e^{in\tau} = 0.$$

where we have used $\cos\tau = \frac{1}{2}(e^{i\tau} + e^{-i\tau})$. These equations will only be satisfied for all τ if

$$\beta c_{n+1} + 2\{\nu - n^2 + i\kappa n\}c_n + \beta c_{n-1} = 0, \quad n = 0, \pm 1, \pm 2, \dots. \tag{9.59}$$

Let

$$\gamma_n = \frac{\beta}{2(\nu - n^2 + i\kappa n)}, \tag{9.60}$$

and express eqns (9.59) in the form

$$\gamma_{n+1}c_{n+1} + c_n + \gamma_{n-1}c_{n-1} = 0. \tag{9.61}$$

There are non-zero solutions for the sequence $\{c_n\}$ if, and only if, the infinite determinant is zero, that is,

$$\begin{vmatrix} \cdots & \cdots & \cdots & \cdots & \cdots & \cdots & \cdots \\ \cdots & \gamma_1 & 1 & \gamma_1 & 0 & 0 & \cdots \\ \cdots & 0 & \gamma_0 & 1 & \gamma_0 & 0 & \cdots \\ \cdots & 0 & 0 & \gamma_{-1} & 1 & \gamma_{-1} & \cdots \\ \cdots & \cdots & \cdots & \cdots & \cdots & \cdots & \cdots \end{vmatrix} = 0,$$

or

$$\begin{vmatrix} \cdots & \cdots & \cdots & \cdots & \cdots & \cdots & \cdots \\ \cdots & \gamma_1 & 1 & \gamma_1 & 0 & 0 & \cdots \\ \cdots & 0 & \gamma_0 & 1 & \gamma_0 & 0 & \cdots \\ \cdots & 0 & 0 & \overline{\gamma}_1 & 1 & \overline{\gamma}_1 & \cdots \\ \cdots & \cdots & \cdots & \cdots & \cdots & \cdots & \cdots \end{vmatrix} = 0,$$

since $\gamma_{-n} = \overline{\gamma}_n$ $(n = 1, 2, \dots)$, the conjugate of γ_n. We can approximate to the determinant by choosing a finite number of rows. Let

$$E_1 = \begin{vmatrix} 1 & \gamma_1 & 0 \\ \gamma_0 & 1 & \gamma_0 \\ 0 & \overline{\gamma}_1 & 1 \end{vmatrix} = 1 - \frac{\beta^2(\nu - 1)}{2\nu[(\nu - 1)^2 + \kappa^2]}.$$

With $\nu = \frac{1}{4} + O(\varepsilon)$,

$$E_1 = 1 + \tfrac{8}{3}\beta_2\varepsilon^2 + o(\varepsilon^2).$$

Hence E_1 cannot be zero for ε small. The implication is that there are no 2π periodic solutions in the variable τ.

To search for 4π periodic solutions, let

$$\xi = \sum_{n=-\infty}^{\infty} d_n e^{\frac{1}{2}in\tau}.$$

Substitution of this series into (9.50) leads to

$$\sum_{n=-\infty}^{\infty} [\beta d_{n+2} + 2\{v - \tfrac{1}{4}n^2 + \tfrac{1}{2}i\kappa n\} d_n + \beta d_{n-2}] e^{\frac{1}{2}in\tau} = 0.$$

These equations will only be satisfied for all τ if

$$\beta d_{n+2} + 2\{v - \tfrac{1}{4}n^2 + \tfrac{1}{2}i\kappa n\} d_n + \beta d_{n-2} = 0. \tag{9.62}$$

As in Section 9.3 there are two independent sets of equations for n even and for n odd. The even case duplicates the previous case for 2π periodic solutions so that we need not consider it. For the case of n odd, let

$$\delta_m = \frac{\beta}{2[v - \tfrac{1}{4}(2m-1)^2 + \tfrac{1}{2}(2m-1)i\kappa]} \qquad (m \geq 1)$$

Elimination of $\{d_n\}$ in (9.62) results in the infinite determinant equation

$$\begin{vmatrix} \cdots & \cdots & \cdots & \cdots & \cdots & \cdots & \cdots & \cdots \\ \cdots & \delta_2 & 1 & \delta_2 & 0 & 0 & 0 & \cdots \\ \cdots & 0 & \delta_1 & 1 & \delta_1 & 0 & 0 & \cdots \\ \cdots & 0 & 0 & \bar{\delta}_1 & 1 & \bar{\delta}_1 & 0 & \cdots \\ \cdots & 0 & 0 & 0 & \bar{\delta}_2 & 1 & \bar{\delta}_2 & \cdots \\ \cdots & \cdots & \cdots & \cdots & \cdots & \cdots & \cdots & \cdots \end{vmatrix} = 0,$$

where $\bar{\delta}_m$ is the conjugate of δ_m. The value of this determinant can be approximated by

$$G_1 = \begin{vmatrix} 1 & \delta_1 \\ \bar{\delta}_1 & 1 \end{vmatrix} = 1 - \delta_1 \bar{\delta}_1 = 1 - \frac{\beta_1^2}{4v_1^2 + \kappa_1^2} + O(\varepsilon).$$

To lowest order, $G_1 = 0$ if

$$1 - \frac{\beta_1^2}{4v_1^2 + \kappa_1^2} = 0, \quad \text{or} \quad v_1 = \pm\tfrac{1}{2}\sqrt{(\beta_1^2 - \kappa_1^2)}.$$

Therefore the stability boundaries are given by

$$v = \tfrac{1}{4} \pm \tfrac{1}{2}\varepsilon\sqrt{(\beta_1^2 - \kappa_1^2)} + O(\varepsilon)$$

Note that if $\kappa_1 = 0$ (no damping), then the stability boundaries given by (9.45) can be recovered. Note also that $\kappa_1 < \beta_1$ is required. Using the stability boundaries, the periodic solution of the Duffing equation is stable in the domain defined by

$$v < \tfrac{1}{4} - \tfrac{1}{2}\varepsilon\sqrt{(\beta_1^2 - \kappa_1^2)}, \quad \text{or} \quad v > \tfrac{1}{4} + \tfrac{1}{2}\varepsilon\sqrt{(\beta_1^2 - \kappa_1^2)},$$

or

$$(v - \tfrac{1}{4})^2 > \tfrac{1}{4}\varepsilon^2(\beta_1^2 - \kappa_1^2).$$

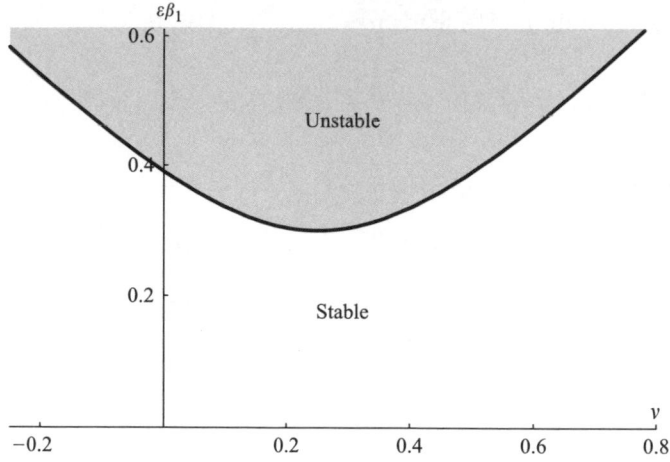

Figure 9.5   The stability for the damped Duffing equation with $\varepsilon\kappa_1 = 0.3$ and $\omega \approx 1$.

A typical stability domain is shown in Fig. 9.5 for $\varepsilon\kappa_1 = 0.3$ in the neighbourhood of resonance near $\omega = 1$. In terms of the original variables, the condition becomes

$$\left(\frac{2 + 3\varepsilon r^2}{8\omega^2} - \frac{1}{4}\right)^2 - \frac{1}{4}\left[\left(\frac{3\varepsilon r^2}{8\omega^2}\right)^2 - \frac{k^2}{4\omega^2}\right] > 0,$$

or

$$(1 - \omega^2)^2 + 3\varepsilon(1 - \omega^2)r^2 + \tfrac{27}{16}\varepsilon^2 r^4 + k^2\omega^2 > 0. \tag{9.63}$$

Since the solutions of the damped Mathieu equation tend to zero in the stable region, asymptotic stability is predicted, confirming the analysis of Chapter 7.

We can also confirm the remark made in Section 5.5 (vi): that stability is to be expected when

$$\frac{d(\Gamma^2)}{d(r^2)} > 0, \tag{9.64}$$

that is, when an increase or decrease in magnitude of Γ results in an increase or decrease respectively in the amplitude. From (9.51) it is easy to verify that $d(\Gamma^2)/d(r^2)$ is equal to the expression on the left of (9.63), and the speculation is therefore confirmed.

In general, when periodic solutions of the original equation are expected the reduced equation (9.4) is the more complicated Hill type (see Problem 9.11). The stability regions for this equation and examples of the corresponding stability estimates may be found in Hayashi (1964).

Problems

9.1 The system

$$\dot{x}_1 = (-\sin 2t)x_1 + (\cos 2t - 1)x_2, \quad \dot{x}_2 = (\cos 2t + 1)x_1 + (\sin 2t)x_2$$

has a fundamental matrix of normal solutions:

$$
\begin{bmatrix}
e^t(\cos t - \sin t) & e^{-t}(\cos t + \sin t) \\
e^t(\cos t + \sin t) & e^{-t}(-\cos t + \sin t)
\end{bmatrix}.
$$

Obtain the corresponding E matrix (Theorem 9.1), the characteristic numbers, and the characteristic exponents.

9.2 Let the system $\dot{x} = P(t)x$ have a matrix of coefficients P with minimal period T (and therefore also with periods $2T, 3T, \ldots$). Follow the argument of Theorem 9.1, using period $mT, m > 1$, to show that $\Phi(t + mT) = \Phi(t)E^m$. Assuming that if the eigenvalues of E are μ_i, then those of E^m are μ_i^m, discuss possible periodic solutions.

9.3 Obtain Wronskians for the following linear systems:
(i) $\dot{x}_1 = x_1 \sin t + x_2 \cos t$, $\dot{x}_2 = -x_1 \cos t + x_2 \sin t$,
(ii) $\dot{x}_1 = f(t)x_2$, $\dot{x}_2 = g(t)x_1$.

9.4 By substituting $x = c + a \cos t + b \sin t$ into Mathieu's equation

$$
\ddot{x} + (\alpha + \beta \cos t)x = 0,
$$

obtain by harmonic balance an approximation to the transition curve near $\alpha = 0$, $\beta = 0$, (compare with Section 9.4).
    By substituting $x = c + a \cos \frac{1}{2}t + b \sin \frac{1}{2}t$, find the transition curves near $\alpha = \frac{1}{4}, \beta = 0$.

9.5 Figure 9.6 represents a particle of mass m attached to two identical linear elastic strings of stiffness λ and natural length l. The ends of the strings pass through frictionless guides A and B at a distance $2L$, $l < L$, apart. The particle is set into lateral motion at the mid-point, and symmetrical displacements $a + b \cos \omega t$, $a > b$, are imposed on the ends of the string. Show that, for $x \ll L$,

$$
\ddot{x} + \left(\frac{2\lambda(L - l + a)}{mL} + \frac{2\lambda b}{mL} \cos \omega t \right) x = 0.
$$

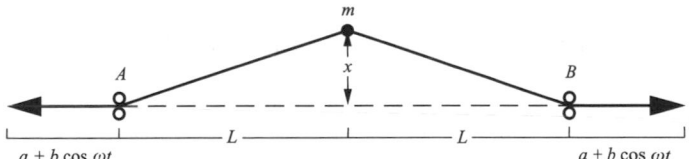

Figure 9.6

Analyse the motion in terms of suitable parameters, using the information of Sections 9.3 and 9.4 on the growth or decay, periodicity and near periodicity of the solutions of Mathieu's equation in the regions of its parameter plane.

9.6 A pendulum with a light, rigid suspension is placed upside-down on end, and the point of suspension is caused to oscillate vertically with displacement y upwards given by $y = \varepsilon \cos \omega t$, $\varepsilon \ll 1$. Show that the equation of motion is

$$
\ddot{\theta} + \left(-\frac{g}{a} - \frac{1}{a}\ddot{y} \right) \sin \theta = 0,
$$

where a is the length of the pendulum, g is gravitational acceleration, and θ the inclination to the vertical. Linearize the equation for small amplitudes and show that the vertical position is stable (that is, the motion of the pendulum restricts itself to the neighbourhood of the vertical: it does not topple over) provided $\varepsilon^2\omega^2/(2ag) > 1$. For further discussion of the inverted pendulum and its stability see Acheson (1997).

9.7 Let $\boldsymbol{\Phi}(t) = (\phi_{ij}(t))$, $i, j = 1, 2$, be the fundamental matrix for the system $\dot{x}_1 = x_2$, $\dot{x}_2 = -(\alpha + \beta\cos t)x_1$, satisfying $\boldsymbol{\Phi}(0) = \boldsymbol{I}$ (Mathieu's equation). Show that the characteristic numbers μ satisfy the equation

$$\mu^2 - \mu\{\phi_{11}(2\pi) + \phi_{22}(2\pi)\} + 1 = 0.$$

9.8 In Section 9.3, for the transition curves of Mathieu's equation for solutions period 2π, let $D_{m,n}$ be the tridiagonal determinant given by

$$D_{m,n} = \begin{vmatrix} 1 & \gamma_m & 0 \\ \gamma_{m-1} & 1 & \gamma_{m-1} \\ & & \ddots \\ & & & \gamma_0 & 1 & \gamma_0 \\ & & & & & \ddots \\ & & & & & & \gamma_{n-1} & 1 & \gamma_{n-1} \\ & & & & & & 0 & \gamma_n & 1 \end{vmatrix}$$

for $m \geq 0, n \geq 0$. Show that

$$D_{m,n} = D_{m-1,n} - \gamma_m\gamma_{m-1}D_{m-2,n}.$$

Let $E_n = D_{n,n}$ and verify that

$$E_0 = 1, \quad E_1 = 1 - 2\gamma_0\gamma_1, \quad E_2 = (1 - \gamma_1\gamma_2)^2 - 2\gamma_0\gamma_1(1 - \gamma_1\gamma_2).$$

Prove that, for $n \geq 1$,

$$E_{n+2} = (1 - \gamma_{n+1}\gamma_{n+2})E_{n+1} - \gamma_{n+1}\gamma_{n+2}(1 - \gamma_{n+1}\gamma_{n+2})E_n$$
$$+ \gamma_n^2\gamma_{n+1}^3\gamma_{n+2}E_{n-1}.$$

9.9 In eqn (9.38), for the transition curves of Mathieu's equation for solutions of period 4π, let

$$F_{m,n} = \begin{vmatrix} 1 & \delta_m & \\ \delta_{m-1} & 1 & \delta_{m-1} \\ & & \ddots \\ & & & \delta_{n-1} & 1 & \delta_{n-1} \\ & & & & \delta_n & 1 \end{vmatrix}.$$

Show as in the previous problem that $G_n = F_{n,n}$ satisfies the same recurrence relation as E_n for $n \geq 2$ (see Problem 9.8). Verify that

$$G_1 = 1 - \delta_1^2,$$
$$G_2 = (1 - \delta_1\delta_2)^2 - \delta_1^2,$$
$$G_3 = (1 - \delta_1\delta_2 - \delta_2\delta_3)^2 - \delta_1^2(1 - \delta_2\delta_3)^2.$$

9.10 Show, by the perturbation method, that the transition curves for Mathieu's equation

$$\ddot{x} + (\alpha + \beta\cos t)x = 0,$$

near $\alpha = 1$, $\beta = 0$, are given approximately by $\alpha = 1 + \frac{1}{12}\beta^2$, $\alpha = 1 - \frac{5}{12}\beta^2$.

9.11 Consider Hill's equation $\ddot{x} + f(t)x = 0$, where f has period 2π, and

$$f(t) = \alpha + \sum_{r=1}^{\infty} \beta_r \cos rt$$

is its Fourier expansion, with $\alpha \approx \frac{1}{4}$ and $|\beta_r| \ll 1, r = 1, 2, \ldots$. Assume an approximate solution $e^{\sigma t}q(t)$, where σ is real and q has period 4π as in (9.34). Show that

$$\ddot{q} + 2\sigma\dot{q} + \left(\sigma^2 + \alpha + \sum_{r=1}^{\infty} \beta_r \cos rt \right) q = 0.$$

Take $q \approx \sin(\frac{1}{2}t + \gamma)$ as the approximate form for q and match terms in $\sin \frac{1}{2}t$, $\cos \frac{1}{2}t$, on the assumption that these terms dominate. Deduce that

$$\sigma^2 = -(\alpha + \tfrac{1}{4}) + \tfrac{1}{2}\sqrt{(4\alpha + \beta_1^2)}$$

and that the transition curves near $\alpha = \frac{1}{4}$ are given by $\alpha = \frac{1}{4} \pm \frac{1}{2}\beta_1$. ($\beta_n$ is similarly the dominant coefficient for transition curves near $\alpha = \frac{1}{4}n^2$, $n \geq 1$.)

9.12 Obtain, as in Section 9.4, the boundary of the stable region in the neighbourhood of $\nu = 1$, $\beta = 0$ for Mathieu's equation with damping,

$$\ddot{x} + \kappa\dot{x} + (\nu + \beta \cos t)x = 0,$$

where $\kappa = O(\beta^2)$.

9.13 Solve Meissner's equation

$$\ddot{x} + (\alpha + \beta f(t))x = 0$$

where $f(t) = 1$, $0 \leq t < \pi$; $f(t) = -1$, $\pi \leq t < 2\pi$ and $f(t + 2\pi) = f(t)$ for all t. Find the conditions on α, β, for periodic solutions by putting $x(0) = x(2\pi)$, $\dot{x}(0) = \dot{x}(2\pi)$ and by making x and $\dot{x}$ continuous at $t = \pi$. Find a determinant equation for α and β.

9.14 By using the harmonic balance method of Chapter 4, show that the van der Pol equation with parametric excitation,

$$\ddot{x} + \varepsilon(x^2 - 1)\dot{x} + (1 + \beta \cos t)x = 0$$

has a 2π-periodic solution with approximately the same amplitude as the unforced van der Pol equation.

9.15 The male population M and female population F for a bird community have a constant death rate k and a variable birth rate $\mu(t)$ which has period T, so that

$$\dot{M} = -kM + \mu(t)F, \quad \dot{F} = -kF + \mu(t)F.$$

The births are seasonal, with rate

$$\mu(t) = \begin{cases} \delta, & 0 < t \leq \varepsilon; \\ 0, & \varepsilon < t \leq T. \end{cases}$$

Show that periodic solutions of period T exist for M and F if $kT = \delta\varepsilon$.

9.16 A pendulum bob is suspended by a light rod of length a, and the support is constrained to move vertically with displacement $\zeta(t)$. Show (by using the Lagrange's equation method or otherwise) that the equation of motion is

$$a\ddot{\theta} + (g + \ddot{\zeta}(t)) \sin \theta = 0,$$

where θ is the angle of inclination to the downward vertical. Examine the stablity of the motion for the case when $\zeta(t) = c \sin \omega t$, on the assumption that it is permissible to put $\sin \theta \approx \theta$.

9.17 A pendulum, with bob of mass m and rigid suspension of length a, hangs from a support which is constrained to move with vertical and horizontal displacements $\zeta(t)$ and $\eta(t)$ respectively. Show that the inclination θ of the pendulum satisfies the equation

$$a\ddot{\theta} + (g + \ddot{\zeta}) \sin \theta + \ddot{\eta} \cos \theta = 0.$$

Let $\zeta = A \sin \omega t$ and $\eta = B \sin 2\omega t$, where $\omega = \sqrt{(g/a)}$. Show that after linearizing this equation for small amplitudes, the resulting equation has a solution

$$\theta = -(8B/A) \cos \omega t.$$

Determine the stability of this solution.

9.18 The equation

$$\ddot{x} + (\tfrac{1}{4} - 2\varepsilon b \cos^2 \tfrac{1}{2}t)x + \varepsilon x^3 = 0$$

has the exact solution $x^*(t) = \sqrt{(2b)} \cos \tfrac{1}{2}t$. Show that the solution is stable by constructing the variational equation.

9.19 Consider the equation $\ddot{x} + (\alpha + \beta \cos t)x = 0$, where $|\beta| \ll 1$ and $\alpha = \tfrac{1}{4} + \beta c$. In the unstable region near $\alpha = \tfrac{1}{4}$ (Section 9.4) this equation has solutions of the form $c_1 e^{\sigma t} q_1(t) + c_2 e^{-\sigma t} q_2(t)$, where σ is real, $\sigma > 0$ and q_1, q_2 have period 4π. Construct the equation for q_1, q_2, and show that $\sigma \approx \pm \beta \sqrt{(\tfrac{1}{4} - c^2)}$.

9.20 By using the method of Section 9.5 show that a solution of the equation

$$\ddot{x} + \varepsilon(x^2 - 1)\dot{x} + x = \Gamma \cos \omega t$$

where $|\varepsilon| \ll 1$, $\omega = 1 + \varepsilon \omega_1$, of the form $x^* = r_0 \cos(\omega t + \alpha)$ (α constant) is asymptotically stable when

$$4\omega_1^2 + \tfrac{3}{16}r_0^4 - r_0^2 + 1 < 0.$$

(Use the result of Problem 9.19.)

9.21 The equation

$$\ddot{x} + \alpha x + \varepsilon x^3 = \varepsilon \gamma \cos \omega t$$

has the exact subharmonic solution

$$x = (4\gamma)^{1/3} \cos \tfrac{1}{3}\omega t,$$

when

$$\omega^2 = 9 \left(\alpha + \frac{3}{4^{1/3}} \varepsilon \gamma^{2/3}\right).$$

If $0 < \varepsilon \ll 1$, show that the solution is stable.

9.22 Analyse the stability of the equation

$$\ddot{x} + \varepsilon x \dot{x}^2 + x = \Gamma \cos \omega t$$

for small ε: assume $\Gamma = \varepsilon \gamma$. (First find approximate solutions of the form $a \cos \omega t + b \cos \omega t$ by the harmonic balance method of Chapter 4, then perturb the solution by the method of Section 9.4.)

9.23 The equation $\ddot{x} + x + \varepsilon x^3 = \Gamma \cos \omega t (\varepsilon \ll 1)$ has an approximate solution $x^*(t) = a \cos \omega t$ where (eqn (7.10)) $\tfrac{3}{4}\varepsilon a^3 - (\omega^2 - 1)a - \Gamma = 0$: Show that the first variational equation (Section 9.4) is $\ddot{\xi} + \{1 + 3\varepsilon x^{*2}(t)\}\xi = 0$. Reduce this to Mathieu's equation and find conditions for stability of $x^*(t)$ if $\Gamma = \varepsilon \gamma$.

9.24 The equation $\ddot{x} + x - \frac{1}{6}x^3 = 0$ has an approximate solution $a \cos \omega t$ where $\omega^2 = 1 - \frac{1}{8}a^2$, $a \ll 1$ (Example 4.10). Use the method of Section 9.4 to show that the solution is unstable.

9.25 Show that a fundamental matrix of the differential equation

$$\dot{x} = A(t)x,$$

where

$$A(t) = \begin{bmatrix} \beta \cos^2 t - \sin^2 t & 1 - (1+\beta) \sin t \cos t \\ -1 - (1+\beta) \sin t \cos t & -1 + (1+\beta) \sin^2 t \end{bmatrix}$$

is

$$\Phi(t) = \begin{bmatrix} e^{\beta t} \cos t & e^{-t} \sin t \\ -e^{\beta t} \sin t & e^{-t} \cos t \end{bmatrix}.$$

Find the characteristic multipliers of the system. For what value of β will periodic solutions exist?

Find the eigenvalues of $A(t)$ and show that they are independent of t. Show that for $0 < \beta < 1$ the eigenvalues have negative real parts. What does this problem indicate about the relationship between the eigenvalues of a linear system with a variable coefficients and the stability of the zero solution?

9.26 Find a fundamental matrix for the system

$$\dot{x} = A(t)x,$$

where

$$A(t) = \begin{bmatrix} \sin t & 1 \\ -\cos t + \cos^2 t & -\sin t \end{bmatrix}.$$

Show that the characteristic multipliers of the system are $\mu_1 = e^{2\pi}$ and $\mu_2 = e^{-2\pi}$. By integration confirm that

$$\exp\left(\int_0^{2\pi} \mathrm{tr}\{A(s)\}\mathrm{d}s \right) = \mu_1 \mu_2 = 1.$$

10 Liapunov methods for determining stability of the zero solution

In Chapter 1, we described the 'energy method' for determining that an equilibrium point at the origin is an attractor, in connection with simple systems that allow a discussion in terms of the change in energy levels along the phase paths as time progresses. The Liapunov theory described in this chapter may be regarded as a development of this idea. It has no necessary connection with energy, concerns solution stability rather than Poincaré stability, and has a broader field of application. We discuss only two-dimensional *autonomous* systems in detail, but the theory can be extended to higher-dimensional and nonautonomous cases (Cesari 1971).

Sections 10.1 to 10.4 give a geometrical treatment for two-dimensional systems based on the Poincaré–Bendixson theorem applied to topographic systems of a simple type. In particular it is shown how a further requirement will frequently allow asymptotic stability to be proved by using a weak Liapunov function. In Section 10.5 the theorems are proved in the usual, and more general, form. With small changes in expression they also apply to *n*-dimensional cases.

10.1 Introducing the Liapunov method

To point the direction in which we are going, consider the following example.

Example 10.1 *Investigate the phase paths of the system*

$$\dot{x} = -y - x^3, \qquad \dot{y} = x - y^3.$$

The system has only one equilibrium point, at the origin. Consider the family of circles

$$V(x, y) = x^2 + y^2 = \alpha, \quad 0 < \alpha < \infty,$$

which are centred on the origin and fill the plane. We shall show that every path crosses these circles in an inward direction; that is to say, the value α of $V(x, y)$ strictly decreases along every phase path as time t increases to infinity. This is illustrated for one path in Fig. 10.1.

To prove this, consider any particular phase path $\mathcal{P}$, corresponding to a solution $x(t), y(t)$. On this path

$$V(x, y) = V\{x(t), y(t)\},$$

so

$$\left(\frac{\mathrm{d}V}{\mathrm{d}t}\right)_{\mathcal{P}} = \left(\frac{\partial V}{\partial x}\dot{x} + \frac{\partial V}{\partial y}\dot{y}\right)_{\mathcal{P}}$$

$$= 2x(-y - x^3) + 2y(x - y^3)$$

$$= -2(x^4 + y^4).$$

Except at the origin, $(\mathrm{d}V/\mathrm{d}t) < 0$ on every phase path $\mathcal{P}$, so V is strictly decreasing along $\mathcal{P}$. In other words, the representative point on $\mathcal{P}$ moves across successive circles of decreasing radius, as in Fig. 10.1. It is plausible to deduce that unless its progress is blocked by a limit cycle, $\mathcal{P}$ enters the equilibrium point at the origin as $t \to \infty$. ●

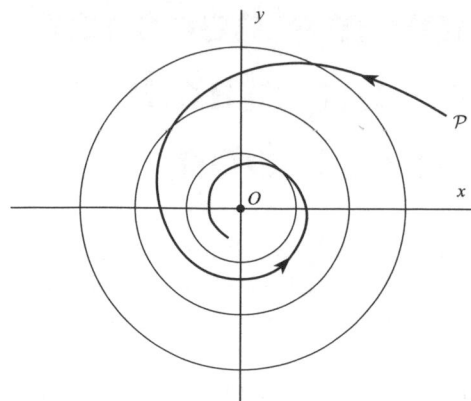

Figure 10.1   Path $\mathcal{P}$ crossing the family of circles in the inward direction.

The family of curves $V(x, y) = x^2 + y^2 = \alpha$ in Example 10.1 is called a **topographic system** for the problem, and it plays a similar role to the **energy levels** which appear in the energy method of Chapter 1.

Some progress has been made beyond the linear approximation at the origin for this system, which predicts a centre rather than a spiral. However, the proof that all paths enter the origin as $t \to \infty$ is incomplete, and we still need to prove Liapunov stability rather than Poincaré stability.

Exercise 10.1
Show that $V(x, y) = x^2 + y^2$ is a topographic system for

$$\begin{bmatrix} \dot{x} \\ \dot{y} \end{bmatrix} = \begin{bmatrix} -2 & -1 \\ 1 & -1 \end{bmatrix} \begin{bmatrix} x \\ y \end{bmatrix}.$$

What type of equilibrium point is the origin?

10.2   Topographic systems and the Poincaré–Bendixson theorem

We need to define a class of general topographic systems which have a topological structure similar to the family of circles used in Example 10.1. They take the form

$$V(x, y) = \alpha, \quad \alpha > 0,$$

and consist of a family of closed curves, surrounding the origin, which converge on to the origin (continuously) as $\alpha \to 0$, as suggested in Fig. 10.2. Formally, we shall adopt the conditions in Definition 10.1, although these may be relaxed in some respects without affecting our results (see Section 10.5). In the theorem statement, a **connected neighbourhood** of the origin can be

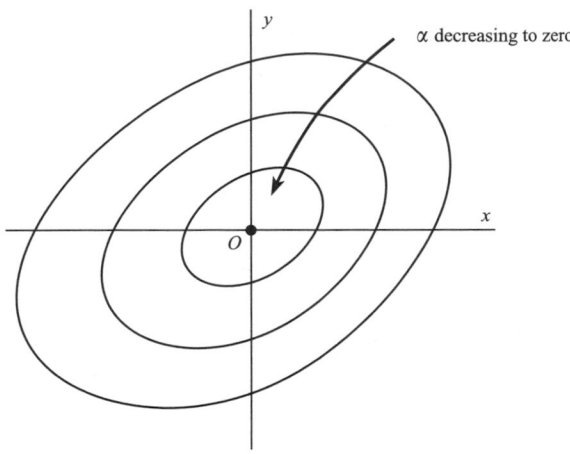

Figure 10.2

regarded as any simple 'patch' on the x, y plane covering the origin that is also an open set, in the sense that its boundary is excluded. Thus the set of points satisfying $x^2 + y^2 < 1$ is an example of a connected neighbourhood of the origin.

Definition 10.1 (Topographic System)   *In some connected neighbourhood $\mathcal{N}$ of the origin, let $V(x, y)$ satisfy:*

(i) *$V(x, y)$ is continuous; $\partial V/\partial x, \partial V/\partial y$ are continuous except possibly at the origin.*

(ii) *$V(0, 0) = 0$ and $V(x, y) > 0$ elsewhere in $\mathcal{N}$.*

(iii) *A value of $\mu > 0$ exists such that, for every value of the parameter α in the interval $0 < \alpha < \mu$, the equation $V(x, y) = \alpha$ for (x, y) in $\mathcal{N}$ defines, uniquely, a simple closed curve T_α in $\mathcal{N}$ which surrounds the origin. Then the family of curves $V(x, y) = \alpha$, $0 < \alpha < \mu$ is called a topographic system on $\mathcal{N}_\mu$, where $\mathcal{N}_\mu$ is a connected neighbourhood of the origin defined by $V(x, y) < \mu$, where $\mathcal{N}_\mu \subseteq \mathcal{N}$.*

Theorem 10.1   *The topographic system of Definition 10.1 has the following properties:*

(i) *$V(x, y) < \alpha$ in the interior of the topographic curve T_α, $0 < \alpha < \mu$.*

(ii) *There is a topographic curve through every point interior to T_α, $0 < \alpha < \mu$ (i.e., $\mathcal{N}_\mu$ is a connected neighbourhood of the origin).*

(iii) *If $0 < \alpha_1 < \alpha_2 < \mu$, then T_{α_1} is interior to T_{α_2}, and conversely.*

(iv) *As $\alpha \to 0$ monotonically, the topographic curves T_α close on to the origin.*

Proof   The proof of Theorem 10.1 is given in Appendix B.   ■

The neighbourhood $\mathcal{N}_\mu$ can be regarded as the largest open set in which the conditions (i) to (iii) hold good, but it is often convenient to use a smaller value of μ, so that $\mathcal{N}_\mu$ is smaller, and a smaller range of topographic curves is utilized.

Despite the technicalities it is usually easy to see when the conditions are satisfied. For example they are satisfied by the system

$$V(x, y) = x^2 + y^2 = \alpha, \quad \mathcal{N}_\mu \text{ is the whole plane;}$$

or we may adopt a smaller $\mathcal{N}_\mu$, say $x^2 + y^2 < 1$.

For the function given by

$$V(x, y) = r^2 e^{-r^2} = \alpha,$$

the greatest $\mathcal{N}_\mu$ allowed is the interior of the circle $r = 1$, corresponding to $\mu = e^{-1}$ (otherwise the uniqueness requirement in Definition 10.1 (iii) is not satisfied: to see this, sketch the curve $V = r^2 e^{-r^2}$, which turns over when $r = 1$).

Consider the surface defined by the equation

$$z = V(x, y).$$

The contours or level curves on the surface, projected on to the x, y plane are given by

$$V(x, y) = \alpha,$$

where the parameter α takes a range of values. If the contours form a topographic system on a range $0 < \alpha < \mu$ then the surface must be bowl shaped or trumpet shaped near enough to the origin, as shown in Fig. 10.3.

Theorem 10.2, proved in Appendix **B**, defines a fairly broad class of eligible functions $V(x, y)$ which suffices for ordinary purposes, and which are usually easy to identify. They are restricted to curves consisting of oval shapes that are concave with repect to the origin.

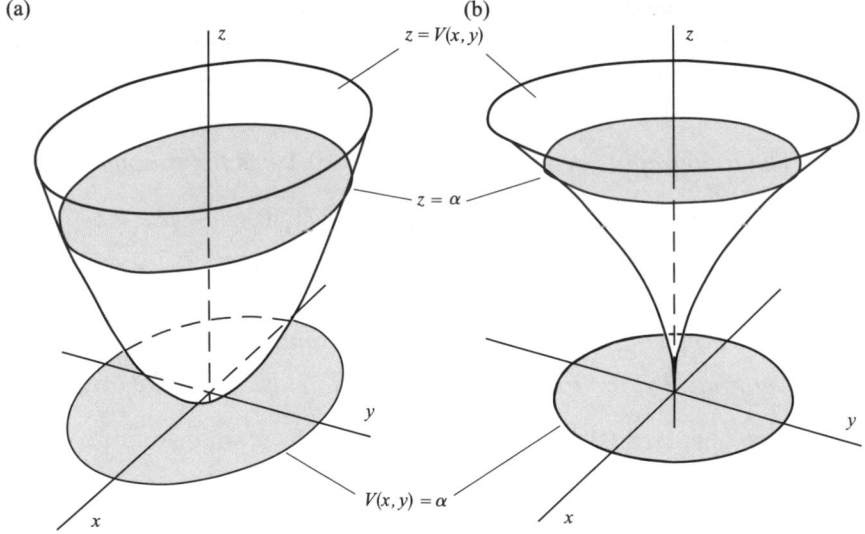

Figure 10.3   Showing typical topographic curves as projections of contour of the function $z = V(x, y)$.

Theorem 10.2   *In some neighbourhood $\mathcal{N}$ of the origin let $V(x, y)$ be continuous, and $\partial V/\partial x$, $\partial V/\partial y$ be continuous except possibly at the origin. Suppose that in $\mathcal{N}$, $V(x, y)$ takes the polar coordinate form*

$$V(x, y) = r^q f(\theta) + E(r, \theta),$$

where

(i) $V(0, 0) = 0$;

(ii) $q > 0$;

(iii) $f(\theta)$ *and* $f'(\theta)$ *are continuous for all values of* θ;

(iv) $f(\theta) > 0$, *and has period* 2π;

(v) $\lim\limits_{r \to 0} r^{-q+1}(\partial E/\partial r) = 0$ *for all* θ.

Then there exists $\mu > 0$ such that

$$V(x, y) = \alpha, \quad 0 < \alpha < \mu$$

defines a topographic system covering a neighbourhood of the origin $\mathcal{N}_\mu$, where $\mathcal{N}_\mu$ lies in $\mathcal{N}$.

Proof   See Appendix B.   ∎

As a special case, if $V(x, y)$ can be represented by a Taylor series about the origin with $V(0, 0) = 0$, and the group of terms of lowest degree present has a minimum at the origin, then the conditions are satisfied. A useful sub-class of this type are functions of the form

$$V(x, y) = f(x) + g(y),$$

where $f(x)$ and $g(y)$ have Taylor series starting with terms Ax^{2n} and By^{2m} respectively with A and B positive (see Fig. 10.3(a)).

The theorem also allows trumpet-shaped surfaces (see Fig. 10.3(b)) arising from expansion in fractional powers, for example the radially symmetric system

$$V(x, y) = r^{1/2} - r = \alpha, \quad 0 < \alpha < \tfrac{1}{4}.$$

Here $\mu = \tfrac{1}{4}$, and $\mathcal{N}_\mu$ is the region $r < \tfrac{1}{4}$ (to secure uniqueness of the topographic curves required by Definition 10.1).

In tracking the progress of the paths in the vicinity of an equilibrium point the following major theorem is frequently required.

Theorem 10.3 (The Poincaré–Bendixson theorem)   *Let $\mathcal{R}$ be a closed, bounded region in the (x, y) plane on which the system $\dot{x} = X(x, y)$, $\dot{y} = Y(x, y)$ is regular (Appendix A). If a positive*

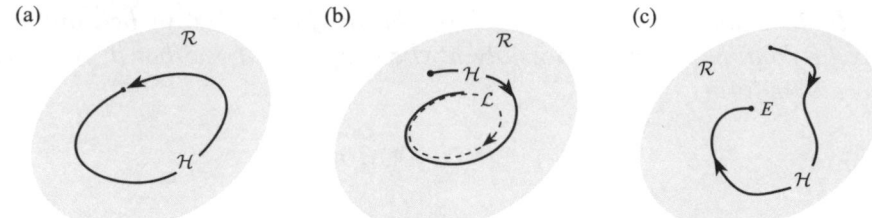

Figure 10.4   (a) $\mathcal{H}$ is a closed paths, (b) $\mathcal{H}$ approaches a limit cycle; (c) $\mathcal{H}$ approaches an equilibrium point E.

half-path $\mathcal{H}$ defined in Section 8.11 lies entirely on $\mathcal{R}$, then either

 (i) *$\mathcal{H}$ consists of a closed phase path on $\mathcal{R}$;*
 (ii) *$\mathcal{H}$ approaches a closed phase path on $\mathcal{R}$;*
(iii) *$\mathcal{H}$ approaches an equilibrium point on $\mathcal{R}$.*   ■

The theorem confirms what we should expect: that for regular systems there are no limiting cases other than closed phase paths and various types of equilibrium point, and that a path cannot wander about at random forever. However, the proof is difficult (see, e.g., Cesari (1971) or Andronov *et al.* (1973)). The three alternatives are illustrated in Fig. 10.4.

Exercise 10.2
Find the time solutions for the positive half-paths of
$$\dot{x} = x(1 - x^2 - y^2) + y, \quad \dot{y} = y(1 - x^2 - y^2) - x,$$
which start at (a) $x=1$, $y=0$, (b) $x=2$, $y=0$, at time $t=0$, and discuss their subsequent behaviour. (Hint: switch to polar coordinates.)

10.3   Liapunov stability of the zero solution

We shall consider regular, autonomous systems (see Appendix A) having the form

$$\dot{x} = X(x, y), \qquad \dot{y} = Y(x, y), \tag{10.1}$$

with an equilibrium point at the origin:

$$X(0, 0) = Y(0, 0) = 0; \tag{10.2}$$

(assuming that, if necessary, the equilibrium point whose stability is being considered has been moved to the origin by a translation of the axes).

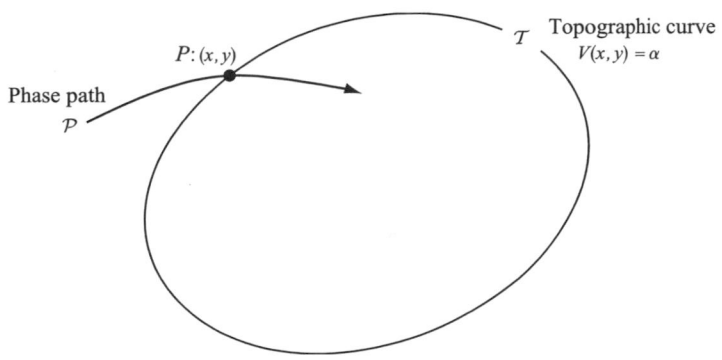

Figure 10.5   A phase path $\mathcal{P}$ crossing a topographic curve defined by $V(x, y) = \alpha$.

We adopt a topographic system (Definition 10.1) defined in a neighbourhood $\mathcal{N}_\mu$ of the origin by a function $V(x, y)$:

$$V(x, y) = \alpha, \quad 0 < \alpha < \mu; \tag{10.3}$$

with $V(0, 0) = 0$.

Now define the function $\dot{V}(x, y)$, whose sign at a point P will determine whether the phase path of (10.1) through P crosses the topographic curve through P inwardly (α decreasing) or outwardly (α increasing). In Fig. 10.5, $\mathcal{T}$ is the topographic curve and $\mathcal{P}$ the phase path through the arbitrary point P: (x, y) in $\mathcal{N}_\mu$. Let the time solutions corresponding to $\mathcal{P}$ be

$$x = x(t), \qquad y = y(t).$$

The time rate of change of $V(x(t), y(t))$ along $\mathcal{P}$ at the point P is given by

$$\frac{\mathrm{d}}{\mathrm{d}t} V(x(t), y(t)) = \dot{x}\frac{\partial V}{\partial x} + \dot{y}\frac{\partial V}{\partial y} = X\frac{\partial V}{\partial x} + Y\frac{\partial V}{\partial y}.$$

This is a *function of position*, denoted by $\dot{V}(x, y)$:

$$\dot{V}(x, y) = X\frac{\partial V}{\partial x} + Y\frac{\partial V}{\partial y}. \tag{10.4}$$

Then, as illustrated in Fig. 10.6.

$$
\begin{aligned}
&\text{if } \dot{V} > 0 \text{ at } P, \quad \mathcal{P} \text{ points outward from } \mathcal{T};\\
&\text{if } \dot{V} < 0 \text{ at } P, \quad \mathcal{P} \text{ points inward across } \mathcal{T};\\
&\text{if } \dot{V} = 0 \text{ at } P, \quad \mathcal{P} \text{ is tangential to } \mathcal{T}.
\end{aligned}
\tag{10.5}
$$

We shall refer to $V(x, y)$ as a **Liapunov test function** for the system (10.1).

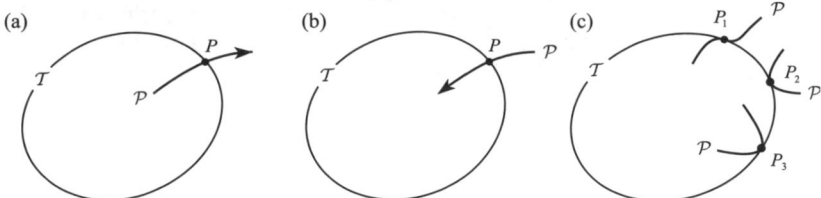

Figure 10.6   Phase path $\mathcal{P}$ crossing topographic curves $\mathcal{T}$. (a) $\dot{V} > 0$ at P. (b) $\dot{V} < 0$ at P. (c) $\dot{V} = 0$ at P: the paths are tangential and the directions undetermined.

Theorem 10.4   *Let $\mathcal{T}_\alpha$ be a topographic curve in $\mathcal{N}_\mu$, defined by*

$$V(x, y) = \alpha < \mu,$$

and suppose that

$$\dot{V}(x, y) \leq 0 \quad in \; \mathcal{N}_\mu. \tag{10.6}$$

Let $\mathcal{H}$ be any half-path that starts at a point P on, or in the interior of, $\mathcal{T}_\alpha$. Then $\mathcal{H}$ can never escape from this region.

Proof   (See Fig. 10.7). Let B be any point exterior to $\mathcal{T}_\alpha$. Since $\alpha < \mu$, there exists α_1 with $\alpha < \alpha_1 < \mu$, such that the topographic curve $\mathcal{T}_{\alpha_1}$ lies between $\mathcal{T}_\alpha$ and the point B. In order to reach B, $\mathcal{P}$ must cross $\mathcal{T}_{\alpha_1}$ outwards, but since $\alpha_1 > \alpha$, this would contradict (10.6). This result is used in Theorem 10.5 and at several points thereafter.   ∎

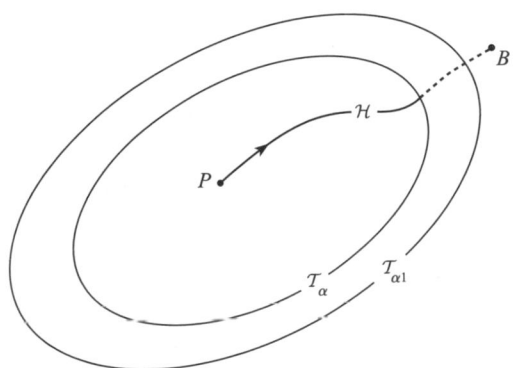

Figure 10.7   If $\dot{V} \leq 0$ on $\mathcal{N}_\mu$ and the topographic curve $\mathcal{T}_\alpha$ lies in $\mathcal{N}_\mu$, then the path from a point P inside $\mathcal{T}_\alpha$ cannot reach B.

Theorem 10.5 (Liapunov stability of the zero solution)   *Let $V(x, y)$ satisfy the conditions of Definition 10.1, and let the system $\dot{x} = X(x, y)$, $\dot{y} = Y(x, y)$ be regular in $\mathcal{N}_\mu$ and have an equilibrium point at the origin. Suppose that*

$$\dot{V}(x, y) \leq 0 \tag{10.7}$$

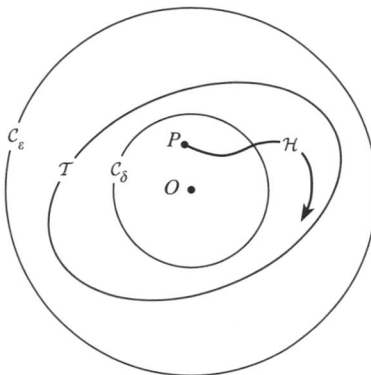

Figure 10.8

in the region consisting of $\mathcal{N}_\mu$ with the origin excluded. Then the zero solution is uniformly stable in the Liapunov sense.

Proof   The following construction (see Fig. 10.8) takes place in $\mathcal{N}_\mu$. $\mathcal{C}_\varepsilon$ is a circle, centre the origin, having arbitrary radius ε, and enclosing no equilibrium points except the origin, $\mathcal{T}$ is any topographic curve interior to $\mathcal{C}_\varepsilon$, and $\mathcal{C}_\delta$ is any circle of radius δ which interior to $\mathcal{T}$. P is any point in the interior of $\mathcal{C}_\delta$ and $\mathcal{H}$ the half-path starting at P, at a time t_0 which is arbitrary. The same value of t_0 is to be used for every such choice of initial point.

Since $\dot{V}(x,y) \leq 0$ in $\mathcal{N}_\mu$, $\mathcal{H}$ can never enter the region exterior to $\mathcal{T}$ (by Theorem 10.4); therefore $\mathcal{H}$ can never reach the circle $\mathcal{C}_\varepsilon$. Stability of the zero time solution $x^*(t) = y^*(t) = 0$, follows immediately. Let the time solution correspoding to the arbitrarily chosen half-path $\mathcal{H}$ be $x(t)$, $y(t)$, $t \geq t_0$. We have shown that given any $\varepsilon > 0$ there exists $\delta > 0$ such that if

$$(x^2(t_0) + y^2(t_0))^{1/2} < \delta$$

(i.e., if $\mathcal{H}$ starts in $\mathcal{C}_\delta$) then

$$(x^2(t) + y^2(t))^{1/2} < \varepsilon, \quad t \geq t_0;$$

(i.e., $\mathcal{H}$ never reaches $\mathcal{C}_\varepsilon$). These are the conditions of Definition 8.2 for Liapunov stability of the zero solution. Also, since δ is independent of t_0, the zero solution is uniformly stable.   ■

Notice the exclusion of the origin from the requirement (10.7). By (10.4), $\dot{V}$, is zero at the origin of $\partial V/\partial x, \partial V/\partial y$ exist there; but this is not required by Definition 10.1. We are only interested in making the distinction between $\dot{V} < 0$ and $\dot{V} \leq 0$ *outside* the origin.

Example 10.2   *Show that the zero solution of the system*

$$\dot{x} = X(x,y) = y, \quad \dot{y} = Y(x,y) = -\beta(x^2 - 1)y - x, \quad \beta < 0 \tag{10.8}$$

is Liapunov stable.

This represents the van der Pol equation $\ddot{x} + \beta(x^2 - 1)\dot{x} + x = 0$ with negative parameter β. Try the topographic system given by the test function

$$V(x,y) = x^2 + y^2 = \alpha, \quad 0 < \alpha < \mu. \tag{10.9}$$

Then

$$\dot{V}(x,y) = X\frac{\partial V}{\partial x} + Y\frac{\partial V}{\partial y} = -2\beta(x^2-1)y^2.$$

Therefore

$$\dot{V}(x,y) \leq 0 \quad \text{when } 0 < |x| < 1.$$

(We require $\mu \leq 1$ in (10.9).) Theorem 10.5 determines that the zero solution, represented by the equilibrium point at the origin of the phase plane, is Liapunov stable.   ●

A function $V(x,y)$ which satisfies the conditions of Theorem 10.5 and which therefore predicts stability of the zero solution, is called a **weak Liapunov function**. The word 'weak' is used because the condition $\dot{V} \leq 0$ is not alone 'strong' enough to distinguish asymptotic stability when it occurs. In Section 10.5 we show how it can be made successful in some cases by adding a further condition.

> **Exercise 10.3**
> Suggest a weak Liapunov function for the zero solution of
>
> $$\ddot{x} + k\dot{x} + x^3 = 0, \quad \dot{x} = y.$$

10.4   Asymptotic stability of the zero solution

Asymptotic stability of the zero solution requires, first, that the solution be Liapunov stable and second, that in the phase plane all half-paths starting sufficiently near to the origin at a time t_0 approach the origin. The conditions of the following theorem exclude the obstacles, namely closed paths and extraneous equilibrium points, that would block the progress of half-paths on their way to the origin.

Theorem 10.6   *Let $V(x,y)$ satisfy the conditions of Definition 10.1 for a topographic system in $\mathcal{N}_\mu$, and let the system $\dot{x} = X(x,y)$, $\dot{y} = Y(x,y)$ be regular in $\mathcal{N}_\mu$. Suppose also that*

$$\dot{V}(x,y) < 0 \tag{10.10}$$

in the region $\mathcal{R}_\mu$, defined to be the neighbourhood $\mathcal{N}_\mu$ with the origin excluded. Then (a) there are no equilibrium points in $\mathcal{R}_\mu$ (the only possible equilibrium point is at the origin); (b) $\mathcal{N}_\mu$ does not contain any closed phase paths (limit cycles).

Proof   (a) Let P be any point in $\mathcal{R}_\mu$. Then P cannot be an equilibrium point since, from (10.4),

$$\dot{V} = \dot{x}\frac{\partial V}{\partial x} + \dot{y}\frac{\partial V}{\partial y}$$

would be zero at P, which would contradict (10.10). Therefore the only possible equilibrium point is at the origin.

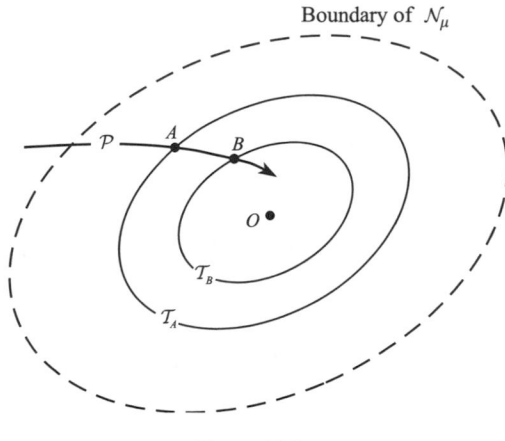

Boundary of $\mathcal{N}_\mu$

Figure 10.9

(b) (See Fig. 10.9) $\mathcal{P}$ is any phase path intersecting $\mathcal{N}_\mu$, and A is any point on $\mathcal{P}$ in $\mathcal{N}_\mu$. B is any point in $\mathcal{N}_\mu$ on $\mathcal{P}$ subsequent to A and $\mathcal{T}_A, \mathcal{T}_B$ are the topographic curves through A and B respectively: by (10.10), the direction of $\mathcal{P}$ is inward across $\mathcal{T}_A$ and $\mathcal{T}_B$. $\mathcal{P}$ can never re-emerge across $\mathcal{T}_B$ to rejoin A, therefore $\mathcal{P}$ cannot be a closed phase path. In particular it cannot be a closed phase path lying in $\mathcal{N}_\mu$. ∎

Theorem 10.7 (Asymptotic stability of the zero solution)   *Let $V(x,y)$ satisfy the conditions of Definition 10.1, of a topographic system, and in the neighbourhood $\mathcal{N}_\mu$ of the definition let the system $\dot{x} = X(x,y)$, $\dot{y} = Y(x,y)$ be regular and have an equilibrium point at the origin. Suppose also that*

$$\dot{V}(x,y) < 0 \tag{10.11}$$

in the region consisting of $\mathcal{N}_\mu$ with the origin excluded. Then the zero solution is (a) *uniformly and* (b) *asymptotically stable.*

Proof   (a) The zero solution is uniformly stable by Theorem 10.5, since the condition (10.11) implies the truth of condition (10.7).

(b) Let P be any point in $\mathcal{N}_\mu, \mathcal{T}$ the topographic curve through P, and $\mathcal{H}$ the half-path starting at P. The region consisting of $\mathcal{T}$ together with its interior is a closed region, so the Poincaré–Bendixson theorem (Theorem 10.3) is applicable to $\mathcal{H}$. Since, by Theorem 10.6, there are no closed paths in $\mathcal{N}_\mu$, or any equilibrium points apart from the origin, $\mathcal{H}$ must approach the origin. Since $\mathcal{H}$ is arbitrary in $\mathcal{N}_\mu$, all half-paths from $\mathcal{N}_\mu$ approach the origin, so the zero solution is asymptotically stable. ∎

A Liapunov test function satisfying the condition (10.11) for the system and therefore determining asymptotic stability is called a **strong Liapunov function**.

Example 10.3   *Investigate the stability of the zero solution of the system $\dot{x} = -y - x^3$, $\dot{y} = x - y^3$.*
Here we return to complete Example 10.1. The family of curves

$$V(x,y) = x^2 + y^2 = \alpha, \quad 0 < \alpha < \infty,$$

is a topographic system. Then

$$\dot{V}(x,y) = X\frac{\partial V}{\partial x} + Y\frac{\partial V}{\partial y} = (-y - x^3)2x + (x - y^3)2y$$
$$= -2(x^4 + y^4)$$

which is negative everywhere except at the origin. Therefore we have found a strong Liapunov function for the system, and by Theorem 10.7 the zero solution is uniformly and asymptotically stable in the Liapunov sense. (This result brings with it the fact that all phase paths approach the origin as $t \to \infty$, which was the question left unresolved in Example 10.1.) ●

Example 10.4 *By using the Liapunov test function $V(x,y) = x^2 + xy + y^2$, show that the zero solution of the linear system $\dot{x} = y$, $\dot{y} = -x - y$ is uniformly and asymptotically stable.*

The proposed test function $V(x,y)$ does define a topographic system, since

$$V(x,y) = x^2 + xy + y^2 = (x + \tfrac{1}{2}y)^2 + \tfrac{3}{4}y^2 > 0$$

for $(x,y) \neq (0,0)$. Also

$$\dot{V}(x,y) = y(2x + y) + (-x - y)(x + 2y)$$
$$= -(x^2 + xy + y^2) = -V(x,y) < 0.$$

for $(x,y) \neq (0,0)$. Therefore, by Theorem 10.7, the zero solution is uniformly and asymptotically stable. (Also, by Theorem 8.1, all the solutions have the same property since the system equations are linear.) ●

Example 10.5 *Investigate the stability of the system $\dot{x} = -x - 2y^2$, $\dot{y} = xy - y^3$.*

The family of curves defined by

$$V(x,y) = x^2 + by^2 = \alpha, \quad \alpha > 0, \ b > 0$$

constitutes a topographic system (of ellipses) everywhere. We obtain

$$\dot{V}(x,y) = -2(x^2 + (2 - b)xy^2 + by^4).$$

It is difficult to reach any conclusion about the sign of $\dot{V}$ by inspection, but if we choose the value $b = 2$ we obtain the simple form

$$\dot{V}(x,y) = -2(x^2 + 2y^4)$$

which is negative everywhere (excluding the origin). Therefore $V(x,y)$ is a strong Liapunov function for the system when

$$V(x,y) = x^2 + 2y^2.$$

By Theorem 10.7 the zero solution is uniformly and asymptotically stable. ●

Exercise 10.4
Show that the system

$$\ddot{x} + (2 + 3x^2)\dot{x} + x = 0$$

is equivalent to the first-order system

$$\dot{x} = y - x^3, \quad \dot{y} = -x + 2x^3 - 2y.$$

Using the Liapunov function $V(x, y) = x^2 + y^2$, show that the origin in the (x, y) plane is asymptotically stable.

In each of the cases of an asymptotically stable zero solution shown in Examples 10.3 to 10.5 we have displayed a neighbourhood $\mathcal{N}_\mu$ from which all half-paths approach the origin as $t \to \infty$. Any neighbourhood having this property is called a **domain of asymptotic stability** or a **domain of attraction** for the origin. $\mathcal{N}_\mu$ will not, in general, be the largest possible domain of attraction, since it depends on the Liapunov function chosen. If the domain of attraction consists of the whole x, y plane, as in Example 10.4, the system is said to be **globally asymptotically stable**.

10.5 Extending weak Liapunov functions to asymptotic stability

In Example 10.2 we showed that the zero solution of the system $\dot{x} = y$, $\dot{y} = -\beta(x^2 - 1)y - x$, with $\beta < 0$ is Liapunov stable, by using a weak Liapunov function $V(x, y) = x^2 + y^2$. However, the linear approximation to the system,

$$\dot{x} = y, \quad \dot{y} = -x + \beta y \ (\beta < 0),$$

predicts a stable spiral or a stable node (in the Poincaré sense) at the origin. It therefore seems likely that the zero solution is not merely stable, but asymptotically stable, so there might exist a *strong* Liapunov function, having the property $\dot{V}(x, y) < 0$ (rather than $\dot{V}(x, y) \leq 0$) in a region exterior to the origin. It is possible to construct such a function as will be shown in Example 10.8; for example,

$$V(x, y) = \{(2 + \beta^2)x^2 - 2\beta xy + 2y^2\}.$$

However, we now show how it is often possible, and simpler, to work with a weak Liapunov function in order to demonstrate asymptotic stability when it does occur.

Theorem 10.8 *Let $V(x, y)$ satisfy the conditions for a topographic system in Definition 10.1, and let the system equations be regular, in $\mathcal{N}_\mu$. Suppose that*

(i) *$\dot{V}(x, y) \leq 0$ in the region consisting of $\mathcal{N}_\mu$ with the origin excluded;*
(ii) *none of the topographic curves in $\mathcal{N}_\mu$ is also a phase path.*

Then $\mathcal{N}_\mu$ does not contain a closed phase path.

Proof In Fig. 10.10, A is any point in $\mathcal{N}_\mu$, $\mathcal{H}$ is the half-path starting at A, and $\mathcal{T}_A$ is the topographic curve through A. $\mathcal{T}_A$ is not a phase path (by the restriction (ii)) so $\mathcal{H}$ must

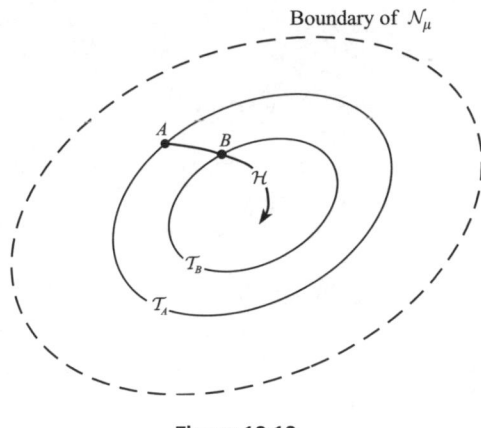

Figure 10.10

separate from $\mathcal{T}_A$, either immediately, or after following $\mathcal{T}_A$ for a time, but before completing the cycle. Also, by (i) and Theorem 10.4, when $\mathcal{H}$ departs from $\mathcal{T}_A$ its direction is inward, as shown.

Let B be a subsequent point on $\mathcal{H}$, and $\mathcal{T}_B$ the topographic curve through B. By the same argument, $\mathcal{H}$ must enter the interior of $\mathcal{T}_B$. Once there, it cannot emerge without violating Theorem 10.4, so it can never rejoin A to form a closed phase path. Since the point A is arbitrary there are, therefore, no closed phase paths in $\mathcal{N}_\mu$. ∎

Theorem 10.9 (Conditions for asymptotic stability from a weak Liapunov function) *Let $V(x, y)$ satisfy the conditions in Definition 10.1 for a topographic system $V(x, y) = \alpha$, $0 < \alpha < \mu$ in a neighbourhood $\mathcal{N}_\mu$ of the origin, in which also the system $\dot{x} = X(x, y)$, $\dot{y} = Y(x, y)$ is regular with a single equilibrium point, situated at the origin. Suppose also that*

(i) *$\dot{V}(x, y) \leq 0$ in the region consisting of $\mathcal{N}_\mu$ with the origin excluded (i.e., $V(x, y)$ is a weak Liapunov function);*

(ii) *no (closed) curve of the topographic system is also a phase path.*
   Then the zero solution is uniformly and asymptotically stable.

Proof   The zero solution is stable, by Theorem 10.5. We have to prove also that all half-paths $\mathcal{H}$ starting in $\mathcal{N}_\mu$ approach the origin as $t \to \infty$.

Choose any point P in $\mathcal{N}_\mu$. Let $\mathcal{H}$ be the half-path starting at P at time t_0, and $\mathcal{T}$ the topographic curve through P. By Theorem 10.4 $\mathcal{H}$ cannot enter the exterior of $\mathcal{T}$, so it remains in the closed region consisting of $\mathcal{T}$ and its interior.

By the conditions in the theorem statement there are no equilibrium points in this region apart from the origin. Also the condition (ii) along with Theorem 10.8 ensures that there are no closed phase paths on the region. Therefore, by the Poincaré–Bendixson theorem (Theorem 10.3) the arbitrarily chosen $\mathcal{H}$, starting in $\mathcal{N}_\mu$, must approach the origin. The zero solution is therefore asymptotically stable. ∎

In Theorem 10.9, notice the importance for the proof of recording firstly that the zero solution is stable. It is not enough to prove that all neighbouring phase paths approach the origin. Such a stable-looking feature of the phase diagram indicates Poincaré stability,   but it does not automatically mean that the representative points remain in step as $t \to \infty$.

It will usually be very easy to determine, in particular cases, whether the condition (b) of Theorem 10.7 is satisfied by inspecting any curves along which $\dot{V}(x, y) = 0$ in the light of the following criteria:

Theorem 10.10 (Satisfaction of condition (b) of Theorem 10.9)   *Let C be any curve along which $\dot{V}(x, y) = 0$, under the conditions of Theorem 10.9. If either* (a) *C is not closed in $\mathcal{N}_\mu$, or* (b) *C is closed, but is not a topographic curve, or* (c) *C is not a phase path of the system, then C is not a curve that need be considered under condition* (b) *of Theorem 10.7.*

Proof   (a) and (b) are obvious. For (c); if C proves to be a topographic curve, then it must be tested against the differential equations since it might, by chance, prove to be a closed phase path. (The possibility that $\dot{V}(x, y)$ is equal to zero on a topographic curve does not imply that it is a phase path: it might merely indicate that at every point on the curve the phase path is tangential to it and then continues inward, as at P_1 in Fig. 10.6(c).)   ∎

In Example 10.2 we showed that the function $V(x, y) = x^2 + y^2$ is a weak Liapunov function for the van der Pol equation for a negative parameter β; this shows only that the zero solution is stable. In Example 10.6 we shall use the reasoning above to show that it is asymptotically stable.

Example 10.6   *Show that the zero solution of the system $\dot{x} = y$, $\dot{y} = -\beta(x^2 - 1)y - x$, with $\beta < 0$, is asymptotically stable.*

Use the test function $V(x, y) = x^2 + y^2 = \alpha, 0 < \alpha < 1$. In Example 10.7 we obtained

$$\dot{V}(x, y) = -2\beta(x^2 - 1)y^2 \leq 0 \quad \text{in } \mathcal{N}_\mu,$$

the neighbourhood $\mathcal{N}_\mu$ being defined by $x^2 + y^2 < 1$. $\dot{V} = 0$ on the curves $y = 0$, $-1 \leq x < 0$ and $y = 0$, $0 < x \leq 1$ (since we exclude the equilibrium point (0,0)). These are not members of the topographic system (neither are they closed; nor, plainly, are they solutions of the system). Therefore Theorem 10.9 applies and the zero solution in asymptotically stable.   ●

10.6   A more general theory for autonomous systems

The following expression of the theory does not depend on the use of the Poincaré–Bendixson theorem and the two-dimensional topographic systems of Definition 10.1, so the conditions on the Liapunov functions are less restrictive. Also, comparatively small changes in wording make the proofs available for n-dimensional systems.

We use the vector formulation. Put

$$x = \begin{bmatrix} x_1 \\ x_2 \end{bmatrix}, \quad \dot{x} = \begin{bmatrix} \dot{x}_1 \\ \dot{x}_2 \end{bmatrix}, \tag{10.12}$$

then the system equations are

$$\dot{x} = X(x), \tag{10.13}$$

where

$$X(x) = \begin{bmatrix} X_1(x) \\ X_2(x) \end{bmatrix},$$

and $X_1(x)$, $X_2(x)$ are scalar functions of position such that (10.13) satisfies the regularity conditions of Appendix A.

A scalar function $f(x)$ is called

positive definite	if $f(0) = 0$, $f(x) > 0$ when $x \neq 0$;
positive semidefinite	if $f(0) = 0$, $f(x) \geq 0$ when $x \neq 0$;
negative definite	if $f(0) = 0$, $f(x) < 0$ when $x \neq 0$;
negative semidefinite	if $f(0) = 0$, $f(x) \leq 0$ when $x \neq 0$.

The functions $V(x)$ used in the previous sections are either positive definite or semidefinite, but they have other restrictions imposed. Notice that if $f(x)$ is definite, then it is also semidefinite.

Theorem 10.11 (Liapunov stability) *Suppose that in a neighbourhood $\mathcal{N}$ of the origin*
(i) *$\dot{x} = X(x)$ is a regular system and $X(0) = 0$;*
(ii) *$V(x)$ is continuous and positive definite;*
(iii) *$\dot{V}(x)$ is continuous and negative semidefinite.*
Then the zero solution of the system is uniformly stable.

Proof Choose any $k > 0$ such that the circular neighbourhood $|x| < k$ lies in $\mathcal{N}$. Without loss of generality, let ε be any number in the range $0 < \varepsilon < k$. In Fig. 10.11, C_ε is the circle $|x| = \varepsilon$ in the phase plane.

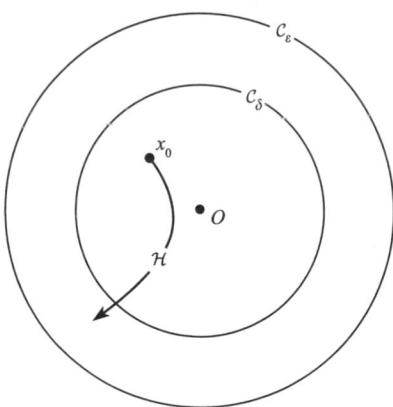

Figure 10.11

Since C_ε is a closed set of points and $V(x)$ is positive and continuous on C_ε, $V(x)$ attains its minimum value m, say, with $m > 0$, at some point on C_ε:

$$\min_{C_\varepsilon} V(x) = m > 0,$$

so that

$$V(x) \geq m > 0 \quad \text{on } C_\varepsilon. \tag{10.14}$$

Since $V(0) = 0$ and $V(x)$ is continuous at the origin

$$\lim_{x \to 0} V(x) = 0.$$

Therefore there exists $\delta > 0$ such that

$$0 \leq V(x) < m \quad \text{for all } |x| < \delta; \tag{10.15}$$

that is, for all x interior to the circle C_δ having radius δ, shown in Fig. 10.11.

Let x_0 be any point (excluding the origin) that is interior to C_δ; $\mathcal{H}$ the half-path starting at x_0 at the arbitrary time t_0; and $x(t)$, $t \geq t_0$, with $x(t_0) = x_0$, is the time solution corresponding to $\mathcal{H}$. By (10.15)

$$0 < V(x_0) < m \tag{10.16}$$

and by condition (iii)

$$\dot{V}(x(t)) \leq 0 \tag{10.17}$$

for all $t \geq t_0$ so long as the path remains in $\mathcal{N}$. Therefore $V(x(t))$ is non-increasing along $\mathcal{H}$, so from (10.16) and (10.17)

$$0 < V(x(t)) < m \quad \text{for all } t \geq t_0.$$

Therefore $\mathcal{H}$ can never meet C_ε, since this would contradict (10.14). The existence of the solution for all $t \geq t_0$ in the domain $0 < |x| < \varepsilon$ is then assured by Theorem A2 of Appendix A.

Therefore, given any $\varepsilon > 0$ there exists $\delta > 0$ such that for any path $x(t)$ the following is true:

$$\text{if } |x(t_0)| < \delta, \quad \text{then } |x(t)| < \varepsilon \text{ for all } t \geq t_0.$$

This is the definition of uniform stability of the zero solution, since δ is independent of t_0. ∎

Theorem 10.11 remains true for systems of dimension $n > 2$.

Theorem 10.12 (Asymptotic stability)   *Suppose that in some neighbourhood $\mathcal{N}$ of the origin*
  (i) *$\dot{x} = X(x)$ is a regular system, and $X(0) = 0$;*
 (ii) *$V(x)$ is positive definite and continuous;*
(iii) *$\dot{V}(x)$ is negative definite and continuous.*
Then the zero solution is uniformly and asymptotically stable.

Proof $\dot{V}(x)$ is negative definite, so it is also negative semidefinite and by Theorem 10.11 the zero solution is uniformly stable. It remains to show that, in terms of phase paths, there exists a neighbourhood of the origin from which every half-path approaches the origin, or that $\lim_{t\to\infty} x(t) = 0$ for every such half-path.

Take any fixed value of $k > 0$ such that the circular neighbourhood

$$|x| < k$$

bounded by $\mathcal{C}_k$ (see Fig. 10.12) lies in $\mathcal{N}$. By Theorem 10.11, there exists a number $\rho > 0$ such that

$$|x(t_0)| = |x_0| < \rho \Rightarrow |x(t)| < k \quad \text{for } t \geq t_0, \tag{10.18}$$

where $x(t)$ is the time solution with initial condition $x(t_0) = x_0$, represented in Fig. 10.12 by the half-path $\mathcal{H}$ from x_0, and where $\mathcal{C}_\rho$ is the circular boundary $|x| = \rho$. Note that ρ depends only on k since the system is autonomous.

Let $\mathcal{C}_\eta$ be any circle, however small, defined by

$$|x| = \eta, \quad 0 < \eta < |x_0| < \rho.$$

It will be shown that $\mathcal{H}$ will enter the interior of $\mathcal{C}_\eta$ at some time $t_\eta > t_0$.

Let $\mathcal{R}$ be the closed annular region defined by

$$\eta \leq |x| \leq k,$$

shown shaded in Fig. 10.12. By condition (iii), $\dot{V}(x)$ attains its maximum value M on $\mathcal{R}$ and this value is negative:

$$M = \max_{\mathcal{R}} \dot{V}(x) < 0,$$

so that

$$\dot{V}(x) \leq M < 0 \quad \text{no } \mathcal{R}. \tag{10.19}$$

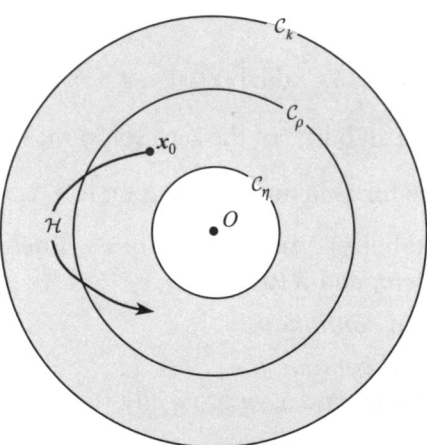

Figure 10.12 $\mathcal{R}$ is shown as shaded.

So long as $\mathcal{H}$ remains on $\mathcal{R}$ the values $V(\boldsymbol{x}(t))$ along $\mathcal{H}$ are given by

$$0 \leq V(\boldsymbol{x}(t)) = V(\boldsymbol{x}_0) + \int_{t_0}^{t} \dot{V}(\boldsymbol{x}(\tau))\,d\tau$$

$$\leq V(\boldsymbol{x}_0) + M(t - t_0),$$

by (10.19). V is non-negative, but the right-hand side becomes negative for large enough t; therefore, $\mathcal{H}$ must exit from $\mathcal{R}$ at some time, through one of the boundaries $|\boldsymbol{x}| = k$ or $|\boldsymbol{x}| = \eta$. By (10.18), $\mathcal{H}$ cannot escape through $\mathcal{C}_k$, so it must cross $\mathcal{C}_\eta$ and enter its interior. Denote the time of its first arrival on $\mathcal{C}_\eta$ by t_η.

Although η is arbitrarily small, $\mathcal{H}$ may, and in general will, emerge again and re-enter $\mathcal{R}$. We now show that, nevertheless, its distance from the origin remains infinitesimally bounded.

Choose any $\varepsilon > 0$ (however small) in the range

$$0 < \varepsilon < |\boldsymbol{x}_0| < \rho.$$

Since the solution is stable, there exists $\delta > 0$ such that any path entering $\mathcal{C}_\delta$ (see Fig. 10.13) remains in the interior of $\mathcal{C}_\varepsilon$ for all subsequent time. Choose the value of η in the foregoing discussion in the range $0 < \eta < \delta$, and let $\mathcal{C}_\eta$ be the circle

$$|\boldsymbol{x}| = \eta \quad \text{with } 0 < \eta < \delta.$$

We have shown above that $\mathcal{H}$, starting in $\mathcal{C}_\rho$ at time t_0, will reach $\mathcal{C}_\eta$ at a certain time t_η. Since $\mathcal{C}_\eta$ is interior to $\mathcal{C}_\delta$, $\mathcal{H}$ will remain interior to $\mathcal{C}_\varepsilon$, for all $t \geq t_\eta$. (The existence of the solution is secured by Theorem A2 of Appendix A.)

In terms of the corresponding time solution $x(t)$ we have, proved that for arbitrary $\eta > 0$,

$$|\boldsymbol{x}(t)| < \varepsilon \quad \text{for all } t \geq t_\eta,$$

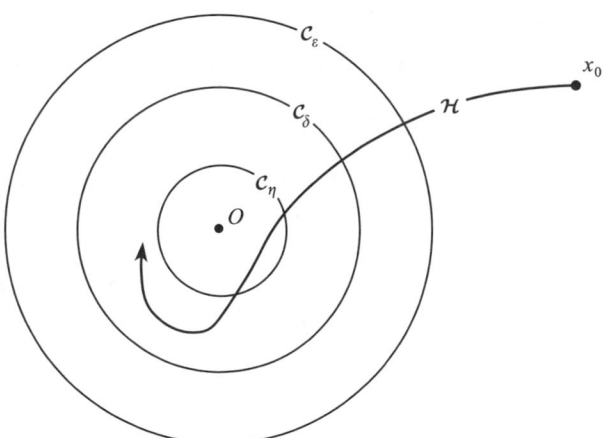

Figure 10.13

or that

$$\lim_{t \to \infty} x(t) = 0,$$

for every solution with $0 < |x(t_0)| < \rho$. The region $0 < |x| < \rho$ is therefore a domain of asymptotic stability. ∎

Theorem 10.12 remains true for systems of dimension $n > 2$.

Exercise 10.5
Using the Liapunov function $V(x, y) = x^2 + y^2$, show that the origin of

$$\dot{x} = -x + y(1 - x^2 - y^2), \quad \dot{y} = -y - x(1 - x^2 - y^2)$$

is asymptotically stable.

10.7   A test for instability of the zero solution: *n* dimensions

The general approach of the Liapunov method can be adapted to test for *instability* of the zero solution of an n-dimensional system. The notation $\|x\|$ denotes any **norm** (a measure of magnitude) of a vector x (see Section 8.2 and Appendix C).

Theorem 10.13   *Let $x(t) = 0$, $t \geq t_0$, be the zero solution of the regular autonomous system $\dot{x} = X(x)$, of dimension n, where $X(0) = 0$. If there exists a function $U(x)$ such that in some neighbourhood $\|x\| \leq k$*

(i) *$U(x)$ and its partial derivatives are continuous;*

(ii) *$U(0) = 0$;*

(iii) *$\dot{U}(x)$ is positive definite for the given system;*

(iv) *in every neighbourhood of the origin there exists at least one point x at which $U(x) > 0$;*

then the zero solution (and therefore every solution) is unstable.

Proof   By (iv), given any δ, $0 < \delta < k$, there exists x_δ such that $0 < \|x_\delta\| < \delta$ and $U(x_\delta) > 0$. Suppose that $x(\delta, t)$, $t \geq t_0$, is the solution satisfying $x(\delta, t_0) = x_\delta$. Its path cannot enter the origin as $t \to \infty$, because $U(0) = 0$ (by (ii)) but $\dot{U}(x) > 0$, $x \neq 0$. Since $\dot{U}$ is positive definite by (iii), and continuous (by (i) and the regularity condition), and since the path is bounded away from the origin there exists a number $m > 0$ such that

$$\dot{U}\{x(\delta, t)\} \geq m > 0, \quad t \geq t_0$$

so long as

$$\|x(\delta, t)\| \leq k.$$

Therefore

$$U\{x(\delta,t)\} - U\{x(\delta,t_0)\} = \int_{t_0}^{t} \dot{U}\{x(\delta,\tau)\}d\tau \geq m(t-t_0). \qquad (10.20)$$

$U(x)$ is continuous by (i) and therefore bounded on $\|x\| \leq k$; but the right-hand side of (10.20) is unbounded. Therefore $x(\delta,t)$ cannot remain in $\|x\| \leq k$ and so the path reaches the boundary $\|x\| = k$.

Therefore, given any ε, $0 < \varepsilon < k$, then for any δ, however small, there is at least one solution $x(\delta,t)$ with $\|x(\delta,t_0)\| < \delta$ but $\|x(\delta,t)\| > \varepsilon$ for some t. Therefore the zero solution is unstable. ∎

Typically (by (iv)), $U(x)$ will take both positive and negative values close to $x = 0$. In the plane, simple functions of the type xy, $x^2 - y^2$, may be successful, as in Example 10.7.

Example 10.7   *Show that $x(t) = 0$, $t \geq t_0$, is an unstable solution of the equation $\ddot{x} - x + \dot{x}\sin x = 0$.*
The equivalent system is $\dot{x} = y, \dot{y} = x - y\sin x$. Consider

$$U(x,y) = xy$$

satisfying (i), (ii), and (iv) for every neighbourhood. Then

$$\dot{U}(x,y) = x^2 + y^2 - xy\sin x.$$

It is easy to show that this has a minimum at the origin and since $\dot{U}(0,0) = 0$, $\dot{U}$ is positive definite in a sufficiently small neighbourhood of the origin. Therefore the zero solution is unstable. (To show by calculus methods that there is a minimum at the origin is sometimes the simplest way to identify such a function.) ●

Exercise 10.6
Show that the origin in the system

$$\dot{x} = y, \quad \dot{y} = -(1-x^2)y + x$$

is an unstable equilibrium point.

10.8   Stability and the linear approximation in two dimensions

The specific question discussed is the following. For the two-dimensional autonomous system

$$\dot{x} = X(x) = Ax + h(x), \qquad (10.21)$$

where A is constant and $h(x)$ is of smaller order of magnitude than Ax, does the instability or asymptotic stability of the zero solution of $\dot{x} = Ax$ imply the same property for the zero solutions of the system $\dot{x} = X(x)$? To answer this question we shall construct explicit Liapunov functions for the linearized system $\dot{x} = Ax$ and show that they also work for the original system.

Suppose that the linear approximate system

$$\dot{x} = Ax, \quad \text{or} \quad \begin{bmatrix} \dot{x} \\ \dot{y} \end{bmatrix} = \begin{bmatrix} a & b \\ c & d \end{bmatrix} \begin{bmatrix} x \\ y \end{bmatrix}, \tag{10.22}$$

is asymptotically stable. We show that a strong Liapunov function $V(x)$ for this system can be constructed which is a quadratic form, namely,

$$V(x) = x^{\mathrm{T}} K x, \tag{10.23}$$

where the constant matrix K has to be determined. It follows that

$$\dot{V}(x) = \dot{x}^{\mathrm{T}} K x + x^{\mathrm{T}} K \dot{x} = x^{\mathrm{T}} A^{\mathrm{T}} K x + x^{\mathrm{T}} K A x$$
$$= x^{\mathrm{T}} [A^{\mathrm{T}} K + K A] x.$$

We now arrange for $V(x)$ to be positive definite and $\dot{V}(x)$ negative definite: the second requirement is satisfied if, for example, we can choose K so that

$$A^{\mathrm{T}} K + K A = -I, \tag{10.24}$$

for then $\dot{V}(x) = -x^2 - y^2$. To find K, in the notation of Chapter 2, let $p = a + d$, $q = ad - bc$. We shall show that there is a solution of (10.24) of the form

$$K = m (A^{\mathrm{T}})^{-1} A^{-1} + n I, \tag{10.25}$$

where m and n are constants, which makes $V(x)$ in (10.23) positive definite. The left of (10.24) becomes

$$A^{\mathrm{T}} K + K A = m \{ A^{-1} + (A^{\mathrm{T}})^{-1} \} + n (A + A^{\mathrm{T}}).$$

Since

$$A = \begin{bmatrix} a & b \\ c & d \end{bmatrix}, \quad A^{-1} = \begin{bmatrix} d & -b \\ -c & a \end{bmatrix} \bigg/ q,$$

then by (10.24) we require

$$A^{\mathrm{T}} K + K A = \frac{1}{q} \begin{bmatrix} 2md + 2naq & (nq - m)(b + c) \\ (nq - m)(b + c) & 2ma + 2ndq \end{bmatrix} = \begin{bmatrix} -1 & 0 \\ 0 & -1 \end{bmatrix}.$$

This equation is satisfied if

$$m = -q/2p, \quad n = -1/2p.$$

Then from (10.25)

$$K = -\frac{1}{2pq} \begin{bmatrix} c^2 + d^2 + q & -ac - bd \\ -ac - bd & a^2 + b^2 + q \end{bmatrix}. \tag{10.26}$$

The associated Liapunov function is given by (10.23):

$$V(\boldsymbol{x}) = -\{(c^2 + d^2 + q)x^2 - 2(ac + bd)xy + (a^2 + b^2 + q)y^2\}/2pq$$
$$= -\{(dx - by)^2 + (cx - ay)^2 + q(x^2 + y^2)\}/2pq, \tag{10.27}$$

which is positive definite at least when

$$p < 0, \quad q > 0. \tag{10.28}$$

Thus V is a strong Liapunov function for the *linearized* system (10.22) when (10.28) is satisfied. Now, the stability regions for (10.22) are already known, and displayed in Fig. 2.10, which is partly reproduced as Fig. 10.14. We see that (10.28) in fact exhausts the regions I of asymptotic stability of eqn (10.22). This fact is used in Theorem 10.14.

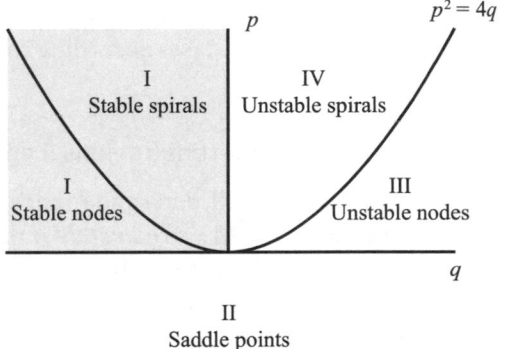

Figure 10.14   Stability regions in the (p, q) plane for $\dot{x} = ax + by, \dot{y} = cx + dy$, where $p = a + d$ and $q = ad - bc$ (see also Fig. 2.10).

Theorem 10.14   *Let* $(0, 0)$ *be an equilibrium point of the regular system*

$$\begin{bmatrix} \dot{x} \\ \dot{y} \end{bmatrix} = \begin{bmatrix} a & b \\ c & d \end{bmatrix} \begin{bmatrix} x \\ y \end{bmatrix} + \begin{bmatrix} h_1(x, y) \\ h_2(x, y) \end{bmatrix}, \tag{10.29}$$

where the order of magnitude of h_1 *and* h_2 *as the origin approached are given by*

$$h_1(x, y) = O(x^2 + y^2), \quad h_2(x, y) = O(x^2 + y^2) \tag{10.30}$$

as $x^2 + y^2 \to 0$. *Then the zero solution of* (10.29) *is asymptotically stable when its linear approximation is asymptotically stable.*

Proof   We shall show that V given by (10.27) is also a strong Liapunov function for the system (10.29). It is a function positive definite when $p < 0, q > 0$; that is, whenever the linearized system is asymptotically stable.

Also, for the given system (10.29),

$$\dot{V} = x^T[A^TK + KA]x + h^T(x)Kx + x^TKh(x)$$

$$= -x^TIx + 2h^T(x)Kx$$

$$= -(x^2 + y^2) + K_{11}h_1^2 + 2K_{12}h_1h_2 + K_{22}h_2^2, \tag{10.31}$$

where the symmetry of K has been used.

By (10.30), for any p, q, there is clearly a neighbourhood of the origin in which the first term of (10.31) predominates, that is to say, where $\dot{V}$ is negative definite. Therefore, V is a strong Liapunov function for (10.29) in the parameter region $p < 0$, $q > 0$; that is, when the *linearized system* is asymptotically stable. The result follows by Theorem 10.12. ∎

The open regions II, III, IV in Fig. 10.14 are, we know, where the solutions of $\dot{x} = Ax$ are *unstable*. To prove a statement in similar terms to the last theorem, only relating to the *instability of the zero solution* of $\dot{x} = Ax + h(x)$, we require a function $U(x)$ with the properties stated in Theorem 10.13. There are two principal cases according to the eigenvalues of A, given as (i) and (ii) below:

(i) Instability: the eigenvalues of A real and different (regions II and III in Fig. 10.14)

Since we hope that the stability of $\dot{x} = Ax$ will determine that of $\dot{x} = Ax + h(x)$, we are interested in the case where the solutions of $\dot{x} = Ax$ are *unstable*: then (Theorem 8.11) *at least one eigenvalue of A is positive*. This covers the regions II and III of Fig. 10.14. (We will not consider the case when the eigenvalues are equal: the line $p^2 = 4q$ in Fig. 10.14; nor when one of the eigenvalues is zero: $q = 0$.) In this case we can reduce the problem to a simpler one by a transformation

$$x = Cu, \tag{10.32}$$

where C is non-singular, and $u = [u_1, u_2]^T$; whence the equation satisfied by u is

$$\dot{u} = C^{-1}ACu. \tag{10.33}$$

It is known that matrices of the form A and $C^{-1}AC$ have the same eigenvalues; hence the stability criteria for $x(t)$ and $u(t)$ in terms of the eigenvalues are the same. When the eigenvalues of A are real and different, it is known that C can then be chosen so that

$$C^{-1}AC = \begin{bmatrix} \lambda_1 & 0 \\ 0 & \lambda_2 \end{bmatrix} = D, \quad \text{say}, \tag{10.34}$$

where the columns of C are eigenvectors corresponding to λ_1, λ_2. Suitable C are given by

$$C = \begin{bmatrix} -b & -b \\ a - \lambda_1 & a - \lambda_2 \end{bmatrix}, \quad b \neq 0;$$

$$C = \begin{bmatrix} -c & -c \\ d - \lambda_1 & d - \lambda_2 \end{bmatrix}, \quad b = 0, \ c \neq 0,$$

and if $b = c = 0$, the equations are already in a suitable form. Then from (10.33) and (10.34)

$$\dot{u} = Du, \tag{10.35}$$

and we wish to establish the instability of the zero solution of this system.
    Consider U defined by

$$U(u) = u^T D^{-1} u = u_1^2/\lambda_1 + u_2^2/\lambda_2. \tag{10.36}$$

Then $U(u) > 0$ at some point in every neighbourhood of $u = 0$ (since instability requires λ_1 or λ_2 positive). $U(u)$ therefore satisfies conditions (i), (ii), and (iv) of Theorem 10.13 in the open regions II and III of Fig. 10.14. Also

$$\dot{U}(u) = \dot{u}^T D^{-1} u + u^T D^{-1} \dot{u} = u^T D D^{-1} u + u^T D^{-1} D u$$

$$= 2(u_1^2 + u_2^2), \tag{10.37}$$

which is positive definite, so that Theorem 10.13 is satisfied.

(ii) Instability: the eigenvalues of A conjugate complex, with positive real part (region IV of Fig. 10.14)

Write

$$\lambda_1 = \alpha + i\beta, \quad \lambda_2 = \alpha - i\beta \qquad (\alpha > 0).$$

The diagonalization process above gives $\dot{u} = Du$ where D and u are complex. Since the theorems refer to real functions, these cannot immediately be used. However, instead of diagonalizing, we can reduce A by a matrix G so that

$$G^{-1} A G = \begin{bmatrix} \alpha & -\beta \\ \beta & \alpha \end{bmatrix}, \tag{10.38}$$

say: an appropriate G is given by

$$G = C \begin{bmatrix} 1 & i \\ 1 & -i \end{bmatrix},$$

where C is as before. The system becomes

$$\dot{u} = \begin{bmatrix} \alpha & -\beta \\ \beta & \alpha \end{bmatrix} u, \tag{10.39}$$

with

$$x = Gu. \tag{10.40}$$

Suppose we define $U(u)$ by

$$U(u) = u^T u. \tag{10.41}$$

This satisfies (i), (ii), and (iv) of Theorem 10.13 in region (IV) of Fig. 10.14. Also

$$\dot{U}(u) = 2\alpha(u_1^2 + u_2^2) \quad \text{with } (\alpha > 0) \tag{10.42}$$

as in case (i), so that $\dot{U}$ is positive definite.

By using the prior knowledge of Chapter 2 as to the regions in which the linearized system is unstable we have determined that wherever $\dot{x} = Ax$ is unstable, U, given by either (10.36) or (10.41) as appropriate, is a function of the type required for Theorem 10.13.

Theorem 10.15   *Let* $(0, 0)$ *be an equilibrium point of the regular two-dimensional system*

$$\dot{x} = Ax + h(x) \tag{10.43}$$

where

$$h(x) = \begin{bmatrix} h_1(x) \\ h_2(x) \end{bmatrix},$$

and $h_1(x), h_2(x) = O(|x|^2)$ *as* $|x| \to 0$. *When the eigenvalues of* A *are different, non-zero, and at least one has positive real part, the zero solution of* (10.43) *is unstable.*

Proof   We shall assume that the eigenvalues of A are real and that one of them is positive (the other case, with λ complex with positive real part, being closely similar). Assume that equation (10.32) is used to reduce (10.43) to the form

$$\dot{u} = Du + g(u); \tag{10.44}$$

then if $u(t) = 0$ is unstable, so is $x(t) = 0$, and conversely. Also, for the components of g,

$$g_1(u), g_2(u) = O(x^2 + y^2) \quad \text{as } x^2 + y^2 \to 0. \tag{10.45}$$

It is sufficient to display a function $U(u)$ for the reduced system (10.44). U given by (10.36) satisfies conditions (i), (ii), and (iv) of Theorem 10.13, namely

$$U(u) = u_1^2/\lambda_1 + u_2^2/\lambda_2.$$

From (10.44) we have

$$\dot{u}_1 = \lambda_1 u_1 + g_1(u), \quad \dot{u}_2 = \lambda_2 u_2 + g_2(u),$$

so that

$$\dot{U}(u) = \dot{u}_1 \frac{\partial U}{\partial u_1} + \dot{u}_2 \frac{\partial U}{\partial u_2}$$

$$= 2(u_1^2 + u_2^2 + \lambda_1^{-1} u_1 g_1(u) + \lambda_2^{-1} u_2 g_2(u)).$$

From (10.45), it is clear that $\dot{U}$ is positive definite in a small enough neighbourhood of the origin, satisfying (iii) of Theorem 10.13. Therefore, by Theorem 10.13, the zero solution $u(t) = 0$ is unstable in regions II and III; thus $x(t) = 0$ is unstable.   ∎

The calculation for region IV of Fig. 10.13 is similar, using U given by (10.41).

Example 10.8   *Show that the zero solution of van der Pol's equation*

$$\ddot{x} + \beta(x^2 - 1)\dot{x} + x = 0$$

is uniformly and asymptotically stable when $\beta < 0$ and unstable when $\beta > 0$. Construct a Liapunov function for the stable case.

The equivalent system is $\dot{x} = y$, $\dot{y} = -x + \beta y - \beta x^2 y$. The linearized system is $\dot{x} = y$, $\dot{y} = -x + \beta y$. The eigenvalues of

$$A = \begin{bmatrix} 0 & 1 \\ -1 & \beta \end{bmatrix}$$

are given by $\lambda_1, \lambda_2 = \frac{1}{2}(\beta \pm \sqrt{(\beta^2 - 4)})$.

Suppose $\beta < 0$. Then the linearized system is asymptotically stable. Also $h^{\mathrm{T}}(x) = [0, -\beta x^2 y]$. Therefore, by Theorem 10.14, the zero solution of the given equation is asymptotically stable.

By (10.23), (10.26), $V = [x, y]K[x, y]^{\mathrm{T}}$ where

$$K = -\frac{1}{2\beta} \begin{bmatrix} 2 + \beta^2 & -\beta \\ -\beta & 2 \end{bmatrix}.$$

Hence

$$V(x, y) = -\{(2 + \beta^2)x^2 - 2\beta xy + 2y^2\}/(2\beta).$$

It can be confirmed that $\dot{V} = -x^2 - y^2 - \beta x^3 y + 2x^2 y^2$, which is negative definite near the origin. The associated topographic family

$$(1 + \tfrac{1}{2}\beta^2)x^2 - \beta xy + y^2 = \text{constant}$$

would be preferable to the simpler family of circles $x^2 + y^2 = \text{constant}$, which is a weak Liapunov function.

Suppose $\beta > 0$. The linearized system is unstable and Theorem 10.15 applies.   ●

Example 10.9   *Investigate the stability of the equilibrium points of*

$$\dot{x} = y(x + 1), \quad \dot{y} = x(1 + y^3).$$

There are two equilibrium points, at $(0, 0)$ and $(-1, -1)$. Near the origin the linear approximation is $\dot{x} = y$, $\dot{y} = x$, with eigenvalues $\lambda = \pm 1$. Hence, the linear approximation has a saddle point at $(0, 0)$. By Theorem 10.13, the zero solution of the given equation is also unstable.

Near the point $(-1, -1)$ put $x = -1 + \xi$, $y = -1 + \eta$. Then

$$\dot{\xi} = \xi(\eta - 1), \quad \dot{\eta} = (\xi - 1)\{1 + (\eta - 1)^3\}.$$

The linear approximation is $\dot{\xi} = -\xi$, $\dot{\eta} = -3\eta$. The eigenvalues are $\lambda = -1, -3$, both negative. By Theorem 10.14 the solution $\xi(t) = \eta(t) = 0$ is therefore asymptotically stable; so is the solution $x(t) = y(t) = -1$ of the original solution.

Figure 10.15 shows the computed phase diagram. The shaded region shows the **domain of attraction** of the point $(-1, -1)$.   ●

The foregoing theorems refer to cases where the linearized system is asymptotically stable, or is unstable, and make no prediction as to the effect of modifying, by nonlinear terms, a case of mere stability. In this case the effect may go either way, as shown by Example 10.10.

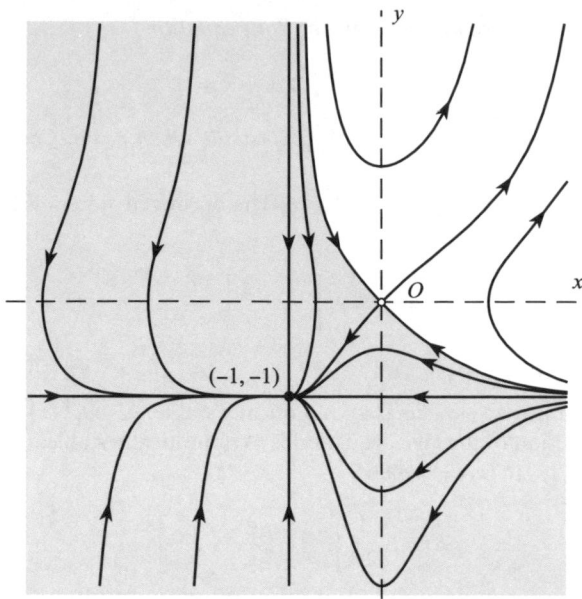

Figure 10.15   Phase diagram for Example 10.9: the domain of attraction of the node at $(-1, -1)$ is shaded.

Example 10.10   *Investigate the stability of the following systems*:
(i) $\dot{x} = y - x(x^2 + y^2)$, $\dot{y} = -x - y(x^2 + y^2)$;
(ii) $\dot{x} = y + x(x^2 + y^2)$, $\dot{y} = -x + y(x^2 + y^2)$.
The linear approximation shows a centre in both cases.

(i) Let $V = x^2 + y^2$. Then

$$\dot{V} = -2(x^2 + y^2)^2$$

which is negative definite so the zero solution is asymptotically stable, by Theorem 10.7 or 10.12.
(ii) Let $U = x^2 + y^2$ (positive in every neighbourhood of the origin). Then

$$\dot{U} = 2(x^2 + y^2)^2$$

which is positive definite. Theorem 10.13 predicts instability.   ●

Exercise 10.7
For the linear system $\dot{x} = Ax$ where

$$A = \begin{bmatrix} 1 & 3 \\ -1 & -2 \end{bmatrix},$$

find the matrix K defined by (10.26). Check that $A^{\mathrm{T}} K + K A = -I$.

10.9   Exponential function of a matrix

It is convenient to have this technical device available in order to solve the equivalent of (10.29) in n dimensions. Let A be a non-singular $n \times n$ matrix. Then e^A is defined by

$$e^A = I + A + \frac{1}{2!}A^2 + \frac{1}{3!}A^3 \cdots, \tag{10.46}$$

whenever the series converges. The partial sums are $n \times n$ matrices and the series converges if the elements of the partial-sum matrices converge, in which case e^A is an $n \times n$ matrix. In fact the series converges for all A, since for any r

$$\left\| I + A + \cdots + \frac{1}{r!}A^r \right\| \leq 1 + \|A\| + \cdots + \frac{1}{r!}\|A\|^r,$$

(where $\| \cdots \|$ denotes the Euclidean matrix norms: see Appendix C) and as $r \to \infty$ the series on the right always converges to $e^{\|A\|}$. It also follows that

(i) $\|e^A\| \leq e^{\|A\|}$.

(ii) $e^{\phi} = I$ where ϕ is the zero matrix.

(iii) $e^A e^B = e^{A+B}$ when $AB = BA$.

(iv) $e^{-A} = (e^A)^{-1}$ ($e^{\pm A}$ are nonsingular).

(v) $\dfrac{d}{dt}e^{At} = Ae^{At} = e^{At}A$.

(vi) $(e^{At})^T = e^{A^T t}$, where T denotes the transpose.

(vii) Let the eigenvalues of A of A be $\lambda_1, \lambda_2, \ldots, \lambda_n$. Then for any γ such that $\gamma > \max_{1 \leq i \leq n}(\lambda_i)$ there exists a constant $c > 0$ such that $\|e^{At}\| < ce^{\gamma t}$.

(viii) The exponential function can be used to represent the solution of the system $\dot{x} = Ax$, where A is constant. Let

$$x = e^{At}c, \tag{10.47}$$

where c is any constant vector. Then $x(0) = e^{\phi}c = Ic = c$ and

$$\dot{x} = Ae^{At}c = Ax,$$

by (v) above, which confirms the solution. In fact, e^{At} is a fundamental matrix for the system, since e^{At} is nonsingular.

(ix) It can be shown (Ferrar 1951) that the exponential series (10.46) can be expressed as a **polynomial** in A of degree less than n, whose coefficients are functions of the eigenvalues of A. We shall not need this result, but it is worth noting that it is not essential to evaluate infinite series.

Example 10.11   *Find the solution of*

$$\dot{x} = \begin{bmatrix} \dot{x} \\ \dot{y} \end{bmatrix} = \begin{bmatrix} 1 & 1 \\ 0 & 1 \end{bmatrix} \begin{bmatrix} x \\ y \end{bmatrix} = Ax,$$

with $x(0) = 1$, $y(0) = 1$.
By (10.47) the solution has the form

$$\begin{bmatrix} x \\ y \end{bmatrix} = e^{At} \begin{bmatrix} c_1 \\ c_2 \end{bmatrix}.$$

Putting $t = 0$ we have $c_1 = 1$, $c_2 = 1$. The representation is obviously a generalization of the one-dimensional solution $ce^{\alpha t}$. To evaluate e^{At}, note that

$$A^r = \begin{bmatrix} 1 & r \\ 0 & 1 \end{bmatrix} = B_r, \quad r \geq 1,$$

say. Therefore

$$e^{At} = \sum_{r=0}^{\infty} \frac{B_r t^r}{r!} = \begin{bmatrix} \sum_{r=0}^{\infty} \dfrac{t^r}{r!} & 1 + \sum_{r=1}^{\infty} \dfrac{t^r}{(r-1)!} \\ 0 & \sum_{r=0}^{\infty} \dfrac{t^r}{r!} \end{bmatrix} = \begin{bmatrix} e^t & 1 + te^t \\ 0 & e^t \end{bmatrix}.$$

Therefore

$$x = e^{At} \begin{bmatrix} 1 \\ 1 \end{bmatrix} = \begin{bmatrix} e^t & 1 + te^t \\ 0 & e^t \end{bmatrix} \begin{bmatrix} 1 \\ 1 \end{bmatrix} = \begin{bmatrix} e^t + 1 + te^t \\ e^t \end{bmatrix}. \qquad \bullet$$

The relation between the exponential matrix solution and the fundamental matrix solution of $\dot{x} = Ax$ can be obtained as follows. Suppose that A is an $n \times n$ matrix with n distinct eigenvalues $\lambda_1, \lambda_2, \ldots, \lambda_n$. Let the eigenvector associated with λ_r be c_r, where

$$[A - \lambda_r I] c_r = 0 \quad (r = 1, 2, \ldots, n).$$

Let

$$C = [c_1, c_2, \ldots, c_n].$$

Then

$$AC = A[c_1, c_2, \ldots, c_n] = [Ac_1, Ac_2, \ldots, Ac_n]$$
$$= [\lambda_1 c_1, \lambda_2 c_2, \ldots, \lambda_n c_n]$$
$$= [c_1, c_2, \ldots, c_n] D = CD,$$

where D is the diagonal matrix of eigenvalues given by

$$D = \begin{bmatrix} \lambda_1 & 0 & \cdots & 0 \\ 0 & \lambda_2 & \cdots & 0 \\ \cdots & \cdots & \cdots & \cdots \\ 0 & 0 & \cdots & \lambda_n \end{bmatrix}.$$

Hence

$$A = A(CC^{-1}) = (AC)C^{-1} = CDC^{-1},$$

and

$$A^n = (CDC^{-1})(CDC^{-1})\cdots(CDC^{-1}) = CD^nC^{-1},$$

since **D** is diagonal.

Finally, the solution of $\dot{x} = Ax$ is

$$x = e^{At} = \sum_{r=0}^{\infty} \frac{A^r t^r}{r!} = \sum_{r=0}^{\infty} \frac{CD^r C^{-1} t^r}{r!} = C \sum_{r=0}^{\infty} \frac{D^r t^r}{r!} C^{-1}$$

$$= C \sum_{r=0}^{\infty} \begin{bmatrix} \lambda_1^r t^r / r! & 0 & \cdots & 0 \\ 0 & \lambda_2^r t^r / r! & \cdots & 0 \\ \cdots & \cdots & \cdots & \cdots \\ 0 & 0 & \cdots & \lambda_n^r t^r / r! \end{bmatrix} C^{-1}$$

$$= C \begin{bmatrix} e^{\lambda_1 t} & 0 & \cdots & 0 \\ 0 & e^{\lambda_2 t} & \cdots & 0 \\ \cdots & \cdots & \cdots & \cdots \\ 0 & 0 & \cdots & e^{\lambda_n t} \end{bmatrix} C^{-1},$$

which is a fundamental matrix $\Phi(t)$ satisfying $\Phi(0) = CIC^{-1} = I$.

Exercise 10.8
Solve the equation

$$\begin{bmatrix} \dot{x}_1 \\ \dot{x}_2 \\ \dot{x}_3 \end{bmatrix} = \begin{bmatrix} 1 & 2 & 1 \\ 2 & 1 & 1 \\ 1 & 1 & 2 \end{bmatrix} \begin{bmatrix} x_1 \\ x_2 \\ x_3 \end{bmatrix}$$

using the exponential matrix method, and find the fundamental matrix $\Phi(t)$ for which $\Phi(0) = I$.

10.10   Stability and the linear approximation for *n*th order autonomous systems

Given the *n*-dimensional nonlinear autonomous system

$$\dot{x} = Ax + h(x), \quad h(0) = 0, \tag{10.48}$$

and its linear approximation near the origin

$$\dot{x} = Ax, \tag{10.49}$$

we wish to identify conditions on h whereby asymptotic stability or instability of (10.49) implies the similar property for the zero solutions of (10.48). We proceed as in Section 10.6 for the two-dimensional case.

Suppose first that the solutions of the linearized system (10.49) are *asymptotically stable*. This implies (Theorem 8.8) that

$$\text{Re}\{\lambda_i\} < 0, \quad i = 1, 2, \ldots, n, \tag{10.50}$$

where λ_i are the eigenvalues of A. We construct a strong Liapunov function for (10.48) which is a quadratic form:

$$V(x) = x^{\text{T}} K x, \tag{10.51}$$

where the matrix K can be determined to make V positive definite and $\dot{V}$ negative definite.

In order also to have any hope of making $\dot{V}$ negative definite for the nonlinear system (10.48), we at least need $\dot{V}$ negative definite for the linearized system (10.49). From (10.51)

$$\dot{V}(x) = x^{\text{T}} (A^{\text{T}} K + K A) x, \tag{10.52}$$

and we shall arrange that

$$A^{\text{T}} K + K A = -I, \tag{10.53}$$

which makes (10.52) negative definite:

$$\dot{V}(x) = -\sum_{i=1}^{n} x_i^2.$$

To achieve this, consider the product $e^{A^{\text{T}} t} e^{A t}$. We have

$$\frac{\text{d}}{\text{d}t}\{e^{A^{\text{T}} t} e^{A t}\} = A^{\text{T}} e^{A^{\text{T}} t} e^{A t} + e^{A^{\text{T}} t} e^{A t} A. \tag{10.54}$$

By (vii) and (viii) of Section 10.9, when $\dot{x} = A x$ is asymptotically stable,

$$\|e^{A t}\| \leq c e^{\gamma t}, \quad c > 0, \ \gamma < 0.$$

Since the eigenvalues of A^{T} are the same as those of A we can choose c so that both

$$\|e^{A t}\| \quad \text{and} \quad \|e^{A^{\text{T}} t}\| \leq c e^{\gamma t}, \quad \text{where } c > 0, \ \gamma < 0. \tag{10.55}$$

This ensures the convergence of the integrals below. From (10.54),

$$\int_0^{\infty} \frac{\text{d}}{\text{d}t}\{e^{A^{\text{T}} t} e^{A t}\}\text{d}t = A^{\text{T}} \left(\int_0^{\infty} e^{A^{\text{T}} t} e^{A t} \, \text{d}t\right) + \left(\int_0^{\infty} e^{A^{\text{T}} t} e^{A t} \, \text{d}t\right) A;$$

and it also equals $[e^{A^{\text{T}} t} e^{A t}]_0^{\infty} = -I$, by (10.55). Comparing this result with (10.53) we see that

$$K = \int_0^{\infty} e^{A^{\text{T}} t} e^{A t} \, \text{d}t \tag{10.56}$$

will satisfy (10.53). The matrix K is symmetrical by (iii) of Section 10.9. Note that (10.56), and hence (10.53), hold whenever the eigenvalues of A are negative, that is, whenever $\dot{x} = Ax$ is asymptotically stable.

We have finally to show that V is positive definite. From (10.51) and (10.56),

$$V(x) = x^{\mathrm{T}} K x = \int_0^\infty (x^{\mathrm{T}} e^{A^{\mathrm{T}} t})(e^{At} x)\, \mathrm{d}t$$

$$= \int_0^\infty (e^{At} x)^{\mathrm{T}} (e^{At} x)\, \mathrm{d}t = \int_0^\infty B(t)^{\mathrm{T}} B(t)\, \mathrm{d}t,$$

where the matrix $B(t) = e^{At} x$. Since $B(t)^{\mathrm{T}} B(t) \geq 0$ for all t (this true for any real square matrix), $V(x)$ is positive definite as required. We have therefore obtained a strong Liapunov function $V(x)$ for the linearized system (10.49).

Theorem 10.16   *If the n-dimensional system $\dot{x} = Ax + h(x)$, with A constant, is regular, and*

 (i) *the zero solution (hence every solution: Theorem 8.1) of $\dot{x} = Ax$ is asymptotically stable;*
 (ii) *$h(0) = 0$, and $\lim_{\|x\| \to 0} \|h(x)\| / \|x\| = 0$,*   (10.57)

then $x(t) = 0$, $t \geq t_0$, for any t_0 is an asymptotically stable solution of

$$\dot{x} = Ax + h(x). \tag{10.58}$$

Proof   We have to show there is a neighbourhood of the origin where $V(x)$ defined by (10.51) and (10.56) is a strong Liapunov function for (10.58). The function V given by

$$V(x) = x^{\mathrm{T}} K x,$$

where

$$K = \int_0^\infty e^{A^{\mathrm{T}} t} e^{At}\, \mathrm{d}t$$

is positive definite when (i) holds. Also for (10.58),

$$\dot{V}(x) = \dot{x}^{\mathrm{T}} K x + x^{\mathrm{T}} K \dot{x}$$

$$= x^{\mathrm{T}} (A^{\mathrm{T}} K + KA) x + h^{\mathrm{T}} K x + x^{\mathrm{T}} K h$$

$$= -x^{\mathrm{T}} x + 2 h^{\mathrm{T}} (x) K x, \tag{10.59}$$

by (10.53) and the symmetry of K.

We have to display a neighbourhood of the origin in which the first term of (10.59) dominates. From the Cauchy–Schwarz inequality (see Appendix C(vi))

$$|2 h^{\mathrm{T}} (x) K x| \leq 2 \|h(x)\| \, \|K\| \, \|x\|. \tag{10.60}$$

By (ii), given any $\varepsilon > 0$ there exists $\delta > 0$ such that

$$\|\boldsymbol{x}\| < \delta \Rightarrow \|\boldsymbol{h}(\boldsymbol{x})\| < \varepsilon\|\boldsymbol{x}\|,$$

so that from (10.60),

$$|2\boldsymbol{h}^{\mathrm{T}}(\boldsymbol{x})\boldsymbol{K}\boldsymbol{x}| \leq 2\varepsilon\|\boldsymbol{K}\|\,\|\boldsymbol{x}\|^2. \tag{10.61}$$

Let ε to be chosen so that

$$\varepsilon < 1/(4\|\boldsymbol{K}\|),$$

then from (10.61),

$$\|\boldsymbol{x}\| < \delta \Rightarrow \|2\boldsymbol{h}^{\mathrm{T}}(\boldsymbol{x})\boldsymbol{K}\boldsymbol{x}\| < \tfrac{1}{2}\|\boldsymbol{x}\|^2 = \tfrac{1}{2}(x_1^2 + x_2^2 + \cdots + x_n^2).$$

Therefore, by (10.59). $\dot{V}(\boldsymbol{x})$ for (10.58) is negative definite on $\|\boldsymbol{x}\| < \delta$. By Theorem 10.12, the zero solution is asymptotically stable. ∎

A similar theorem can be formulated relating to **instability**. Consider the regular system

$$\dot{\boldsymbol{x}} = \boldsymbol{A}\boldsymbol{x} + \boldsymbol{h}(\boldsymbol{x}), \quad \boldsymbol{h}(0) = 0, \tag{10.62}$$

where $\boldsymbol{A}$ is constant. Let $\boldsymbol{C}$ be a nonsingular matrix, and change the variables to $\boldsymbol{u}$ by writing

$$\boldsymbol{x} = \boldsymbol{C}\boldsymbol{u}; \tag{10.63}$$

then the original equation becomes

$$\dot{\boldsymbol{u}} = \boldsymbol{C}^{-1}\boldsymbol{A}\boldsymbol{C}\boldsymbol{u} + \boldsymbol{h}(\boldsymbol{C}\boldsymbol{u}). \tag{10.64}$$

Let $\boldsymbol{C}$ be chosen to reduce $\boldsymbol{A}$ to canonical form, as in Section 10.8. The eigenvalues are unchanged, so whenever a solution $\boldsymbol{x}(t)$ of (10.62) is stable, the corresponding solution $\boldsymbol{u}(t)$ of (10.64) is stable. Since the converse is also true, the same must apply to instability. The zero solution of (10.62) corresponds with that of (10.64): we may therefore investigate the stability of (10.64) in place of (10.62). The same applies to the linearized pair

$$\dot{\boldsymbol{x}} = \boldsymbol{A}\boldsymbol{x}, \quad \dot{\boldsymbol{u}} = \boldsymbol{C}^{-1}\boldsymbol{A}\boldsymbol{C}\boldsymbol{u}. \tag{10.65}$$

Suppose, for simplicity, that the eigenvalues of $\boldsymbol{A}$ are distinct, and that at least one of them has positive real part. Then all solutions of (10.65) are unstable. There are two cases.

(i) Eigenvalues of A real, distinct, with at least one positive

It is known that a real $\boldsymbol{C}$ may be chosen so that

$$\boldsymbol{C}^{-1}\boldsymbol{A}\boldsymbol{C} = \boldsymbol{D}, \tag{10.66}$$

where $\boldsymbol{D}$ is diagonal with the elements equal to the eigenvalues of $\boldsymbol{A}$. Then (10.64) becomes

$$\dot{\boldsymbol{u}} = \boldsymbol{D}\boldsymbol{u} + \boldsymbol{h}(\boldsymbol{C}\boldsymbol{u}). \tag{10.67}$$

(ii) When the eigenvalues of A are distinct, not all real, and at least one has positive real part

It is then known that a real nonsingular matrix G may be chosen so that

$$G^{-1}AG = D^*, \tag{10.68}$$

where D^* is block-diagonal: in place of the pair of complex roots λ, $\bar{\lambda}$ in diagonal positions which would be delivered by a transformation of type (10.62) and (10.65), we have a 'block' of the form

$$D^* = \begin{bmatrix} \cdots & & 0 \\ & \alpha & -\beta \\ & \beta & \alpha \\ 0 & & \cdots \end{bmatrix}, \tag{10.69}$$

where

$$\lambda = \alpha + \mathrm{i}\beta.$$

The argument in the following theorem is hardly altered in this case.

Theorem 10.17   *If A is constant, and*

(i) *the eigenvalues of A are distinct, none are zero, and at least one has positive real part;*
(ii) $\lim_{\|x\|\to 0} \|h(x)\|/\|x\| = 0;$ $\qquad\qquad\qquad\qquad\qquad\qquad\qquad\qquad$ (10.70)

then the zero solutions $x(t) = 0$, $t \ge t_0$, of the regular system

$$\dot{x} = Ax + h(x)$$

are unstable.

Proof   We shall carry out the proof only for the case where the eigenvalues of A are all real. Reduce (10.70) to the form (10.67):

$$\dot{u} = Du + h(Cu),$$

where C is nonsingular and D is diagonal with the eigenvalues of A as its elements, at least one being positive. As explained, it is only necessary to determine the stability of $u(t) = 0$, $t \ge t_0$. Write

$$U(u) = u^{\mathrm{T}} D^{-1} u = \sum_{i=1}^{n} \frac{u_i^2}{\lambda_i}. \tag{10.71}$$

If $\lambda_k > 0$, then $U(u) > 0$ when $u_k \neq 0$ and $u_1 = u_2 = \cdots = u_{k-l} = u_{k+1} = \cdots = u_n = 0$. Therefore (i), (ii), and (iv) of Theorem 10.13 are satisfied.

Also, for (10.70),

$$\dot{U}(u) = \dot{u}^{\mathrm{T}} D^{-1} u = u^{\mathrm{T}} D D^{-1} u + u^{\mathrm{T}} D^{-1} D u + h^{\mathrm{T}} D^{-1} u + u^{\mathrm{T}} D^{-1} h$$

$$= 2(u_1^2 + u_2^2 + \cdots + u_n^2) + 2u^{\mathrm{T}} D^{-1} h(Cu), \tag{10.72}$$

since $DD^{-1} = D^{-1}D = I$ and D is symmetrical. Theorem 10.13 requires $\dot{U}$ to be positive definite in a neighbourhood of the origin. It is clearly sufficient to show that the second term in (10.72) is smaller than the first in a neighbourhood of the origin.

By (iii), given $\varepsilon > 0$ there exists $\delta > 0$ such that

$$\|C\| \, \|u\| < \delta \Rightarrow \|Cu\| < \delta \Rightarrow \|h(Cu)\| < \varepsilon \|Cu\|, \tag{10.73}$$

or alternatively

$$\|u\| < \delta / \|C\| \Rightarrow \|h(Cu)\| < \varepsilon \|C\| \, \|u\|.$$

Therefore (see (10.72)), $\|u\| < \delta / \|C\|$ implies that

$$|2u^{\mathrm{T}} D^{-1} h(Cu)| \leq 2\|u^{\mathrm{T}}\| \, \|D^{-1}\| \, \|h(Cu)\|,$$

$$\leq 2\varepsilon \|D^{-1}\| \, \|C\| \, \|u\|^2. \tag{10.74}$$

If we choose

$$\varepsilon < 1/(2\|D^{-1}\| \, \|C\|),$$

then (10.74) becomes

$$|2u^{\mathrm{T}} D^{-1} h(Cu)| < \|u\|^2$$

for all u close enough to the origin. By referring to (10.72) we see that $\dot{U}(u)$ is positive definite in a neighbourhood of the origin, as required by (iii) of Theorem 10.11. Therefore the zero solution is unstable. ∎

Example 10.12 *Prove that the zero solution of the system*

$$\dot{x}_1 = -x_1 + x_2^2 + x_3^2, \quad \dot{x}_2 = x_1 - 2x_2 + x_1^2, \quad \dot{x}_3 = x_1 + 2x_2 - 3x_3 + x_2 x_3,$$

is uniformly and asymptotically stable.

Here we have

$$A = \begin{bmatrix} -1 & 0 & 0 \\ 1 & -2 & 0 \\ 1 & 2 & -3 \end{bmatrix}$$

and

$$h(x) = [x_2^2 + x_3^2, \ x_1^2, \ x_2 x_3]^{\mathrm{T}}.$$

The eigenvalues of A are $-1, -2, -3$; therefore the zero solution of $\dot{x} = Ax$ is uniformly and asymptotically stable. Also

$$\|h(x)\| = x_2^2 + x_3^2 + x_1^2 + |x_2 x_3|,$$

and
$$\|h(x)\|/\|x\| = (x_2^2 + x_3^2 + x_1^2 + |x_2 x_3|)/(|x_1| + |x_2| + |x_3|),$$

tending to zero with $\|x\|$. By Theorem 10.16 the zero solution of the given system is therefore uniformly asymptotically stable. ●

Example 10.13 *Show that the zero solution of the equation*
$$\dddot{x} - 2\ddot{x} - \dot{x} + 2x = \ddot{x}(x + \dot{x})$$

is unstable.

Write as the equivalent system
$$\dot{x} = y, \quad \dot{y} = z, \quad \dot{z} = -2x + y + 2z + z(x + y).$$

Then
$$A = \begin{bmatrix} 0 & 1 & 0 \\ 0 & 0 & 1 \\ -2 & 1 & 2 \end{bmatrix},$$

with eigenvalues 2, 1, -1, two of which are positive. Also
$$\|h(x)\|/\|x\| = |z(x + y)|/(|x| + |y| + |z|)$$

which tends to zero with $\|x\|$. Therefore, by Theorem 10.17, the zero solution of the given equation is unstable. ●

Exercise 10.9
For the linear system $\dot{x} = Ax$ where
$$A = \begin{bmatrix} -2 & 1 \\ 1 & -2 \end{bmatrix},$$

find the matrix K using
$$K = \int_0^\infty e^{A^T t} e^{At} \, dt.$$

Find the Liapunov function $V(x) = x^T K x$, and check that
$$A^T K + K A = -I.$$

10.11   Special systems

Quadratic systems

Consider the strongly nonlinear system whose lowest-order terms are of the second degree:
$$\dot{x} = X(x, y) \approx a_1 x^2 + 2b_1 xy + c_1 y^2, \tag{10.75}$$
$$\dot{y} = Y(x, y) \approx a_2 x^2 + 2b_2 xy + c_2 y^2. \tag{10.76}$$

Neglecting any higher-order terms, the equations can be written as

$$\dot{x} = [x, y]A_1 \begin{bmatrix} x \\ y \end{bmatrix}, \qquad \dot{y} = [x, y]A_2 \begin{bmatrix} x \\ y \end{bmatrix}, \tag{10.77}$$

where

$$A_1 = \begin{bmatrix} a_1 & b_1 \\ b_1 & c_1 \end{bmatrix}, \qquad A_2 = \begin{bmatrix} a_2 & b_2 \\ b_2 & c_2 \end{bmatrix}.$$

For Theorem 10.13 let

$$U(x, y) = \alpha x + \beta y. \tag{10.78}$$

Then for any non-zero α and β, $U > 0$ for some points in every neighbourhood of the origin. Thus (i), (ii), and (iv) of Theorem 10.13 are satisfied. For (10.77), $\dot{U}$ is given by

$$\dot{U}(x, y) = \alpha\dot{x} + \beta\dot{y} = \alpha[x, y]A_1 \begin{bmatrix} x \\ y \end{bmatrix} + \beta[x, y]A_2 \begin{bmatrix} x \\ y \end{bmatrix}$$

$$= [x, y] \begin{bmatrix} \alpha a_1 + \beta a_2 & \alpha b_1 + \beta b_2 \\ \alpha b_1 + \beta b_2 & \alpha c_1 + \beta c_2 \end{bmatrix} \begin{bmatrix} x \\ y \end{bmatrix}.$$

This quadratic form is positive definite if the conditions

$$\alpha a_1 + \beta a_2 > 0, \Delta(\alpha, \beta) > 0 \tag{10.79}$$

hold, where

$$\Delta(\alpha, \beta) = \alpha^2(a_1c_1 - b_1^2) + \alpha\beta(a_1c_2 + a_2c_1 - 2b_1b_2) + \beta^2(a_2c_2 - b_2^2),$$

$$= (\alpha, \beta) \begin{bmatrix} a_1c_1 - b_1^2 & \frac{1}{2}(a_1c_2 + a_2c_1 - 2b_1b_2) \\ \frac{1}{2}(a_1c_2 + a_2c_1 - 2b_1b_2) & a_2c_2 - b_2^2 \end{bmatrix} \begin{bmatrix} \alpha \\ \beta \end{bmatrix}. \tag{10.80}$$

Consider whether α, β exist satisfying (10.79) and (10.80). On a plane for $\alpha, \beta, a_1\alpha + b_1\beta = 0$ is a straight line, and (10.79) is represented by a half-plane, like that shown in Fig. 10.16(a). Since $\Delta(\alpha, \beta)$ is a quadratic form it is either positive definite, positive on a ray or in a sector as in Fig. 10.16(b), or negative definite. Unless $\Delta(\alpha, \beta)$ is negative definite, or unless it is negative except on the line $a_1\alpha + b_1\beta = 0$ where it is zero, there exist *some* values of α, β making $U(x)$ positive definite and therefore satisfying the final conditions of Theorem 10.13. The zero solution of (10.75) and (10.79) is therefore unstable, except possibly in the particular cases, depending on exact relations between the coefficients, mentioned above. We shall not investigate these cases.

Hamiltonian problems in dynamics

Conservative problems, particularly in dynamics, can be expressed in the form

$$\dot{p}_i = -\frac{\partial \mathcal{H}}{\partial q_i}, \quad \dot{q}_i = \frac{\partial \mathcal{H}}{\partial p_i}, \quad i = 1, 2, \ldots, n, \tag{10.81}$$

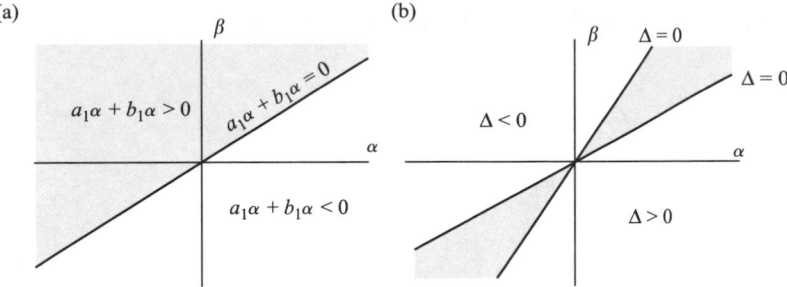

Figure 10.16

where $\mathcal{H}$ is a given function called the Hamiltonian of the system, q_i is a generalized coordinate, and p_i a generalized momentum. The Hamiltonian is defined by

$$\mathcal{H}(\boldsymbol{p},\boldsymbol{q}) = \mathcal{T}(\boldsymbol{p},\boldsymbol{q}) + \mathcal{V}(\boldsymbol{q}),$$

where $\mathcal{T}$ is the kinetic energy and $\mathcal{V}$ the potential energy: it is assumed that $\mathcal{V}(0) = 0$. $\mathcal{T}$, being kinetic energy, is a positive definite quadratic form in p_i, so $\mathcal{H}(0,0) = 0$, since $\mathcal{T}(\boldsymbol{p},0) = 0$.

Suppose that $\boldsymbol{q} = 0$ is a minimum of $\mathcal{V}(\boldsymbol{q})$ so that $\mathcal{V}$, and hence $\mathcal{H}$, is positive definite in a neighbourhood of the origin. Then

$$\dot{\mathcal{H}}(\boldsymbol{p},\boldsymbol{q}) = \sum_{i=1}^{n} \frac{\partial \mathcal{H}}{\partial p_i}\dot{p}_i + \sum_{i=1}^{n} \frac{\partial \mathcal{H}}{\partial q_i}\dot{q}_i = 0$$

by (10.81). Thus $\mathcal{H}$ is a weak Liapunov function for (10.81). Therefore, by Theorem 10.19 the zero solution $\boldsymbol{p} = 0$, $\boldsymbol{q} = 0$, *a position of equilibrium, is stable when it is at a minimum of $\mathcal{V}$.*

Now, suppose the origin is a *maximum* of $\mathcal{V}$, and that $\mathcal{V}$ has the expansion

$$\mathcal{V}(\boldsymbol{q}) = P_N(\boldsymbol{q}) + P_{N+1}(\boldsymbol{q}) + \cdots, \quad N \geq 2$$

where $P_N(\boldsymbol{q})$ is a homogeneous polynomial of degree $N \geq 1$ in the q_i. Consider a function $U(\boldsymbol{p},\boldsymbol{q})$, of the type in Theorem 10.13, defined by

$$U(\boldsymbol{p},\boldsymbol{q}) = \sum_{i=1}^{n} p_i q_i.$$

Then U satisfies conditions (i), (ii), and (iv) of Theorem 10.13. Also

$$\dot{U}(p,q) = \sum_{i=1}^{n} p_i \dot{q}_i + \sum_{i=1}^{n} \dot{p}_i q_i$$

$$= \sum_{i=1}^{n} p_i \frac{\partial \mathcal{H}}{\partial p_i} - \sum_{i=1}^{n} q_i \frac{\partial \mathcal{H}}{\partial q_i}$$

$$= \sum_{i=1}^{n} p_i \frac{\partial \mathcal{T}}{\partial p_i} - \sum_{i=1}^{n} q_i \frac{\partial \mathcal{T}}{\partial q_i} - \sum_{i=1}^{n} q_i \frac{\partial \mathcal{V}}{\partial q_i}$$

$$= 2\mathcal{T} - \sum_{i=1}^{n} \frac{\partial \mathcal{T}}{\partial q_i} - N P_N(q) - (N+1) P_{N+1}(q) + \cdots,$$

where Euler's theorem on homogeneous functions has been applied to $\mathcal{T}, P_N, P_{N+1}, \ldots$. The dominating terms in the series near $p = q = 0$ are $2\mathcal{T} - N P_N(q)$. Since, by hypothesis, P_N has a maximum at $q = 0$ and $\mathcal{T}$ is positive definite, $\dot{U}$ is positive definite in a neighbourhood of the origin. Therefore, by Theorem 10.13, *the equilibrium is unstable when it is at a maximum of* $\mathcal{V}$.

The Liénard equation

This is the equation

$$\ddot{x} + f(x)\dot{x} + g(x) = 0, \tag{10.82}$$

or the equivalent system in the Liénard plane

$$\dot{x} = y - F(x), \qquad \dot{y} = -g(x); \tag{10.83}$$

where

$$F(x) = \int_0^x f(u)\, du.$$

Suppose that f and g are continuous and that

(i) $f(x)$ is positive in a neighbourhood of the origin, except at $x = 0$, where it is zero;

(ii) $g(x)$ is positive / negative when x is positive / negative (implying $g(0) = 0$).

   Now let

$$G(x) = \int_0^x g(u)\, du, \tag{10.84}$$

and consider the function

$$V(x, y) = G(x) + \tfrac{1}{2} y^2,$$

as a possible weak Liapunov function. It is clearly positive definite since $G(x) > 0$ for $x \neq 0$. Also, by (10.83),

$$\dot{V}(x, y) = g(x)\dot{x} + y\dot{y} = -g(x)F(x).$$

This is negative semidefinite. The zero solution is therefore stable.

Exercise 10.10
Using the Liapunov function $V(x, y) = G(x) + \frac{1}{2}y^2$, show that the origin of the system

$$\ddot{x} + [1 + f(x)]\dot{x} + g(x) + F(x) = 0, \qquad \dot{x} = y - F(x),$$

is asymptotically stable: $f(x)$, $F(x)$, and $g(x)$ satisfy the conditions of the Liénard equation above. Apply the result to

$$\ddot{x} + (1 + x^2)\dot{x} + x + \tfrac{1}{3}x^3 = 0.$$

Problems

10.1 Find a simple V or U function (Theorems 10.5, 10.11 or 10.13) to establish the stability or instability respectively of the zero solution of the following equations:

   (i) $\dot{x} = -x + y - xy^2$, $\dot{y} = -2x - y - x^2y$;

  (ii) $\dot{x} = y^3 + x^2y$, $\dot{y} = x^3 - xy^2$;

 (iii) $\dot{x} = 2x + y + xy$, $\dot{y} = x - 2y + x^2 + y^2$;

 (iv) $\dot{x} = -x^3 + y^4$, $\dot{y} = -y^3 + y^4$;

  (v) $\dot{x} = \sin y$, $\dot{y} = -2x - 3y$;

 (vi) $\dot{x} = x + e^{-y} - 1$, $\dot{y} = x$;

 (vii) $\dot{x} = e^x - \cos y$, $\dot{y} = y$;

(viii) $\dot{x} = \sin(y + x)$, $\dot{y} = -\sin(y - x)$;

 (ix) $\ddot{x} = x^3$;

  (x) $\dot{x} = x + 4y$, $\dot{y} = -2x - 5y$;

 (xi) $\dot{x} = -x + 6y$, $\dot{y} = 4x + y$.

10.2 Show that α may be chosen so that $V = x^2 + \alpha y^2$ is a strong Liapunov function for the system

$$\dot{x} = y - \sin^3 x, \qquad \dot{y} = -4x - \sin^3 y.$$

10.3 Find domains of asymptotic stability for the following systems, using $V = x^2 + y^2$:

 (i) $\dot{x} = -\frac{1}{2}x(1 - y^2)$, $\dot{y} = -\frac{1}{2}y(1 - x^2)$;

(ii) $\dot{x} = y - x(1 - x)$, $\dot{y} = -x$.

10.4 Find a strong Liapunov function at $(0, 0)$ for the system

$$\dot{x} = x(y - b), \qquad \dot{y} = y(x - a)$$

and confirm that all solutions starting in the domain $(x/a)^2 + (y/b)^2 < 1$ approach the origin.

10.5 Show that the origin of the system

$$\dot{x} = xP(x, y), \qquad \dot{y} = yQ(x, y)$$

is asymptotically stable when $P(x, y) < 0$, $Q(x, y) < 0$ in a neighbourhood of the origin.

10.6 Show that the zero solution of

$$\dot{x} = y + xy^2, \qquad \dot{y} = x + x^2y$$

is unstable.

10.7 Investigate the stability of the zero solution of

$$\dot{x} = x^2 - y^2, \qquad \dot{y} = -2xy$$

by using the function $U(x, y) = \alpha xy^2 + \beta x^3$ for suitable constants α and β.

10.8 Show that the origin of the system

$$\dot{x} = -y - x\sqrt{(x^2 + y^2)}, \qquad \dot{y} = x - y\sqrt{(x^2 + y^2)}$$

is a centre in the linear approximation, but in fact is a stable spiral. Find a Liapunov function for the zero solution.

10.9 Euler's equations for a body spinning freely about a fixed point under no forces are

$$A\dot{\omega}_1 - (B - C)\omega_2\omega_3 = 0,$$
$$B\dot{\omega}_2 - (C - A)\omega_3\omega_1 = 0,$$
$$C\dot{\omega}_3 - (A - B)\omega_1\omega_2 = 0,$$

where A, B, and C (all different) are the principal moments of inertia, and $(\omega_1, \omega_2, \omega_3)$ is the spin of the body in principal axes fixed in the body. Find all the states of steady spin of the body.

Consider perturbations about the steady state $(\omega_0, 0, 0)$ by putting $\omega_1 = \omega_0 + x_1$, $\omega_2 = x_2$, $\omega_3 = x_3$, and show that the linear approximation is

$$\dot{x}_1 = 0, \quad \dot{x}_2 = \{(C - A)/B\}\omega_0 x_3, \quad \dot{x}_3 = \{(A - B)/C\}\omega_0 x_2.$$

Deduce that this state is unstable if $C < A < B$ or $B < A < C$.

Show that

$$V = \{B(A - B)x_2^2 + C(A - C)x_3^2\} + \{Bx_2^2 + Cx_3^2 + A(x_1^2 + 2\omega_0 x_1)\}^2$$

is a Liapunov function for the case when A is the largest moment of inertia, so that this state is stable. Suggest a Liapunov function which will establish the stability of the case in which A is the smallest moment of inertia. Are these states asymptotically stable?

Why would you expect V as given above to be a first integral of the Euler equations? Show that each of the terms in braces is such an integral.

10.10 Show that the zero solution of the equation

$$\ddot{x} + h(x, \dot{x})\dot{x} + x = 0$$

is stable if $h(x, y) \geq 0$ in a neighbourhood of the origin.

10.11 The n-dimensional system

$$\dot{x} = \mathbf{grad}\ W(x)$$

has an isolated equilibrium point at $x = 0$. Show that the zero solution is asymptotically stable if W has a local minimum at $x = 0$. Give a condition for instability of zero solution.

10.12 A particle of mass m and position vector $r = (x, y, z)$ moves in a potential field $W(x, y, z)$, so that its equation of motion is $m\ddot{r} = -\text{grad } W$. By putting $\dot{x} = u$, $\dot{y} = v$, $\dot{z} = w$, express this in terms of first-order derivatives. Suppose that W has a minimum at $r = 0$. Show that the origin of the system is stable, by using the Liapunov function

$$V = W + \tfrac{1}{2}m(u^2 + v^2 + w^2).$$

What do the level curves of V represent physically? Is the origin asymptotically stable?
   An additional non-conservative force $f(u, v, w)$ is introduced, so that

$$m\ddot{r} = -\text{grad } W + f.$$

Use the same Liapunov function to give a sufficient condition for f to be of frictional type.

10.13 Use the test for instability to show that if $\dot{x} = X(x, y)$, $\dot{y} = Y(x, y)$ has an equilibrium point at the origin, then the zero solution is unstable if there exist constants α and β such that

$$\alpha X(x, y) + \beta Y(x, y) > 0$$

in a neighbourhood of the origin and is zero at the origin.

10.14 Use the result of Problem 10.13 to show that the origin is unstable for each of the following:

   (i) $\dot{x} = x^2 + y^2$, $\dot{y} = x + y$;
   (ii) $\dot{x} = y \sin y$, $\dot{y} = xy + x^2$;
   (iii) $\dot{x} = y^{2m}$, $\dot{y} = x^{2n}$ (m, n positive integers).

10.15 For the system $\dot{x} = y$, $\dot{y} = f(x, y)$, where $f(0, 0) = 0$, show that V given by

$$V(x, y) = \tfrac{1}{2}y^2 - \int_0^x f(u, 0)\, du$$

is a weak Liapunov function for the zero solution when

$$\{f(x, y) - f(x, 0)\}y \le 0, \quad \int_0^x f(u, 0)\,du < 0,$$

in a neighbourhood of the origin.

10.16 Use the result of Problem 10.15 to show the stability of the zero solutions of the following:

   (i) $\ddot{x} = -x^3 - x^2\dot{x}$;
   (ii) $\ddot{x} = -x^3/(1 - x\dot{x})$;
   (iii) $\ddot{x} = -x + x^3 - x^2\dot{x}$.

10.17 Let

$$\dot{x} = -\alpha x + \beta f(y), \qquad \dot{y} = \gamma x - \delta f(y),$$

where $f(0) = 0$, $yf(y) > 0 (y \ne 0)$, and $\alpha\delta > 4\beta\gamma$, where $\alpha, \beta, \gamma, \delta$ are positive. Show that, for suitable values of A and B,

$$V = \tfrac{1}{2}Ax^2 + B\int_0^y f(u)\, du$$

is a strong Liapunov function for the zero solutions (and hence that these are asymptotically stable).

10.18 A particle moving under a central attractive force $f(r)$ per unit mass has the equations of motion

$$\dot{r} - r\dot{\theta}^2 = f(r), \quad \frac{d}{dt}(r^2\dot{\theta}) = 0.$$

For a circular orbit, $r = a$, show that $r^2 \dot{\theta} = h$, a constant, and $h^2 + a^3 f(a) = 0$. The orbit is subjected to a small radial perturbation $r = a + \rho$, in which h is kept constant. Show that the equation for ρ is

$$\ddot{\rho} - \frac{h^2}{(a+\rho)^3} - f(a+\rho) = 0.$$

Show that

$$V(\rho, \dot{\rho}) = \frac{1}{2}\dot{\rho}^2 + \frac{h^2}{2(a+\rho)^2} - \int_0^\rho f(a+u)\,du - \frac{h^2}{2a^2}$$

is a Liapunov function for the zero solution of this equation provided that $3h^2 > a^4 f'(a)$, and that the gravitational orbit is stable in this sense.

10.19 Show that the following Liénard-type equations have zero solutions which are asymptotically stable:

(i) $\ddot{x} + |x|(\dot{x} + x) = 0$;

(ii) $\ddot{x} + (\sin x / x)\dot{x} + x^3 = 0$;

(iii) $\dot{x} = y - x^3, \dot{y} = -x^3$.

10.20 Give a geometrical account of Theorem 10.13.

10.21 For the system

$$\dot{x} = f(x) + \beta y, \quad \dot{y} = \gamma x + \delta y, \quad (f(0) = 0),$$

establish that V given by

$$V(x, y) = (\delta x - \beta y)^2 + 2\delta \int_0^x f(u)\,du - \beta\gamma x^2$$

is a strong Liapunov function for the zero solution when, in some neighbourhood of the origin,

$$\delta \frac{f(x)}{x} - \beta\gamma > 0, \quad \frac{f(x)}{x} + \delta < 0$$

for $x \neq 0$ (Barbashin 1970).

Deduce that for initial conditions in the circle $x^2 + y^2 < 1$, the solutions of the system

$$\dot{x} = -x^3 + x^4 + y, \quad \dot{y} = -x,$$

tend to zero.

10.22 For the system $\dot{x} = f(x) + \beta y, \dot{y} = g(x) + \delta y, f(0) = g(0) = 0$, show that V given by

$$V(x, y) = (\delta x - \beta y)^2 + 2 \int_0^x \{\delta f(u) - \beta g(u)\}du$$

is a strong Liapunov function for the zero solution when, in some neighbourhood of the origin,

$$\{\delta f(x) - \beta g(x)\}x > 0, \quad xf(x) + \delta x^2 < 0$$

for $x \neq 0$ (Barbashin 1970).

Deduce that the zero solution of the system $\dot{x} = -x^3 + 2x^4 + y, \dot{y} = -x^4 - y$ is asymptotically stable. Show how to find a domain of initial conditions from which the solutions tend to the origin. Sketch phase paths and a domain of asymptotic stability of the origin.

10.23 Consider van der Pol's equation, $\dot{x} + \varepsilon(x^2 - 1)\dot{x} + x = 0$, for $\varepsilon < 0$, in the Liénard phase plane, eqn (10.83):

$$\dot{x} = y - \varepsilon(\tfrac{1}{3}x^3 - x), \quad \dot{y} = -x.$$

Show that, in this plane, $V = \frac{1}{2}(x^2 + y^2)$ is a strong Liapunov function for the zero solution, which is therefore asymptotically stable. Show that all solutions starting from initial conditions inside the circle $x^2 + y^2 = 3$ tend to the origin (and hence the limit cycle lies outside this region for every $\varepsilon < 0$. Sketch this 'domain of asymptotic stability' in the ordinary phase plane with $\dot{x} = y$.

10.24 Show that the system $\dot{x} = -x - xy^2$, $\dot{y} = -y - x^2y$ is globally asymptotically stable, by guessing a suitable Liapunov function.

10.25 Assuming that the conditions of Problem 10.22 are satisfied, obtain further conditions which ensure that the system is globally asymptotically stable.
Show that the system $\dot{x} = y - x^3$, $\dot{y} = -x - y$ is globally asymptotically stable.

10.26 Assuming that the conditions of Problem 10.23 are satisfied, obtain further conditions which ensure that the system is globally asymptotically stable.
Show that the system $\dot{x} = -x^3 - x + y$, $\dot{y} = -x^3 - y$ is globally asymptotically stable.

10.27 Give conditions on the functions f and g of the Liénard equation, $\ddot{x} + f(x)\dot{x} + g(x) = 0$ which ensure that the corresponding system $\dot{x} = y - F(x)$, $\dot{y} = -g(x)$ (Section 10.11) is globally asymptotically stable.
Show that all solutions of the equation $\ddot{x} + x^2\dot{x} + x^3 = 0$ tend to zero.

10.28 (Zubov's method.) Suppose that a function $W(x, y)$, negative definite in the whole plane, is chosen as the time derivative $\dot{V}$ of a possible Liapunov function for a system $\dot{x} = X(x, y)$, $\dot{y} = Y(x, y)$, for which the origin is an asymptotically stable equilibrium point. Show that $V(x, y)$ satisfies the linear partial differential equation

$$X\frac{\partial V}{\partial x} + Y\frac{\partial V}{\partial y} = W$$

with $V(0, 0) = 0$.
Show also that for the path $x(t)$, $y(t)$ starting at (x_0, y_0) at time t_0

$$V\{x(t), y(t)\} - V(x_0, y_0) = \int_{t_0}^{t} W\{x(u), y(u)\}du.$$

Deduce that the boundary of the domain of asymptotic stability (the domain of initial conditions from which the solutions go into the origin) is the set of points (x, y) for which $V(x, y)$ is infinite, by considering the behaviour of the integral as $t \to \infty$, firstly when (x_0, y_0) is inside this domain and then when it is outside. (Therefore the solution $V(x, y)$ of the partial differential equation above could be used to give the boundary of the domain directly. However, solving this equation is equivalent in difficulty to finding the paths: the characteristics are in fact the paths themselves.)

10.29 For the system

$$\dot{x} = X(x, y) = -\tfrac{1}{2}x(1 - x^2)(1 - y^2),$$

$$\dot{y} = Y(x, y) = -\tfrac{1}{2}y(1 - x^2)(1 - y^2)$$

show that the Liapunov function $V = x^2 + y^2$ leads to $\dot{V} = -(x^2 + y^2)(1 - x^2)(1 - y^2)$ and explain why the domain of asymptotic stability (see Problem 10.30) contains at least the unit circle $x^2 + y^2 = 1$.
Alternatively, start with $\dot{V} = -x^2 - y^2 - 2x^2y^2$ and obtain V from the equation

$$X\frac{\partial V}{\partial x} + Y\frac{\partial V}{\partial y} = \dot{V}, \quad V(0, 0) = 0$$

(see Problem 10.30). It is sufficient to verify that $V = -\log\{(1 - x^2)(1 - y^2)\}$. Explain why the square $|x| < 1$, $|y| < 1$ is the complete domain of asymptotic stability for the zero solution.

10.30 Use the series definition of e^{At} to prove the following properties of the exponential function of a matrix:

   (i) $e^{A+B} = e^A e^B$ if $AB = BA$;

   (ii) e^A is nonsingular and $(e^A)^{-1} = e^{-A}$;

   (iii) $\dfrac{d}{dt} e^{At} = Ae^{At} = e^{At}A$;

   (iv) $(e^{At})^T = e^{A^T t}$.

10.31 Let the distinct eigenvalues of the $n \times n$ matrix A be $\lambda_1, \lambda_2, \ldots, \lambda_n$. Show that, whenever $\gamma > \max_{1 \le i \le n} \operatorname{Re}(\lambda_i)$, there exists a constant $c > 0$ such that $\|e^{At}\| \le c e^{\gamma t}$.

10.32 Express the solution of

$$\begin{bmatrix} \dot{x} \\ \dot{y} \end{bmatrix} = \begin{bmatrix} 0 & 1 \\ 1 & 0 \end{bmatrix} \begin{bmatrix} x \\ y \end{bmatrix}, \quad x(0) = 0,\ \dot{x}(0) = 1,$$

   in matrix form, and, by evaluating the exponential obtain the ordinary form from it.

10.33 Evaluate

$$K = \int_0^\infty e^{A^T t} e^{At}\, dt, \quad \text{where } A = \frac{1}{2} \begin{bmatrix} -3 & 1 \\ 1 & -3 \end{bmatrix},$$

   and confirm that $A^T K + KA = -I$.

10.34 Show that, if B is an $n \times n$ matrix, $A = e^B$ and C is nonsingular, then $C^{-1}AC = e^{C^{-1}BC}$.

10.35 (i) Let $L = \operatorname{diag}(\lambda_1, \lambda_2, \ldots, \lambda_n)$, where λ_i are distinct and $\lambda_i \ne 0$ for any i. Show that $L = e^D$, where $D = \operatorname{diag}(\ln \lambda_1, \ln \lambda_2, \ldots, \ln \lambda_n)$. Deduce that for nonsingular A with distinct eigenvalues, $A = e^B$ for some matrix B.

   (ii) Show that, for the system $\dot{x} = P(t)x$, where $P(t)$ has period T, and E (eqn (9.15)) is nonsingular with distinct eigenvalues, every fundamental matrix has the form $\Phi(t) = R(t)e^{Mt}$, where $R(t)$ has period T, and M is a constant matrix. (See the result of Problem 10.35.)

10.36 Using the results from Problem 10.36, show that the transformation $x = R(t)y$ reduces the system $\dot{x} = P(t)x$, where $P(t)$ has period T, to the form $\dot{y} = My$, where M is a constant matrix.

The existence of periodic solutions

Suppose that the phase diagram for a differential equation contains a single, unstable equilibrium point and a limit cycle surrounding it, as in the case of the van der Pol equation. Then in practice all initial states lead to the periodic oscillation represented by the limit cycle. In such cases the limit cycle, is the principal feature of the system from the practical point of view, and it is desirable to be able to decide with certainty whether it is there or not. Hitherto our attitude to this question has been intuitive; we assemble qualitative evidence supporting the existence of a limit cycle, from energy considerations or geometrical arguments, then attempt to estimate the radius by the methods of Chapters 4, 5, and 7, a definite result being taken as further confirmation that the limit cycle is really there. The present chapter contains theorems and methods for proving positively the existence or non-existence of limit cycles and centres for certain types of equation. The cases chosen can be interpreted physically as involving a balance between energy gain and loss on various regions in the phase plane; they include the type discussed in Sections 1.5 and 4.1; also others with different symmetry. The latter serve as a guide to intuition for related types which do not quite fit the conditions of the theorems. Further theorems can be found, for example, in Andronov *et al.* (1973a) and Cesari (1971).

11.1 The Poincaré–Bendixson theorem and periodic solutions

We shall be concerned with second-order autonomous differential equations. A limit cycle is an isolated periodic motion, which appears as an isolated closed curve in the phase plane. We take Fig. 3.18 as showing the general nature of a limit cycle; neighbouring paths resemble spirals which approach, or recede from, the limit cycle, though the appearance of actual limit cycles may be somewhat different (see, e.g., Fig. 3.24). In Section 3.4, we gave criteria whereby it can sometimes be shown that *no* limit cycle can exist in a certain region, and also a necessary condition for a limit cycle (the index requirement), but we have no tests giving *sufficient* conditions for existence. In particular we lack information when the differential equations do not contain a small parameter.

We require the **Poincaré–Bendixson theorem**, which describes the ultimate behaviour on $t \to \infty$ of a phase path which enters and remains on a closed bounded region. This theorem was stated and described in Section 10.2 (Theorem 10.3), but in order to preserve continuity we state it again here:

Theorem 11.1 (The Poincaré–Bendixson theorem) *Let $\mathcal{R}$ be a closed, bounded region consisting of nonsingular points of a plane system $\dot{x} = X(x)$ such that some positive half-path $\mathcal{H}$ of the system lies entirely within $\mathcal{R}$. Then either $\mathcal{H}$ is itself a closed path, or it approaches a closed path, or it terminates at an equilibrium point.* ∎

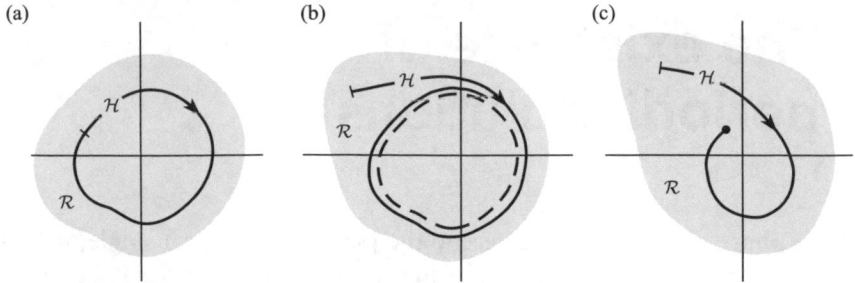

Figure 11.1   Possible behaviour of half-paths restricted to a bounded region $\mathcal{R}$ shown shaded: (a) closed path; (b) path approaching a closed path; (c) path approaching an equilibrium point.

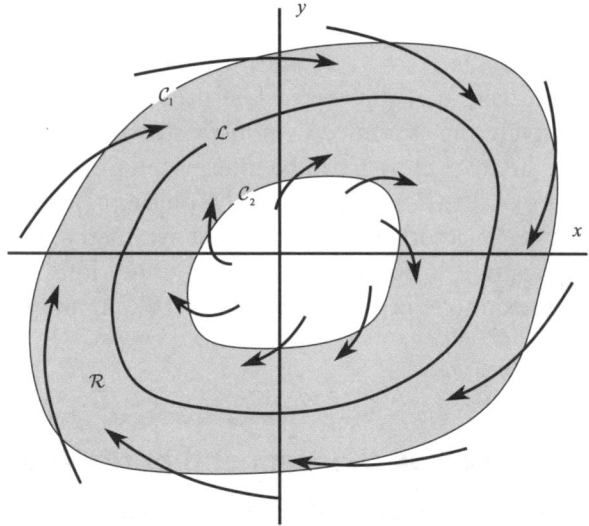

Figure 11.2   All paths are directed into the shaded region $\mathcal{R}$ as they cross $\mathcal{C}_1$ and $\mathcal{C}_2$. There exists at least one limit cycle $\mathcal{L}$ in $\mathcal{R}$.

If we can isolate a region from which some path cannot escape, the theorem describes what may happen to it. The possibilities are illustrated in Fig. 11.1. The theorem implies, in particular, that if $\mathcal{R}$ contains no equilibrium points, and some half-path $\mathcal{H}$ remains in $\mathcal{R}$, then $\mathcal{R}$ must contain a periodic solution.

The theorem can be used in the following way. Suppose (Fig. 11.2) that we can find two closed curves, $\mathcal{C}_1$ and $\mathcal{C}_2$, with $\mathcal{C}_2$ inside $\mathcal{C}_1$, such that *all* paths crossing $\mathcal{C}_1$ and $\mathcal{C}_2$ enter the annular region $\mathcal{R}$ between them. Then no path which enters $\mathcal{R}$ can ever escape from $\mathcal{R}$. If, further, we know that $\mathcal{R}$ has no equilibrium points in it, then the theorem predicts at least one closed path $\mathcal{L}$ somewhere in $\mathcal{R}$. Evidently $\mathcal{L}$ must wrap round the inner curve $\mathcal{C}_2$ as shown, for the index of a closed path is 1 and it must therefore have an equilibrium point interior to it, but $\mathcal{R}$ contains no equilibrium points. There must exist suitable equilibrium points interior to $\mathcal{C}_2$ for all this to be possible.

The same result is true if paths are all outward from $\mathcal{R}$ across $\mathcal{C}_1$ and $\mathcal{C}_2$ (by reversing the time variable).

The practical difficulty is in finding, for a given system, a suitable $\mathcal{C}_1$ and $\mathcal{C}_2$ to substantiate in this way the existence of a limit cycle. There is a similarity between this process and that of finding Liapunov functions (Chapter 10).

Example 11.1 *Show that the system*

$$\dot{x} = x - y - (x^2 + \tfrac{3}{2}y^2)x, \quad \dot{y} = x + y - (x^2 + \tfrac{1}{2}y^2)y$$

has a periodic solution.

The system has an equilibrium point, at $(0, 0)$. Any other equilibrium points must satisfy

$$y = x - (x^2 + \tfrac{3}{2}y^2)x, \qquad x = -y + (x^2 + \tfrac{1}{2}y^2)y.$$

A sketch of these curves indicates that they cannot intersect.

We shall try to find two circles, centred on the origin, with the required properties. In Fig. 11.3(a), $\mathbf{n} = (x, y)$ is a normal, pointing outward at P from the circle radius r, and $\mathbf{X} = (X, Y)$ is in the direction of the path through P. Also $\cos\phi = \mathbf{n} \cdot \mathbf{X}/|\mathbf{n}||\mathbf{X}|$ and therefore $\mathbf{n} \cdot \mathbf{X}$ is positive or negative according to whether $\mathbf{X}$ is pointing away from, or towards, the interior of the circle. We have

$$\begin{aligned}
\mathbf{n} \cdot \mathbf{X} &= xX + yY = x^2 + y^2 - x^4 - \tfrac{1}{2}y^4 - \tfrac{5}{2}x^2y^2 \\
&= r^2 - r^4 + \tfrac{1}{2}y^2(y^2 - x^2) \\
&= r^2 - r^4(1 + \tfrac{1}{4}\cos 2\theta - \tfrac{1}{4}\cos^2 2\theta).
\end{aligned} \qquad (11.1)$$

When, for example, $r = \tfrac{1}{2}$, this is positive for all θ and so all paths are directed outwards on this circle, and when $r = 2$ it is negative, with all paths directed inwards. Therefore, somewhere between $r = \tfrac{1}{2}$ and $r = 2$, there is at least one closed path, corresponding to periodic solution, since there are no equilibrium points in this annulus. Figure 11.3(b) shows the computed stable limit cycle.

We can look for the pair of circles which pin down *the narrowest annular region in which we can predict that a closed path exists*. The interval of r in which a closed path might lie is that for which, given a value of r, $\mathbf{X}$ points inward on some points of the circle, outward on others; that is to say, values of r in this region give

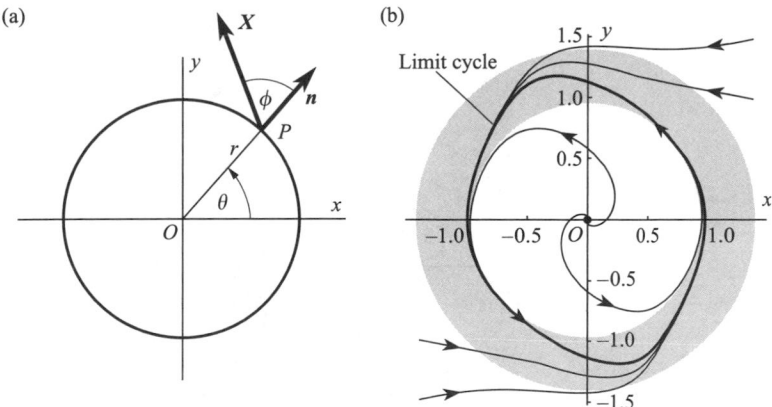

Figure 11.3 (a) Notation for Example 11.1. (b) Showing the computed limit cycle, and the region between the inner and outer 'cycles with contact', $0.97 \le r \le 1.41$, shown shaded.

circles to which X is tangential at some point ('cycles with contact': Andronov *et al.* 1973). For such values of r the equation

$$r^2 - r^4 \left(1 + \tfrac{1}{4}\cos 2\theta - \tfrac{1}{4}\cos^2 2\theta \right) = 0$$

has a solution (r, θ). Write this in the form

$$4 \left(\frac{1}{r^2} - 1 \right) = \cos 2\theta - \cos^2 2\theta.$$

Investigation of the maximum and minimum values of the right-hand side shows that

$$-2 \le \cos 2\theta - \cos^2 2\theta \le \tfrac{1}{4}$$

so that solutions will exist for values of the left-hand side which lie in this range, that is, when

$$\tfrac{4}{\sqrt{17}} \le r \le \sqrt{2}, \quad \text{or} \quad 0.97 \le r \le 1.41,$$

which is shown as the shaded annular region in Fig. 11.3(b). This figure also shows the computed limit cycle and neighbouring phase paths. ●

Another point of view, closely resembling that of the Liapunov functions of Chapter 10, is the following. Let

$$v(x, y) = c > 0, \quad c_1 < c < c_2,$$

be a 'band' of closed curves for some range of c including $c = c_0$ (the curve $\mathcal{C}_0$), the outer curves corresponding to larger c (Fig. 11.4). If, for every P,

$$\left[\frac{dv}{dt} \right]_P = \dot{v}(x, y) = X \frac{\partial v}{\partial x} + Y \frac{\partial v}{\partial y} > 0, \quad P \in \mathcal{C}_0,$$

then all paths point outwards on $\mathcal{C}_0$ (see Fig. 11.4). Similarly, we may seek a closed curve over which all paths are inward.

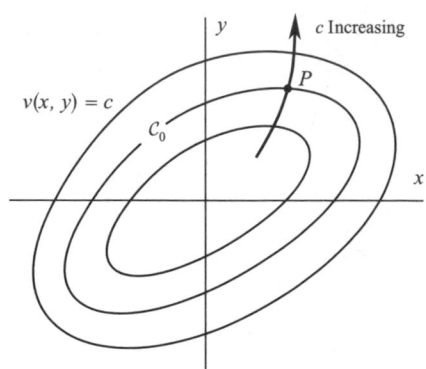

Figure 11.4

Example 11.2   *Show that the system*

$$\dot{x} = y - x^3 + x, \quad \dot{y} = -x - y^3 + y$$

has at least one closed path in the phase plane.

A sketch of the cubic curves $y = -x + x^3$, $x = y + y^3$ indicates that they intersect only at $(0, 0)$ which implies that the origin is the only equilibrium point.

Put $v(x, y) = x^2 + y^2$. Then

$$\frac{dv}{dt} = 2x(y - x^3 + x) + 2y(-x - y^3 + y) = 2(x^2 + y^2 - x^4 - y^4).$$

If $x^2 + y^2 < 1$ then

$$x^2 + y^2 > (x^2 + y^2)^2 = x^4 + 2x^2 y^2 + y^4 \geq x^4 + y^4.$$

Hence dv/dt is strictly positive on the circle $x^2 + y^2 = c$ for any c such that $0 < c < 1$. On the other hand, for $x^2 + y^2 > 2$,

$$2(x^4 + y^4) \geq (x^2 + y^2)^2 > 2(x^2 + y^2).$$

As a result dv/dt is strictly negative on the circle $x^2 + y^2 = c$ for any $c > 2$. Therefore the system has at least one closed path lying between the two concentric circles.   ●

It is generally difficult to find curves $v(x, y) = c$ with the required properties. It may be possible, however, to construct supporting arguments which are not rigorous, as in the following example.

Example 11.3   *Show that the system*

$$\dot{x} = y, \quad \dot{y} = -4x - 5y + \frac{6y}{1 + x^2}$$

has a periodic solution.

These equations are a simplified version of equations for a tuned-grid vacuum-tube circuit (Andronov and Chaikin 1949). The only equilibrium point on the finite plane is at $x = 0, y = 0$. Near the origin, the linear approximation is $\dot{x} = y, \dot{y} = -4x + y$, an unstable spiral. As in the previous example, the existence of a Liapunov function for the stable spiral (see Chapter 10) guarantees that there is in the present case a closed curve surrounding the origin over which the paths point outward.

To find what happens at a great distance use, for example, the mapping $z = 1/x, u = y/x$ (eqn (3.15)); the system becomes

$$\dot{z} = -vz, \quad \dot{u} = -4 - 5u - u^2 + 6uz^2/(z^2 + 1).$$

The equilibrium points on the horizon $z = 0$ are at $u = -1$ and $u = -4$ and these are unstable.

The situation is better viewed in the diametrical plane $\mathcal{P}^*$ of Section 3.3. The picture is as in Fig. 11.5, giving a reasonable assurance of a periodic solution, shown by the heavy line in the figure.   ●

The Poincaré–Bendixson principle can be employed to obtain theorems covering broad types of differential equation, of which the following is an example.

Theorem 11.2   *The differential equation*

$$\ddot{x} + f(x, \dot{x})\dot{x} + g(x) = 0 \qquad (11.2)$$

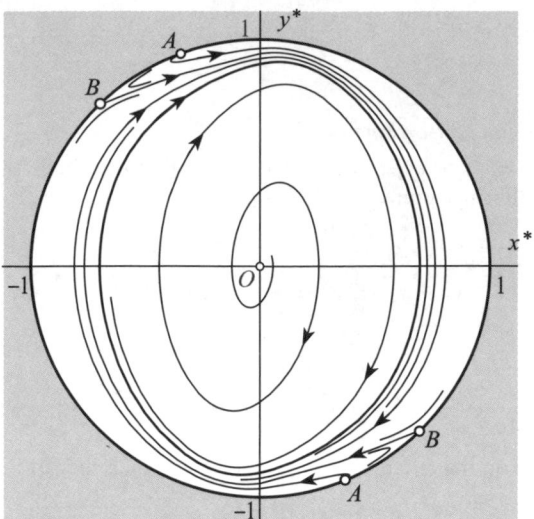

Figure 11.5 Phase diagram plotted on to a disc in the (x^*, y^*) plane (see Section 3.3) showing a stable limit cycle. At infinity the equilibrium points marked A are saddle point and those marked B unstable nodes.

(*the Liénard equation*), *or the equivalent system*

$$\dot{x} = y, \quad \dot{y} = -f(x, y)y - g(x).$$

where f and g are continuous, has at least one periodic solution under the following conditions:

(i) *there exists $a > 0$ such that $f(x, y) > 0$ when $x^2 + y^2 > a^2$;*
(ii) *$f(0, 0) < 0$ (hence $f(x, y) < 0$ in a neighbourhood of the origin);*
(iii) *$g(0) = 0, g(x) > 0$ when $x > 0$, and $g(x) < 0$ when $x < 0$;*
(iv) *$G(x) = \int_0^x g(u)\, du \to \infty$ as $x \to \infty$.*

Proof (iii) implies that there is a single equilibrium point, at the origin.
  Consider the function

$$\mathcal{E}(x, y) = \tfrac{1}{2}y^2 + G(x). \tag{11.3}$$

This represents the energy of the system (potential plus kinetic) when it is regarded, say, as representing a spring–particle system with external forces. Clearly, $G(0) = 0, G(x) > 0$ when $x \neq 0$, and G is monotonic increasing to infinity (by (iv)); and is continuous. Therefore $\mathcal{E}(0, 0) = 0$, and $\mathcal{E}(x, y) > 0$ for $x \neq 0$ and $y \neq 0$ ($\mathcal{E}$ is positive definite). Also $\mathcal{E}$ is continuous and increases monotonically in every radial direction from the origin. Therefore (Fig. 11.6) the family of contours of $\mathcal{E}$ with parameter $c > 0$:

$$\mathcal{E}(x, y) = c \tag{11.4}$$

consists of simple closed curves encircling the origin. As c tends to zero they approach the origin and as $c \to \infty$ they become infinitely remote (the principal consequence of (iv)).

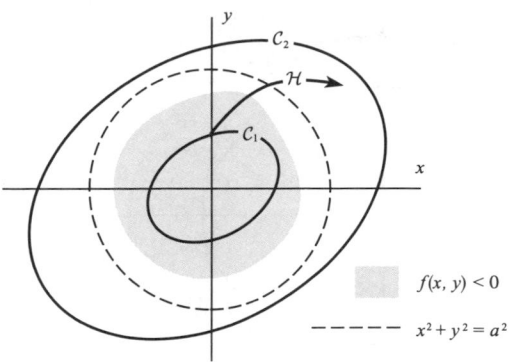

Figure 11.6

We can choose $c, c = c_1$, small enough for the corresponding contour, $\mathcal{C}_1$, to be entirely within the neighbourhood of the origin where, by (ii), $f(x, y) < 0$. We will examine a half-path $\mathcal{H}$ starting at a point on $\mathcal{C}_1$.

Consider $\dot{\mathcal{E}}(x, y)$ on $\mathcal{H}$:

$$\dot{\mathcal{E}}(x, y) = g(x)\dot{x} + y\dot{y} = g(x)y + y\{-f(x, y)y - g(x)\}$$
$$= -y^2 f(x, y). \tag{11.5}$$

This is positive, except at $y = 0$, on $\mathcal{C}_1$. Choose $\mathcal{H}$ to start at a point on $\mathcal{C}$ other than $y = 0$. Then it leaves $\mathcal{C}_1$ in the outward direction. It can never reappear inside $\mathcal{C}_1$, since to do so it must cross some interior contours in the inward direction, which is impossible since, by (11.5), $\dot{\mathcal{E}} \geq 0$ on all contours near to $\mathcal{C}_1$, as well as on $\mathcal{C}_1$.

Now consider a contour $\mathcal{C}_2$ for large $c, c = c_2$ say. $\mathcal{C}_2$ can be chosen, by (iv), to lie entirely outside the circle $x^2 + y^2 = a^2$, so that, by (i), $f(x, y) > 0$ on $\mathcal{C}_2$. By (11.5), with $f(x, y) > 0$, all paths crossing $\mathcal{C}_2$ cross inwardly, or are tangential (at $y = 0$), and by a similar argument to the above, no positive half-path, once inside $\mathcal{C}_2$, can escape.

Therefore $\mathcal{H}$ remains in the region bounded by $\mathcal{C}_1$ and $\mathcal{C}_2$, and by Theorem 11.1, there is a periodic solution in this region. ∎

The theorem can be interpreted as in Section 1.5: near to the origin the 'damping' coefficient, f, is negative, and we expect paths to spiral outwards due to intake of energy. Further out, $f(x, y) > 0$, so there is loss of energy and paths spiral inwards. Between the families we expect a closed path.

Example 11.4 *Show that the equation $\ddot{x} + (x^2 + \dot{x}^2 - 1)\dot{x} + x^3 = 0$ has a limit cycle, and locate it between two curves $\mathcal{E}(x, y) = $ constant.*

With $\dot{x} = y$, the only equilibrium point is at $(0, 0)$. In Theorem 11.2, $f(x, y) = x^2 + y^2 - 1$, $g(x) = x^3$, $G(x) = \frac{1}{4}x^4$. Therefore (11.4) gives the contours of $\mathcal{E}$: $\frac{1}{4}x^4 + \frac{1}{2}y^2 = c$. The closed path located by the theorem lies between two such contours, one inside, the other outside, of the curve $f(x, y) = 0$, or $x^2 + y^2 = 1$, and is most closely fixed by finding the smallest contour lying outside this circle and the largest lying inside. We require respectively min/max of $x^2 + y^2$ subject to $\frac{1}{4}x^4 + \frac{1}{2}y^2 = c$, c being then chosen so that the min/max is

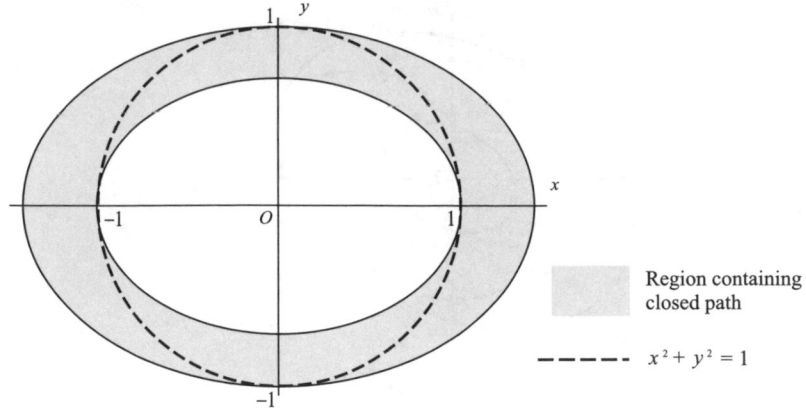

Figure 11.7   The shaded region is bounded by the curves $\frac{1}{4}x^4 + \frac{1}{2}y^2 = \frac{1}{4}$ and $\frac{1}{4}x^4 + \frac{1}{2}y^2 = \frac{1}{2}$.

equal to 1. The calculation by means of a Lagrange multiplier gives $c = \frac{1}{2}$ and $c = \frac{1}{4}$ respectively (see Fig. 11.7). ●

Exercise 11.1
Using the function $v(x, y) = x^2 + \frac{1}{2}y^2$, show that the system

$$\dot{x} = x + y - xr^2, \quad \dot{y} = -2x + 2y - yr^2, \quad r^2 = x^2 + y^2$$

has a periodic solution which lies between

$$v(x, y) = \frac{1}{2} \quad \text{and} \quad v(x, y) = 1.$$

11.2   A theorem on the existence of a centre

The following theorem involves ingredients different from those previously considered. (For a theorem of similar form but with less restrictive conditions see Minorsky 1962, p. 113.)

Theorem 11.3   *The origin is a centre for the equation*

$$\ddot{x} + f(x)\dot{x} + g(x) = 0,$$

or for the equivalent system

$$\dot{x} = y, \quad \dot{y} = -f(x)y - g(x), \tag{11.6}$$

when, in some neighbourhood of the origin, f and g are continuous, and

(i) *$f(x)$ is odd, and of one sign in the half-plane $x > 0$;*

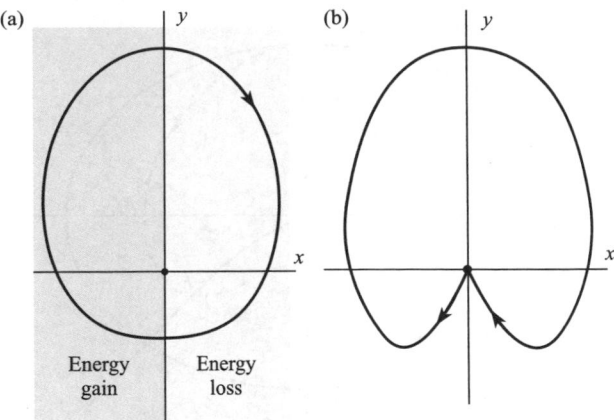

Figure 11.8

(ii) $g(x) > 0$, $x > 0$, and $g(x)$ is odd (implying $g(0) = 0$);
(iii) $g(x) > \alpha f(x) F(x)$ for $x > 0$, where $F(x) = \int_0^x f(u)\,\mathrm{d}u$, and $\alpha > 1$.

Proof   We may assume that $f(x) > 0$ when $x > 0$, since the other case, $f(x) < 0$, reduces to this by putting $-t$ for t. The equation may describe a particle–spring system, with positive damping (loss of energy) for $x > 0$, and negative damping (gain in energy) for $x < 0$. Since f and g are odd, the paths are symmetrical about the y axis (put $x = -z$, $t = -\tau$ into (11.6)). As in Theorem 11.2, (ii) ensures that there is a single equilibrium point, at the origin. Figure 11.8 shows possible types of closed path satisfying these conditions: we have to exclude possibilities such as (*b*).
  Let

$$\mathcal{E}(x, y) = \tfrac{1}{2}y^2 + G(x), \tag{11.7}$$

where

$$G(x) = \int_0^x g(u)\,\mathrm{d}u.$$

As in Theorem 11.2, the family of contours

$$\mathcal{E}(x, y) = c > 0, \tag{11.8}$$

where c is a parameter, define a family of closed curves about the origin (but this time only in some neighbourhood of the origin where the conditions hold, or for small enough c), and as $c \to 0$ the closed curves approach the origin.
  Let $\mathcal{C}_0$ be an arbitrary member of the family (11.8) in the prescribed neighbourhood of the origin, and consider the half-path $\mathcal{H}$ starting at A on the intersection of $\mathcal{C}_0$ with the y axis (Fig. 11.9). On $\mathcal{H}$,

$$\dot{\mathcal{E}}(x, y) = \frac{\partial \mathcal{E}}{\partial x}\dot{x} + \frac{\partial \mathcal{E}}{\partial x}\dot{y} = g(x)y + y[-f(x)y - g(x)] = -y^2 f(x). \tag{11.9}$$

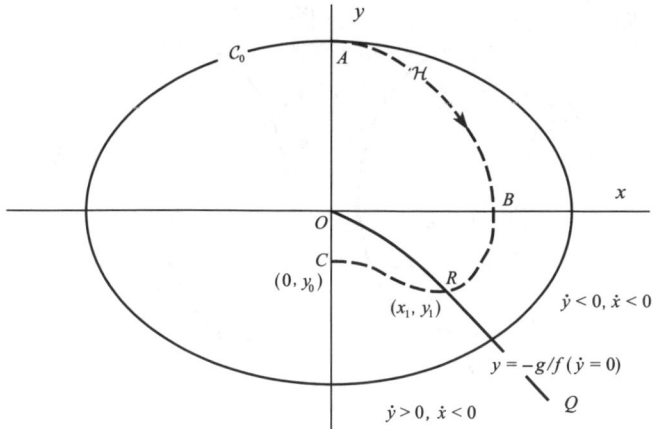

Figure 11.9

While $y > 0$, $\dot{x} > 0$, so $\mathcal{H}$ constantly advances to the right in the first quadrant. Also, by (11.9), $\dot{\mathcal{E}}(x, y) < 0$, so y moves continuously to curves with smaller $\mathcal{E}$. Since $\mathcal{H}$ remains within C_0 it leaves the quadrant at some point B on the positive x axis, inside C_0.

On the axis $y = 0$, $x > 0$ we have, by (ii),

$$\dot{x} = 0, \quad \dot{y} = -g(x) < 0.$$

Therefore all paths cut the positive x axis vertically downward, and so cannot re-enter the first quadrant from the fourth. Moreover, since $\dot{\mathcal{E}}(x, y) < 0$ in the fourth quadrant, $\mathcal{H}$ continues to lie inside C_0; therefore it must either enter the origin (the only equilibrium point), or cross the y axis at some point C for which $y < 0$.

We shall show that it does not enter the origin. In the fourth quadrant,

$$\dot{x} < 0 \tag{11.10}$$

but

$$\dot{y} = -f(x)y - g(x), \tag{11.11}$$

which is sometimes negative and sometimes positive. We can deal with (11.11) as follows (Fig. 11.9). OQ is the curve or **isocline** on which $\dot{y} = 0$ (and $\mathrm{d}y/\mathrm{d}x = 0$), dividing the region where $\dot{y} > 0$ from that where $\dot{y} < 0$ (eqn (11.11)). Assume initially that this curve approaches the origin, that is, that

$$\lim_{x \to 0} \{-g(x)/f(x)\} = 0. \tag{11.12}$$

For this case, $\mathcal{H}$ turns over in the manner indicated; is horizontal at R, and has negative slope up to C, where $\mathrm{d}y/\mathrm{d}x = 0$ again. We have to show that $y_0 < 0$. On RC

$$y_1 - y_0 = \int_0^{x_1} \frac{\mathrm{d}y}{\mathrm{d}x}\,\mathrm{d}x.$$

Therefore,

$$y_0 = y_1 + \int_0^{x_1} \{f(x) + y^{-1}g(x)\}dx \quad \text{(by (11.6))},$$

$$= -\frac{g(x_1)}{f(x_1)} + F(x_1) + \int_0^{x_1} y^{-1}g(x)dx \quad \text{(since } F(0) = 0\text{)},$$

$$< -\frac{g(x_1)}{f(x_1)} + F(x_1) \quad \text{(since } g(x) > 0, y \le 0 \text{ on } 0 \le x \le x_1\text{)},$$

$$< -\alpha F(x_1) + F(x_1) < 0 \quad \text{(by (iii))}.$$

Therefore $y_0 < 0$, and $ABRC$ is half of a closed path which is its own reflection in the y axis.

Now return to the possibility that the curve $y = -g(x)/f(x)$ does not pass through the origin. Since $g(x) > 0$ and $f(x) > 0$ when $x > 0$, it lies entirely below the x axis, and Fig. 11.10 shows how $\mathcal{H}$ must in this case also remain below the x axis, and therefore cannot enter the origin.

Finally, since the argument is independent of A near enough to the origin, the origin is a centre.   ∎

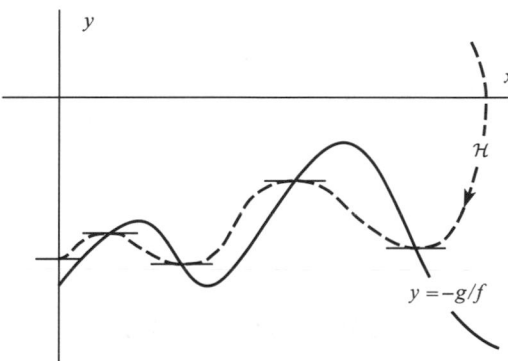

Figure 11.10

Example 11.5   *The equation $\ddot{x} + x\dot{x} + x^3 = 0$ has a centre at the origin.*

With $\dot{x} = y$, the system had only one equilibrium point at the origin. It cannot be classified by linearization. Theorem 11.3 will be applied with $f(x) = x$ and $g(x) = x^3$, so that $F(x) = \frac{1}{2}x^2$. Conditions (i) and (ii) of Theorem 11.3 are satisfied. Also

$$f(x)F(x) = \frac{1}{2}x^3.$$

Therefore,

$$g(x) - \alpha f(x)F(x) = \frac{1}{2}x^3(2 - \alpha)$$

so (iii) is satisfied with $1 < \alpha < 2$ (see Fig. 11.11).   ●

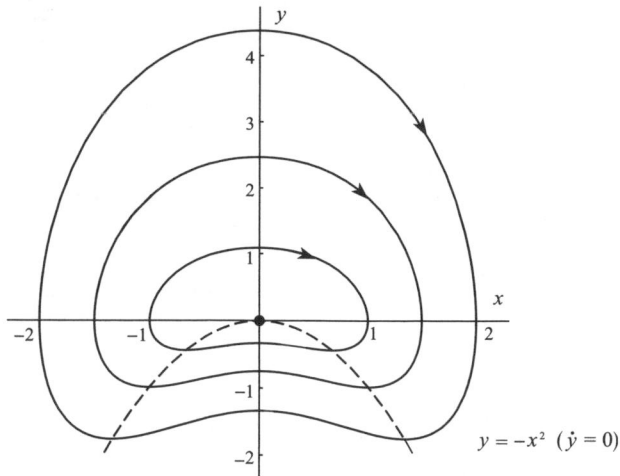

Figure 11.11   Computed paths for $\ddot{x}+x\dot{x}+x^3=0$ together with the isocline $\dot{y}=0(y=-x^2)$.

Exercise 11.2
In equation $\ddot{x}+f(x)\dot{x}+f(x)=0$ $f(x)$ satisfies the conditions of Theorem 11.3. Show that the origin is a centre if

$$\int_0^x f(u)\,\mathrm{d}u < 1.$$

If $f(x)=xe^{-x^2}$, find the equation of the phase paths of the centre.

11.3   A theorem on the existence of a limit cycle

We consider the equation

$$\ddot{x}+f(x)\dot{x}+g(x)=0, \tag{11.13}$$

where, broadly speaking, $f(x)$ is positive when $|x|$ is large, and negative when $|x|$ is small, and where g is such that, in the absence of the damping term $f(x)\dot{x}$, we expect periodic solutions for small x. **Van der Pol's equation,** $\ddot{x}+\beta(x^2-1)\dot{x}+x=0,\ \beta>0$, is of this type. Effectively the theorem demonstrates a pattern of expanding and contracting spirals about a limit cycle. Paths far from the origin spend part of their time in regions of energy input and part in regions of energy loss, so the physical argument for a limit cycle is less compelling than in the case of Theorem 11.2.

   The proof is carried out on a different phase plane from the usual one:

$$\dot{x}=y-F(x),\quad \dot{y}=-g(x), \tag{11.14}$$

(the Liénard plane) where

$$F(x) = \int_0^x f(u)\,du.$$

The use of this plane enables the shape of the paths to be simplified (under the conditions of Theorem 11.4, $\dot{y} = 0$ only on $x = 0$) without losing symmetry, and, additionally, allows the burden of the conditions to rest on F rather than f, f being thereby less restricted.

Theorem 11.4   *The equation $\ddot{x} + f(x)\dot{x} + g(x) = 0$ has a unique periodic solution if f and g are continuous, and*
  (i) *$F(x)$ is an odd function;*
 (ii) *$F(x)$ is zero only at $x = 0$, $x = a$, $x = -a$, for some $a > 0$;*
(iii) *$F(x) \to \infty$ as $x \to \infty$ monotonically for $x > a$;*
(iv) *$g(x)$ is an odd function, and $g(x) > 0$ for $x > 0$.*
(Conditions (i) to (iii) imply that $f(x)$ is even, $f(0) < 0$ and $f(x) > 0$ for $x > a$.)

Proof   The characteristics of $f(x)$ and $F(x)$ are shown schematically in (Fig. 11.12). The general pattern of the paths can be obtained from the following considerations.

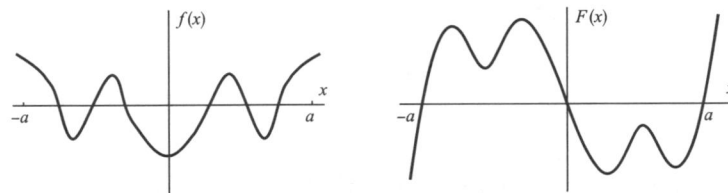

Figure 11.12   Relation between $f(x)$ and $F(x)$.

(a) If $x(t)$, $y(t) = \dot{x} - F(x)$ is a solution, so is $-x(t)$, $-y(t)$ (since F and g are odd); therefore the *whole phase diagram* is symmetrical about the origin (but not necessarily the individual phase paths).

(b) The slope of a path is given by

$$\frac{dy}{dx} = \frac{-g(x)}{y - F(x)}$$

so the paths are horizontal only on $x = 0$ (from (iv)), and are vertical only on the curve $y = F(x)$. Above $y = F(x)$, $\dot{x} > 0$, and below, $\dot{x} < 0$.

(c) $\dot{y} < 0$ for $x > 0$, and $\dot{y} > 0$ for $x < 0$, by (iv).

In what follows, the *distance* between any two points, say A and B, is denoted by AB. The (directed) *phase path* joining two points C and D, or any three points C, D, and E, are denoted by $\widehat{CD}$ and $\widehat{CDE}$ respectively.

Some typical phase paths are shown in Fig. 11.13. A path $\widehat{YY'Y''}$ is closed if and only if Y and Y'' coincide. The symmetry condition (a) implies that this is the case if and only if

$$OY = OY'. \tag{11.15}$$

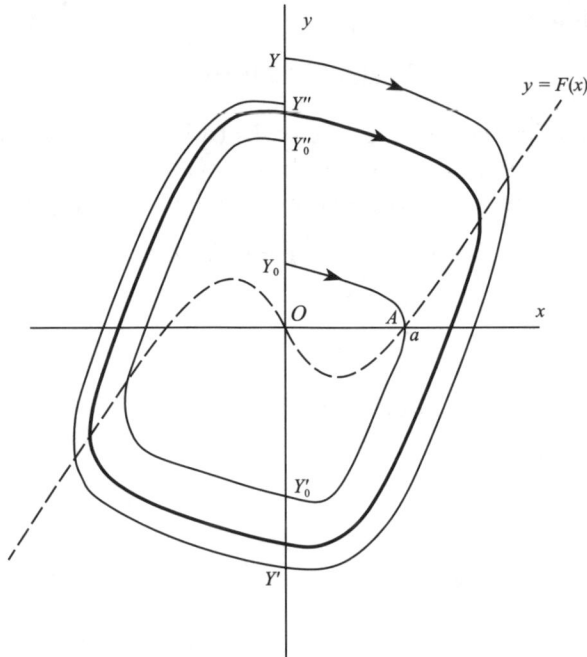

Figure 11.13

We prove the theorem by showing that for the path $\overset{\frown}{Y_0Y_0'Y_0''}$ through A: $(0,a)$, where $F(\alpha)=0$ (condition (ii) and Fig. 11.12),

$$OY_0 - OY_0' < 0,$$

and that as Y recedes to infinity along the positive y axis

$$OY - OY' \to \infty$$

monotonically. There will then be exactly one point Y for which $OY - OY'$ is zero, and this identifies the (single) closed path. Thus we are really confirming a pattern of spirals converging on a limit cycle.

Now let

$$v(x, y) = \int_0^x g(u)\, du + \tfrac{1}{2}y^2, \qquad (11.16)$$

and for any two points, S and T say, on a phase path write (Fig. 11.14)

$$v_{\widehat{ST}} = v_T - v_S = \int_{\widehat{ST}} dv, \qquad (11.17)$$

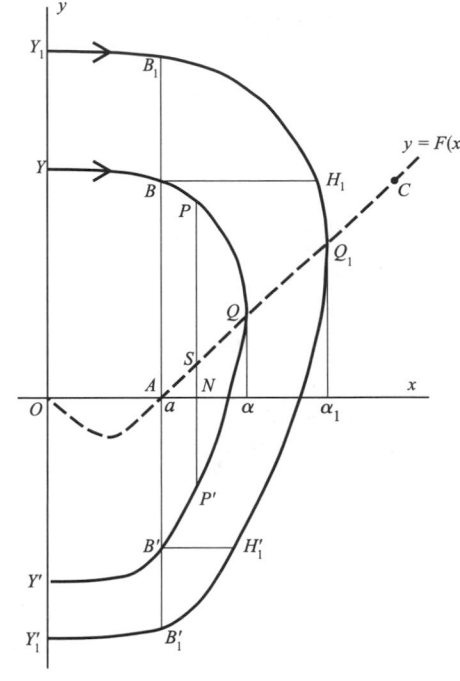

Figure 11.14 $\overset{\frown}{YQY'}$ and $\overset{\frown}{Y_1Q_1Y_1'}$ are the representative paths considered in the text.

where the integral is a line integral. Along an element of a path we have

$$dv = y\,dy + g\,dx = y\,dy + g\,\frac{dx}{dy}\,dy \quad \text{(from (11.16))}$$

$$= y\,dy + g\frac{y - F}{-g}dy \quad \text{(from (11.14))}$$

$$= F\,dy. \tag{11.18}$$

Let the ordinate BB' (through $x = a$) separate the parts of the right-hand half plane where $F(x)$ is positive from where it is negative. We consider these parts separately in estimating $V_{YY'}$ expressed as the sum

$$V_{\overset{\frown}{YY'}} = V_{\overset{\frown}{YB'}} + V_{\overset{\frown}{BB'}} + V_{\overset{\frown}{B'Y'}}. \tag{11.19}$$

The proof is carried out through the steps (A) to (F) below.

(A) As Q moves out from A along AC, $V_{YB} + V_{B'Y'}$ is positive and monotonic decreasing

Choose any fixed path $\overset{\frown}{YQY'}$, with Q at the point $(\alpha, F(\alpha))$ as shown, and another, $\overset{\frown}{Y_1Q_1Y_1'}$, with Q_1 at the point $(\alpha_1, F(\alpha_1))$, where $\alpha_1 > \alpha$.

By (11.14), since $y > 0$ and $F(x) < 0$ on $\widehat{YB}$, $\widehat{Y_1 B_1}$,

$$0 < \left(\frac{dy(x)}{dx}\right)_{\widehat{YB}} < \left(\frac{dy(x)}{dx}\right)_{\widehat{Y_1 B_1}} \tag{11.20}$$

for every x. Similarly, since $y < F(x)$ (from (b)) and $F(x) < 0$ on $\widehat{B'Y'}$, $\widehat{B_1' Y_1'}$,

$$0 < \left(\frac{dy(x)}{dx}\right)_{\widehat{B_1' Y_1'}} < \left(\frac{dy(x)}{dx}\right)_{\widehat{B'Y'}} \tag{11.21}$$

(That is to say, $\widehat{Y_1 B_1}$ and $\widehat{B_1' Y_1'}$ are shallower than $\widehat{YB}$ and $\widehat{B'Y'}$, respectively.) Therefore from (11.18), with $F(x) < 0$,

$$V_{YB} = \int_{\widehat{YB}} F \, dy = \int_{\widehat{YB}} (-F) \left(-\frac{dy}{dx}\right) dx > \int_{\widehat{Y_1 B_1}} (-F) \left(-\frac{dy}{dx}\right) dx$$

$$= V_{\widehat{Y_1 B_1}} > 0 \tag{11.22}$$

using (11.20). Similarly,

$$V_{\widehat{B'Y'}} > V_{\widehat{B_1' Y_1'}} > 0. \tag{11.23}$$

The result (A) follows from (11.21) and (11.23).

(B) As Q moves out from A along AC, $V_{BB'}$ is monotonic decreasing

Choose $\widehat{BQB'}$ arbitrarily to the right of A and let $\widehat{B_1 Q_1 B_1'}$ be another path with Q_1 to the right of Q, as before. Then $F(x) > 0$ on these paths, and we write

$$V_{\widehat{B_1 B_1'}} = -V_{\widehat{B_1' B_1}} - \int_{\widehat{B_1' B_1}} F(x) \, dy \le - \int_{\widehat{H_1' H_1}} F(x) \, dy,$$

(where $\widehat{BH_1}$, $\widehat{B'H_1'}$ are parallel to the x axis)

$$\le - \int_{\widehat{B'B}} F(x) \, dy = V_{\widehat{BB'}} \tag{11.24}$$

(since, for the same values of y, $F(x)$ on $B'B$ is less than or equal to $F(x)$ on $H_1' H_1$).

(C) From (A) and (B) we deduce that $V_{YY'}$ is monotonic decreasing as Q moves from A to infinity in the direction of C.

(D) $V_{BB'}$ tends to $-\infty$ as the paths recede to infinity

Let S be a point on the curve $y = F(x)$, to the right of A, and let $\widehat{BQB'}$ be an arbitrary path with Q to the right of S. The line $PSNP'$ is parallel to the y axis. Then, as before, referring to Fig. 11.14,

$$V_{BB'} = - \int_{\widehat{B'B}} F(x) \, dy \le \int_{\widehat{P'P}} F(x) \, dy.$$

Also

$$F(x) \geq NS \quad \text{on } PP'$$

since $F(x)$ is monotonic increasing by (iii); therefore

$$V_{\widehat{BB'}} \leq -NS \int_{\widehat{P'P}} dy = -NS \cdot PP' \leq -NS \cdot NP. \tag{11.25}$$

But as Q goes to infinity towards the right, $NP \to \infty$.

(E) From (C) and (D), $V_{YY'}$ is monotonic decreasing to $-\infty$, to the right of A

(F) $V_{YY'} > 0$ when Q is at A or to the left of A (For then $F(x) < 0$ and $dy < 0$.)

From (E) and (F), there is one and only one path for which $V_{YY'} = 0$, and this, by eqn (11.17) and the symmetry of the paths, is closed. ■

Example 11.6   *The van der Pol equation $\ddot{x} + \beta(x^2 - 1)\dot{x} + x = 0, \beta > 0$, has a unique limit cycle. (The case $\beta < 0$ is exactly similar (put $-t$ for t). Also note that this proof holds even when β is not small.)*
The phase plane x, y is defined by

$$\dot{x} = y - \beta x(\tfrac{1}{3}x^2 - 1), \quad \dot{y} = x$$

in this first-order form of the van der Pol equation.
Here,

$$f(x) = \beta(x^2 - 1), \quad g(x) = x$$

so that

$$F(x) = \beta(\tfrac{1}{3}x^3 - x).$$

The conditions of the theorem are satisfied, with $a = \sqrt{3}$. The x-extremities of the limit cycle must be beyond $x = \pm\sqrt{3}$. ●

Example 11.7   *Show that the equation*

$$\ddot{x} + \frac{x^2 + |x| - 1}{x^2 - |x| + 1}\dot{x} + x^3 = 0$$

has a unique periodic solution.
We apply Theorem 11.4 with $f(x) = (x^2 + |x| - 1)/(x^2 - |x| + 1)$ and $g(x) = x^3$. The graph of $f(x)$ versus x is shown in Fig. 11.15(a). Since $f(x) \to 1$ as $x \to \infty$, (iii) is satisfied, and (i), (ii), and (iv) are obviously satisfied. Therefore a limit cycle exists, its extreme x-values being beyond the nonzero roots of $F(x) = 0$. The computed limit cycle is shown in Fig. 11.15(b). ●

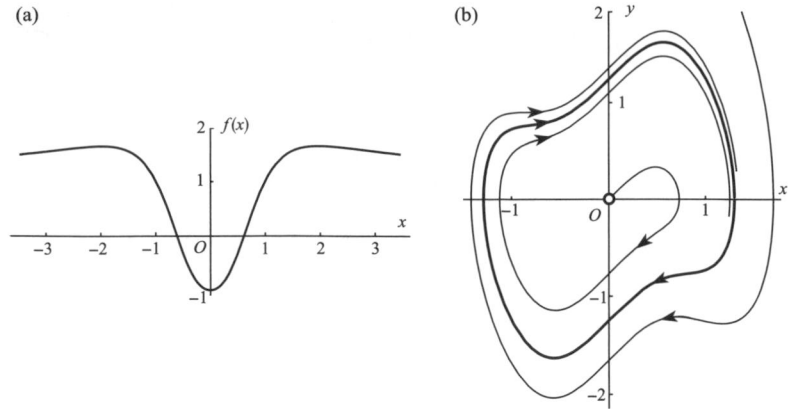

Figure 11.15 (a) Graph of $f(x)$ versus x; (b) phase diagram showing the limit cycle.

Exercise 11.3
Compute the phase diagram for the van der Pol equation

$$\ddot{x} + \beta(x^2 - 1)\dot{x} + x = 0 \quad (\beta > 0),$$

in the phase plane defined by

$$\dot{x} = y - F(x), \quad \dot{y} = -x$$

(see Theorem 11.4 and Example 11.6). Possible parameter values are $\beta = 0.5, 1, 2$. Compare with Fig. 3.24.

11.4   Van der Pol's equation with large parameter

Van der Pol's equation with large parameter β

$$\ddot{x} + \beta(x^2 - 1)\dot{x} + x = 0$$

is equivalent (by putting $\delta = 1/\beta$) to the equation

$$\delta\ddot{x} + (x^2 - 1)\dot{x} + \delta x = 0 \tag{11.26}$$

with δ small. This can be regarded as a singular perturbation problem in terms of Chapter 6 (the small parameter is the coefficient of the highest order derivative), in which the influence of the various terms is differently disposed on different parts of the solution curves, or on the phase paths. The existence of a limit cycle is confirmed in Example 11.6. Here we shall look briefly at the analytical construction of the limit cycle of (11.26) for small δ in the usual phase plane with $\dot{x} = y$, and not the Liénard plane. It is necessary to piece together the limit cycle from several approximate solutions corresponding to different balances between the terms. The approximations are made for $\delta > 0$; the phase path is then completed by invoking its symmetry in the origin.

The equation for the phase path is

$$\frac{dy}{dx} = -\frac{x^2 - 1}{\delta} - \frac{x}{y}, \tag{11.27}$$

and the isocline for zero slope is the curve $y = \delta x/(1 - x^2)$.

In eqn (11.26) put $t = \mu\tau$, so that

$$\frac{\delta}{\mu^2}x'' + \frac{1}{\mu}(x^2 - 1)x' + \delta x = 0. \tag{11.28}$$

When $\mu = O(\delta)$ the first two terms predominate, and to the first order

$$x'' + (x^2 - 1)x' = 0. \tag{11.29}$$

If $x = a$ (the amplitude) when $y = 0$, then integration of (11.29) gives

$$x' = x - \tfrac{1}{3}x^3 - a + \tfrac{1}{3}a^3, \tag{11.30}$$

or

$$x' = \tfrac{1}{3}(x - a)(3 - x^2 - ax - a^2).$$

If $a > 1$, then on $x' > 0$ (the upper half of the phase plane)

$$\tfrac{1}{2}\{-a + \sqrt{(12 - 3a^2)}\} < x < a,$$

and (11.30) represents the phase path in this interval: the amplitude a will be determined later.

When we put $\mu = O(\delta^{-1})$ in (11.28), the second and third terms predominate and are of the same order, so that

$$x' = x/(1 - x^2), \tag{11.31}$$

which is essentially the zero-slope isocline of (11.27). This equation can only be valid for $y > 0$ and $-a < x < -1$ because of the singularity at $x = -1$.

The two approximate solutions given by (11.30) and (11.31) must be connected by a third equation in the neighbourhood of $x = -1$. To the required order the solution given by (11.30) must pass through $y = 0$, $x = -1$: whence

$$\tfrac{1}{2}\{-a + \sqrt{(12 - 3a^2)}\} = -1, \quad \text{or} \quad a = 2.$$

For the transition, put $z = x + 1$; then (11.28) becomes

$$\frac{\delta}{\mu^2}z'' + \frac{1}{\mu}\{(z - 1)^2 - 1\}z' + \delta z - \delta = 0.$$

Now put $\mu = O(\delta^\alpha)$, and let $z = \delta^\beta u$; then a balance is achieved between the terms containing z'', z', z and the constant if

$$1 - 2\alpha + \beta = -\alpha + 2\beta = 1;$$

that is, if $\alpha = \frac{1}{3}$ and $\beta = \frac{2}{3}$. The differential equation for the transition region is

$$u'' - 2uu' - 1 = 0, \tag{11.32}$$

an equation without an elementary solution, relating u' and u.

There remains the solution around $x = -2$ to be constructed. Put $z = x + 2$ in (11.28):

$$\frac{\delta}{\mu^2} z'' + \frac{1}{\mu}\{(z-2)^2 - 1\}z' + \delta z - 2\delta = 0.$$

Again, put $\mu = O(\delta^\alpha)$ and $z = \delta^\beta u$. The equation

$$u'' + 3u' = 2 \tag{11.33}$$

follows if $1 - 2\alpha + \beta = \beta - \alpha = 1$, or if $\alpha = 1$, $\beta = 2$. From (11.33)

$$\tfrac{1}{3}u' + \tfrac{2}{9}\log(2 - 3u') + u = \text{constant}.$$

When $u = 0$, $u' = 0$, so finally

$$\tfrac{1}{3}u' + \tfrac{2}{9}\log(2 - 3u') + u = \tfrac{2}{9}\log 2.$$

After restoring the original variables, x and t, the sections of the limit cycle on the phase plane for $y > 0$ are as follows:

(i) $-2 \leq x < -2 + O(\delta^2)$:

$$\tfrac{1}{3}\delta y + \tfrac{2}{9}\delta^2 \log(1 - \tfrac{3}{2}\delta^{-1}y) = -(x + 2);$$

(ii) $-2 + O(\delta^2) < x < -1 - O(\delta^{2/3})$:

$$y = \delta x / (1 - x^2);$$

(iii) $-1 - O(\delta^{2/3}) < x < -1 + O(\delta^{2/3})$: the appropriate solution of

$$\delta \ddot{x} - 2(x+1)\dot{x} = \delta;$$

(iv) $-1 + O(\delta^{2/3}) < x \leq 2$:

$$x = \tfrac{1}{3}y^3 + \delta y - \tfrac{2}{3}.$$

The complete limit cycle can now be constructed since it is symmetrical about the origin. Figure 11.16(a) shows the computed limit cycle for $\beta = 10$, and the corresponding time solution is shown in Fig. 11.16(b).

The van der Pol equation for large parameter β (or small δ) is an example of a **relaxation oscillation,** in which typically, as Fig. 11.16(b) shows, the system displays a slow build-up followed by a sudden discharge, repeated periodically.

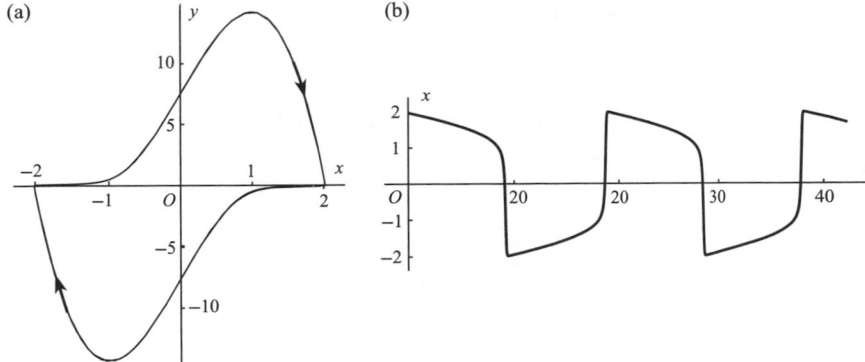

(a)

(b)

Figure 11.16 (a) Limit cycle for van der Pol's equation (11.26) with $\delta = 1/\beta = 0.1$; (b) time-solution curve corresponding to the limit cycle

Problems

11.1 Prove that the equilibrium point of

$$\ddot{x} + \frac{x}{1+x^2}\dot{x} + x\ln(1+x^2) = 0, \quad \dot{x} = y$$

is a centre in the (x, y) plane. Compute the phase diagram in the neighbourhood of $(0, 0)$.

11.2 A system has exactly one equilibrium point, n limit cycles, and no other periodic solutions. Explain why an asymptotically stable limit cycle must be adjacent to unstable limit cycles, but an unstable limit cycle may have stable or unstable cycles adjacent to it.

Let c_n be the number of possible configurations, with respect to stability, of n nested limit cycles. Show that $c_1 = 2$, $c_2 = 3$, $c_3 = 5$, and that in general

$$c_n = c_{n-1} + c_{n-2}.$$

(This recurrence relation generates the Fibonacci sequence.) Deduce that

$$c_n = \{(2 + \sqrt{5})(1 + \sqrt{5})^{n-1} + (-2 + \sqrt{5})(1 - \sqrt{5})^{n-1}\}/(2^{n-1}\sqrt{5}).$$

11.3 By considering the path directions across each of the suggested topographic systems show that in each of the cases given there exists a limit cycle. Locate the region in which a limit cycle might exist as closely as possible. Show that in each case only one limit cycle exists:

(i) $\dot{x} = 2x + 2y - x(2x^2 + y^2)$, $\dot{y} = -2x + y - y(2x^2 + y^2)$
(topographic system $x^2 + y^2$ = constant);

(ii) $\dot{x} = -x - y + x(x^2 + 2y^2)$, $\dot{y} = x - y + y(x^2 + 2y^2)$
(topographic system $x^2 + y^2$ = constant);

(iii) $\dot{x} = x + y - x^3 - 6xy^2$, $\dot{y} = -\frac{1}{2}x + 2y - 8y^3 - x^2y$
(topographic system $x^2 + 2y^2$ = constant); compute the phase diagram, and show the topographic system.

(iv) $\dot{x} = 2x + y - 2x^3 - 3xy^2$, $\dot{y} = -2x + 4y - 4y^3 - 2x^2y$
(topographic system $2x^2 + y^2$ = constant).

11.4 Show that the equation $\ddot{x} + \beta(x^2 + \dot{x}^2 - 1)\dot{x} + x^3 = 0$ $(\beta > 0)$ has at least one periodic solution.

11.5 Show that the origin is a centre for the equations
   (i) $\ddot{x} - x\dot{x} + x = 0$;
   (ii) $\ddot{x} + x\dot{x} + \sin x = 0$.

11.6 Suppose that $f(x)$ in the equation $\ddot{x} + f(x)\dot{x} + x = 0$ is given by $f(x) = x^n$. Show that the origin is a centre if n is an odd, positive integer.

11.7 Show that the equation

$$\ddot{x} + \beta(x^2 - 1)\dot{x} + \tanh kx = 0$$

has exactly one periodic solution when $k > 0$, $\beta > 0$. Decide on its stability.
   The 'restoring force' resembles a step function when k is large. Is the conclusion the same when it is exactly a step function?

11.8 Show that

$$\ddot{x} + \beta(x^2 - 1)\dot{x} + x^3 = 0$$

has exactly one periodic solution.

11.9 Show that

$$\ddot{x} + (|x| + |\dot{x}| - 1)\dot{x} + x|x| = 0$$

has at least one periodic solution.

11.10 Show that the origin is a centre for the equation

$$\ddot{x} + (k\dot{x} + 1)\sin x = 0.$$

11.11 Using the method of Section 11.4, show that the amplitude of the limit cycle of

$$\varepsilon\ddot{x} + (|x| - 1)\dot{x} + \varepsilon x = 0, \quad \dot{x} = y \ (0 < \varepsilon \ll 1),$$

is approximately $a = 1 + \sqrt{2}$ to order ε. Show also that the solution for $y > 0$ is approximately

$$\varepsilon y = (x - a) - \tfrac{1}{2}x^2\text{sgn}(x) + \tfrac{1}{2}a^2, \quad (-1 < x < a).$$

Compare the curve with the computed phase path for $\varepsilon = 0.1$.

11.12 Let F and g be functions satisfying the conditions of Theorem 11.4. Show that the equation

$$\ddot{u} + F(\dot{u}) + g(u) = 0$$

has a unique periodic solution (put $\dot{u} = -z$). Deduce that Rayleigh's equation $\ddot{u} + \beta(\tfrac{1}{3}\dot{u}^3 - \dot{u}) + u = 0$ has a unique limit cycle.

11.13 Show that the equation

$$\ddot{x} + \beta(x^2 + \dot{x}^2 - 1)\dot{x} + x = 0,$$

unlike the van der Pol equation, does not have a relaxation oscillation for large positive β.

11.14 For the van der Pol oscillator

$$\delta\ddot{x} + (x^2 - 1)\dot{x} + \delta x = 0$$

for small positive δ, use the formula for the period, eqn (1.13), to show that the period of the limit cycle is approximately $(3 - 2\ln 2)\delta^{-1}$. (Hint: the principal contribution arises from that part of the limit cycle given by (ii) in Section 11.4.)

11.15 Use the Poincaré–Bendixson theorem to show that the system

$$\dot{x} = x - y - x(x^2 + 2y^2), \quad \dot{y} = x + y - y(x^2 + 2y^2)$$

has at least one periodic solution in the annulus $1/\sqrt{2} < r < 1$, where $r = \sqrt{(x^2 + y^2)}$.

12 Bifurcations and manifolds

A characteristic of nonlinear oscillating systems which has become a subject of considerable recent interest is the great variety of types of response of which they are capable as initial conditions or parameter values change. The passage from one set of responses to another often occurs very suddenly, or 'catastrophically': instances of this were shown in Section 1.7 in connection with a parameter-dependent conservative system, and in Section 7.3 on the jump phenomenon of Duffing's equation. Further examples of such changes are given in the present chapter, the linking concept being the idea of bifurcation, where the sudden change in behaviour occurs as a parameter passes through a critical value, called a bifurcation point. A system may contain more than one parameter, each with its own bifurcation points, so that it can display extremely complex behaviour, and computer studies play an important part in providing a taxonomy for the behaviour of such systems. We shall look at some of the elementary characteristics of bifurcations as they arise in the fold and cusp catastrophes, and the Hopf bifurcation. More discussion of bifurcations with many examples is given by Hubbard and West (1995). A **manifold** is a subspace of a phase or solution space on which a characteristic property such as stability can be associated.

12.1 Examples of simple bifurcations

In this section we take an intuitive view of some simple bifurcations. Consider the system

$$\dot{x} = y, \qquad \dot{y} = -\lambda x$$

containing a parameter λ with values in $(-\infty, \infty)$. The phase diagram contains a centre for $\lambda > 0$ and a saddle for $\lambda < 0$, these classifications representing radically different types of stable and unstable system behaviour. The change in stability occurs as λ passes through $\lambda = 0$: a bifurcation is said to occur at $\lambda = 0$, and $\lambda = 0$ is called the **bifurcation point**.

The behaviour of the damped system

$$\dot{x} = y, \qquad \dot{y} = -ky - \omega^2 x, \tag{12.1}$$

with $\omega > 0$ given and k the **bifurcation parameter**, depends on the roots of the characteristic equation

$$m^2 + km + \omega^2 = 0.$$

The phase diagram is an unstable node for $k < -2\omega$, an unstable spiral for $-2\omega < k < 0$, a stable spiral for $0 < k < 2\omega$, and a stable node for $k > 2\omega$. It might appear at first sight that, for example, the transition from stable spiral to stable node should indicate a bifurcation, but an important common feature is that both are asymptotically stable, so that even though the stable spiral looks different from a stable node we do not call the value $k = 2\omega$ a bifurcation point. However, the transition of k from negative k to positive k through $k = 0$ is accompanied by a *change of stability*. Therefore we regard the system as having a bifurcation point at $k = 0$. We shall say more about the general linear problem in Section 12.3.

The equation

$$\ddot{x} - k(x^2 + \dot{x}^2 - 1)\dot{x} + x = 0$$

(see Example 1.9) has a limit cycle $x^2 + y^2 = 1$ in the phase plane with $\dot{x} = y$. Let k be the bifurcation parameter. The behaviour close to the origin is indicated by the linearized equation

$$\ddot{x} + k\dot{x} + x = 0$$

which is essentially the same as (12.1) with $\omega = 1$. The equilibrium point at $(0, 0)$ passes through the same bifurcation as that of (12.1). The limit cycle $x^2 + y^2 = 1$ is always present but its stability changes from stable to unstable as k increases through zero, so in this example the change in the phase diagram is not limited just to the equilibrium point but also to other features.

Example 12.1   *Find the bifurcation points of the system $\dot{x} = -\lambda x + y$, $\dot{y} = -\lambda x - 3y$.*

Let

$$x = \begin{bmatrix} x \\ y \end{bmatrix}, \quad A(\lambda) = \begin{bmatrix} -\lambda & 1 \\ -\lambda & -3 \end{bmatrix}$$

so that, in matrix form, the system is equivalent to

$$\dot{x} = A(\lambda)x.$$

If $\lambda \neq 0$, the system has only one equilibrium point, at the origin: if $\lambda = 0$, equilibrium occurs at all points on $y = 0$. Following the method of Chapter 2, we find the eigenvalues m of $A(\lambda)$ from

$$|A(\lambda) - mI| = 0,$$

or

$$\begin{vmatrix} -\lambda - m & 1 \\ -\lambda & -3 - m \end{vmatrix} = 0.$$

Hence m satisfies

$$m^2 + (3 + \lambda)m + 4\lambda = 0,$$

which has the roots

$$m_1, m_2 = \tfrac{1}{2}[-\lambda - 3 \pm \sqrt{\{(\lambda - 1)(\lambda - 9)\}}].$$

We look for bifurcations where the character of the roots changes—having regard to whether they are real or complex, and to the sign when they are real or the sign of the real part when they are complex.

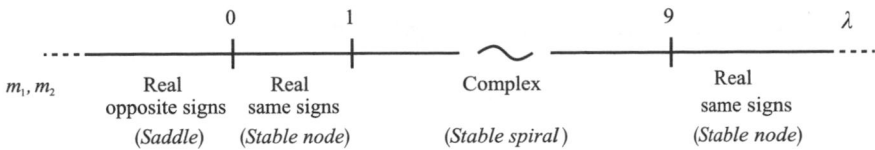

Figure 12.1

The classification of the origin is:

λ	m_1, m_2	Type
$\lambda < 0$	Real, opposite signs	Saddle
$0 < \lambda < 1$	Real, negative	Stable node
$1 < \lambda < 9$	Complex, negative real part	Stable spiral
$\lambda > 9$	Real negative	Stable node

The transitions are shown in Fig. 12.1. The system has a single bifurcation point $\lambda = 0$, where there is a change from a stable node to a saddle. ●

Exercise 12.1
In Example 12.1, $\dot{x} = Ax$ where

$$A = \begin{bmatrix} -\lambda & 1 \\ -\lambda & -3 \end{bmatrix}.$$

Referring back to Fig. 2.10, express p and q in terms of λ. Show that $q = -4(p + 3)$, and draw this line on Fig. 2.10. Discuss the bifurcations of the system.

12.2 The fold and the cusp

Bifurcations of systems are closely linked to **catastrophes**. In general terms a system experiences a **catastrophe** when a smooth change in the values of a parameter results in a sudden change in the response of the system. A comprehensive account of catastrophe theory is given by Poston and Stewart (1978). Here we shall in a general way relate the two simplest catastrophes and their associated bifurcation sets of parameters as they arise in the 'conservative' systems of Section 1.3, which take the standard form

$$\ddot{x} = f(x) = -\mathcal{V}'(x),$$

where $\mathcal{V}$ is called the potential energy, or potential, of the system.
  Suppose we have a potential containing a parameter λ, defined by

$$\mathcal{V}(\lambda, x) = \tfrac{1}{3}x^3 + \lambda x. \tag{12.2}$$

The equilibrium points occur where $\partial V/\partial x = 0$, or where

$$x^2 + \lambda = 0. \tag{12.3}$$

If $\lambda < 0$ there are two equilibrium points, at $x = \pm\sqrt{\lambda}$, one if $\lambda = 0$, and none if $\lambda > 0$. The phase diagrams may be obtained by using the method of Section 1.3: a plot of V against x for constant λ is made, and must take one of the forms shown in Fig. 12.2, depending on the value of λ. If $\lambda < 0$ then V has a minimum at $x = \sqrt{\lambda}$, indicating a centre at this point, and a maximum at $x = -\sqrt{\lambda}$, indicating a saddle there. The case $\lambda = 0$ and $\lambda > 0$ may be discussed similarly.

Equation (12.3) defines the so-called 'catastrophe manifold' for this problem: this is the curve shown in Fig. 12.3. In effect it shows the positions of the equilibrium points in terms of the values of λ, but its real significance is to display the fact that $\lambda = 0$ is a bifurcation point for the problem, where the nature of the solutions suddenly changes.

The stability of the equilibrium points to which the two branches of the curve relate can be deduced from the method of Section 1.7. If the region where $(-\partial V/\partial x) > 0$ is shaded, then equilibrium points which form the boundary above this region are stable and the rest are unstable.

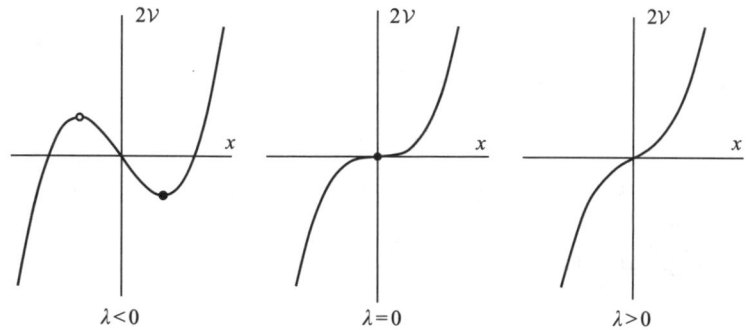

Figure 12.2

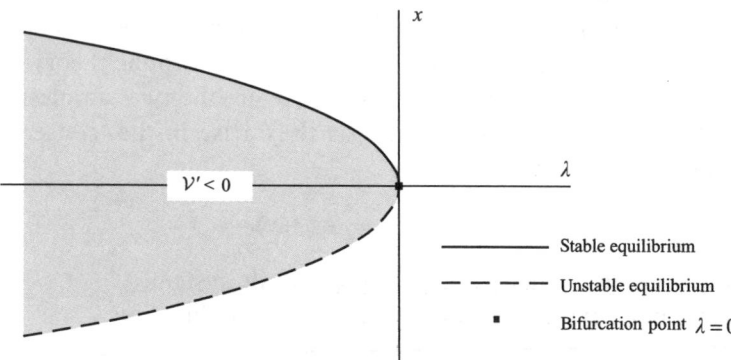

Figure 12.3   Manifold curve for the fold catastrophe.

Potentials may contain two or more parameters. Suppose now that the potential consists of a fourth-degree polynomial in x of the form

$$\mathcal{V}(\lambda, \mu, x) = \tfrac{1}{4}x^4 - \tfrac{1}{2}\lambda x^2 + \mu x \tag{12.4}$$

where λ and μ are two parameters. Equilibrium points occur where $\partial \mathcal{V}/\partial x = 0$, or where

$$M(\lambda, \mu, x) \equiv x^3 - \lambda x + \mu = 0. \tag{12.5}$$

Regarding (12.5) as a cubic equation for x, the nature of its solutions depends on the sign of the discriminant

$$D(\lambda, \mu) = \mu^2 + 4\left(\tfrac{-\lambda}{3}\right)^3 = \mu^2 - \tfrac{4}{27}\lambda^3$$

(see, e.g., Ferrar, 1950). If $D > 0$ there is one real solution and two complex ones; if $D = 0$ there are two real roots one of which is a repeated root; and if $D < 0$ there are three real roots, all different. The potential $\mathcal{V}$ then takes one of the forms shown in Fig. 12.4.

The catastrophe manifold defined by (12.5) is this time a surface in λ, μ, x space (Fig. 12.5). A fold grows from the origin in the general direction of the λ axis. The two curves radiating

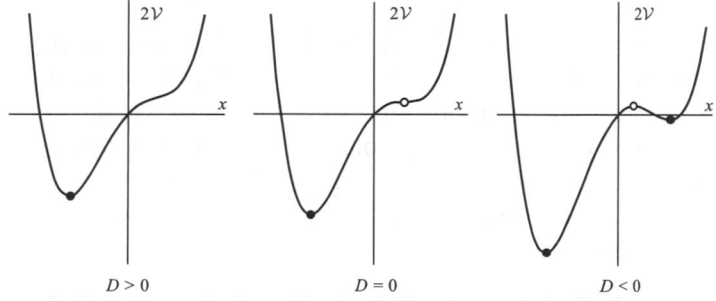

Figure 12.4

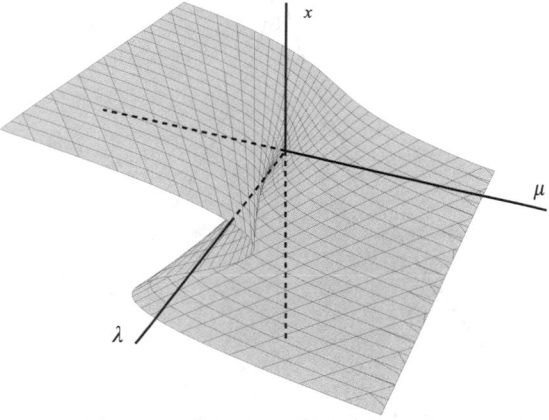

Figure 12.5   Manifold for the cusp catastrophe.

from the origin along which the fold turns over are of particular interest, since they form a demarcation between the parameter region where there is a single equilibrium point and the region in which there are three. We therefore project the latter region on to the λ, μ, plane. At the edge of the fold

$$\frac{\partial M}{\partial x} = 3x^2 - \lambda = 0. \tag{12.6}$$

The projection of the boundary of the fold is obtained by eliminating x between (12.5) and (12.6), and the resulting curves in the $\lambda, \mu = 1$ plane are given by

$$\mu = \pm \frac{2}{3\sqrt{3}} \lambda^{3/2}$$

as shown in Fig. 12.6. (This expression is equivalent to $D(\lambda, \mu) = 0$.)

Figure 12.6 is called a **catastrophe map** (of a **cusp catastrophe**): it indicates the critical values of the parameters at which sudden change in the nature of the equation's output can be expected. The **bifurcation set** consists of the cusp itself. In the shaded region between the two branches of the curve there are three equilibrium points, and outside there is one.

The technique of Section 1.7 can again be used to establish which parts of the surface of Fig. 12.5 represent stable equilibrium points, and which unstable, by noticing that the surface divides the space into two parts, and that in the 'lower' half $(\partial V/\partial x)$ is negative and in the 'upper' half positive (the separation occurring when $(-\partial V/\partial x) = 0$, which is the equation of the surface itself). The connecting portion of the fold between the upper and lower parts therefore represents unstable equilibrium points, and the rest are stable. The unstable set lies between the branches of the cusp in Fig. 12.6. Thus if the parameter values undergo a change in which they cross the cusp a marked change in response of the system is to be expected, this change taking place 'catastrophically'.

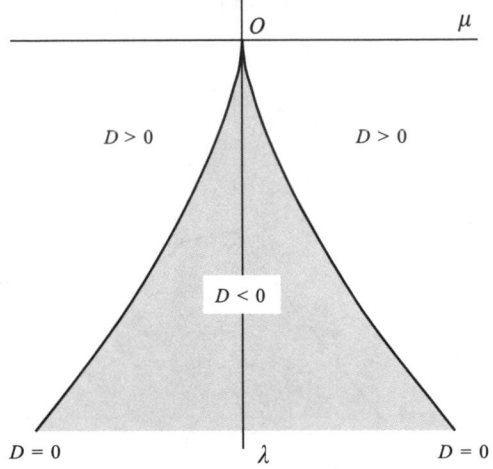

Figure 12.6 Bifurcation set for the cusp catastrophe.

It is shown in Section 7.3 how Duffing's equation generates a similar cusp catastrophe when the parameters are taken to be the forcing amplitude and the forcing frequency: the observed variables in that case are not equilibrium points and their stability, but amplitudes of steady forced oscillations and their stability, as parametrized in the van der Pol plane.

Exercise 12.2
In the conservative system

$$\ddot{x} = -\frac{\partial \mathcal{V}(\lambda, \mu, x)}{\partial x},$$

let

$$\mathcal{V}(\lambda, \mu, x) = \tfrac{1}{4}x^4 - \tfrac{1}{3}\lambda x^3 + \mu x.$$

Show that equilibrium points occur where

$$M(\lambda, \mu, x) \equiv x^3 - \lambda x^2 + \mu = 0.$$

Reduce this equation to standard form (12.5) by the transformation $x = z + \tfrac{1}{3}\lambda$. Show that in the equation for z the discriminant is

$$D(\lambda, \mu) = \mu^2 - \tfrac{4}{27}\mu\lambda^3.$$

12.3   Further types of bifurcation

In this section we examine further parametric bifurcations of first-order systems, mainly in two variables. The approach is through examples, and no attempt is made to cover general bifurcation theory. However, we can first make some general observations about the first-order autonomous system in n real variables which contains m real parameters. This can be expressed in the form

$$\dot{x} = X(\mu, x), \quad x \in \mathbb{R}^n, \ \mu \in \mathbb{R}^m, \tag{12.7}$$

where μ is an m-dimensional vector of real parameters (the letter $\mathbb{R}$ stands for the set of real numbers). Equilibrium points occur at the solutions for x of the n scalar equations expressed in vector form by

$$X(\mu, x) = 0 \tag{12.8}$$

for any given μ.

Suppose that (μ_0, x_0) is a solution of this equation. Then $\mu = \mu_0$ is a **bifurcation point** if the structure of the phase diagram changes as μ passes through μ_0. This rather imprecise definition covers a number of possibilities including a change in the number of equilibrium points as μ passes through μ_0, or a change in their stability. There are many ways in which μ can pass through $\boldsymbol{\mu}_0$: if $m = 3$ then we would look at changes in all straight lines through $\boldsymbol{\mu}_0$. Some directions might produce a bifurcation and some might not, but if at least one does so then $\boldsymbol{\mu}_0$ is a bifurcation point. Another possibility is that equilibrium points appear and disappear although their number remains fixed, causing the phase diagram to suffer a discontinuous change (see Example 12.5).

Geometrically, the equilibrium points occur where the n surfaces given by (12.8) in the n-dimensional $\boldsymbol{x}$ space intersect, but obviously the locations of these surfaces will vary with changes in the parameter vector $\boldsymbol{\mu}$. Alternatively, we can view the equilibrium points as the intersections of the surfaces in the $(m+n)$ dimensional $(\boldsymbol{x}, \boldsymbol{\mu})$ space. Any projection $\boldsymbol{\mu} = constant$ $vector$ in this space will give the n surfaces in the $\boldsymbol{x}$ subspace, whose intersections determine the equilibrium points of the system.

We shall look at some examples of further common bifurcations which occur in plane systems containing a single parameter.

Saddle-node bifurcation (or fold bifurcation)

Consider the equations

$$\dot{x} = y, \quad \dot{y} = x^2 - y - \mu. \tag{12.9}$$

Equilibrium points occur where $y = 0$, $x^2 - y - \mu = 0$. Geometrically they lie on the intersection of the plane $y = 0$ and the surface $y = x^2 - \mu$ in (x, y, μ) space. In this case, all the equilibrium points lie in the plane $y = 0$, so that we need only show the curve $x^2 = \mu$ of equilibrium points in the (x, μ) plane as shown in Fig. 12.7(a). There is a bifurcation point at $\mu = \mu_0 = 0$ where $x = x_0 = 0$, $y = y_0 = 0$. For $\mu < 0$ there are no equilibrium points whilst for $\mu > 0$ there are two at $x = \pm\sqrt{\mu}$.

For $\mu > 0$, let $x = \pm\sqrt{\mu} + x'$, $y = y'$ in (12.12). Hence, to the first order

$$\dot{x}' = y', \quad \dot{y}' = (\pm\sqrt{\mu} + x')^2 - y' - \mu \approx \pm 2x'\sqrt{\mu} - y'. \tag{12.10}$$

In the notation of the linear approximation (Section 2.4)

$$p = -1, \quad q = \mp 2\sqrt{\mu}, \quad \Delta = 1 \pm 8\sqrt{\mu}. \tag{12.11}$$

Hence, for $x > 0$ the equilibrium point is a saddle; whilst for $x < 0$, the equilibrium point is a stable node, which becomes a stable spiral for $\sqrt{\mu} > \frac{1}{8}$. However the immediate bifurcation as μ increases through zero is of **saddle-node** type. Figure 12.7(b) shows the saddle-node phase diagram for $\mu = 0.01$.

Transcritical bifurcation

Consider the parametric system

$$\dot{x} = y, \quad \dot{y} = \mu x - x^2 - y.$$

It has equilibrium points at $(0, 0)$ and $(\mu, 0)$. Again all the equilibrium points lie only in the plane $y = 0$ in the (x, y, μ) space. The bifurcation curves are given by $x(x - \mu) = 0$, which are two straight lines intersecting at the origin (see Fig. 12.8).

(a)

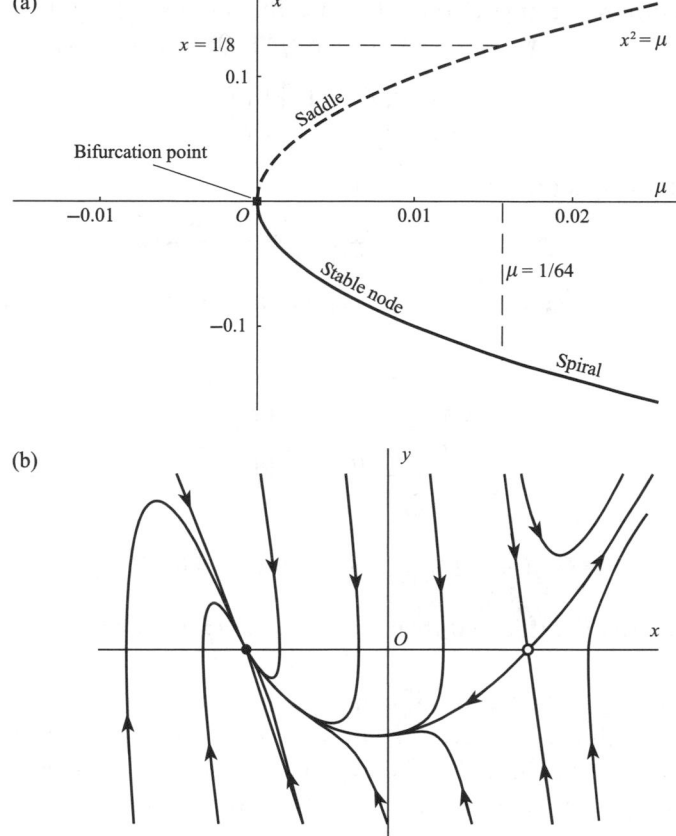

(b)

Figure 12.7   (a) A saddle-node bifurcation; (b) phase diagram for $\mu = 0.01$.

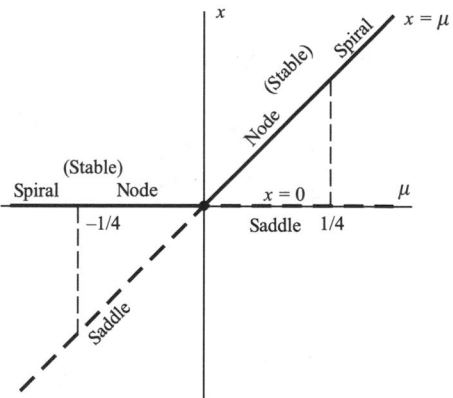

Figure 12.8   A transcritical bifurcation.

There is a bifurcation point at $\mu = \mu_0 = 0$, since the number of equilibrium points changes from two ($\mu < 0$) to one ($\mu = 0$) and back to two ($\mu > 0$) as μ increases. Near the origin

$$\begin{bmatrix} \dot{x} \\ \dot{y} \end{bmatrix} \approx \begin{bmatrix} 0 & 1 \\ \mu & -1 \end{bmatrix} \begin{bmatrix} x \\ y \end{bmatrix}.$$

Thus in the notation of eqns (2.35) and (2.36),

$$p = -1, \quad q = -\mu, \quad \Delta = 1 + 4\mu.$$

For $\mu < 0$, the origin is *stable*; a *node* if $-\frac{1}{4} < \mu < 0$, and a *spiral* if $\mu < -\frac{1}{4}$. If $\mu > 0$ then the origin is a *saddle*.

Near $x = \mu$, with $x = \mu + x'$ and $y = y'$,

$$\begin{bmatrix} \dot{x}' \\ \dot{y}' \end{bmatrix} \approx \begin{bmatrix} 0 & 1 \\ -\mu & -1 \end{bmatrix} \begin{bmatrix} x' \\ y' \end{bmatrix}.$$

Hence

$$p = -1, \quad q = \mu, \quad \Delta = 1 - 4\mu.$$

For $\mu < 0$, $x = \mu$ is a *saddle*; if $\mu > 0$ then $x = \mu$ is a *stable node* for $0 < \mu < \frac{1}{4}$ and a *stable spiral* for $\mu > \frac{1}{4}$.

This is an example of a **transcritical bifurcation** where, at the intersection of the two bifurcation curves, stable equilibrium switches from one curve to the other at the bifurcation point. As μ increases through zero, the saddle point collides with the node at the origin, and then remains there whilst the stable node moves away from the origin.

Pitchfork (or flip) bifurcation

Consider the system

$$\dot{x} = y, \qquad \dot{y} = \mu x - x^3 - y. \tag{12.12}$$

The (x, μ) bifurcation diagram, shown in Fig. 12.9, has a bifurcation point at $\mu = \mu_0 = 0$. Near the origin

$$\begin{bmatrix} \dot{x} \\ \dot{y} \end{bmatrix} \approx \begin{bmatrix} 0 & 1 \\ \mu & -1 \end{bmatrix} \begin{bmatrix} x \\ y \end{bmatrix}.$$

Hence

$$p = -1, \quad q = -\mu, \quad \Delta = 1 + 4\mu.$$

If $\mu < 0$, then the origin is *stable*, a *node* if $-\frac{1}{4} < \mu < 0$ and a *spiral* if $\mu < -\frac{1}{4}$; if $\mu > 0$ then the origin is a *saddle point*.

The additional equilibrium points are at $x = \pm\sqrt{\mu}$ for $\mu > 0$. Linearization about these points shows that $x = \pm\sqrt{\mu}$ are *stable nodes* for $0 < \mu < \frac{1}{8}$, and *stable spirals* for $\mu > \frac{1}{8}$. As μ increases through zero the stable node for $\mu < 0$ bifurcates into two stable nodes and a saddle point for $\mu > 0$. This is an example of a **pitchfork bifurcation** named after its shape in the (x, μ) plane.

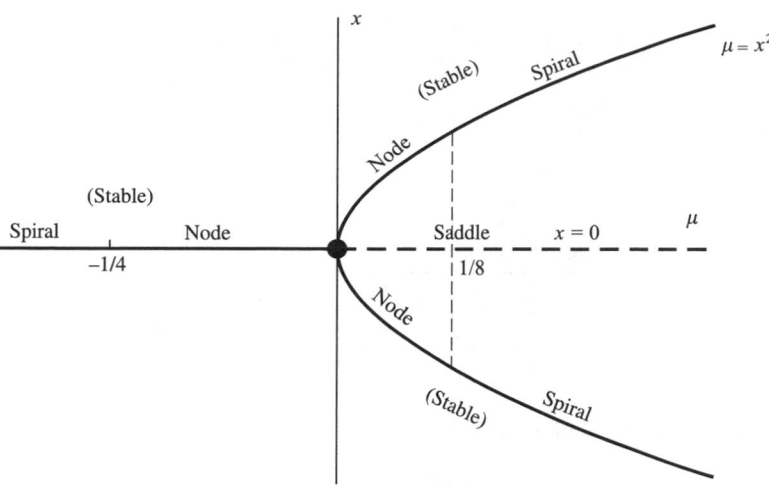

Figure 12.9 A pitchfork bifurcation.

The bead on the rotating circular wire (Example 1.12) is a possible mechanical model for a pitchfork bifurcation.

Pitchfork bifurcations are often associated with **symmetry breaking**. In the system above, $x = 0$ is the only (stable) equilibrium point for $\mu < 0$, but as μ increases through zero the system could be disturbed into either stable mode, thus destroying symmetry.

Example 12.2 *Investigate the bifurcation point of*

$$\dot{x} = x - 2y - \mu, \qquad \dot{y} = y - x^2 + \mu. \tag{12.13}$$

Equilibrium occurs where

$$x - 2y - \mu = 0, \qquad y - x^2 + \mu = 0.$$

In the (x, y, μ) space, the equilibrium points lie on the curve where plane $x - 2y - \mu = 0$ intersects the surface $y - x^2 + \mu = 0$. Elimination of μ leads to $y = x - x^2$. Hence parametically the curve of intersection is given by

$$x = w, \qquad y = w - w^2, \qquad \mu = -w + 2w^2, \tag{12.14}$$

where w is the parameter. The projection of this three-dimensional curve onto the (μ, y) plane is shown in Fig. 12.10. Bifurcation occurs where μ has a maximum or a minimum in terms of w. Thus $d\mu/dw = 0$ where $w = \frac{1}{4}$, for which value $\mu = -\frac{1}{8}$. Hence there are no equilibrium points for $\mu < -\frac{1}{8}$ and two for $\mu > -\frac{1}{8}$.

Suppose that (μ_1, x_1, y_1) is an equilibrium state with $\mu_1 \geq -\frac{1}{8}$. Let $x = x_1 + x'$ and $y = y_1 + y'$. The linear approximations to (12.13) are

$$\dot{x}' = x_1 + x' - 2y_1 - 2y' - \mu_1 = x' - 2y', \tag{12.15}$$

$$\dot{y}' = y_1 + y' - (x_1 + x')^2 + \mu_1 \approx -2x_1x' + y'. \tag{12.16}$$

Hence

$$p = 2, \quad q = 1 - 4x_1, \quad \Delta = p^2 - 4q = 16x_1.$$

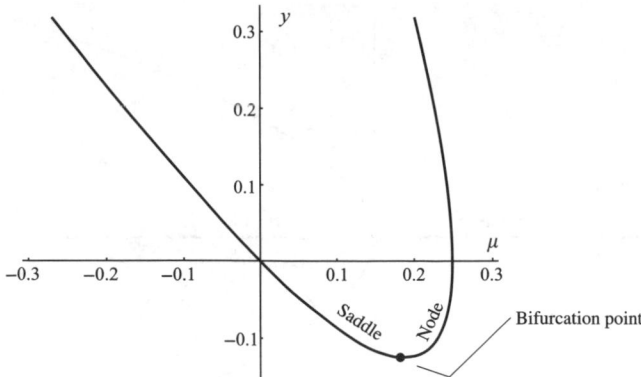

Figure 12.10   Projection of the equilibrium curve in the (x, y, μ) space on to the (μ, y) plane: the bifurcation point occurs at $\mu_1 = \frac{1}{8}$.

From (12.14), $2x_1^2 - x_1 - \mu_1 = 0$, so

$$x_1 = \tfrac{1}{4}[1 \pm \sqrt{(1 + 8\mu_1)}] \quad (\mu_1 \geq -\tfrac{1}{8}).$$

The upper sign corresponds to a *saddle* since $q < 0$, whilst the lower sign implies that $q > 0$, $\Delta > 0$, which means that this branch is an *unstable node* for $\mu_1 \geq 0$ and small (see Fig. 12.10).   ●

Example 12.3   *Show that*

$$\dot{x} = y, \qquad \dot{y} = [(x + 1)^2 - \mu + y][(x - 1)^2 + \mu + y]$$

has two equilibrium points for all μ. Discuss the bifurcation which takes place at $\mu = 0$.

Equilibrium occurs where $y = 0$ and

$$(x + 1)^2 - \mu = 0, \quad \text{or,} \quad (x - 1)^2 + \mu = 0.$$

The equilibrium curves are shown in Fig. 12.11. There are always two equilibrium points for all μ, but one point disappears for $\mu > 0$ and another appears at $\mu = 0$ (see Fig. 12.11). They are both saddle-node bifurcations with bifurcation points at $(0, -1)$ and $(0, 1)$ in the (μ, x) plane. (See Exercise 12.3 at the end of this section.)   ●

As a final example in elementary bifurcation theory, we shall consider in detail a one-parameter plane system which has several bifurcations including a **Hopf bifurcation** (see Section 12.5), and requires a three-dimensional view.

Example 12.4   *Investigate the bifurcations of the system*

$$\dot{x} = X(x, y) = 2x(\mu - x) - (x + 1)y^2, \quad \dot{y} = Y(x, y) = y(x - 1).$$

Equilibrium occurs where

$$2x(\mu - x) - (x + 1)y^2 = 0, \quad y(x - 1) = 0,$$

which can be separated into three curves in (x, y, μ) space:

$$\mathcal{C}_1 \colon x = 0, \ y = 0; \quad \mathcal{C}_2 \colon x = \mu, \ y = 0; \quad \mathcal{C}_3 \colon x = 1, \ y^2 = \mu - 1(\mu \geq 1).$$

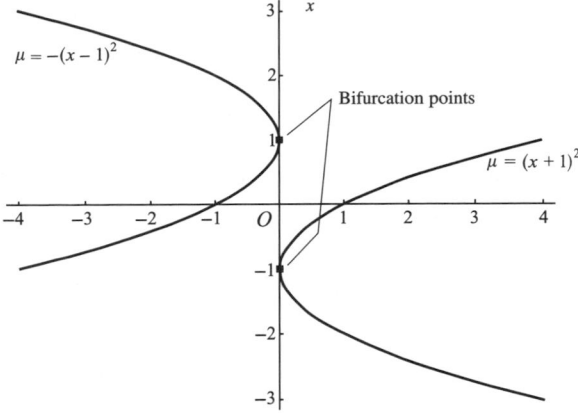

Figure 12.11

The equilibrium states are shown in Fig. 12.12. The Jacobian of $[X(x, y), Y(x, y)]^{\mathrm{T}}$ is

$$
J(x, y, \mu) = \begin{bmatrix} \dfrac{\partial X}{\partial x} & \dfrac{\partial X}{\partial y} \\[2mm] \dfrac{\partial Y}{\partial x} & \dfrac{\partial Y}{\partial y} \end{bmatrix} = \begin{bmatrix} 2\mu - 4x - y^2 & -2(x+1)y \\ y & x-1 \end{bmatrix}.
$$

We need to investigate the eigenvalues of $J(x, y, \mu)$ in each equilibrium state since this will decide the type of each linear approximation.

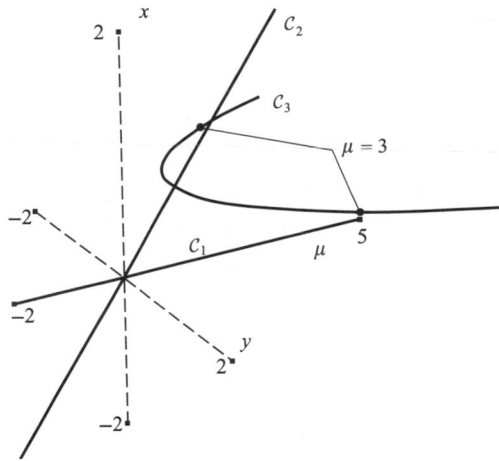

Figure 12.12   Equilibrium curves for $\dot{x} = 2x(\mu - x) - (x+1)y^2$, $\dot{y} = y(x-1)$.

(a) *Equilibrium points on* C_1. Here

$$
J(0, 0, \mu) = \begin{bmatrix} 2\mu & 0 \\ 0 & -1 \end{bmatrix}
$$

and the eigenvalues of $J(0, 0, \mu)$ are given by

$$\det[J(0, 0, \mu) - mI_2] = \begin{vmatrix} 2\mu - m & 0 \\ 0 & -1 - m \end{vmatrix} = 0,$$

so that $m = -1$ or 2μ. Hence for $\mu < 0$, the equilibrium point is a *stable node* and for $\mu > 0$ a *saddle point*.

(b) *Equilibrium points on* C_2. In this case

$$J(\mu, 0, \mu) = \begin{bmatrix} -2\mu & 0 \\ 0 & \mu - 1 \end{bmatrix}.$$

The eigenvalues of $J(\mu, 0, \mu)$ are $m = -2\mu$ or $\mu - 1$. Hence for $\mu < 0$ and $\mu > 1$, the eigenvalues are real and of opposite signs, indicating a *saddle*, whilst for $0 < \mu < 1$ the eigenvalues are both real and negative indicating a *stable node*.

From (a) and (b) it follows that there is a *transcritical bifurcation* at $\mu = 0$. The situation at $\mu = 1$ is more complicated since curve C_3 intersects C_2 there.

(c) *Equilibrium points on* C_3. On this curve

$$J(1, \sqrt{(\mu - 1)}, \mu) = \begin{bmatrix} \mu - 3 & -4\sqrt{(\mu - 1)} \\ \sqrt{(\mu - 1)} & 0 \end{bmatrix}.$$

Hence the eigenvalues are given by

$$\begin{vmatrix} \mu - 3 - m & -4\sqrt{(\mu - 1)} \\ \sqrt{(\mu - 1)} & -m \end{vmatrix} = 0, \quad \text{or} \quad m^2 - (\mu - 3)m + 4(\mu - 1) = 0.$$

The eigenvalues can be defined as

$$m_1, m_2 = \tfrac{1}{2}[\mu - 3 \pm \sqrt{\{(\mu - 11)^2 - 96\}}].$$

The eigenvalues for $\mu > 1$ are of the following types:

(I) $1 < \mu < 11 - 4\sqrt{6}$ (≈ 1.202): m_1, m_2 real and negative (stable node);

(II) $11 - 4\sqrt{6} < \mu < 3$: m_1, m_2 complex conjugates with negative real part (stable spiral);

(III) $3 < \mu < 11 + 4\sqrt{6}$ (≈ 20.798): m_1, m_2 complex conjugates with positive real part (unstable spiral);

(IV) $11 + 4\sqrt{6} < \mu$: m_1, m_2 both real and positive (unstable node).

Results (b) and (c)(I) above show that a *pitchfork bifurcation* occurs at the bifurcation point at $\mu = 1$. As μ increases through $\mu = 3$, the stable spirals at $x = 1$, $y = \sqrt{2}$ become unstable spirals. Hence $\mu = 3$ is a bifurcation point indicating a change of stability. A similar bifurcation occurs at the symmetric point $x = 1$, $y = -\sqrt{(\mu - 1)}$. These are often referred to as **Hopf bifurcations**: as we shall see in the next section this stability change of a spiral can generate a limit cycle under some circumstances, but not in this example.

To summarize, the system has three bifurcation points: a transcritical bifurcation at $\mu = 0$, a pitchfork bifurcation at $\mu = 1$ and Hopf bifurcations at $\mu = 3$.   ●

Further elementary discussion on bifurcation theory can be found in the books by Hubbard and West (1995) and Grimshaw (1990); a more advanced treatment is given by Hale and Kocak (1991).

Exercise 12.3
In Example 12.3, x satisfies

$$\ddot{x} = [(x + 1)^2 - \mu + \dot{x}][(x - 1)^2 + \mu + \dot{x}].$$

Equilibrium occurs where

$$(x + 1)^2 - \mu = 0, \quad (x - 1)^2 + \mu = 0.$$

Investigate the linear approximations for $\mu > 0$. For the equilibrium point at $x = -1 + \sqrt{\mu}$, let $x = -1 + \sqrt{\mu} + Z$, and show that Z satisfies approximately

$$\ddot{Z} - h(\mu)\dot{Z} - 2\sqrt{\mu}h(\mu)Z = 0,$$

where $h(\mu) = (\sqrt{\mu} - 2)^2 + \mu$. Confirm that this equilibrium point is a saddle. Similarly show that $x = -1 - \sqrt{\mu}$ is an unstable node.

Exercise 12.4
Discuss all bifurcations of the system

$$\dot{x} = x^2 + y^2 - 2, \quad \dot{y} = y - x^2 + \mu.$$

Compute phase diagrams for typical parameter values.

12.4 Hopf bifurcations

Some bifurcations generate limit cycles or other periodic solutions. Consider the system

$$\dot{x} = \mu x + y - x(x^2 + y^2), \tag{12.17}$$

$$\dot{y} = -x + \mu y - y(x^2 + y^2), \tag{12.18}$$

where μ is the bifurcation parameter. The system has a single equilibrium point, at the origin.
  In polar coordinates the equations become

$$\dot{r} = r(\mu - r^2), \quad \dot{\theta} = -1.$$

If $\mu \leq 0$ then the entire diagram consists of a stable spiral. If $\mu > 0$ then there is an unstable spiral at the origin surrounded by a stable limit cycle which grows out of the origin—the steps in its development are shown in Fig. 12.13. This is an example of a **Hopf bifurcation** which generates a limit cycle. Typically for such cases the linearization of (12.17) and (12.18) predicts a centre at the origin, which proves to be incorrect: the origin changes from being asymptotically stable to being unstable without passing through the stage of being a centre.

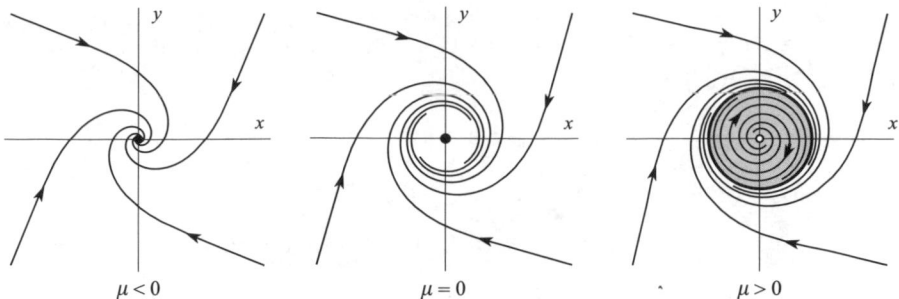

Figure 12.13 Development of a limit cycle in a Hopf bifurcation.

The following is a simple version, restricted to polar-type equations, of a more general result.

Theorem 12.1 *Given the equations* $\dot{x} = \mu x + y - xf(r)$, $\dot{y} = -x + \mu y - yf(r)$, *where* $r = \sqrt{(x^2 + y^2)}$, $f(r)$ *and* $f'(r)$ *are continuous for* $r \geq 0$, $f(0) = 0$, *and* $f(r) > 0$ *for* $r > 0$. *The origin is the only equilibrium point. Then*

 (i) *for* $\mu < 0$ *the origin is a stable spiral covering the whole plane;*

 (ii) *for* $\mu = 0$ *the origin is a stable spiral;*

(iii) *for* $\mu > 0$ *there is a stable limit cycle whose radius increases from zero as* μ *increases from zero.*

Proof   Choose the Liapunov function $V(x, y) = \frac{1}{2}(x^2 + y^2)$ (see Section 10.5 for the Liapunov method). Then

$$\dot{V}(x, y) = \frac{\partial V}{\partial x}\dot{x} + \frac{\partial V}{\partial y}\dot{y}$$

$$= x[\mu x + y - xf(r)] + y[-x + \mu y - yf(r)]$$

$$= r^2(\mu - f(r)) < 0.$$

Suppose $\mu \leq 0$. By Theorem 10.7 the origin is asymptotically stable and its domain of attraction is the entire plane. That it is a spiral can be seen from the polar form of the equations

$$\dot{r} = r(\mu - f(r)), \quad \dot{\theta} = -1.$$

Note that the case $\mu = 0$ is included.

Suppose $\mu > 0$. Reverse the time, so t is replaced by $(-t)$ in the equations. Select the same Liapunov function. Then

$$\dot{V}(x, y) = r^2(f(r) - \mu) < 0$$

for $\mu > 0$ in some interval $0 < r < r_1$ by continuity of $f(r)$ and the inequality $f(r) > 0$ for $r > 0$. This proves that the origin of the time-reversed system is stable, so the original system is unstable, and the polar equation confirms that it is a spiral in $0 < r < r_1$.

Finally in some interval $0 < \mu < \mu_1$, $\mu - f(r) = 0$ must have exactly one solution for μ_1 sufficiently small. For this value of r, the corresponding circular path is a limit cycle. The polar equations confirm that this limit cycle must be stable. ∎

Example 12.5   *Show that the equation*

$$\ddot{x} + (x^2 + \dot{x}^2 - \mu)\dot{x} + x = 0$$

exhibits a Hopf bifurcation as μ increases through zero.

We can apply Theorem 3.5 to show that the system has no periodic solutions for $\mu < 0$. Write the equation as

$$\dot{x} = X(x, y) = y,$$

$$\dot{y} = Y(x, y) = -(x^2 + y^2 - \mu)y - x.$$

Then

$$\frac{\partial X}{\partial x} + \frac{\partial Y}{\partial y} = -x^2 - 3y^2 + \mu < 0$$

for all (x, y) and all $\mu < 0$. Hence by Theorem 3.5, there can be no closed paths in the phase plane.

For the critical case $\mu = 0$, we can use the Liapunov function

$$V(x, y) = \tfrac{1}{2}(x^2 + y^2)$$

which results in

$$\dot{V}(x, y) = -(x^2 + y^2)y^2.$$

Since $\dot{V}(x, y)$ vanishes along the line $y = 0$, but is otherwise strictly negative, the origin is an asymptotically stable equilibrium point (see Section 10.5). For $\mu > 0$ the system has a stable limit cycle with path $x^2 + y^2 = \mu$ (cf. Example 1.6) which evidently emerges from the origin at $\mu = 0$ with a radius which increases with μ. ●

Exercise 12.5
A system is given by

$$\dot{r} = -\mu r + r^2 - r^3, \quad \dot{\theta} = -1, \quad 1 > \mu > 0$$

in polar coordinates. As μ decreases from 1, show that a limit cycle of radius $\tfrac{1}{2}$ appears when $\mu = \tfrac{1}{4}$, which then bifurcates into stable and unstable limit cycles. Sketch the phase diagrams for $\tfrac{1}{4} < \mu < 1$, $\mu = \tfrac{1}{4}$ and $\mu < \tfrac{1}{4}$.

Exercise 12.6
Discuss the bifurcations of the limit cycles of the system

$$\dot{r} = (r^4 - 2r^2 + \mu)r, \quad \dot{\theta} = -1.$$

12.5   Higher-order systems: manifolds

This section contains further development of n-th order autonomous systems introduced in Section 8.9. There we considered the real system

$$\dot{x} = X(x), \quad x = [x_1, x_2, \ldots, x_n]^{\mathrm{T}}, \tag{12.19}$$

$$X = [X_1, X_2, \ldots, X_n]^{\mathrm{T}} \in \mathbb{R}^n,$$

which has equilibrium points where

$$X(x) = 0.$$

If $x = x_0$ is an equilibrium point then the linear approximation to (12.19) will be, after putting $x = x_0 + x'$,

$$\dot{x}' = J(x_0)x' = Ax', \quad \text{say,}$$

where J is the Jacobian matrix of derivatives of X evaluated at $x = x_0$, namely

$$J(x_0) = [J_{ij}(x_0)] = \left[\frac{\partial X_i(x)}{\partial x_j} \right]_{x=x_0} \quad (i, j = 1, 2, \ldots, n).$$

The stability classification of equilibrium points of linear approximations will depend on the eigenvalues of A as we explained in Section 8.9. If all the eigenvalues of A have negative real part then the linear approximation is asymptotically stable and so may be is the nonlinear system. Conditions under which this holds good are provided in Section 10.8. If at least one eigenvalue has positive real part then the equilibrium point will be unstable. (There are also critical case where the eigenvalues have either negative real part or zero real part (imaginary eigenvalues) which require further investigation regarding stability.)

In this section we shall look specifically at the solutions which approach the equilibrium point either as $t \to \infty$ or $t \to -\infty$. In the case of a two-dimensional saddle point, such time solutions correspond to the four asymptotes or separatrices (see, e.g., Fig. 2.6). Higher-order systems will be introduced by way of the following three-dimensional example.

Example 12.6   *Find the eigenvalues and eigenvectors associated with*

$$\dot{x} = \begin{bmatrix} \dot{x} \\ \dot{y} \\ \dot{z} \end{bmatrix} = \begin{bmatrix} 1 & 2 & 1 \\ 2 & 1 & 1 \\ 1 & 1 & 2 \end{bmatrix} \begin{bmatrix} x \\ y \\ z \end{bmatrix},$$

and identify stable and unstable solutions.

The eigenvalues of A are given by

$$\det[A - mI_3] = \begin{vmatrix} 1-m & 2 & 1 \\ 2 & 1-m & 1 \\ 1 & 1 & 2-m \end{vmatrix} = (4-m)(-1-m)(1-m) = 0.$$

Let the eigenvalues be denoted by $m_1 = 4$, $m_2 = 1$, $m_3 = -1$. The eigenvalues are real with two positive and one negative so that the origin will be an unstable equilibrium point. The corresponding eigenvectors r_1, r_2, r_3 are given by

$$r_1 = \begin{bmatrix} 1 \\ 1 \\ 1 \end{bmatrix}, \quad r_2 = \begin{bmatrix} -1 \\ -1 \\ 2 \end{bmatrix}, \quad r_3 = \begin{bmatrix} 1 \\ -1 \\ 0 \end{bmatrix},$$

so that the general solution is

$$\begin{bmatrix} x \\ y \\ z \end{bmatrix} = \alpha \begin{bmatrix} 1 \\ 1 \\ 1 \end{bmatrix} e^{4t} + \beta \begin{bmatrix} -1 \\ -1 \\ 2 \end{bmatrix} e^{t} + \gamma \begin{bmatrix} 1 \\ -1 \\ 0 \end{bmatrix} e^{-t},$$

where α, β, and γ are constants.

The equilibrium point is unstable since two eigenvalues are positive and one is negative. It can be described as being of *three-dimensional saddle* type. Any solutions for which $\alpha = \beta = 0$ initially will always lie on the *line* given parametrically by

$$x = \gamma e^{-t}, \ y = -\gamma e^{-t}, \ z = 0, \text{ or by } x + y = 0, \ z = 0.$$

Any solution which starts on this straight line will approach the origin as $t \to \infty$. This line of initial points is an example of a **stable manifold** of the equilibrium point. Any solution for which $\gamma = 0$ will lie in the *plane* $x - y = 0$ (obtained by eliminating α and β in the solutions for x, y and z) for all t, and approach the origin as $t \to -\infty$. Every solution which starts on the plane $x - y = 0$ $((x, y) \neq (0, 0))$ approaches the origin as $t \to -\infty$. This collection of initial points defines the **unstable manifold** of the equilibrium point.   ●

We have used the term **manifold** rather loosely in the previous example. Technically a manifold is a subspace of dimension $m \leq n$ in $\mathbb{R}^n$ usually satisfying continuity and differentiability conditions. Thus the sphere surface $x^2 + y^2 + z^2 = 1$ is a manifold of dimension 2 in $\mathbb{R}^3$, the solid sphere $x^2 + y^2 + z^2 < 1$ is a manifold of dimension 3 in $\mathbb{R}^3$; and the parabola $y = x^2$ is a manifold of dimension 1 in $\mathbb{R}^2$. If a solution of a differential equation starts on a given space, surface or curve (that is, a manifold) remains within it *for all time*, then the manifold is said to be **invariant**. For example, van der Pol's equation has a limit cycle (Section 11.3). Any solution which starts on the limit cycle will remain on it for all time. Hence this closed curve in the phase plane is an **invariant manifold** (the term **set** is also used instead of manifold). Equilibrium points are invariant manifolds, as are any complete phase paths, or sets of phase paths. Since the context is clear in this book, and since we only discuss invariant manifolds, the term invariant has been dropped. Further details of the mathematical background can be found in the book by Arrowsmith and Place (1990).

We shall use the result of the following Example to illustrate how a homoclinic path lies on both the stable and unstable manifolds on the equilibrium point.

Example 12.7 *Consider the system*

$$\dot{x} = -\tfrac{1}{2}x - y + \tfrac{1}{2}z + 2y^3, \tag{12.20}$$

$$\dot{y} = -\tfrac{1}{2}x + \tfrac{1}{2}z, \tag{12.21}$$

$$\dot{z} = z - (x + z)y^2. \tag{12.22}$$

Where is the system in equilibrium? Investigate the linear approximation in the neighbourhood of the origin. Confirm that

$$\dot{x} - 2\dot{y} + \dot{z} = (x - 2y + z)(\tfrac{1}{2} - y^2), \tag{12.23}$$

and deduce that time-solutions exist that lie in the plane $x - 2y + z = 0$. *Hence show that there exists a homoclinic solution of the origin given by*

$$x = (1 + \tanh t)\operatorname{sech} t, \quad y = \operatorname{sech} t, \quad z = (1 - \tanh t)\operatorname{sech} t.$$

For equilibrium, we must have

$$-\tfrac{1}{2}x - y + \tfrac{1}{2}z + 2y^3 = -\tfrac{1}{2}x + \tfrac{1}{2}z = z - (x + z)y^2 = 0.$$

Hence

$$z = x, \quad y(1 - 2y^2) = 0, \quad x(1 - 2y^2) = 0.$$

Thus equilibrium occurs at the origin and at all points on the line of intersection of the planes $z = x, y = \pm 1/\sqrt{2}$. Near the origin the linear approximation of the differential equations is

$$\begin{bmatrix} \dot{x} \\ \dot{y} \\ \dot{z} \end{bmatrix} = \begin{bmatrix} -\tfrac{1}{2} & -1 & \tfrac{1}{2} \\ -\tfrac{1}{2} & 0 & \tfrac{1}{2} \\ 0 & 0 & 1 \end{bmatrix} \begin{bmatrix} x \\ y \\ z \end{bmatrix}.$$

Hence the eigenvalues are given by

$$\begin{vmatrix} -\tfrac{1}{2} - m & -1 & \tfrac{1}{2} \\ -\tfrac{1}{2} & -m & \tfrac{1}{2} \\ 0 & 0 & 1 - m \end{vmatrix} = 0,$$

which reduces to

$$(m - 1)(2m - 1)(m + 1) = 0.$$

Denote the solutions by $m_1 = -1, m_2 = \tfrac{1}{2}$ and $m_3 = 1$. The eigenvectors are given by

$$r_1 = \begin{bmatrix} 2 \\ 1 \\ 0 \end{bmatrix}, \quad r_2 = \begin{bmatrix} -1 \\ 1 \\ 0 \end{bmatrix}, \quad r_3 = \begin{bmatrix} 0 \\ \tfrac{1}{2} \\ 1 \end{bmatrix},$$

from which it follows that the general solution is

$$\begin{bmatrix} x \\ y \\ z \end{bmatrix} = \alpha r_1 e^{-t} + \beta r_2 e^{\frac{1}{2}t} + \gamma r_3 e^t, \tag{12.24}$$

where α, β, and γ are constants. Since the eigenvalues are real but not the same sign the origin is an unstable equilibrium point of saddle type.

From the original differential equations (12.20), (12.21) and (12.22)

$$\frac{\mathrm{d}}{\mathrm{d}t}(x - 2y + z) = (x - 2y + z)\left(\frac{1}{2} - y^2\right).$$

Therefore there exist time-solutions which lie in the plane

$$x - 2y + z = 0. \tag{12.25}$$

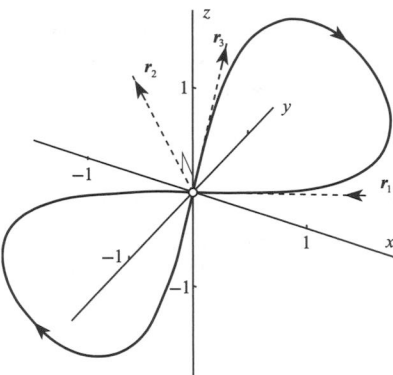

Figure 12.14 Homoclinic paths of the origin and the eigenvector directions in Example 12.9.

To obtain these solutions use (12.25) to eliminate x from (12.21) and (12.22), leading respectively to

$$\dot{y} = z - y, \tag{12.26}$$

$$\dot{z} = z - 2y^3. \tag{12.27}$$

By eliminating z and $\dot{z}$ between (12.26) and (12.27) we obtain $\ddot{y} = y - 2y^3$, which has a particular solution

$$y(t) = \operatorname{sech} t. \tag{12.28}$$

By substituting (12.28) into (12.27) we obtain $\dot{z} = z - 2\operatorname{sech}^3 t$, which has a solution

$$z(t) = \operatorname{sech} t (1 - \tanh t). \tag{12.29}$$

Finally, from (12.25) again, we have

$$x(t) = 2y(t) - z(t) = \operatorname{sech} t (1 + \tanh t). \tag{12.30}$$

The functions $x(t)$, $y(t)$, $z(t) \to 0$ as $t \to \pm\infty$, confirming that the corresponding phase path is homoclinic to the origin. This path (which lies in the quadrant $x > 0$, $y > 0$, $z > 0$ on the plane (12.25)), together with its reflection in the origin, which is also homoclinic there, are shown in Fig. 12.14. ●

In Example 12.7, the eigenvalues associated with the origin are $-1, \frac{1}{2}, 1$: two are positive and one is negative. The homoclinic path in $x > 0$, which is known *exactly* by (12.28), (12.29) and (12.30), emerges from the origin in the direction of the eigenvector $\boldsymbol{r}_3 (\alpha = \beta = 0)$ (see eqn (12.24)), and approaches the origin in the direction of $\boldsymbol{r}_1 (\beta = \gamma = 0)$. Since the origin has only one negative eigenvalue *all* initial coordinates which lie on this homoclinic path are the *only* points whose solutions approach the origin as $t \to \infty$. This curve defines explicitly the **stable manifold** of the origin.

Now consider all initial points in the phase space whose solutions approach the origin as $t \to -\infty$. Since the origin has *two* positive eigenvalues the points will lie on a two-dimensional manifold which is the **unstable manifold**. This surface must contain the stable manifold as a curve embedded in it. The **tangent plane to the unstable manifold** as $t \to \infty$ is defined locally near the origin by the plane through the eigenvectors $\boldsymbol{r}_2$ and $\boldsymbol{r}_3$.

Generally the **notations** W^s and W^u will be used to denote stable and unstable manifolds. Stable and unstable manifolds can be computed (as we shall see in Section 13.5) by considering appropriate initial points close to the origin and then extrapolating solutions either forwards or backwards in time. However they can be quite complex structures which are difficult to visualize. The following rather contrived example shows explicit manifolds.

Example 12.8   *Find the stable and unstable manifolds of*

$$\dot{x} = y, \quad \dot{y} = x - x^3, \quad \dot{z} = 2z.$$

This is really a plane system with an added separable equation for z. There are equilibrium points at $(0, 0, 0)$, $(1, 0, 0)$, and $(-1, 0, 0)$. The eigenvalues of the linear approximation at the origin are $m_1 = -1, m_2 = 1, m_3 = 2$ with corresponding eigenvectors

$$\boldsymbol{r}_1 = [-1, 1, 0]^{\mathrm{T}}, \quad \boldsymbol{r}_2 = [1, 1, 0]^{\mathrm{T}}, \quad \boldsymbol{r}_3 = [0, 0, 1]^{\mathrm{T}}.$$

Two eigenvectors lie in the (x, y) plane and one points along the z axis. The homoclinic paths associated with the origin are given by (see, Example 3.10)

$$x = \pm\sqrt{2}\,\mathrm{sech}\,t, \quad y = \mp\sqrt{2}\,\mathrm{sech}\,t\,\tanh t, \quad z = 0.$$

Only solutions with initial points on these curves, which define the *stable manifolds* W^{s} of the origin, approach the origin as $t \to \infty$ since $z = Ce^{2t}$. On the other hand, solutions with initial points on the *cylinders*

$$x = \pm\sqrt{2}\,\mathrm{sech}\,t, \qquad y = \mp\sqrt{2}\,\mathrm{sech}\,t\,\tanh t$$

approach the origin as $t \to -\infty$ since $z = Ce^{2t}$. These cylinders define the *unstable manifolds* W^{u} of the origin (see Fig. 12.15).   ●

Of course the previous example is a special case in which the stable manifold is embedded in the unstable manifold as shown in Fig. 12.15. If friction is added to the system as in the *damped* Duffing equation

$$\dot{x} = y, \quad \dot{y} = -ky + x - x^3, \quad \dot{z} = 2z,$$

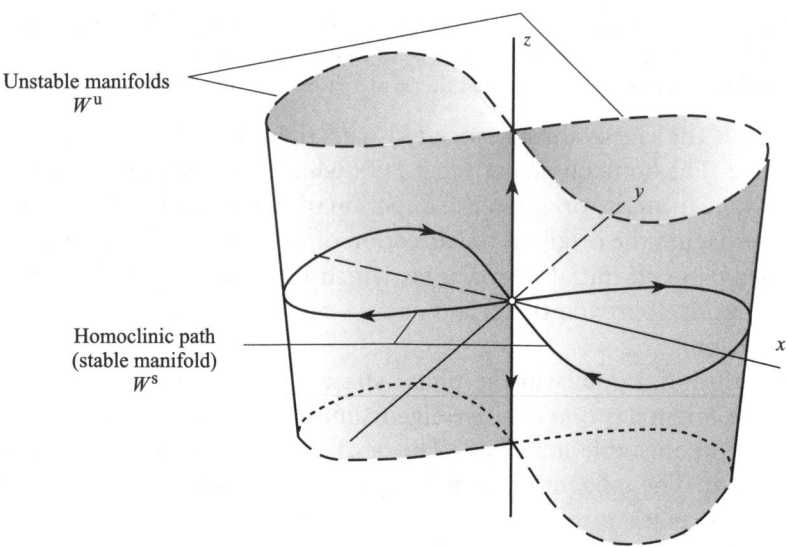

Figure 12.15   Stable and unstable manifolds for $\dot{x} = y$, $\dot{y} = x - x^3$, $\dot{z} = 2z$.

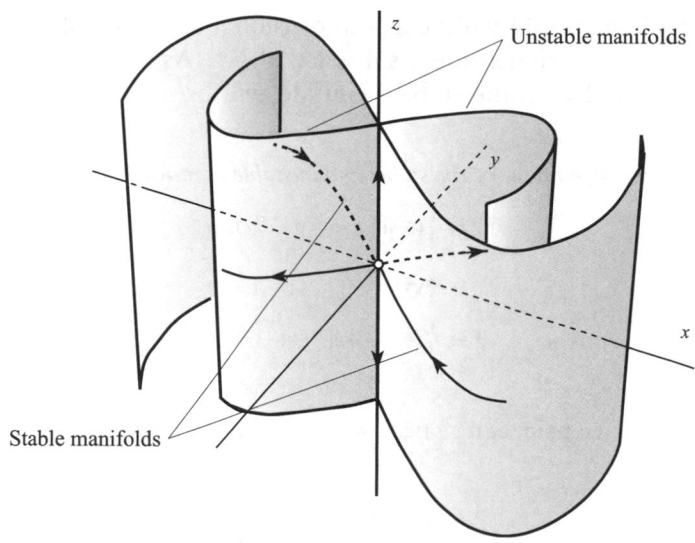

Figure 12.16   Stable and unstable manifolds for $\dot{x} = y$, $\dot{y} = -ky + x - x^3$, $\dot{z} = 2z$.

with separable z equation, then, for $k > 0$, the origin still has two positive eigenvalues and one negative eigenvalue. Here there is no homoclinic path. The stable manifolds are still curves, and the unstable manifolds remain as cylindrical surfaces as shown in Fig. 12.16, but the two no longer meet except at the origin.

Stable and unstable manifolds must be associated with either equilibrium points or limit cycles as we shall see in the next chapter. For example, we can discuss manifolds of an equilibrium point of a system, but manifolds without such a reference point or limit cycle would be meaningless. They do not exist as separate entities.

12.6   Linear approximation: centre manifolds

Suppose that an nth order nonlinear autonomous system has an equilibrium point at the origin (if it is at some other point, we can always move it to the origin by a translation of $\boldsymbol{x}$), with linear approximation

$$\dot{\boldsymbol{x}} = \boldsymbol{A}\boldsymbol{x},$$

where $\boldsymbol{A}$ is a constant $n \times n$ matrix. The manifolds of the origin are determined by the signs of the real parts of the eigenvalues of $\boldsymbol{A}$. If all the eigenvalues of $\boldsymbol{A}$ have negative real part then the stable manifold will be the whole of $\mathbb{R}^n$ for the linear system, and there will be no unstable manifold. For a nonlinear system the stable manifold will occupy a subset of $\mathbb{R}^n$ including a neighbourhood of the origin.

We shall concentrate on the local behaviour of the nonlinear system by looking at its linear approximation. If $\boldsymbol{A}$ has $p < n$ eigenvalues with negative and $n - p$ eigenvalues with positive real part then the stable manifold will be a p dimensional subspace of $\mathbb{R}^n$, whilst the unstable manifold will be a $n - p$ dimensional subspace of $\mathbb{R}^n$. As we have seen in the previous section if $n = 3$ and $p = 2$ then the stable manifold will be a plane and the unstable one a straight line.

Example 12.9   *Find the local behaviour of the stable and unstable manifolds of*

$$\dot{x} = \tfrac{1}{4}(-\sin x + 5y - 9z),$$

$$\dot{y} = \tfrac{1}{4}(4x - e^z + 1),$$

$$\dot{z} = \tfrac{1}{4}(4x - 4\sin y + 3z)$$

near the origin.

The system has an equilibrium point at the origin, where its linear approximation is $\dot{x}' = Ax'$, and

$$A = \tfrac{1}{4}\begin{bmatrix} -1 & 5 & -9 \\ 4 & 0 & -1 \\ 4 & -4 & 3 \end{bmatrix}.$$

The eigenvalues of A are $m_1 = -\tfrac{1}{4} - i$, $m_2 = -\tfrac{1}{4} + i$, $m_3 = 1$, and its corresponding eigenvectors are

$$r_1 = \begin{bmatrix} -i \\ 1 \\ 1 \end{bmatrix}, \quad r_2 = \begin{bmatrix} i \\ 1 \\ 1 \end{bmatrix}, \quad r_3 = \begin{bmatrix} 1 \\ 1 \\ 0 \end{bmatrix}.$$

Hence the general linear approximation is

$$\begin{bmatrix} x' \\ y' \\ z' \end{bmatrix} = \alpha r_1 e^{(-\frac{1}{4} - i)t} + \bar{\alpha} r_2 e^{(-\frac{1}{4} + i)t} + \beta r_3 e^t,$$

where α is a complex constant and β is a real constant. Solutions for which $\beta = 0$ lie in the plane $y = z$, which defines the tangent plane to the *stable manifold* of the nonlinear system at the origin. Solutions for which $\alpha = 0$ are given by

$$x = \beta e^t, \quad y = \beta e^t, \quad z = 0,$$

which is a straight line defining the tangent to the *unstable manifold*. The manifolds of the linear approximation are shown in Fig. 12.17.   ●

This is a useful stage to summarize properties of manifolds for equilibrium points of higher order systems which have linearized approximations with eigenvalues having nonzero real parts. As stated in Section 2.5, such equilibrium points are said to be **hyperbolic**. Consider the autonomous system.

$$\dot{x} = f(x) = Ax + g(x), \quad x \in \mathbb{R}^n, \tag{12.31}$$

where $f(0) = g(0) = 0$. In (12.31) it is assumed that $x = 0$ is an isolated equilibrium point (any equilibrium point can always be translated to the origin). It is also assumed that

$$\|g\| = o(\|x\|) \quad \text{as } \|x\| \to 0,$$

where the norms are defined in Section 8.3.

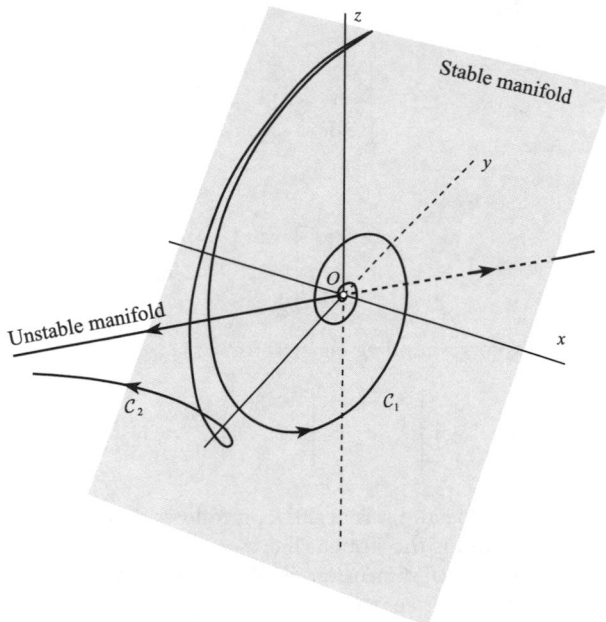

Figure 12.17 Stable and unstable manifolds of the linear approximations at the origin in Example 12.9. C_1 is a phase path in the stable manifold, whilst C_2 is a path starting from a neighbouring initial point to that of C_1. Ultimately this path approaches the unstable manifold.

The stable and unstable manifolds of the linear approximation

$$\dot{x}' = Ax' \tag{12.32}$$

are known as **linear manifolds** of the original system (12.31) and their spaces are denoted by E^s and E^u. Here E^s is a subspace defined by the k, ($0 \le k \le n$) eigenvectors whose eigenvalues have negative real parts, and E^u is the subspace defined by the eigenvectors whose eigenvalues have positive real parts. The full nonlinear equations (12.31) will have stable and unstable manifolds denoted by W^s and W^u of the origin, to which the subspaces E^s and E^u are tangential at $x = 0$.

So far the case in which pure imaginary eigenvalues occur has been excluded. If any eigenvalues are pure imaginary, then a third manifold E^c arises for the linear approximation. This is known as the **centre manifold**.

The definitions imply that, in general, any linear approximation at the origin may contain any or all of the three manifolds E^s, E^u, and E^c. This is illustrated in the following example of a linear system.

Example 12.10 *Find the manifolds for the linear system*

$$\dot{x} = -x + 3y,$$
$$\dot{y} = -x + y - z,$$
$$\dot{z} = -y - z.$$

The eigenvalues of

$$A = \begin{bmatrix} -1 & 3 & 0 \\ -1 & 1 & -1 \\ 0 & -1 & -1 \end{bmatrix}$$

are given by

$$\begin{bmatrix} -1-m & 3 & 0 \\ -1 & 1-m & -1 \\ 0 & -1 & -1-m \end{bmatrix} = -(m+1)(m^2+1) = 0.$$

Let $m_1 = -i$, $m_2 = i$, $m_3 = -1$: the corresponding eigenvectors are

$$r_1 = \begin{bmatrix} -3 \\ -1+i \\ 1 \end{bmatrix}, \quad r_2 = \begin{bmatrix} -3 \\ -1-i \\ 1 \end{bmatrix}, \quad r_3 = \begin{bmatrix} -1 \\ 0 \\ 1 \end{bmatrix}.$$

Since $m_3 = -1$ is the only real eigenvalue and it is negative, it follows that there is no unstable manifold, and that, parametrically, the stable manifold is the straight line $x = -u, y = 0, z = u (-\infty < u < \infty)$. The centre manifold E^c (which must always have even dimension) is the plane $x + 3z = 0$. In algebraic terminology, we say that E^s is **spanned** by r_3, and that E^c is spanned by r_1 and r_2, written as

$$E^s = \text{span}\{r_3\}, \quad E^u = \text{span}\{0\}, \quad E^c = \text{span}\{r_1, r_2\}.$$

This terminology entails the linear structure of the associated solution sets. For example, E^c consists of the (real) solution set $\alpha r_1 e^{m_1 t} + \beta r_2 e^{m_2 t}$, where $m_2 = \bar{m}_1$, $r_2 = \bar{r}_1$ and $\beta = \bar{\alpha}$, α being an arbitrary complex constant. Some phase paths of the system are shown in Fig. 12.18.   ●

For a nonlinear system having an equilibrium point with *linear* manifolds E^s, E^u, and E^c, the actual manifolds W^s, W^u, and W^c are locally *tangential* to E^s, E^u, and E^c. Whilst solutions on W^s and W^u behave asymptotically as solutions on E^s and E^u as $t \to \pm\infty$ respectively, the same

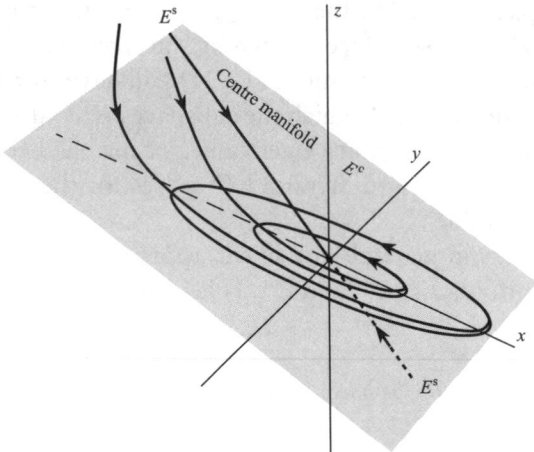

Figure 12.18   The centre manifold E^c lies in the plane $x + 3z = 0$, and the stable manifold E^s is the line $x = -t$, $y = 0$, $z = t$.

is not true of W^c. Solutions on W^c can be stable, unstable, or oscillatory (more information on the mathematics of manifolds can be found in the book by Wiggins 1990).

In the *linear* problem in Example 12.10, the origin is a stable equilibrium point but it is *not asymptotically stable*, due to the presence of E^c. The question arises: if, at an equilibrium point, the linear approximation to a nonlinear problem has a centre manifold, what can we say anything about the stability of the equilibrium point? The subject of pure imaginary eigenvalues and their relation with stability is known as **centre manifold theory**, and a detailed account of it can be found in the book by Carr (1981). Here we shall look at an illustrative example.

Consider the system

$$\dot{x} = -y - xz, \tag{12.33}$$

$$\dot{y} = x - y^3, \tag{12.34}$$

$$\dot{z} = -z - 2xy - 2x^4 + x^2. \tag{12.35}$$

The system has three equilibrium points, at $(0,0,0)$, $(2^{\frac{1}{2}}, 2^{-\frac{5}{6}}, 2^{\frac{1}{3}})$, $(-2^{\frac{1}{2}}, -2^{-\frac{5}{6}}, 2^{\frac{1}{3}})$. The equilibrium point at the origin has the linear approximation given by

$$\dot{x}' = \begin{bmatrix} \dot{x}' \\ \dot{y}' \\ \dot{z}' \end{bmatrix} = \begin{bmatrix} 0 & -1 & 0 \\ 1 & 0 & 0 \\ 0 & 0 & -1 \end{bmatrix} \begin{bmatrix} x' \\ y' \\ z' \end{bmatrix} = Ax'.$$

The eigenvalues of A are $m_1 = i, m_2 = -i, m_1 = -1$ with corresponding eigenvectors

$$r_1 = \begin{bmatrix} 1 \\ -i \\ 0 \end{bmatrix}, \quad r_2 = \begin{bmatrix} 1 \\ i \\ 0 \end{bmatrix}, \quad r_3 = \begin{bmatrix} 0 \\ 0 \\ 1 \end{bmatrix},$$

Since A has two pure imaginary eigenvalues, it follows that the center manifold E^c is given by the coordinate plane $z = 0$, and the stable manifold E^s by the z axis.

System (12.33)–(12.35) has a *surface* through the origin on which solutions exist. From (12.33) and (12.35) after elimination of y,

$$\dot{z} = -z + x^2 + 2x\dot{x} - 2x^4 + 2x^2z,$$

or

$$\frac{d}{dt}(z - x^2) = -(z - x^2)(1 - 2x^2).$$

Thus there are solutions for which $z = x^2$: this is a manifold to which E^c, given by $z = 0$, is *tangential* at the origin as shown in Fig. 12.19. On this surface x and y satisfy

$$\dot{x} = -y - x^3, \tag{12.36}$$

$$\dot{y} = x - y^3, \tag{12.37}$$

from (12.33) and (12.34). Thus solutions which start on this surface remain on it.

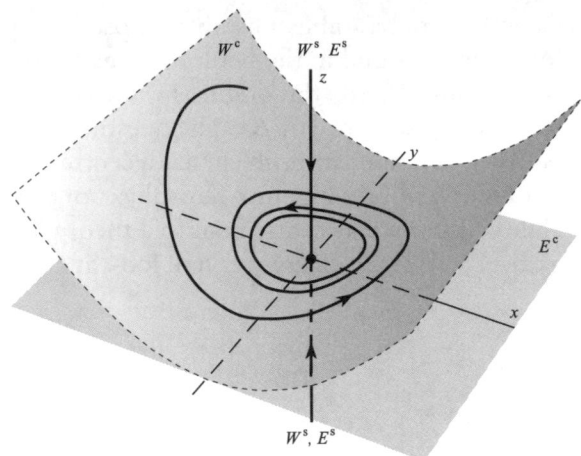

Figure 12.19 The centre manifolds E^c and W^c of the origin for the system $\dot{x} = -y - xz$, $\dot{y} = x - y^3$, $\dot{z} = -z - 2xy - 2x^4 + x^2$. A stable spiral solution is shown on W^c which is the surface $z = x^2$. The origin is asymptotically stable.

For (12.36) and (12.37) introduce a Liapunov function $V(x, y) = x^2 + y^2$ (see Chapter 10). It follows that

$$\frac{\mathrm{d}V}{\mathrm{d}t} = -\frac{\partial V}{\partial x}\dot{x} - \frac{\partial V}{\partial y}\dot{y} = -2x^4 - 2y^4 < 0$$

for $(x, y) \neq (0, 0)$. Hence by Theorems 10.7 or 10.12, the system given by (12.36) and (12.37) has an asymptotically stable equilibrium point at the origin. In fact all paths which start on the surface $z = x^2$ remain on it for all subsequent times, and approach the origin at $t \to \infty$. Also there is an *exact* solution $x = y = 0, z = \beta \mathrm{e}^{-t}$. All this suggests that the origin is, overall, asymptotically stable.

The **centre manifold** W^c is defined to be a set of paths in the nonlinear system which has E^c as its tangent plane at the origin. Thus W^c in the case considered above is the surface $z = x^2$. A computed spiral path on W^c is shown in Fig. 12.19.

This example illustrates an essential feature of a **centre manifold theorem** which can be summarized as follows.

Consider an n-dimensional system expressed in the form

$$\dot{x} = Ax + h(x), \tag{12.38}$$

where *all* linear terms on the right-hand side are included in the term Ax. Suppose that the constant matrix A can be represented in the **block-diagonal form**

$$A = \begin{bmatrix} B & 0 \\ 0 & C \end{bmatrix}, \tag{12.39}$$

where C is an $m \times m$ matrix, and B is an $(n - m) \times (n - m)$ matrix. All other elements are zero. (A block-diagonal matrix is one in which the leading diagonal consists of square matrices with zeros everywhere else. Also, a linear transformation may be required to formulate the system in block-diagonal form.) It is assumed that $h(0) = 0$, so that the origin is an equilibrium point.

Let x be partitioned in the form

$$x = \begin{bmatrix} u \\ v \end{bmatrix},$$

where u is an $n - m$ vector and v an m vector. Then (12.38) can be replaced by the two vector equations

$$\dot{u} = Bu + f(u, v), \quad (n - m \text{ equations}),$$
$$\dot{v} = Cv + g(u, v), \quad (m \text{ equations}),$$

where

$$h(x) = \begin{bmatrix} f(u, v) \\ g(u, v) \end{bmatrix}.$$

It is assumed that the origin is an isolated equilibrium point, and that f and g are sufficiently smooth for any required derivatives and expansions to exist.

The n eigenvalues of A (assumed to be all different) comprise the $(n-m)$ eigenvalues of B and the m eigenvalues of C. centre manifold theory is concerned with the stability of equilibrium points which have some imaginary eigenvalues.

Theorem 12.2 (centre manifold)   *Suppose that the matrix B has only imaginary eigenvalues (which implies that $n - m$ must be an even integer) and that the eigenvalues of C have negative real part. The stable manifold W^s has dimension m, and the centre manifold W^c has dimension $n - m$. Suppose that W^c can be represented locally by the manifold $v = p(u)$.*
If $u = 0$ is an asymptotically stable equilibrium point of the $(n - m)$-dimensional system

$$\dot{u} = Bu + f(u, p(u)),$$

then the origin of

$$\dot{x} = Ax + h(x)$$

is asymptotically stable. ■

In other words what happens to the solutions on $v = p(u)$ projected onto W^c decides the asymptotic stability of the origin. More details and proofs are given by Carr (1981) and Guckenheimer and Holmes (1983).

Problems

12.1 Find the bifurcation points of the linear system $\dot{x} = A(\lambda)x$ with $x = [x_1, x_2]^T$ and $A(\lambda)$ given by

(i) $A(\lambda) = \begin{bmatrix} -2 & \frac{1}{4} \\ -1 & \lambda \end{bmatrix}$;

(ii) $A(\lambda) = \begin{bmatrix} \lambda & \lambda - 1 \\ 1 & \lambda \end{bmatrix}$.

12.2 In a conservative system, the potential is given by $\mathcal{V}(x, \lambda) = \frac{1}{3}x^3 + \lambda x^2 + \lambda x$ (cf. eqn (12.2)). Find the equilibrium points of the system, and show that it has bifurcation points at $\lambda = 0$ and at $\lambda = 1$. What type of bifurcations occur for $\lambda < 0$ and $\lambda > 1$?

12.3 Let $V(x, \lambda, \mu) = \frac{1}{4}x^4 - \frac{1}{2}\lambda x^2 + \mu x$ as in eqn (12.4). Draw projections of the bifurcations given by the cusp surface $x^3 - \lambda x + \mu = 0$ on to both the (x, λ)-plane and the (x, μ)-plane. Sketch the projection of the cusp on to the (μ, λ) plane.

12.4 Discuss the stability and bifurcation of the equilibrium points of the parameter-dependent conservative system

$$\ddot{x} = -V_x(x, \lambda)$$

where $V(x, \lambda) = \frac{1}{4}x^4 - \frac{1}{2}\lambda x^2 + \lambda x$.

12.5 Discuss bifurcations of the system $\dot{x} = y^2 - \lambda$, $\dot{y} = x + \lambda$.

12.6 Find the bifurcation points of $\dot{x} = y^2 - \lambda$, $\dot{y} = x + \lambda$.

12.7 Consider the system $\dot{x} = y, \dot{y} = x(\lambda - x^2)$, $-\infty < \lambda < \infty$. Investigate the phase diagrams for $\lambda < 0, \lambda = 0$, and $\lambda > 0$. Describe the bifurcation of the system as λ increases through zero.

12.8 Discuss the bifurcations of $\dot{x} = (y^2 - \lambda)y$, $\dot{y} = x + \lambda$.

12.9 Investigate the bifurcation of the system

$$\dot{x} = x, \qquad \dot{y} = y^2 - \lambda,$$

at $\lambda = 0$. Show that, for $\lambda > 0$, the system has an unstable node at $(0, \sqrt{\lambda})$, and a saddle point at $(0, -\sqrt{\lambda})$. Sketch the phase diagrams for $\lambda < 0, \lambda = 0$. and $\lambda > 0$.

12.10 A homoclinic path (Section 3.6) is a phase path which joins an equilibrium point to itself in an autonomous system. Show that $\dot{x} = y, \dot{y} = x - x^2$ has such a path and find its equation. Sketch the phase paths for the perturbed system

$$\dot{x} = y + \lambda x, \qquad \dot{y} = x - x^2,$$

for both $\lambda > 0$ and $\lambda < 0$. (The homoclinic saddle connection is destroyed by the perturbation; the system undergoes what is known as a **homoclinic bifurcation** (Section 3.6) at $\lambda = 0$.)

12.11 A heteroclinic path (Section 3.6) is a phase path which joins two different equilibrium points. Find the heteroclinic saddle connection for the system $\dot{x} = xy, \dot{y} = 1 - y^2$. Sketch the phase paths of the perturbed system $\dot{x} = xy + \lambda, \dot{y} = 1 - y^2$ for both $\lambda > 0$ and $\lambda < 0$.

12.12 Let

$$\dot{x} = -\mu x - y + x/(1 + x^2 + y^2),$$

$$\dot{y} = x - \mu y + y/(1 + x^2 + y^2).$$

Show that the equations display a Hopf bifurcation as $\mu > 0$ decreases through $\mu = 1$. Find the radius of the periodic path for $0 < \mu < 1$.

12.13 Show that the system

$$\dot{x} = x - \gamma y - x(x^2 + y^2), \quad \dot{y} = \gamma x + y - y(x^2 + y^2) - \gamma \quad (\gamma > 0),$$

has a bifurcation point at $\gamma = \frac{1}{2}$, by investigating the numbers of equilibrium points for $\gamma > 0$. Compute the phase diagram for $\gamma = \frac{1}{4}$.

12.14 Let $\dot{x} = Ax$, where $x = [x, y, z]^T$. Find the eigenvalues and eigenvectors of A in each of the following cases. Describe the stable and unstable manifolds of the origin.

$$\text{(a) } A = \begin{bmatrix} 1 & 1 & 2 \\ 1 & 2 & 1 \\ 2 & 1 & 1 \end{bmatrix} \quad \text{(b) } A = \begin{bmatrix} 3 & 0 & -1 \\ 0 & 1 & 0 \\ 2 & 0 & 0 \end{bmatrix}$$

$$(c) \ A = \begin{bmatrix} 2 & 0 & 0 \\ 0 & 2 & 2 \\ 0 & 2 & -1 \end{bmatrix} \quad (d) \ A = \begin{bmatrix} 6 & 5 & 5 \\ 5 & 6 & 5 \\ 5 & 5 & 6 \end{bmatrix}$$

12.15 Show that $\dot{x} = Ax$ where $x = [x, y, z]^T$ and

$$A = \begin{bmatrix} -3 & 0 & -2 \\ -4 & -1 & -4 \\ 3 & 1 & 3 \end{bmatrix}$$

has two imaginary eigenvalues. Find the equation of the centre manifold of the origin. Is the remaining manifold stable or unstable?

12.16 Show that the centre manifold of

$$\begin{bmatrix} \dot{x} \\ \dot{y} \\ \dot{z} \end{bmatrix} = \begin{bmatrix} -1 & 0 & 1 \\ 0 & 1 & -2 \\ 0 & 1 & -1 \end{bmatrix} \begin{bmatrix} x \\ y \\ z \end{bmatrix},$$

is given by the plane $2x + y - 2z = 0$.

12.17 Show that the phase paths of $\dot{x} = y(x+1)$, $\dot{y} = x(1 - y^2)$ are given by

$$y = \pm\sqrt{[1 - Ae^{-2x}(1+x)^2]},$$

with singular solutions $x = -1$ and $y = \pm 1$. Describe the domains in the (x, y) plane of the stable and unstable manifolds of each of the three equilibrium points of the system.

12.18 Show that the linear approximation at $(0, 0, 0)$ of

$$\dot{x} = -y + yz + (y - x)(x^2 + y^2),$$
$$\dot{y} = x - xz - (x + y)(x^2 + y^2),$$
$$\dot{z} = -z + (1 - 2z)(x^2 + y^2),$$

has a centre manifold there. Show that $z = x^2 + y^2$ is a solution of this system of equations. To which manifold of the origin is this surface tangential? Show also that, on this surface, x and y satisfy

$$\dot{x} = -y + (2y - x)(x^2 + y^2), \quad \dot{y} = x - (2x + y)(x^2 + y^2),$$

Using polar coordinates determine the stability of solutions on this surface and the stability of the origin.

12.19 Investigate the stability of the equilibrium points of

$$\dot{x} = \mu x - x^2, \quad \dot{y} = y(\mu - 2x)$$

in terms of the parameter μ. Draw a stability diagram in the (μ, x) plane for $y = 0$. What type of bifurcation occurs at $\mu = 0$? Obtain the equations of the phase paths, and sketch the phase diagrams in the cases $\mu = -1, \mu = 0$ and $\mu = 1$.

12.20 Where is the bifurcation point of the parameter dependent-system

$$\dot{x} = x^2 + y^2 - \mu, \quad \dot{y} = 2\mu - 5xy?$$

Discuss how the system changes as μ increases. For $\mu = 5$ find all linear approximations for all equilibrium points, and classify them.

12.21 Obtain the polar equations for (r, θ) of

$$\dot{x} = y + x[\mu - (x^2 + y^2 - 1)^2],$$
$$\dot{y} = -x + y[\mu - (x^2 + y^2 - 1)^2],$$

where $|\mu| < 1$. Show that, for $0 < \mu < 1$, the system has two limit cycles, one stable and one unstable, which collide at $\mu = 0$, and disappear for $\mu < 0$. This is an example of a **blue sky catastrophe** in which a finite-amplitude stable limit cycle simply disappears as a parameter is changed incrementally.

12.22 Discuss the bifurcations of the equilibrium points of

$$\dot{x} = y, \quad \dot{y} = -x - 2x^2 - \mu x^3$$

for $-\infty < \mu < \infty$. Sketch the bifurcation diagram in the (μ, x) plane. Confirm that there is a fold bifurcation at $\mu = 1$. What happens at $\mu = 0$?

12.23 Consider the system

$$\dot{x} = y - x(x^2 + y^2 - \mu), \quad \dot{y} = -x - y(x^2 + y^2 - \mu),$$

where μ is a parameter. Express the equations in polar form in terms of (r, θ). Show that the origin is a stable spiral for $\mu < 0$, and an unstable spiral for $\mu > 0$. What type of bifurcation occurs at $\mu = 0$?

12.24 In polar form a system is given

$$\dot{r} = r(r^2 - \mu r + 1), \quad \dot{\theta} = -1,$$

where μ is a parameter. Discuss the bifurcations which occur as μ increases through $\mu = 2$.

12.25 The equations of a displaced van der Pol oscillator are given by

$$\dot{x} = y - a, \quad \dot{y} = -x + \delta(1 - x^2)y,$$

where $a > 0$ and $\delta > 0$. If the parameter $a = 0$ then the usual equations for the van der Pol oscillator appear. Suppose that a is increased from zero. Show that the system has two equilibrium points one of which is a saddle point at $x \approx -x1/(a\delta), y = a$ for small a. Compute phase paths for $\delta = 2$, and $a = 0.1, 0.2, 0.4$, and observe that the saddle point approaches the stable limit cycle of the van der Pol equation. Show that at $a \approx 0.31$ the saddle point collides with the limit cycle, which then disappears.

12.26 Find the stable and unstable manifolds of the equilibrium points of

$$\dot{x} = x^2 + \mu, \quad \dot{y} = -y, \quad \dot{z} = z,$$

for $\mu < 0$. What type of bifurcation occurs at $\mu = 0$?

12.27 Consider the system

$$\dot{x} = \mu x - y - x^3, \quad \dot{y} = x + \mu y - y^3.$$

By putting $z = \mu - x^2$, show that any equilibrium points away from the origin are given by solutions of

$$z^4 - \mu z^3 + \mu z + 1 = 0.$$

Plot the graph of μ against z and show that there is only one equilibruim point at the origin if $\mu < 2\sqrt{2}$, approximately, and 9 equilibrium points if $\mu > 2\sqrt{2}$.

Investigate the linear approximation for the equilibrium point at the origin and show that the system has a Hopf bifurcation there at $\mu = 0$. Compute the phase diagram for $\mu = 1.5$.

12.28 Show that the system

$$\dot{x} = x^2 + y + z + 1, \quad \dot{y} = z - xy, \quad \dot{z} = x - 1$$

has one equilibrium point at $(1, -1, -1)$. Determine the linear approximation

$$\dot{x}' = A x'$$

to the system at this point. Find the eigenvalues and eigenvectors of A, and the equations of the stable and unstable manifolds E^s and E^c of the linear approximation.

12.29 Consider the equation

$$\dot{z} = \lambda z - |z|^2 z,$$

where $z = x + iy$ is a complex variable, and $\lambda = \alpha + i\beta$ is a complex constant. Classify the equilibruim point at the origin, and show that the system has a Hopf bifurcation as α increases through zero for $\beta \neq 0$. How does the system behave if $\beta = 0$?.

13 Poincaré sequences, homoclinic bifurcation, and chaos

13.1 Poincaré sequences

Consider the *autonomous* system

$$\dot{x} = X(x, y), \qquad \dot{y} = Y(x, y)$$

and its phase diagram in the (x, y)-plane. Let Σ be a curve or *cross-section* of the plane with the property that it cuts each phase path transversely in some region of the phase diagram, that is, it is nowhere tangential to a phase path. It is called a **Poincaré section** of the phase diagram. Consider a point $A_0 : (x_0, y_0)$ on the Poincaré section Σ shown in Fig. 13.1. If we follow the phase path through A_0 in its direction of flow then it next cuts Σ in the same sense at $A_1 : (x_1, y_1)$. This point is the **first return**, or **Poincaré map**, of the point A_0. We are not implying that such a point must exist, but if it does then it is called a first return. If we continue on the phase path, then the first return of A_1 is $A_2 : (x_2, y_2)$ (see Fig. 13.1). We can represent this process as a **mapping** or **function** by an **operator** P_Σ which is such that for the particular Σ chosen, and for every (x, y) on Σ,

$$(x', y') = P_\Sigma(x, y),$$

where (x', y') is the point of first return of the path from (x, y). For successive returns starting from (x_0, y_0) we use the notation

$$(x_2, y_2) = P_\Sigma(P_\Sigma(x_0, y_0)) = P_\Sigma^2(x_0, y_0); \quad (x_n, y_n) = P_\Sigma^n(x_0, y_0).$$

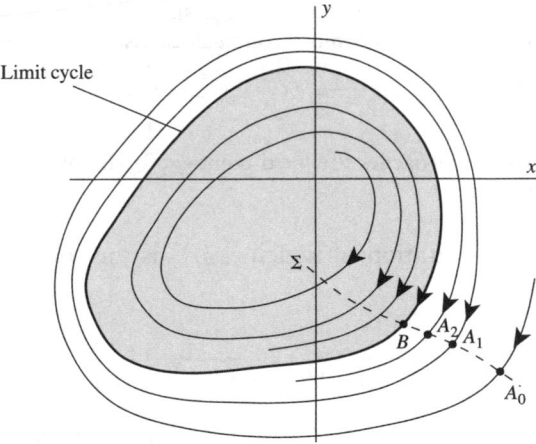

Figure 13.1 First returns associated with a spiral and a limit cycle, and the Poincaré section Σ; $A_1, A_2, \ldots$ are the first returns of $A_0, A_1, \ldots$ **in the same sense** across Σ, and B is the fixed points which they are approaching.

The starting time for a sequence of first returns is immaterial since the system is autonomous. Note that the time lapse between returns is not in general constant.

Figure 13.1 shows a stable limit cycle with interior and exterior phase paths spiralling towards the cycle. The returns $A_1, A_2, \ldots$ starting at A_0 on Σ approach the point B which is the intersection of the limit cycle and Σ. A similar approach of returns to B occurs if A_0 is on Σ, but inside the limit cycle. If B is chosen as the initial point on Σ, then all returns are at B: B is a **fixed point of the map** P_Σ. All sequences which start on Σ but sufficiently close to B approach B. The behaviour of these returns indicates that B is a *stable* fixed point. In general we call a sequence $A_0, A_1, A_2, \ldots$ a **Poincaré sequence** (it need not necessarily be stable).

We can see from Fig. 13.1 that B could have alternative section Σ, and also that every point on the limit cycle is a possible fixed point of such a section.

Example 13.1   *Obtain the map of first returns P_Σ for the differential equations*

$$\dot{x} = \mu x + y - x\sqrt{(x^2 + y^2)},$$
$$\dot{y} = -x + \mu y - y\sqrt{(x^2 + y^2)},$$

for the section Σ given by $y = 0$, $x > 0$ with initial point $(x_0, 0)$ $(x_0 < \mu)$.

Since the equations are autonomous the starting time t_0 is immaterial—assume that $t_0 = 0$. In polar coordinates the equations become

$$\dot{r} = r(\mu - r), \quad \dot{\theta} = -1$$

with solutions

$$r = \mu r_0 / \{r_0 + (\mu - r_0)e^{-\mu t}\}, \quad \theta = -t + \theta_0,$$

where $r(0) = r_0$ and $\theta(0) = \theta_0$. $r = \mu$ is a limit cycle, and the approach directions indicated by the sign of $\dot{r}$ imply that it is stable. By eliminating t the paths are given by

$$r = \mu r_0 / \{r_0 + (\mu - r_0)e^{\mu(\theta - \theta_0)}\}.$$

The section given corresponds to $\theta_0 = 0$, and the required successive returns occur for $\theta = -2\pi, -4\pi, \ldots$ (bearing in mind that the representative points move round clockwise); with initial point $(r_0, 0)$, so

$$r_n = \mu r_0 / \{r_0 + (\mu - r_0)e^{-\mu 2n\pi}\}, \quad \theta_n = -2n\pi \quad (n = 1, 2, \ldots). \tag{13.1}$$

As $n \to \infty$ the sequence of points approaches the fixed point $(\mu, 0)$, as shown in Fig. 13.2, corresponding to the intersection with the limit cycle.   ●

We can find the **difference equation** of which (13.1) is the solution. It follows from (13.1) that

$$r_{n+1} = \frac{\mu r_0}{r_0 + (\mu - r_0)e^{-2(n+1)\mu\pi}} = \frac{\mu r_n}{r_n(1 - e^{-2\mu\pi}) + \mu e^{-2\mu\pi}} = f(r_n), \tag{13.2}$$

say, after eliminating r_0 and using (13.1) again. Equation (13.2) is a first-order difference equation for r_n. Whilst we know the solution r_n, $n = 0, 1, \ldots$ in this case, it is instructive to

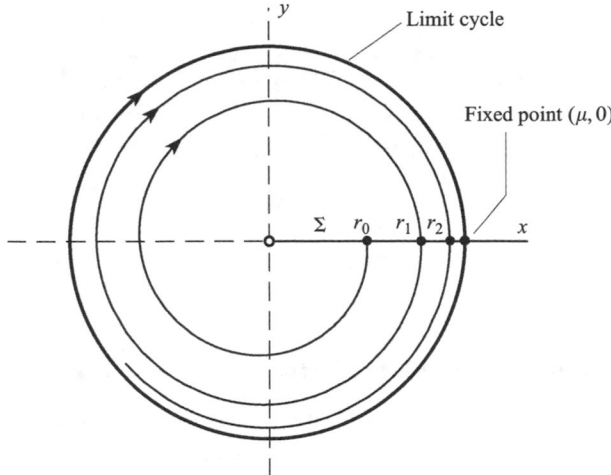

Figure 13.2 First returns approaching $(\mu, 0)$ on Σ: $\theta_0 = 0$.

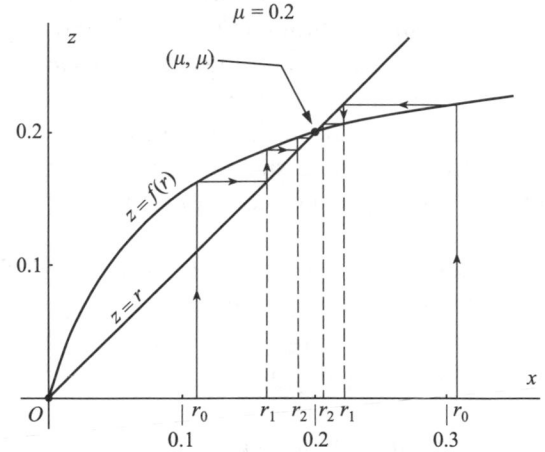

Figure 13.3 Cobweb diagram for the sequence $\{r_n\}$ given by (13.2) with $\mu = 0.2$.

see how we can represent it geometrically (see, Jordan and Smith 1997, Ch. 37). As shown in Fig. 13.3, plot the curve $z = f(r)$ and the line $z = r$. The **fixed points** of the Poincaré section or map occur where the curve and the line intersect, namely where r satisfies $r = f(r)$. As we expect, the solution are $r = 0$ and $r = \mu$. The sequence of values of r starting from $r = r_0$ are also shown in Fig. 13.3, which has been constructed by the **cobweb** sequence of lines reflected between $z = r$ and $z = f(r)$. Representatives cobwebs for $r_0 < \mu$ and $r_0 > \mu$ are shown in the figure. The stability of the fixed point is indicated by the cobwebs converging to the point (μ, μ). The sequential mapping described by (13.2) corresponds to the operator P_Σ mentioned at the beginning of Section 13.1.

In this particular case the time lapses between returns are all equal, the interval being 2π, but this will not usually be true.

Exercise 13.1
Consider the polar system $\dot{r} = r(1 - r)$, $\dot{\theta} = -1$ (as in Example 13.1 with $\mu = 1$) and the section $y = \frac{1}{2}(-\infty < y < \infty)$. Discuss the returns which occur from all initial points on the line with particular reference to the intersections with the limit cycle. What happens to the returns if the section is $y = 2$?

13.2 Poincaré sections for nonautonomous systems

Important features of a system can be revealed by Poincaré maps or sequences. For example, as we have seen, in autonomous systems periodic solutions may exist when fixed points appear, so some useful information is retained if the phase paths of Figs 13.2 or 13.3 are deleted and only the dots of the Poincaré map retained. But for autonomous systems there is no particular advantage in using the map of first returns since the paths can be plotted (numerically if necessary) and the phase diagrams interpreted without difficulty. For *nonautonomous systems*, on the other hand, the diagram of solution curves projected on to the (x, y)-plane which corresponds to the phase diagram in the autonomous case appears as a tangle of intersecting and self-intersecting curves, since each initial state or point in the plane generates an infinite number of curves corresponding to the various initial times t_0: see, for example, Fig. 7.11(b). Important features can be totally obscured in such a diagram, but Poincaré maps can be used to detect underlying structure, such as periodic solutions having the forcing or a subharmonic frequency.

In this context the investigation of periodic solutions, nearly periodic solutions, and similar phenomena is to a considerable extent an exploratory matter in which computation plays at the present time a very significant part. A search for hidden periodicities, such as subharmonic periods, is best carried out by starting with a period in mind and then looking for solutions with this period.

In that case a variant of the Poincaré map is usually more profitable—if the solutions sought are expected to have period T, then we should plot on the x, y plane a sequence of points calculated at times $T, 2T, 3T, \ldots$ along the phase paths starting from various states, and see whether any of these sequences indicate that we are approaching a periodic solution. This does not quite fit the definition of a Poincaré map, which does not involve any mention of time intervals but picks out intersections of a phase path under investigation with another given curve (the 'section'). The two procedures can with advantage for the analysis (especially in multidimensional cases) be brought together in a manner suggested by the following example.

Consider the first-order nonautonomous system of dimension unity:

$$\dot{x} = -\tfrac{1}{8}x + \cos t \tag{13.3}$$

which has the general solution

$$x = c\mathrm{e}^{-t/8} + \tfrac{8}{65}(\cos t + 8 \sin t).$$

The right-hand side of (13.3) has time period 2π. In fact if t is replaced by $t + 2\pi$ throughout, the equation remains unchanged; but this does not mean that all the solutions have period 2π. If we start with a given initial state and proceed through an interval of length 2π, then the state arrived at is the initial state for the next interval, but generally this second state will be different from the first.

By the following artifice, eqn (13.3) can be written as an *autonomous* system of *one dimension higher*. We nominate a second variable, θ, to take the place of t, and rewrite the relations as

$$\dot{x} = -\tfrac{1}{8}x + \cos\theta, \tag{13.4}$$

$$\dot{\theta} = 1, \tag{13.5}$$

(the independent variable is still t). It is also necessary to complete the identification of θ with t by requiring

$$\theta(0) = 0. \tag{13.6}$$

The two-dimensional systems (13.4) and (13.5) now has the property that if we substitute $\theta + 2n\pi$ for θ, n being an integer, then the system is unchanged. By taking advantage of this property we can relate a plot of calculated values of x at equal time steps 2π to a Poincaré map associated with (13.4) and (13.5), not in a plane space (x, y), but on a cylindrical surface. The space is constructed by picking out the strip

$$-\infty < x < \infty, \quad 0 < \theta \leq 2\pi$$

from the (x, y) plane, and wrapping it round a cylinder of circumference 2π, so that the side $\theta = 0$ becomes attached to the side $\theta = 2\pi$. On this space, and for all times, eqns (13.4) and (13.5) hold good, and the solutions are represented by curves which wrap round the cylinder and, in the case of periodic solutions whose period is a multiple of 2π, wrap round it more than once and join up smoothly (see Fig. 13.4). The periodic variable θ is an angle, which explains the choice of notation.

In a similar way the two-dimensional system

$$\dot{x} = X(x, y, t), \qquad \dot{y} = Y(x, y, t),$$

where X and Y are periodic in with period T in t, is equivalent to the system

$$\dot{x} = X(x, y, (2\pi/T)\theta), \quad \dot{y} = Y(x, t, (2\pi/T)\theta), \quad \dot{\theta} = \frac{2\pi}{T},$$

where X and Y are now 2π-periodic in θ. Effectively, the time is rescaled to ensure that t is still an angular variable. This may be visualized as a 'toroidal space' for which the planes $\theta = 0, 2\pi, 4\pi, \ldots$ (equivalent to $t = 0, T, 2T, \ldots$) are bent round to coincide with $\theta = 0$, which serves as the section for a Poincaré sequence.

The θ curve in which the values of x and y for θ between $2\pi n$ and $2\pi(n + 1)$ are mapped on to θ between 0 and 2π is usually denoted by the symbol $\mathcal{S}$. The toroidal phase space is then $\mathcal{R}^2 \times \mathcal{S}$. Mathematically, we say that $\theta = 2\pi t/T \,(\mathrm{mod}\, 2\pi)$. Generally, a Poincaré sequence for a section $\theta = \theta_0$ with an initial starting coordinates $x(t_0) = x_0$, $y(t_0) = y_0$ would consist of the sequence of pairs of coordinates $(x(t_0 + nT), (y(t_0 + nT))$ in the (x, y) plane.

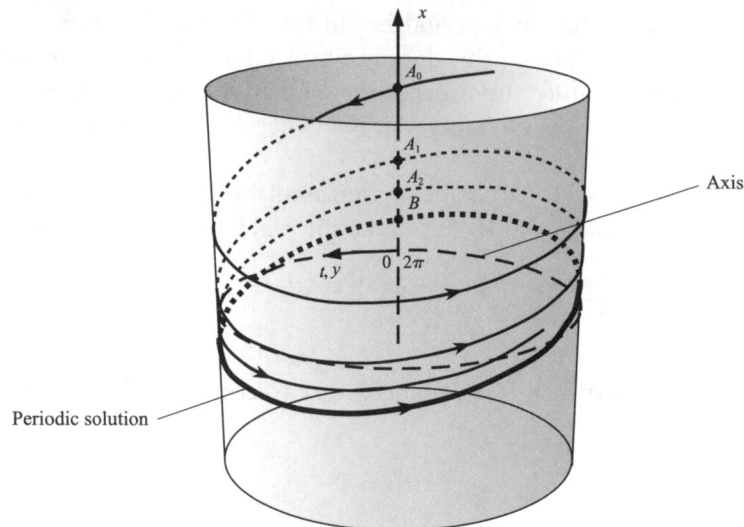

Figure 13.4   Solution of (13.3) mapped on to a cylinder of circumference 2π. On the section $y = 0$ (which includes $t = 2\pi, 4\pi, \dots$), the Poincaré map is the sequence $A_0, A_1, A_2, \dots$ with limiting point B on the periodic solution $x = \frac{8}{65}(\cos t + 8 \sin t)$.

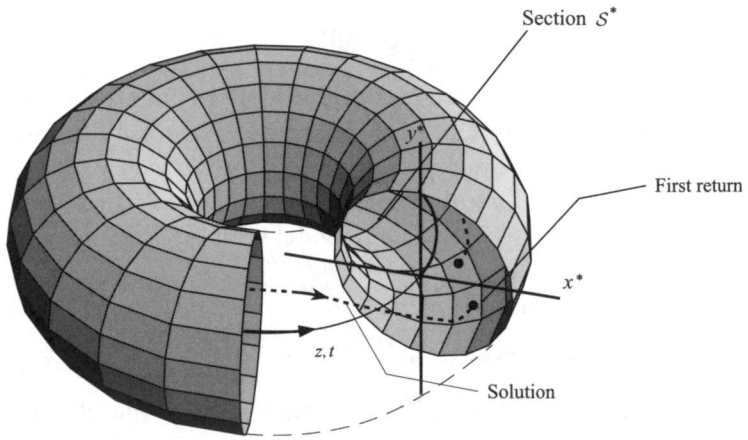

Figure 13.5   A torus cut open to show the projection $\mathcal{S}^*$ in the (x^*, y^*) plane.

If the phase diagram occupies the whole of the (x, y) plane, then we cannot draw a representation of the torus. However, it can be drawn if the (x, y) plane is mapped onto the diametrical plane (x^*, y^*) using the hemispherical transformation of Section 3.3 (see Fig. 3.13). The Poincaré map then appears as a cross section $\mathcal{S}^*$ of a torus, which is a disc of unit radius as shown in Fig. 13.5.

An alternative method of viewing a sequence of Poincaré points taken at a constant time interval T along a phase path is to let the time t be the third (vertical) axis. Take sections at $t = 0, t = T, t = 2T, \dots$, and then project the points for a particular phase path back onto the

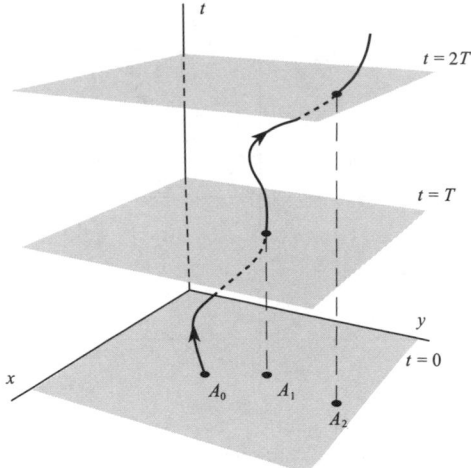

Figure 13.6   First returns obtained from the sections $t = 0, t = T, t = 2T, \ldots$ in the (x, y, t) space.

plane $t = 0$ as shown in Fig. 13.6. Obviously, any other start time $t = t_0$ with time steps of T can be chosen as described above.

Example 13.2   *Find the Poincaré sequences with $T = 2\pi$ for the section Σ: $\theta = 0$ for arbitrary initial states, and locate the corresponding fixed point of the nonautonomous equation*

$$\ddot{x} + 3\dot{x} + 2x = 10 \cos t$$

in the phase plane for which $\dot{x} = y$.

When we introduce a new variable $\theta = t$ (in this case period $T = 2\pi$), the system takes the autonomous form

$$\dot{x} = y, \quad \dot{y} = -2x - 3y + 10 \cos \theta, \quad \dot{\theta} = 1, \text{ with } \theta(0) = 0,$$

and a Poincaré sequence corresponding to $t = 0, 2\pi, \ldots$ appears as a sequence of points following the phase path $(x(t), y(t), \theta(t))$ as it intersects $\theta = 0$, or the (x, y) plane. They can be found by referring to the known solution with initial condition $x = x_0$, $y = y_0$ at $t = 0$:

$$x = (-5 + 2x_0 + y_0)e^{-t} + (4 - x_0 - y_0)e^{-2t} + \cos t + 3 \sin t.$$
$$y = -(-5 + 2x_0 + y_0)e^{-t} - 2(4 - x_0 - y_0)e^{-2t} - \sin t + 3 \cos t.$$

The sequence starting at (x_0, y_0) in the plane $\theta = 0$ is given by

$$x_n = (-5 + 2x_0 + y_0)e^{-2n\pi} + (4 - x_0 - y_0)e^{-4n\pi} + 1,$$
$$y_n = -(-5 + 2x_0 + y_0)e^{-2n\pi} - 2(4 - x_0 - y_0)e^{-4n\pi} + 3$$

for $n = 1, 2, \ldots$. As $n \to \infty$, $(x_n, y_n) \to (1, 3)$ irrespective of the initial coordinates (x_0, y_0). This is the *fixed point* of the Poincaré sequence of the limiting periodic solution in which $t = 0, 2\pi, 4\pi, \ldots$, as shown in Fig. 13.7. It would not turn out to be the same point, $(1, 3)$, if we had considered an initial time $t = t_0$ having a different value from zero or a multiple of 2π, but it might not be necessary to explore all these possibilities since the

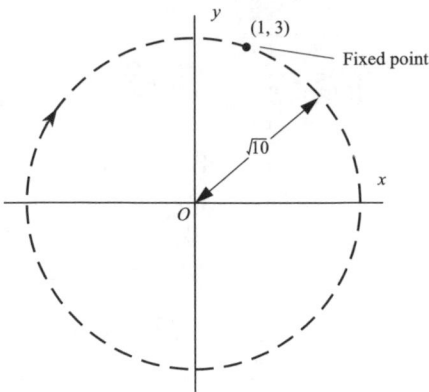

Figure 13.7 Fixed point at $(1, 3)$ of the periodic solution of $\ddot{x} + 3\dot{x} + 2x = 10\cos t$ at $t = 0, 2\pi, \ldots$. The circle is the stable limit cycle $x = \cos t + 3\sin t$, $y = -\sin t + 3\cos t$. Starting from (x_0, y_0), the sequence follows on approaching spiral path.

existence of a single point, independent of x_0, y_0, suggests the possibility of either an equilibrium point, or a closed path of period 2π or an integral multiple of 2π. In the latter case it provides an initial condition for calculating the whole cycle.   ●

Example 13.3 *Find the Poincaré sequence having constant time interval 2π for the solution of*

$$64\ddot{x} + 16\dot{x} + 65x = 64\cos t$$

which starts from $x(0) = 0$, $\dot{x}(0) = 0$. Calculate also the coordinates of the resulting fixed point of the periodic solution.

The roots of the characteristic equation of the differential equation are $p_1 = -\frac{1}{8} + i$ and $p_2 = -\frac{1}{8} - i$. To find the forced periodic response let

$$x = A\cos t + B\sin t.$$

By direct substitution it follows that $A = 64/257$ and $B = 1024/257$. The general solution is therefore

$$x = e^{-t/8}(C\cos t + D\sin t) + (64\cos t + 1024\sin t)/257.$$

The initial conditions imply that $C = -64/257$ and $D = -1032/257$. The sequence of first returns of this solution is given by

$$x_n = A(1 - e^{-\frac{1}{4}n\pi}), \quad y_n = B(1 - e^{-\frac{1}{4}n\pi}), \quad n = 1, 2, \ldots.$$

Figure 13.8 shows the sequence of first returns from the origin. As $n \to \infty$ the sequence of dots approaches the fixed point at (A, B). This is a point on the phase path corresponding to the forced periodic response $A\cos t + B\sin t$.   ●

Exercise 13.2
Find the general solution of $\ddot{x} + 3\dot{x} + 2x = 10\cos t$. If $x(0) = \dot{x}(0) = 0$, find the sequence of 2π returns in the (x, y) plane and the limits of the sequence.

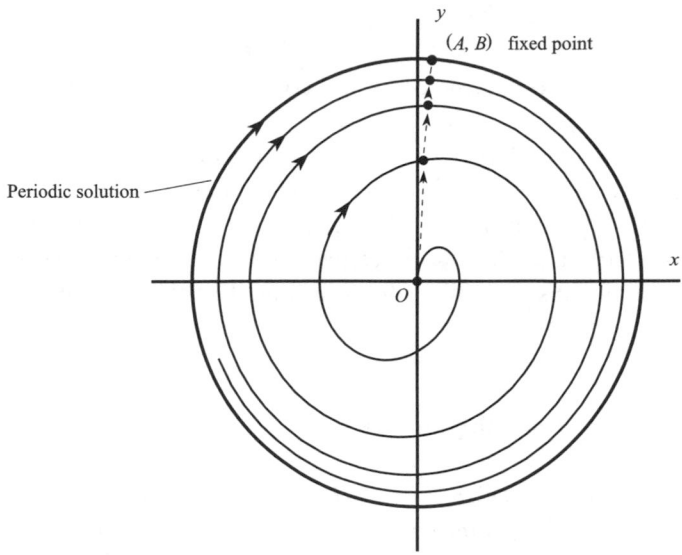

Figure 13.8 First returns for $64\ddot{x} + 16\dot{x} + 65x = 64\cos t$, $\dot{x} = y$ taken at $t = 0, 2\pi, \ldots$ starting at $(0, 0)$.

13.3 Subharmonics and period doubling

As we have seen, the appearance of fixed points is closely connected with periodic solutions. However, some solutions have fixed points arising from some sections but not others. For example, a particular solution of

$$\ddot{x} - x = 2e^{-t}(\cos 2t - \sin 2t)$$

is given by

$$x = e^{-t}\sin^2 t, \quad y = \dot{x} = e^{-t}(-\sin^2 t + 2\sin t \cos t).$$

Then $x(t), y(t)$ are zero at $t = 0, 2\pi, \ldots$, so that $x = 0$, $y = 0$ is a fixed point, but $x(t)$ and $y(t)$ are not periodic in t, nor is $(0, 0)$ an equilibrium point. The Poincaré sequence will only indicate a periodic solution with certainty if a fixed point exists for *all* sections.

When the forcing term has period T, a sequence with interval T may detect *possible* solutions of period $T, 2T, \ldots$ Some sequences for possible periodic solutions of systems with 2π forcing period are discussed below and shown in Fig. 13.9. The Poincaré sequences shown on the phase paths as dots consist of the one or more points with coordinates $(x(t), \dot{x}(t))$ taken at $t = 0, 2\pi, 4\pi, \ldots$

The linear equation

$$\ddot{x} + \tfrac{1}{4}x = \tfrac{3}{4}\cos t, \quad \dot{x} = y$$

has a particular solution $x = \cos t$ which has period 2π. The time solution and phase path are shown in Fig. 13.9(a) with fixed point at $(1, 0)$ in the (x, y) plane. The equation also has the solution

$$x = 0.5 \sin \tfrac{1}{2}t - 0.3 \cos \tfrac{1}{2}t + \cos t,$$

which is a **subharmonic** of period 4π which results in a double-loop phase path, and *two* fixed points, at $(0.7, 0.25)$ and $(1.3, -0.25)$, between which the first returns alternate (Fig. 13.9(b)). The differential equation is undamped so that there can be no converging Poincaré sequences (all solutions are periodic). The alternating returns will only be revealed if the initial point is chosen to be one of them. This example, shown in Fig. 13.9(b), displays **period doubling** or a **period-2 map**.

A solution such as

$$x = -0.4 \sin \tfrac{1}{3}t + 0.8 \sin t - 0.4 \cos t$$

has a subharmonic of period 6π, which leads to the triple-loop phase path shown in Fig. 13.9(c). The Poincaré map at interval $2a\pi$ alternates between *three* fixed points, and is called a period-3map.

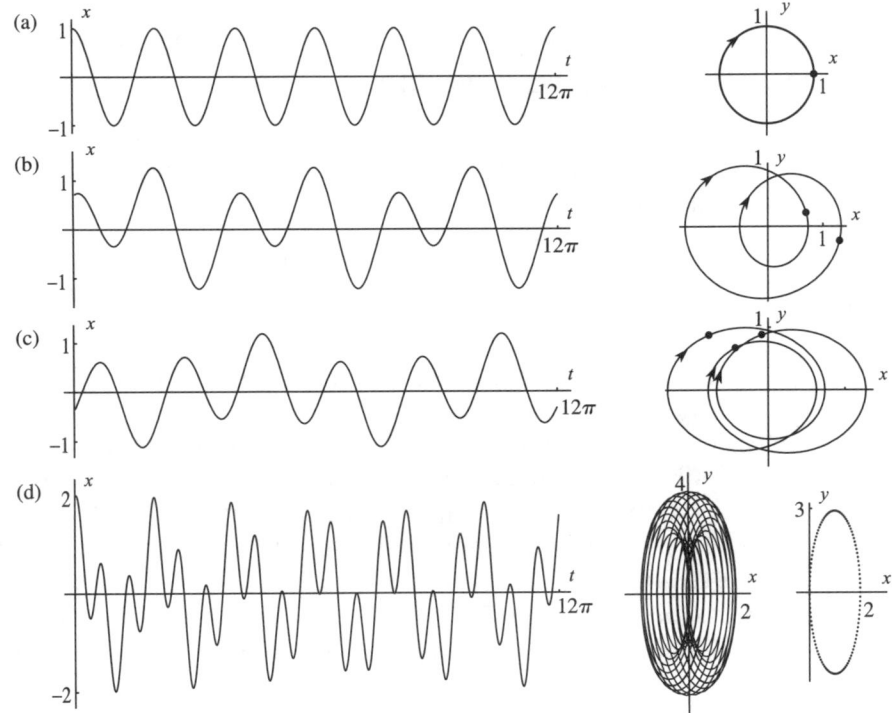

Figure 13.9 Poincaré maps at $t = 0, 2\pi, \ldots$ for a sample of signals. (a) $x = \cos t$ which has one fixed point; (b) $x = -0.3 \cos \tfrac{1}{2}t + 0.5 \sin \tfrac{1}{2}t + \cos t$ which has a subharmonic of period 4π resulting in two fixed points, characteristic of period-2 oscillations; (c) $x = -0.4 \sin \tfrac{1}{3}t + 0.8 \sin t - 0.4 \cos t$ showing the Poincaré map of a period-3 oscillation; (d) $x = \cos t + \cos \pi t$, which is the sum of two oscillations, with frequency ratio irrational, showing that the returns (250 displayed) of the Poincaré map accumulate on the ellipse shown separately.

The solution shown in Fig. 13.9(d) is that for

$$x = \cos(\pi t) + \cos t,$$

which is the sum of two periodic functions, the ratio of whose frequencies is not a rational number. The phase paths of this **quasi-periodic** solution fill an oval-shaped region, and the first returns lie on an ellipse.

When the exact solution to the differential equation is not obtainable, as will be the case in all instances of significance, computation must take the place of an algebraic derivation of the Poincaré map. This is fairly easy to compute and display graphically—a step-by-step solution of the equations is organized to print out a result at the end of every period, and the points are normally displayed on continuous linking curves with the direction of movement indicated. The technique is sometimes referred to as the **stroboscopic method** since the representative point is, as it were, illuminated once in every cycle (see Minorsky 1962).

In seeking stable periodic solutions via the approach to fixed points of a Poincaré map it is advantageous to be able to start the calculation at an initial point as close as possible to the closed cycle in the x, y plane which is sought. This is especially true if the evolution of periodic solutions as a parameter varies is being traced, as in Section 13.2, since otherwise spurious connections may be identified. For this purpose any of the approximate methods for finding periodic solutions described in the earlier chapters may be used. In the following Example we indicate the use of the van der Pol plane for this purpose.

Example 13.4 *Find approximate positions for the fixed points of the Poincaré map of time interval $2\pi/\omega$ for the Duffing equation*

$$\ddot{x} + k\dot{x} + x + \beta x^3 = \Gamma \cos \omega t \quad (k > 0, \beta < 0). \tag{13.7}$$

In Section 7.2 it is shown that there exists a stable oscillation of the approximate form

$$x(t) = a_0 \cos \omega t + b_0 \sin \omega t \tag{13.8}$$

where a_0 and b_0 are obtained by solving the equations

$$b\{\omega^2 - 1 - \tfrac{3}{4}\beta(a^2 + b^2)\} + k\omega a = 0, \tag{13.9}$$

$$a\{\omega^2 - 1 - \tfrac{3}{4}\beta(a^2 + b^2)\} - k\omega b = \Gamma. \tag{13.10}$$

On the phase plane with $y = \dot{x}$ the approximate solution (13.8) traces out an ellipse with its centre at the origin. The true path should be a closed curve close to this ellipse. If we solve (13.9), (13.10) for a_0 and b_0 numerically and substitute the values into (13.8), we may then use any point on the ellipse as a starting point for an exact Poincaré map—say the point where $t = 0$:

$$x_0 = a_0, \quad y_0 = \omega b_0.$$

This leads comparatively quickly to a good estimate for the position of a fixed point, which will lie somewhere on the true oscillation and serve as an initial condition for plotting the complete cycle in the (x, y)-plane. Similar techniques were used to assist in obtaining the later figures in this chapter.   ●

Exercise 13.3
Verify that the equation

$$\ddot{x} - 24\delta^{-3}(2 - x^2 - \dot{x}^2)\dot{x} + \tfrac{8}{9}\dot{x} + \tfrac{1}{9}x = -\tfrac{16}{9}\sin t$$

has the 2π-periodic solution $x = \cos t + \sin t$ for all $\delta \neq 0$. Show that this solution co-exists with a subharmonic $x = \delta \cos \tfrac{1}{3} t$ if $\delta = -3$ or $\delta = 3(-1 \pm \sqrt{3})$. Sketch their phase paths in the (x, y) plane, where $y = \dot{x}$, and their fixed points of period 2π.
Using a computer investigation show that if $\delta = -3$, then $x = \cos t + \sin t$ is unstable, but it has a neighbouring stable 2π-periodic solution.

13.4 Homoclinic paths, strange attractors and chaos

Studies of computer-generated solutions of forced second-order equations such as Duffing's equation and three-dimensional autonomous systems have revealed unexpectedly complex solution structures arising from what might appear to be relatively simple nonlinear differential equations. The system

$$\dot{x} = a(y - x), \quad \dot{y} = bx - y - xz, \quad \dot{z} = xy - cz \qquad (13.11)$$

$(a, b, c$ constant$)$ was first investigated by Lorenz (1963). The Lorenz equations, as they are now known, arose in a model for convective motion in the atmosphere. The solutions calculated by computer display very complex behaviour for a wide range of parameter values. By complex behaviour we include seemingly random or 'chaotic' output from the system although the solutions remain deterministic in terms of their initial values and there is no 'random' input. These phenomena arise specifically with third- and higher-order equations, or in forced second-order systems (they have no obvious counterpart in autonomous second-order systems). The Lorenz equations and similar models have been put forward as explanations of turbulent flow in fluid mechanics. However, certain aspects of the long-run behaviour of such systems show systematic characteristics with a degree of independence of initial conditions and parameter values, which is opening the way to a rational study even of their 'chaotic' behaviour.

We shall look in some detail at the Rössler system (see Nicolis 1995)

$$\dot{x} = -y - z, \quad \dot{y} = x + ay, \quad \dot{z} = bx - cz + xz \quad (a, b, c > 0),$$

which is simpler than the Lorenz equations in that it has just one nonlinear quadratic term. The system has equilibrium points where

$$-y - z = 0, \quad x + ay = 0, \quad bx - cz + xz = 0.$$

Obviously the origin is always an equilibrium point, and there is a second one at

$$x = x_1 = c - ab, \quad y = y_1 = -(c - ab)/a, \quad z = z_1 = (c - ab)/a. \qquad (13.12)$$

In this three-parameter system, we shall look at the special case in which $a = 0.4$ and $b = 0.3$ but allow the parameter c to vary (the values for a and b were chosen after numerical investigations indicated domains in the parameter space (a, b, c) where interesting chaotic phenomena might occur). The two equilibrium points coincide when $c = ab$: we shall assume that $c > ab$. For fixed a and b, the equilibrium points move out on a straight line through the origin, given parametrically by (13.12), as c increases from the value ab. The linear approximation of the Rössler system near the origin is $\dot{x} = Ax$, where

$$A = \begin{bmatrix} 0 & -1 & -1 \\ 1 & a & 0 \\ b & 0 & -c \end{bmatrix}.$$

The eigenvalues of A satisfy the cubic

$$\lambda^3 + (c - a)\lambda^2 + (b + 1 - ac)\lambda + c - ab = 0$$

or, with $a = 0.4$ and $b = 0.3$,

$$f(\lambda, c) \equiv \lambda^3 + (c - 0.4)\lambda^2 - (0.4c - 1.3)\lambda + (c - 0.12) = 0.$$

It follows that $f(0, c) = c - 0.12 > 0$ for $c > ab = 0.12$. Since $f(\lambda, c) \to \pm\infty$ as $\lambda \to \pm\infty$, there is at least one real negative eigenvalue. The other eigenvalues are complex with positive real part. This equilibrium point is called a **saddle-spiral**.

In the neighbourhood of the other equilibrium point, at (x_1, y_1, z_1) say, let $x = x_1 + x', y = y_1 + y', z = z_1 + z'$. Then the linear approximation at (x_1, y_1, z_1) is $\dot{x}' = Bx'$ where

$$B = \begin{bmatrix} 0 & -1 & -1 \\ 1 & a & 0 \\ b + z_1 & 0 & -c + x_1 \end{bmatrix}.$$

The eigenvalues of B satisfy (with $a = 0.4$, $b = 0.3$)

$$g(\lambda, c) \equiv \lambda^3 - 0.28\lambda^2 + (0.952 + 2.5c)\lambda - (c - 0.12) = 0.$$

The derivative with respect to λ is

$$\frac{\partial g(\lambda, c)}{\partial \lambda} = 3\lambda^2 - 0.56\lambda + (0.952 + 2.5c).$$

For $c > ab = 0.12$, one root of $g(\lambda, c) = 0$ is real and positive (since $g(0, c) < 0$), and two roots are complex (since $\partial g(\lambda, c)/\partial \lambda = 0$ has no real solutions for λ) with negative real part if $c < 1.287$ and positive real part if $c > 1.287$ approximately. Again the equilibrium point is always unstable. Hence for all $c > 0.12$, both equilibrium points are unstable.

There exist homoclinic paths which spiral out from the origin lying approximately on the plane unstable manifold associated with the linear approximation at the origin. The direction of the normal to this plane will be determined by the two complex eigenvalues at the origin. Eventually the solution starts to turn in the z direction as the nonlinear term begins to take effect. It is then possible for this path to return to into the origin along its stable manifold. There also exist *near-homoclinic* orbits as shown in Fig. 13.10 for $a = 0.4$, $b = 0.3$ and $c = 4.449$.

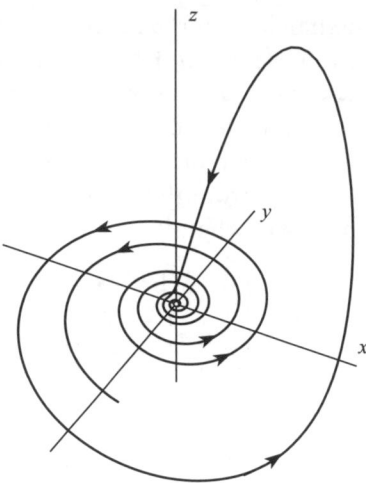

Figure 13.10   A near-homoclinic path for the Rössler attractor $\dot{x} = -y - z$, $\dot{y} = x + ay$, $\dot{z} = bx - cz + xz$ for $a = 0.4$, $b = 0.3$, $c = 4.449$.

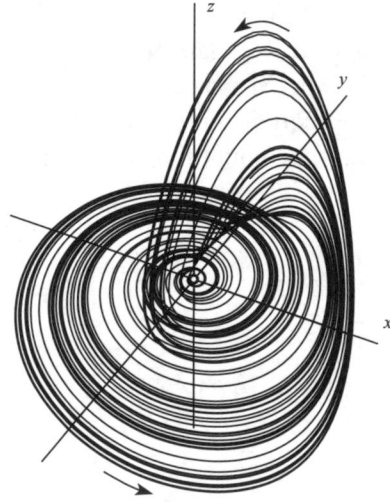

Figure 13.11   The Rössler attractor $\dot{x} = -y - z$, $\dot{y} = x + ay$, $\dot{z} = bx - cz + xz$ for $a = 0.4$, $b = 0.3$, $c = 4.449$; a long-time solution.

Such paths are drawn towards O by the stabilizing effect of the stable manifold, and then spun out again by the influence of the unstable manifold. The result is a solution which seems to wander in a non-repeating manner in some bounded attracting set in $\mathbb{R}^3$: see Fig. 13.11 which shows a long-time solution for the same parameter values of a, b, and c. How could this have arisen?

Whilst there are no stable equilibrium points for this Rössler system, it is possible that there could exist a stable limit cycle which could attract all solutions. For $c = 1.4$, a simple numerical search reveals a stable periodic solution which is shown in Fig. 13.12(a). If c increases

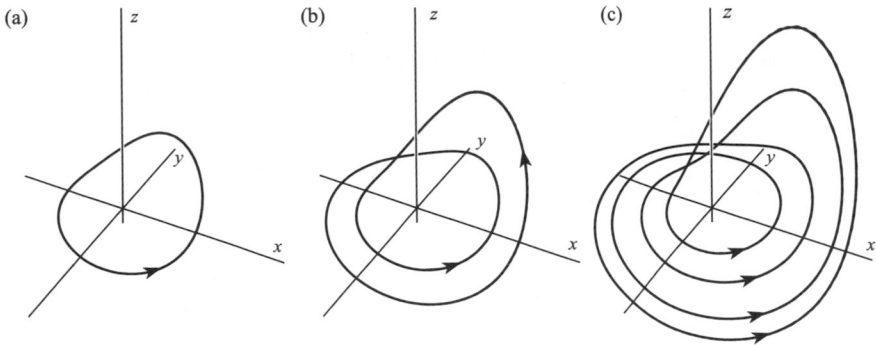

Figure 13.12 Rössler attractor $\dot{x} = -y - z$, $\dot{y} = x + ay$, $\dot{z} = bx - cz + xz$ for $a = 0.4$, $b = 0.3$ and (a) $c = 1.4$; (b) $c = 2$, period-2 solution; (c) $c = 2.63$, period-4 solution.

to $c = 2$, this solution bifurcates at an intermediate value of c into a solution which circuits the origin twice when its phase path is projected onto the (x, y) plane (see Fig. 13.12(b)). A further increase to $c = 2.63$ shows that a further bifurcation has taken place which results in a projected path which circuits the origin four times, giving a period-4 solution as shown in Fig. 13.12(c). The process continues with further **period doubling** to 8, 16, 32, ... circuits. The parameter sequence converges to a value of c beyond which all these periods are present, and we have a **chaotic attracting set** as shown in Fig. 13.11 for $c = 4.449$. This is known as a **strange attractor** which has developed parametrically through a process of period doubling. At each (pitchfork) bifurcation, an unstable periodic orbit will always remain so that the final attractor must include an unbounded set of unstable limit cycles.

Of course this description does not really indicate why these limit cycles become unstable and bifurcate. If we look at the period doubling development from $c = 1.4$ (Fig. 13.12(a)), the closed curve is close to the tangent plane to the unstable mainfold of the origin. As c increases parts of path become closer to the origin whilst a loop grows in the z direction. In other words the periodic solution becomes closer to the destablizing effect of the homoclinic paths which appear as c increases.

13.5 The Duffing oscillator

Phenomena such as period doubling and chaos cannot arise from plane autonomous systems: autonomous systems must be nonlinear and of third-order or above for the possibility of such behaviour, as with the Rössler oscillator in the previous section. However, *forced plane* systems can show chaotic responses. Such a system can be viewed as a third-order system simply by introducing a new variable z. Thus the forced system

$$\dot{x} = X(x, y, t), \quad \dot{y} = Y(x, y, t)$$

could be replaced by the third-order autonomous system

$$\dot{x} = X(x, y, z), \quad \dot{y} = Y(x, y, z), \quad \dot{z} = 1, \quad z(0) = 0.$$

In most practical applications the forcing is periodic in time, and the main interest is in harmonic and subharmonic responses. For such systems it is convenient to compute the Poincaré sequence at the periodic times of the forcing period to detect period doubling and chaos as we explained in Section 13.1. We shall pursue this approach rather than the third-order autonomous analogy.

An equation whose solutions can exhibit some of these interesting phenomena is the Duffing equation

$$\ddot{x} + k\dot{x} - x + x^3 = \Gamma \cos \omega t. \tag{13.13}$$

Note that this version has negative linear and positive cubic restoring terms. The corresponding autonomous system ($\Gamma = 0$) has equilibrium points at $x = \pm 1$, (stable spirals when $0 < k < 2\sqrt{2}$; stable nodes when $k > 2\sqrt{2}$), and at $x = 0$ (a saddle point). As Γ increases from zero we might expect stable forced periodic solutions to develop from $x = \pm 1$ rather than from $x = 0$ as was the case in Section 7.2.

With this in view we shall look for shifted $2\pi/\omega$-periodic solutions of the form

$$x = c(t) + a(t) \cos \omega t + b(t) \sin \omega t. \tag{13.14}$$

As with earlier applications of harmonic balance we assume that the amplitudes $a(t)$ and $b(t)$ are 'slowly varying', so that in the subsequent working their second derivatives can be neglected (it will shortly be found that this approximation would not be appropriate for $c(t)$). Also, harmonics of order higher than the first will be neglected in the approximations as before. With these assumptions, when (13.14) is substituted into (13.13) and the coefficients of $\cos \omega t$ and $\sin \omega t$, and the constant term, are matched on either side of (13.13) we find that

$$\ddot{c} + k\dot{c} = -c \left(c^2 - 1 + \tfrac{3}{2}r^2 \right), \tag{13.15}$$

$$k\dot{a} + 2\omega\dot{b} = -a \left(-1 - \omega^2 + 3c^2 + \tfrac{3}{4}r^2 \right) - k\omega b + \Gamma, \tag{13.16}$$

$$-2\omega\dot{a} + k\dot{b} = -b \left(-1 - \omega^2 + 3c^2 + \tfrac{3}{4}r^2 \right) + k\omega a, \tag{13.17}$$

where $r^2 = a^2 + b^2$. If we further put

$$\dot{c} = d, \tag{13.18}$$

then (13.15) becomes

$$\dot{d} = -c \left(c^2 - 1 + \tfrac{3}{2}r^2 \right) - kd. \tag{13.19}$$

Paths defined by the system (13.16)–(13.19) lie in a van der Pol 4-space (a, b, c, d).

The equilibrium points of the system, which correspond to steady oscillations, are obtained by equating the right-hand sides of (13.16) to (13.19) to zero.

It follows that $d = 0$ and

$$c\left(c^2 - 1 + \tfrac{3}{2}r^2\right) = 0, \tag{13.20}$$

$$a\left(-1 - \omega^2 + 3c^2 + \tfrac{3}{4}r^2\right) + k\omega b = \Gamma, \tag{13.21}$$

$$b\left(-1 - \omega^2 + 3c^2 + \tfrac{3}{4}r^2\right) - k\omega a = 0. \tag{13.22}$$

By squaring and adding (13.21) and (13.22) we find that r^2 satisfies the cubic equation

$$r^2\left[\left(-1 - \omega^2 + 3c^2 + \tfrac{3}{4}r^2\right)^2 + k^2\omega^2\right] = \Gamma^2. \tag{13.23}$$

There are two sets of solutions of (13.23) to be considered:

$$TYPE\,I \quad c = 0, \quad r^2\left[\left(-1 - \omega^2 + \tfrac{3}{4}r^2\right)^2 + k^2\omega^2\right] = \Gamma^2; \tag{13.24}$$

$TYPE\,II \quad$ For $r \le \sqrt{(2/3)}$ only:

$$c^2 = 1 - \tfrac{3}{2}r^2, \quad r^2\left[\left(2 - \omega^2 - \tfrac{15}{4}r^2\right)^2 + k^2\omega^2\right] = \Gamma^2. \tag{13.25}$$

Figure 13.13 shows plots of r against $|\Gamma|$ using (13.24) (the curve C_1) and (13.25) (the curve C_2). The results are plotted for the case $k = 0.3$, $\omega = 1.2$, these values being chosen to display particularly interesting properties of the solutions. The best choice for the parameters is a matter of experiment with computed solutions and graphic displays, but for this value of k, the

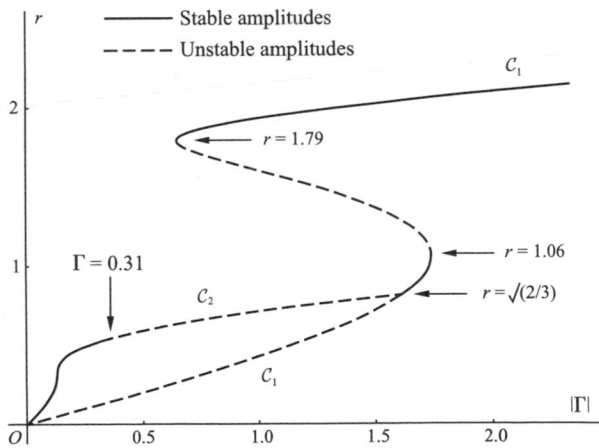

Figure 13.13 Curves showing the response amplitude r in terms of the forcing amplitude Γ, for $k = 0.3$ and $\omega = 1.2$, plotted from eqns (13.24) and (13.25). Type I solution is shown as C_1, and Type II solution is shown as C_2. Linear analysis indicates the stability or instability of the periodic solutions. Remember that whilst C_2 shows the response amplitude, the oscillation, limit cycle is centred about $x = c$.

damping coefficient, the phenomena to be discussed occur over a wide range of frequency ω and amplitude Γ.

TYPE I (eqn (13.24)) Let $a = a_0$, $b = b_0$, $c = 0$, $d = 0$ be an equilibrium point of (13.16)–(13.19). Let $a = a_0 + \xi$, $b = b_0 + \eta$. Then for small $|\xi|, |\eta|, |c|, |d|$, the linearized approximations of (13.16)–(13.19) are

$$k\dot{\xi} + 2\omega\dot{\eta} + A\xi + B\eta = 0, \tag{13.26}$$

$$-2\omega\dot{\xi} + k\dot{\eta} + C\xi + D\eta = 0; \tag{13.27}$$

$$\dot{c} = d \tag{13.28}$$

$$\dot{d} = c(1 - \tfrac{3}{2}r_0^2) - kd \tag{13.29}$$

where $r_0^2 = a_0^2 + b_0^2$ and

$$A = -1 - \omega^2 + \tfrac{9}{4}a_0^2 + \tfrac{3}{4}b_0^2, \quad B = k\omega + \tfrac{3}{2}a_0b_0, \tag{13.30}$$

$$C = -k\omega + \tfrac{3}{2}a_0b_0, \quad D = -1 - \omega^2 + \tfrac{3}{4}a_0^2 + \tfrac{9}{4}b_0^2. \tag{13.31}$$

The equations for ξ, η, c, d uncouple, with ξ, η determined by (13.26) and (13.27), and c and d by (13.28) and (13.29).

Given that $k > 0$, (13.28) and (13.29) have asymptotically stable solutions if and only if

$$r_0^2 > \tfrac{2}{3}.$$

(13.26) and (13.27) can be written in the form

$$\dot{\boldsymbol{\xi}} = \boldsymbol{P}\boldsymbol{\xi},$$

where $\boldsymbol{\xi} = [\xi, \eta]^{\mathrm{T}}$ and

$$\boldsymbol{P} = \frac{1}{k^2 + 4\omega^2} \begin{bmatrix} -Ak + 2\omega C & -Bk + 2\omega D \\ -Ck - 2\omega A & -Dk - 2\omega B \end{bmatrix}.$$

The eigenvalues of $\boldsymbol{P}$ have negative real part, which implies asymptotic stability, if in the usual notation, (see (2.62))

$$p = 2\omega(C - B) - k(A + D) < 0, \tag{13.32}$$

and

$$q = (-Ak + 2\omega C)(-Dk - 2\omega B) - (-Bk + 2\omega D)(-Ck - 2\omega A)$$
$$= -(4\omega^2 + k^2)(BC - AD) > 0 \tag{13.33}$$

(see Fig. 2.8). Using (13.30) and (13.31), inequalities (13.32) and (13.33) become

$$p = k(2 - 2\omega^2 - 3r_0^2) < 0 \quad \text{or} \quad r_0^2 > \max[\tfrac{2}{3}(1 - \omega^2)], \tag{13.34}$$

and

$$q = \tfrac{1}{16}(4\omega^2 + k^2)[27r_0^4 - 48(1 + \omega^2)r_0^2 + 16(1 + \omega^2)^2] > 0,$$

or

$$[r_0^2 - \tfrac{24}{27}(1 + \omega^2)]^2 + \tfrac{16}{81}[3k^2\omega^2 - (1 + \omega^2)^2] > 0. \tag{13.35}$$

These conditions are necessary and sufficient for asymptotic stability. For the case of Fig. 13.13, where $k = 0.3$, $\omega = 1.2$, conditions (13.34) and (13.35) are satisfied if any d only if

$$r_0 > 1.794 \quad \text{or} \quad 0.816 < r_0 < 1.058.$$

The boundaries are shown in Fig. 13.13. It can also be shown that these critical values occur where the (Γ, r_0)-curve turns over and a sudden change happens.

TYPE II (eqn (13.25)) The centre of oscillation is not at the origin, which complicates the analysis.

The steps in the stability calculation are as follows. With $c^2 = 1 - \tfrac{3}{2}r^2$ and a selected value for Γ, we solve (13.21) and (13.32) numerically for a_0 and b_0. Equations (13.16)–(13.19) are linearized in the neighbourhood of

$$a = a_0, \quad b = b_0, \quad c = c_0 = \sqrt{[1 - \tfrac{3}{2}(a_0^2 + b_0^2)]}, \quad d = d_0 = 0. \tag{13.36}$$

Let their perturbations be respectively a', b', c' and d'. After some algebra, it can be shown that they satisfy

$$\dot{u} = Qu,$$

where $\mathbf{u} = [a', b', c', d']$,

$$Q = \begin{bmatrix} q_{11} & q_{12} & q_{13} & 0 \\ q_{21} & q_{22} & q_{23} & 0 \\ 0 & 0 & 0 & 1 \\ -3a_0c_0 & -3b_0c_0 & -2c_0^2 & -k \end{bmatrix}, \tag{13.37}$$

and

$$q_{11} = [(-2 - \omega^2 + \tfrac{15}{4}r_0^2 - \tfrac{3}{2}a_0^2)k + 3a_0b_0\omega]/(k^2 + 4\omega^2)$$

$$q_{12} = [(4 - k^2 - 2\omega^2 - \tfrac{15}{2}r_0^2 + 3b_0^2)\omega - \tfrac{3}{2}a_0b_0k]/(k^2 + 4\omega^2)$$

$$q_{13} = 6c_0(2\omega b_0 - a_0k)/(k^2 + 4\omega^2)$$

$$q_{21} = [(-4 + k^2 + 2\omega^2 + \tfrac{15}{2}r_0^2 - 3a_0^2)\omega - \tfrac{3}{2}a_0b_0k]/(k^2 + 4\omega^2)$$

$$q_{22} = [(-2 - \omega^2 + \tfrac{15}{4}r_0^2 - \tfrac{3}{2}b_0^2)k - 3a_0b_0\omega]/(k^2 + 4\omega^2)$$

$$q_{23} = 6c_0(-2\omega a_0 - b_0k)/(k^2 + 4\omega^2).$$

The eigenvalues of Q can be computed as follows. The parameters ω, k and r_0 are specified; Γ (assumed positive) is then calculated from (13.25); a_0, b_0 and c_0 are then given by (13.36);

finally the eigenvalues of Q, of which there will be four, can be computed. Stability can then be determined by the signs of the real parts of the eigenvalues. Boundaries of stability occur where the eigenvalues where two eigenvalues have zero real parts and two have negative real parts. This can be achieved numerically by trying a sequence of trial values for r_0 (for the same ω and k) until zeros are approximated to within an acceptable error.

For comparison with Fig. 13.13, we select the same parameter values $k = 0.3$ and $\omega = 1.2$ in the calculation of eigenvalues. For $r_0 \approx 0.526$, computations give

$$\Gamma \approx 0.314, \quad a_0 \approx -0.419, \quad b_0 \approx 0.317,$$

and the eigenvalues of Q (all complex)

$$\{-0.276 \pm 0.58\mathrm{i}, 0 \pm 0.527\}$$

to 3 decimal places. For $\gamma < 0.314$, approximately, at least one eigenvalue has positive real part, which indicates instability on the curve C_2 shown in Fig. 13.13.

The remaining solutions on C_2 for $r_0 < \sqrt{\frac{2}{3}}$ are unstable according to this method. However at $\Gamma \approx 0.65$, a stable solution of amplitude $r_0 \approx 1.79$ appears on the curve C_2 in Fig. 13.13. This analysis using the van der Pol space indicates stable 2π-periodic solutions for $0 < \Gamma < 0.314$, and for $\Gamma > 0.65$, approximately for the specified k and ω. Behaviour in the 'stability gap' is complicated as we shall illustrate below.

A numerical search of the Duffing equation (13.13) for periodic solutions confirms the existence of stable $2\pi/\omega$-periodic solutions for $0 < \Gamma < 0.27$ approximately, compared with the theoretical estimate of $\Gamma \approx 0.31$ just obtained. A computed solution for $\Gamma = 0.2$ for $k = 0.3$ and $\omega = 1.2$ is shown in Fig. 13.14(a) together with its phase path and the corresponding Poincaré map for the times $t = 0, 2\pi, \ldots$. For $\Gamma = 0.28$, in the unstable interval, a stable subharmonic of period $4\pi/\omega$ exists and is shown in Fig. l3.14(b). The two-point Poincaré map is shown on the right. A further increase in Γ to 0.29 reveals a stable periodic subharmonic of period $8\pi/\omega$, a period-4 solution (see Fig. 13.14(c)). This **period doubling cascade** is a rapidly accelerating sequence of bifurcations which by $\Gamma = 0.3$, approximately, has doubled to infinity leaving an 'oscillation' without any obvious periodic behaviour. The solution is bounded but not periodic. The Poincaré map becomes a bounded set of returns without any obvious repetitions, and is another example of a **strange attractor**. A sample solution and corresponding phase diagram are shown in Fig. 13.15. Any initial transient behaviour was eliminated before the solution was plotted.

Between $\Gamma = 0.3$ and 0.36 no obvious regularity is observable: the solution plots reveal wandering solutions of an irregularly oscillating type without any uniform pattern. Such solutions are said to display chaotic behaviour. Close to $\Gamma = 0.37$, however, stable solutions of period 5 appear (see Fig. 13.16(a))—these are centred on $x = 0$. Another chaotic regime follows between $\Gamma = 0.37$ and about 0.65 (a segment of a solution plot is shown in Fig. 13.15 for $\Gamma = 0.5$). In an interval around $\Gamma = 0.65$ more stable period-2 solution appear, centred somewhere close to $x = 0$ but of a very unsymmetrical type. At $\Gamma \approx 0.73$ (theory predicts $\Gamma \approx 0.65$), a stable period-2 solution is now available as shown in Fig. 13.16(c) (see also the stability diagram, Fig. 13.13).

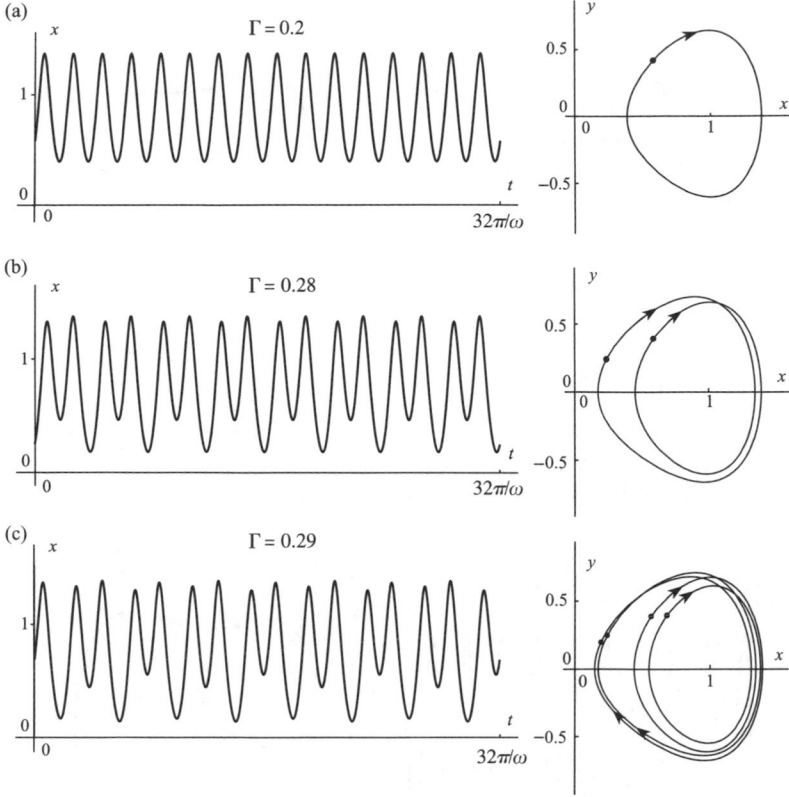

Figure 13.14 Time solutions, phase paths, and Poincaré maps of the Duffing oscillator for $k = 0.3$ and $\omega = 1.2$: (a) $\Gamma = 0.20$ (period-1); (b) $\Gamma = 0.28$ (period-2); (c) $\Gamma = 0.29$ (period-4).

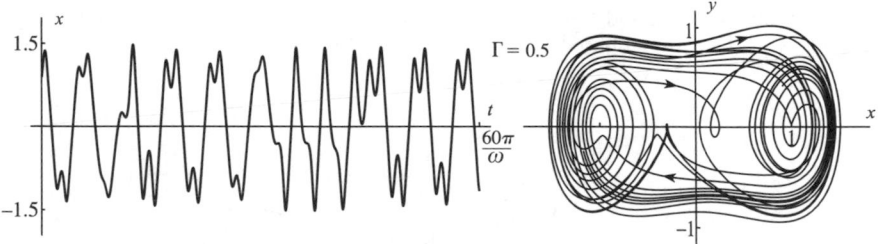

Figure 13.15 Sample chaotic response for $k = 0.3, \omega = 1.2$ and $\Gamma = 0.5$: the bounded time solution and corresponding phase path are shown.

The search for periodicities can be assisted by plotting Poincaré maps appropriate to the period being looked for: $2\pi/\omega$ for the forcing frequency, $4\pi/\omega$ for period doubling, and so on. Periodic solutions of period $2\pi/\omega$ produce maps in the (x, y)-plane having a single fixed point, period doubling two fixed points, etc. (see Fig. 13.9). If we apply the same procedure to the 'chaotic' output, say for $\Gamma = 0.5$, we find that after all there does appear to be some degree of underlying structure. Figure 13.17 shows a Poincaré map for this case. The sequence shown is for $t = 2\pi n$ ($n = 0, 1, 2, \ldots$) and nearly 2000 points are displayed. Not all the

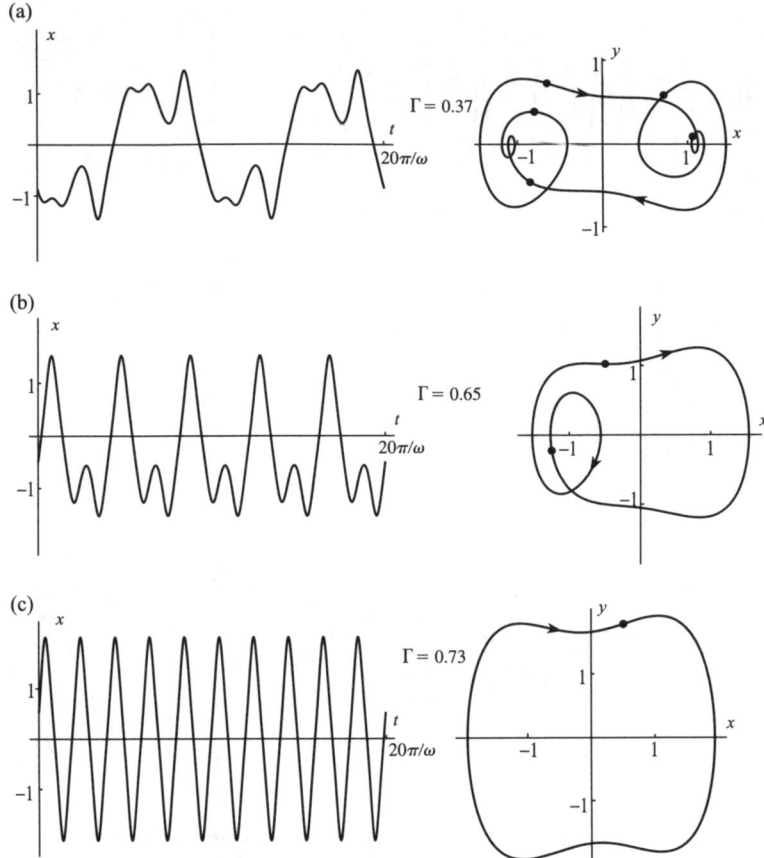

Figure 13.16   Time solutions and phase diagrams for Duffing's equation (13.8) for $k = 0.3$ and $\omega = 1.2$ in the cases: (a) $\Gamma = 0.37$, a large amplitude, period-5 response; (b) $\Gamma = 0.65$, a large amplitude, period-2 response, on of a symmetric pair; (c) $\Gamma = 0.73$, a large amplitude, period-1 response. The Poincaré sections for $t = 0, 2\pi, \ldots$ are shown by the dots.

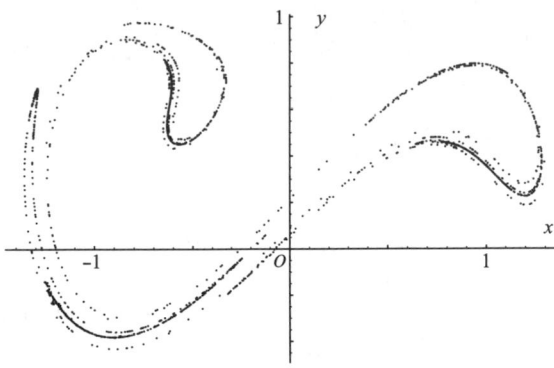

Figure 13.17   Poincaré returns for Duffing's equation (13.8) for $k = 0.3$, $\omega = 1.2$ and $\Gamma = 0.5$: 2000 returns are shown for the section $t = 0, 2\pi/\omega, 4\pi/\omega, \ldots$.

points calculated are plotted—working from a given initial condition the earlier points are subject to a form of **transient behaviour**, and not until they find their way into the set displayed are they recorded. There are no fixed points observed in the chaotic regime: instead the set plays a similar role. The set is to a large extent independent of the initial values, and has the property that any point once in the set generates a sequence of first returns all of which lie in the set: it is another example of a **strange attractor**. Computer studies point also to the fact that, given any point in the attracting set, some future first return will come arbitrarily close to it. The shape of the strange attractor depends upon which sections are chosen: for example the returns for the sections $t = (2n + 1)\pi/\omega$ ($n = 0, 1, 2, \ldots$) will fill a differently shaped region when projected on to the (x, y) plane. In the x, y, t space the solutions (after the disappearance of 'transients' as above) will remain in a tube in the direction of the t axis. Alternatively, periodicity of the forcing term can used to design a space in which the t axis is turned so that the points $x = 0$, $y = 0$, and $t = 2n\pi/\omega$ coincide with the origin (0, 0, 0) to give a **toroidal** space looking like that shown in Fig. 13.5. The Poincaré sequence then becomes simply a cross section of the torus. The three-dimensional diagram is difficult to draw unless some projection such as that given in Section 3.3 is used.

In the Duffing oscillator, as Γ increases from zero to about 0.27 (for the parameter values $k = 0.3$, $\omega = 1.2$), the amplitude of the response increases, tending to carry the representative point in the phase plane into the neighbourhood of the unstable limit cycle growing from the origin (Type I above). It seems to be this interaction between the growing incursion of the stable limit cycle into the domain of the unstable limit cycle which destroys the stability of the former which shows itself as period doubling.

As we shall see in Section 13.6, homoclinic bifurcation is also a destabilizing influence for this Duffing oscillator.

The period-doubling sequence of bifurcations of the periodic solution emerging from the equilibrium point (1,0) as Γ increases from zero, can be represented by Poincaré fixed points in the (x, y) plane. Figure 13.18(a) shows a schematic view of these returns (not to scale). At $\Gamma \approx 0.27$ a pitchfork bifurcation occurs resulting in a stable period-2 subharmonic, whilst the original harmonic becomes unstable. The Poincaré map reveals the pitchfork structure. At $\Gamma \approx 0.28$ this subharmonic suffers a further pitchfork bifurcation to a period-4 solution, and

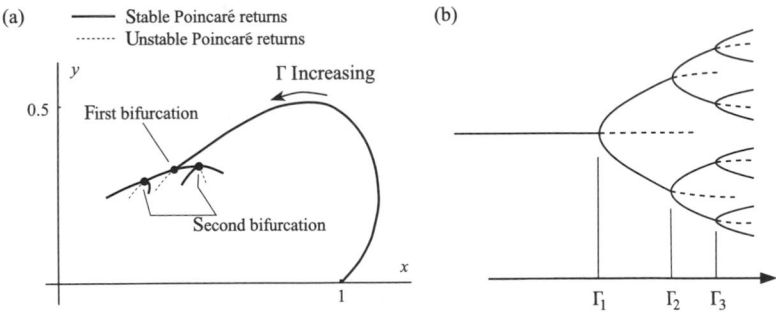

Figure 13.18 Track of the fixed points of the periodic solution in $x > 0$ as Γ increases from zero. (a) The first pitchfork bifurcation takes place at $\Gamma \approx 0.27$. The diagram is a representation and not an accurate figure near the bifurcations. (b) Scheme showing a period doubling cascade through a sequence of pitchfork bifurcations.

so on. If these period doubling bifurcations occur at $\Gamma = \Gamma_1, \Gamma_2, \Gamma_3, \ldots$, then it is known the progression of this sequence obeys a universal law

$$\lim_{k \to \infty} \frac{\Gamma_k - \Gamma_{k-1}}{\Gamma_{k+1} - \Gamma_k} = \delta = 4.66292 \ldots.$$

The constant δ is a universal constant for period doubling sequences, and is known as the **Feigenbaum constant** (see Nicolis 1995 and Addison 1997 for more details).

A representation of a sequence of pitchfork bifurcations is shown in Fig. 13.18(b). The period doubling cascade arises from bifurcations of a discrete system consisting of Poincaré first returns. A simple model of a discrete system which exhibits a period doubling cascade is the **logistic map** which will be described in the next section.

13.6 A discrete system: the logistic difference equation

Probably the simplest nonlinear difference equation is the logistic equation defined by

$$u_{n+1} = \alpha u_n (1 - u_n).$$

or, by putting $f(u) = \alpha u (1 - u)$,

$$u_{n+1} = f(u_n), \quad n = 0, 1, 2, \ldots$$

Fixed points of a difference equation occur where $u_n = u$ say, for all n, that is, where $f(u) = u$, or

$$u = \alpha u (1 - u) \quad (\alpha > 0)$$

for the logistic equation. Thus there are two fixed points, at $u = 0$ and at $u = (\alpha - 1)/\alpha$. Let $v = f(u)$. Then in the (u, v) plane equilibrium occurs at $(0, 0)$ and $((\alpha - 1)/\alpha, (\alpha - 1)/\alpha)$, at the intersections of the line $v = u$ and the parabola $v = f(u) = \alpha u (1 - u)$, namely the origin and the point P in Fig. 13.19. This figure is drawn in the case $\alpha = 2.8$. A **cobweb** construction can

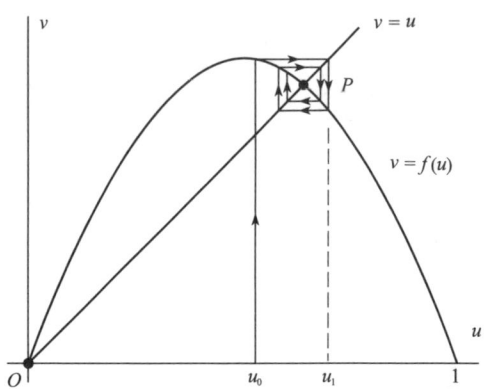

Figure 13.19   Cobweb construction showing the stability of the fixed point P for $\alpha = 2.8$.

indicate the stability of P. Suppose that $u = u_0$ initially. Then $u_1 = \alpha u_0 (1 - u_0)$, and value of u_1 can be constructed by drawing a line vertically to meet $v = f(u)$, and then a horizontal line to $v = u$ as shown in Fig. 13.19. The value of u there is u_1. The process can be repeated to find $u_1, u_2, \ldots$. It seems evident graphically that P will be stable since the cobweb converges to P, where $u = (\alpha - 1)/\alpha = 9/14$ for $\alpha = 2.8$.

The stability of P depends critically on the slope of the tangent to $v = f(u)$ at P. Its slope there is

$$f'((\alpha - 1)/\alpha) = 2 - \alpha.$$

It can be proved that if this slope is *greater* than -1, then the solution is stable, and $u_n \to (\alpha - 1)/\alpha$ as $n \to \infty$. If the slope is less than -1 then the cobweb will spiral away from the fixed point. The critical slope separating stability and instability is

$$f'((\alpha - 1)/\alpha) = 2 - \alpha = -1, \text{ that is, where } \alpha = 3.$$

If $\alpha > 3$ (the unstable interval), what happens to the iterations of the logistic equation? Consider the curve

$$v = f(f(u)) = \alpha[\alpha u(1 - u)]\,[1 - \alpha u(1 - u)] = \alpha^2 u(1 - u)[1 - \alpha u(1 - u)].$$

The line $v = u$ intersects this curve where

$$u = \alpha^2 u(1 - u)[1 - \alpha u(1 - u)].$$

Hence the fixed points of $v = f(f(u))$ occur where

$$u = 0, \quad \text{or} \quad [\alpha u^2 - (\alpha - 1)]\,[\alpha^2 u^2 - \alpha(1 + \alpha)u + (1 + \alpha)] = 0.$$

Note that the equation $u = f(f(u))$ automatically includes the solutions of $u = f(u)$. Hence the second equation always has the solution $u = (\alpha - 1)/\alpha$, and two further real solutions if $\alpha > 3$. Figure 13.20(a) shows the curve $v = f(f(u))$ for the critical case $\alpha = 3$ and for $\alpha = 3.4$. For $\alpha = 3.4$ let the points of intersection with the line $v = u$ be O, A, B, C as shown in Fig. 13.20(b). Since

$$u_A = f(f(u_A)), \quad u_B = f(f(u_B)),$$

it is possible to insert a square with diagonal AC such that its other corners lie on $v = f(u)$. This follows since the pair of equations above always has the solution

$$u_B = f(u_A), \quad u_A = f(u_B).$$

The existence of the square indicates the presence of a **period-2 solution** of the logistic equation such that u_n alternates between u_A and u_B, where u_A and u_B are the roots of

$$\alpha^2 u^2 - \alpha(1 + \alpha)u + (1 + \alpha) = 0$$

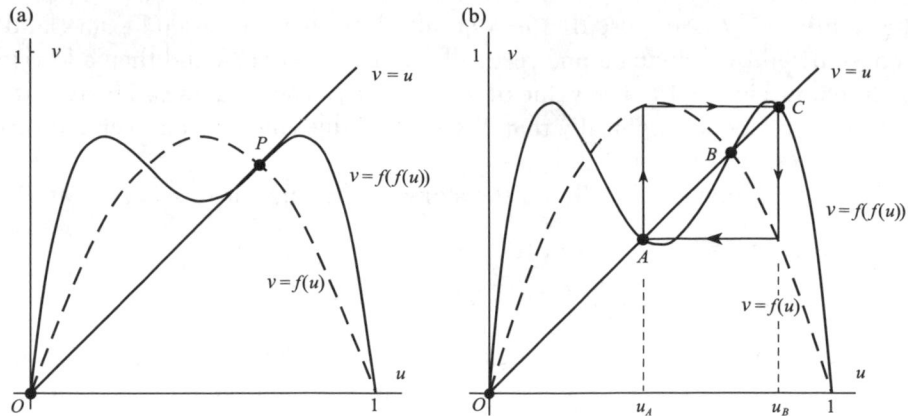

Figure 13.20 (a) Graph of $v = f(f(u))$ for the critical case $\alpha = 3$. (b) Graph of $v = f(f(u))$ for $\alpha = 3.4$ showing the period-2 cycle between A and B.

for $\alpha > 3$. At $\alpha = 3.4$, the period-2 solution is stable since the slopes at u_A and u_B to the curve $u = f(f(u))$ both exceed -1. The value of α at which the critical slope of -1 occurs can be found as follows. The slope is -1 if

$$\frac{\mathrm{d}}{\mathrm{d}u} f(f(u)) = \alpha^2 - 2\alpha^2(1+\alpha)u + 6\alpha^3 u^2 - 4\alpha^3 u^3 = -1, \tag{13.38}$$

where u satisfies

$$\alpha^2 u^2 - \alpha(1+\alpha)u + (1+\alpha) = 0. \tag{13.39}$$

Multiply (13.39) by $4\alpha u$ and eliminate u^3 in (13.38), which results in

$$u^2 - \frac{\alpha+1}{\alpha}u + \frac{\alpha^2+1}{2\alpha^2(\alpha-2)} = 0, \tag{13.40}$$

whilst (13.39) can be written as

$$u^2 - \frac{\alpha+1}{\alpha}u + \frac{\alpha+1}{\alpha^2} = 0. \tag{13.41}$$

Finally, eliminate u between (13.40) and (13.41) so that α must satisfy

$$\alpha^2 - 2\alpha - 5 = 0.$$

Since $\alpha > 3$, the required root is $\alpha = 1 + \sqrt{6} = 3.449\ldots$ Hence there exists a stable period-2 solution for $3 < \alpha < 1 + \sqrt{6}$. Beyond this value we must look at the curve $v = f(f(f(f(u))))$. Its graph for $\alpha = 3.54$ is shown in Fig. 13.21: there are now 8 fixed points on $v = u$. The only stable solution on this curve is a period-4 one. The next doubling occurs at $\alpha \approx 3.544$. The interval between period doubling rapidly decreases until a limit is reached at $\alpha \approx 3.570$, beyond which irregular chaotic behaviour occurs.

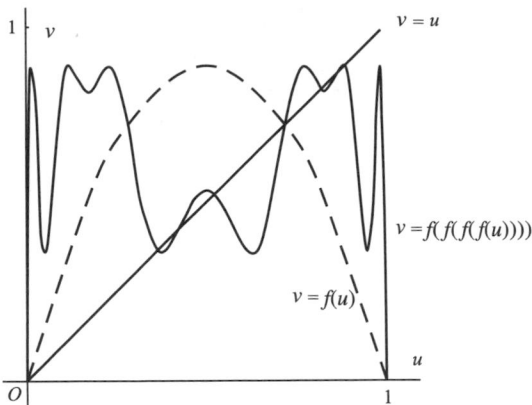

Figure 13.21   The eight fixed points of $v = f(f(f(f(u))))$ for $\alpha = 3.54$ lying on the intersection with the line $v = u$.

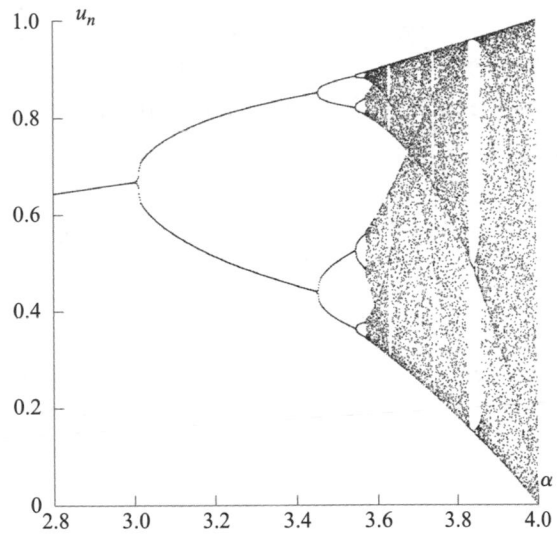

Figure 13.22   Period doubling for the logistic equation as α increases from $\alpha = 2.8$ to $\alpha = 4$: chaotic intersections occur beyond $\alpha \approx 3.57$.

The development of the attractor is shown in Fig. 13.22 for $2.8 < \alpha < 4$. Each bifurcation is of pitchfork type (Fig. 13.23) in which a doubled stable cycle is created leaving behind an unstable cycle.

Exercise 13.4
Find the fixed points of the difference equation $u_{n+1} = \alpha u_n(1 - u_n^3)$. Find the value of α for which the non-zero fixed point becomes unstable.

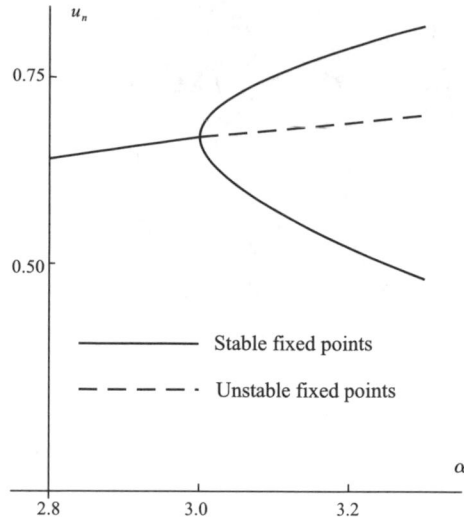

Figure 13.23　First pitchfork bifurcation for the logistic equation at $\alpha = 3$.

13.7　Liapunov exponents and difference equations

Whilst Fig. 13.22 indicates numerically period doubling and chaotic behaviour, how can we identify domains of the parameter α where such phenomena occur? This analysis refers to difference equations but it gives clues to chaotic behaviour in differential equations which will be considered later. Consider the difference equation

$$u_{n+1} = f(u_n).$$

Suppose that the initial value is A_0: $(0, u_0)$ with subsequent iterations A_1:$(1, u_1)$ where $u_1 = f(u_0)$, A_2:$(2, u_2)$ where $u_2 = f(u_1)$, and so on. The sequence of iterates denoted by $A_0, A_1, A_2 \ldots$, is shown in Fig. 13.24. Consider another initial value M_0: $(0, u_0 + \varepsilon)$ which is close to u_0 assuming that $|\varepsilon|$ is small. This maps into M_1: $(1, f(u_0 + \varepsilon))$. Approximately

$$f(u_0 + \varepsilon) \approx f(u_0) + f'(u_0)\varepsilon = u_1 + f'(u_0)\varepsilon$$

using the Taylor expansion.

The distance between the initial values (at $n = 0$) is $|\varepsilon|$, and the distance between A_1 and M_1 at $n = 1$ is approximately $|f'(u_0)||\varepsilon|$. Chaotic behaviour is associated with exponential divergence of neighbouring solutions. In the expectation of exponential divergence the distances can be converted to linear growth by taking their logarithms. Thus the increase in the logarithmic distance between $n = 0$ and $n = 1$ is

$$\ln(|f'(u_0)|\,|\varepsilon|) - \ln|\varepsilon| = \ln|f'(u_0)|.$$

We should continue this process between successive steps but the exponential growth is numerically unsustainable from a computing position, so instead restart the process at N_1: $(1, u_1 + \varepsilon)$ (see Fig. 13.24). Also, although the divergence can be locally exponential the system can still

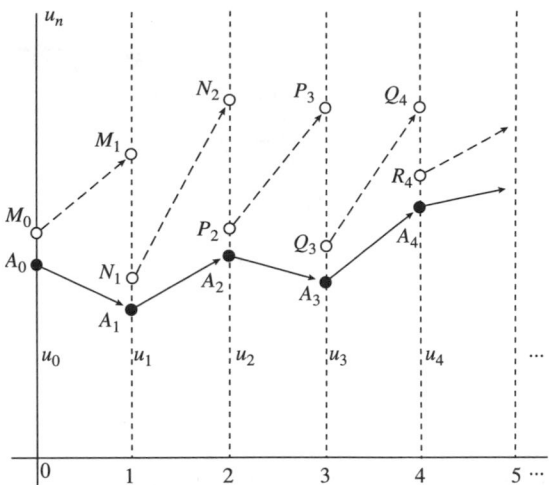

Figure 13.24 The exact iterated sequence is $A_0, A_1, A_2, \ldots$, where u_n takes the values $u_0, u_1, u_2 \ldots$. The approximating sequence is $M_0 M_1; N_1 N_2; P_2 P_3 : \ldots$, where $A_0 M_0 = A_1 N_1 = A_2 P_2 = \cdots \varepsilon$.

be bounded. The point N_1 maps into N_2 (say), where

$$f(u_1 + \varepsilon) \approx f(u_1) + f'(u_1)\varepsilon = u_2 + f'(u_1)\varepsilon.$$

The increase over this step is $\ln |f'(u_1)|$. We now repeat restart this process over each step. The *average* of the total growth over N steps is therefore

$$\frac{1}{N} \sum_{k=0}^{N} \ln |f'(u_k)|.$$

The limit of this sum as $N \to \infty$ is known as the **Liapunov exponent** Λ of this difference equation, namely

$$\Lambda = \lim_{N \to \infty} \frac{1}{N} \sum_{k=0}^{N} \ln |f'(u_k)|. \tag{13.42}$$

which is, incidently, independent of ε. Chaos occurs if the Liapunov exponent Λ is positive. Further discussion and examples are given by Moon (1987).

Consider again the logistic equation

$$u_{n+1} = \alpha u_n (1 - u_n) \tag{13.43}$$

discussed in the previous section. In this case $f(x) = \alpha x(1 - x)$. Hence eqn (13.42) becomes

$$\Lambda(\alpha) = \lim_{N \to \infty} \frac{1}{N} \sum_{k=0}^{N} \ln |\alpha(1 - 2u_k)|, \tag{13.44}$$

in terms of the parameter α. The distribution of Liapunov exponents in terms of the parameter α is shown in Fig. 13.25 for $3.3 < \alpha < 3.9$. The positive values indicate chaotic output.

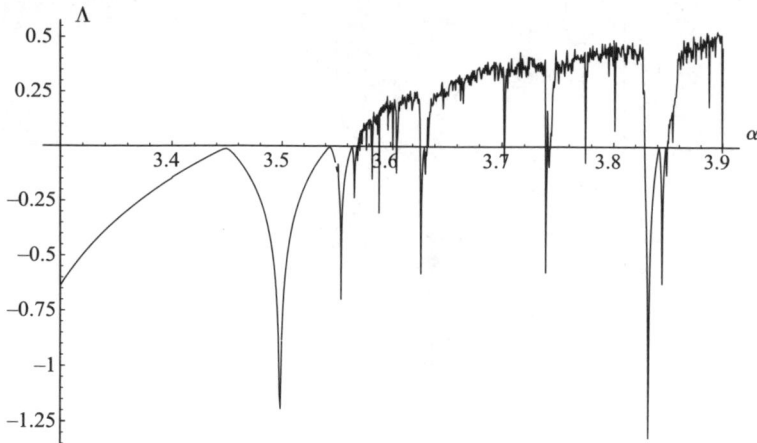

Figure 13.25 Λ has been computed from (13.44) for about 1100 values of α between $\alpha = 3.3$ and $\alpha = 3.9$ with $N = 250$: convergence is slow and the diagram gives the general shape of the Liapunov exponents. For greater accuracy for a particular value of α, a larger value of N should be chosen.

This figure should be compared with Fig. 13.22. In Fig. 13.25, there are visible 'negative' spikes which indicate windows of periodic (non-chaotic) solutions, which are also just visible in Fig. 13.22.

Example 13.5   *Verify that $u_n = \sin^2(2^n)$ is a solution of $u_{n+1} = 4u_n(1 - u_n)$. Find the Liapunov exponent of this solution. Is the solution chaotic?*

Using a simple identity

$$4u_n(1 - u_n) = 4\sin^2(2^n)(1 - \sin^2(2^n))$$

$$= 4\sin^2(2^n)\cos^2(2^n) = \sin^2(2^{n+1}) = u_{n+1},$$

which verifies the solution.

Using (13.42) with $f(u) = 4u(1 - u)$, the Liapunov exponent of the solution is

$$\Lambda = \lim_{N \to \infty} \frac{1}{N} \sum_{k=0}^{N} \ln|f'(u_k)| = \lim_{N \to \infty} \frac{1}{N} \sum_{k=0}^{N} \ln|4 - 8u_k|$$

$$= \lim_{N \to \infty} \frac{1}{N} \sum_{k=0}^{N} \ln|4 - 8\sin^2(2^k)|$$

$$= \lim_{N \to \infty} \frac{1}{N} \sum_{k=0}^{N} \ln|4\cos(2^{k+1})|.$$

Whilst this is the *exact* formula for the Liapunov exponent, there are computational problems which arise from the exponential growth of 2^k as k increases. It is easier to calculate Λ by computing the sequence $\{u_k\}$ from an appropriate initial value u_0, and then using (13.44). Thus, if $u_0 = 0.9$, then

$$u_1 = 4u_0(1 - u_0) = 0.36, \quad u_2 = 4u_1(1 - u_1) = 0.9216,\ldots,$$

which are easy to compute compared with the trigonometric function $\cos(2^k)$. From (13.44), it follows that $\Lambda = 0.693\ldots$, which is positive indicating chaotic behaviour.

Note that even though the solution $u_n = \sin^2(2^n)$ is bounded it can still be chaotic: chaos does not necessarily imply unboundedness.   ●

13.8   Homoclinic bifurcation for forced systems

We have referred earlier in this chapter (Section 13.3) to the significance of homoclinic bifurcation as a possible trigger for chaotic output in certain third-order systems such as the Rössler attractor. For the autonomous Rössler system homoclinic paths are associated with an equilibrium point, but they may equally occur for a limit cycle. We shall specifically discuss such behaviour for systems of the form

$$\dot{x} = y, \quad \dot{y} = f(x, y, t), \tag{13.45}$$

where $f(x, y, t + (2\pi/\omega)) = f(x, y, t)$ for all t: this is a $2\pi/\omega$-periodically forced system. Suppose that the system has an unstable limit cycle *with the same period* $2\pi/\omega$ as the forcing period. Suppose also that, associated with the limit cycle, there are two families of solutions, an attracting one which approaches the limit cycle as $t \to \infty$, and a repelling one which approaches the limit cycle as $t \to -\infty$. We take Poincaré sequences with time-steps equal to the forcing period, namely $2\pi/\omega$, since the differential equation (13.45) is unaffected by time translations $t \to t + 2\pi n/\omega$ for any integer n. Consider Poincaré sequences which start from time $t = 0$. Since the limit cycle has period $2\pi/\omega$, the point on it for which $t = 0$ will be fixed point.

Now consider the set of initial points $(x(0), y(0))$ from which solutions approach the limit cycle, but only record the points $(x(2n\pi/\omega), y(2n\pi/\omega))$ on the phase paths for $n = 1, 2, \ldots$, rather like a series of periodic snapshots. On the (x, y) phase plane the accumulation of these points appears as a curve which approaches the fixed point on the limit cycle. This resulting curve of Poincaré sequences is known as a **stable manifold** (with reference time $t = 0$) of the fixed point of the limit cycle. We could start with a different initial time, say, $t = t_0$ which would lead to a different stable manifold with a different fixed point but still on the limit cycle.

Similarly we can consider an alternative set of initial values (x_0, y_0) for which the backward Poincaré sequences $(x(2\pi n/\omega), y(2\pi n/\omega))$, $n = -1, -2, \ldots$ approaches the fixed point on the limit cycle. This is known as the **unstable manifold** with reference to the initial time $t = 0$. There is a strong similarity with saddle points of equilibrium points in the plane for autonomous systems. The two manifolds appear as two pairs of incoming and outgoing accumulations of sequences.

Homoclinic bifurcation occurs if the stable and unstable manifolds intersect, since through any such point of intersection there is a solution which approaches the limit cycle as $t \to \infty$ and as $t \to -\infty$. Such solutions are said to be **homoclinic to the limit cycle**.

Generally for nonlinear systems it is difficult to create examples in which explicit solutions can be found for the manifolds, but, for some linear systems, it is possible to generate them which is helpful in understanding their construction.

Consider the linear equation

$$\ddot{x} + \dot{x} - 2x = 10\cos t, \quad \dot{x} = y. \tag{13.46}$$

This equation has the general solution

$$x(t) = Ae^t + Be^{-2t} - 3\cos t + \sin t.$$

This equation has been chosen so that the characteristic equation has real solutions of different signs. If $x(0) = x_0$ and $y(0) = y_0$ then, in terms of the initial values,

$$x(t) = \tfrac{1}{3}(2x_0 + y_0 + 5)e^t + \tfrac{1}{3}(x_0 - y_0 + 4)e^{-2t} - 3\cos t + \sin t. \tag{13.47}$$

The differential equation has an isolated, 2π-periodic solution

$$x(t) = -3\cos t + \sin t,$$

with phase path $x^2 + y^2 = 10$, corresponding to the case $A = B = 0$ in the general solution. From (13.47) we see that this solution is selected if the initial condition (x_0, y_0) at $t = 0$ lies on the intersection of the two straight lines L_1 and L_2

$$L_1: 2x_0 + y_0 + 5 = 0, \quad L_2: x_0 - y_0 + 4 = 0,$$

shown in Fig. 13.26. This is the point $Q:(-3, 1)$ and it is a fixed point on the limit cycle for the Poincaré sequence $t = 0, 2\pi, 4\pi, \ldots$

From (13.47), all solutions starting on L_1: ($2x_0 + y_0 + 5 = 0$) approach the limit cycle as $t \to +\infty$ (these include solutions starting from both inside and outside the limit cycle). This is the **attracting family** of solutions.

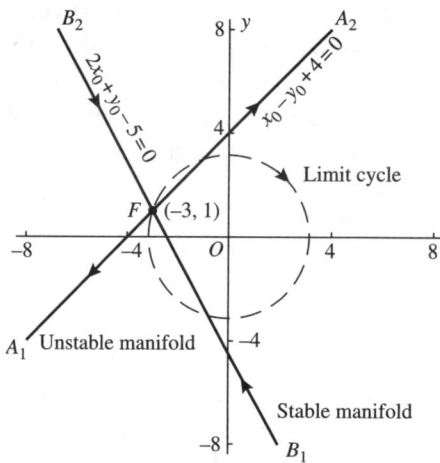

Figure 13.26   Stable and unstable manifolds of the fixed point at $(-3, 1)$, from the Poincaré sequences derived from initial values at $t = 0$, of $\ddot{x} + \dot{x} - 2x = 10\cos t, \dot{x} = y$.

Take any point (x_0, y_0) on L_1, the attracting time-solution selected being denoted by $X(t)$, and consider the Poincaré sequence (x_n, y_n) for $t = 0, 2\pi, 4\pi, \ldots$. The first return $(x_1, y_1) = (X(2\pi), X'(2\pi))$, is also selected from $X(t)$, and (x_1, y_1) serves as the initial conditions for subsequent steps. But the differential equation is of the form (13.45) and *is unaffected by the time translation $t \to t + 2\pi$* due to the 2π-periodicity of the forcing term, so eqn (13.47), with (x_1, y_1) in place (x_0, y_0), together with the requirement $2x_1 + y_1 + 5 = 0$, also hold good. Therefore (x_1, y_1) lies on L_1, and similarly for all subsequent members (x_n, y_n) of the sequence, so the Poincaré sequence with steps 2π converges to Q. The line L_1 is called a stable manifold.

Similarly, the set of points L_2: $(x_0 - y_0 + 4 = 0)$, consisting of initial values of the solutions that do *not* approach the limit cycle, and whose Poincaré sequences (x_n, y_n) for $t = 0, 2\pi, 4\pi, \ldots$ therefore diverge from Q as $n \to \infty$, is called the **unstable manifold** at Q of the limit cycle.

For initial times differing from $t = 0$ a similar construction leads to fixed points other than Q on the limit cycle, since (13.47) takes a different form in such cases. Therefore each point on the limit cycle has a stable and an unstable manifold associated with it.

The stable and unstable manifolds are straight lines, and intersections between them away from the fixed point are not possible. Homoclinic bifurcation cannot occur for forced linear systems.

The Duffing oscillator

We shall now look at the stable and unstable manifolds for the model Duffing oscillator. This a continuation of the analysis started in Section 13.3. Inevitably, the approach is mainly numerical. For nonlinear systems we need to estimate the position of a fixed point of a limit cycle, and the local directions of the manifolds at the fixed point. Consider again the Duffing equation with small damping and forcing

$$\ddot{x} + \varepsilon k \dot{x} - x + x^3 = \varepsilon \gamma \cos \omega t, \quad \dot{x} = y, \quad 0 < \varepsilon \ll 1. \tag{13.48}$$

We are now interested in the unstable limit cycle about the origin in the phase plane, and the stable and unstable manifolds associated with points on it. For small $|x|$, x satisfies

$$\ddot{x} + \varepsilon k \dot{x} - x = \varepsilon \gamma \cos \omega t. \tag{13.49}$$

The periodic solution of this equation is

$$x_p = C \cos \omega t + D \sin \omega t,$$

where

$$C = \frac{-\varepsilon \gamma (1 + \omega^2)}{(1 + \omega^2)^2 + \varepsilon^2 k^2 \omega^2} = \frac{-\varepsilon \gamma}{1 + \omega^2} + O(\varepsilon^3),$$

$$D = \frac{\varepsilon^2 \gamma k \omega}{(1 + \omega^2)^2 + \varepsilon^2 k^2 \omega^2} = \frac{\varepsilon^2 \gamma k \omega}{(1 + \omega^2)^2} + O(\varepsilon^3),$$

which has the fixed point

$$
\left(\frac{-\varepsilon\gamma}{1+\omega^2}, \frac{\varepsilon^2\gamma\kappa\omega^2}{(1+\omega^2)^2} \right),
$$

to order ε^3.

The characteristic equation of (13.49) has roots

$$
m_1, m_2 = \tfrac{1}{2}[-\varepsilon\kappa \pm \sqrt{(\varepsilon^2\kappa^2 + 4)}],
$$

so that the general solution is

$$
x = A e^{m_1 t} + B e^{m_2 t} + x_p.
$$

If $x(0) = x_0$ and $y(0) = y_0$, then

$$
A = [(C - x_0)m_2 - (D\omega - y_0]/(m_2 - m_1),
$$
$$
B = [-(C - x_0)m_2 - (D\omega - y_0)]/(m_2 - m_1).
$$

Since $m_1 > 0$ and $m_2 < 0$ for small $\varepsilon > 0$, the local directions of the stable and unstable manifolds are given by the straight lines

$$
(C - x)m_2 - (D\omega - y) = 0, \quad -(C - x)m_1 - (D\omega - y) = 0,
$$

respectively. Hence the slopes of these lines through the fixed point are m_2 for the stable manifold and m_1 for the unstable manifold, where

$$
m_1, m_2 = \tfrac{1}{2}[-\varepsilon\kappa \pm \sqrt{(\varepsilon^2\kappa^2 + 4)}] \approx \pm 1 - \tfrac{1}{2}\varepsilon\kappa,
$$

for small ε. Thus m_1 and m_2 give the local directions of the stable and unstable manifolds in a similar manner to that shown in Fig. 13.26.

It is fairly easy to compute the manifolds for the fixed point of the Duffing equation (13.48). As in the previous work on period doubling we shall assume the parameter values $k = \varepsilon\kappa = 0.3$ and $\omega = 1.2$ for comparison purposes. As we saw in Section 13.3, period doubling for these parameter values first starts for increasing amplitude Γ at $\Gamma = \varepsilon\kappa \approx 0.27$. However before this happens homoclinic bifurcation has occurred. We can show just how it develops by computing the manifolds for selected values of Γ. Practically this is done by locating the approximate position of the fixed point and the directions of the manifolds there by the method outlined above. Let (x_c, y_c) be the coordinates of the fixed point. Then consider solutions which start at $(x_c \pm \delta, y_c \pm m_1\delta)$ at $t = 0$ where δ is a small positive parameter. The first returns can then be computed at $t = 2\pi/\omega, 4\pi/\omega, \ldots$. This process is repeated for a selection of incremental increases in δ, to cover one cycle of the forcing period. This is partly a trial and error method, but it is helped by the stability of adjacent manifolds. Interpolation of the points generated by this computation results in the manifold. For the stable manifolds we consider solutions which start at $(x_c \pm \delta, y_c \pm m_2\delta)$ at $t = 0$, *but reverse time*. This procedure was used to plot the manifolds shown in Fig. 13.27. These and later figures were first computed by Holmes (1979).

Figure 13.27(a) shows the manifolds for $\Gamma = 0.2$: for this value they do not intersect. At $\Gamma \approx 0.256$, the manifolds touch at A_0 in Fig. 13.27(b). From the periodicity of the Duffing

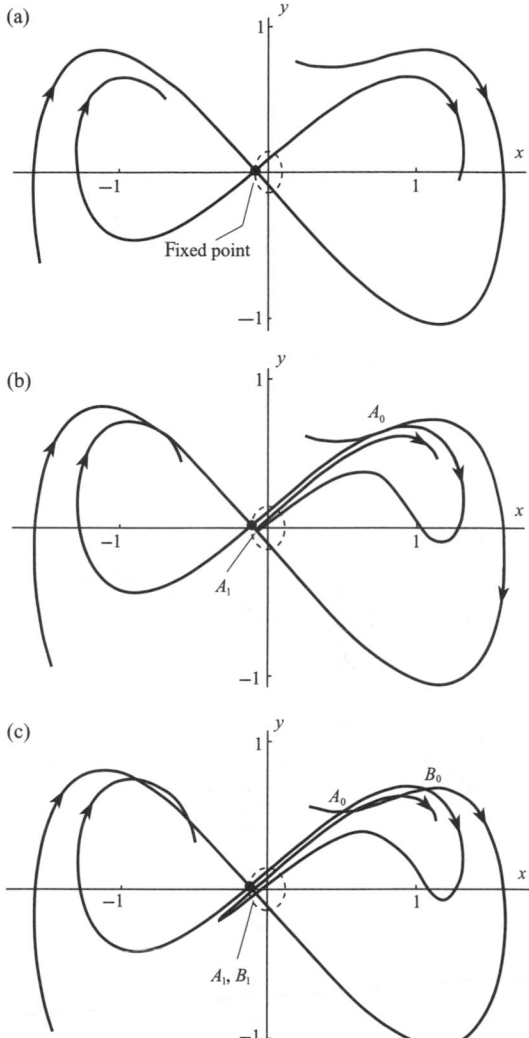

Figure 13.27 Stable and unstable manifolds for Duffing's equation with $k = 0.3$, $\omega = 1.2$ and (a) $\Gamma = 0.20$; (b) $\Gamma = 0.256$; (c) $\Gamma = 0.28$.

equation in t, if the manifolds touch at $t = 0$, then they must inevitably touch at $t = 2n\pi/\omega$ where $n = \ldots, -2, -1, 0, 1, 2, \ldots$, that is, periodically both forward and backward in time. For example, at $t = 2\pi/\omega$ the next tangential contact occurs at A_1 shown in Fig. 13.27(b). Further future contacts will all be in the gap on the stable manifold between the A_1 and the fixed point. The *solution* which passes through A_0 will look something like the curve shown in Fig. 13.28(a) becoming periodic as $t \to \pm\infty$. Its phase diagram is illustrated in Fig. 13.28(b). This solution is highly unstable, and the diagrams in Fig. 13.28 are not exact.

For $\Gamma > 0.256$ transverse intersections of the manifolds occur: the case $\Gamma = 0.28$ is shown in Fig. 13.27(c). The two transverse intersection points A_0, B_0 return after time $2\pi/\omega$ to the points A_1, B_1, and so on. Part of the complexity which arises can be seen in the further intersection

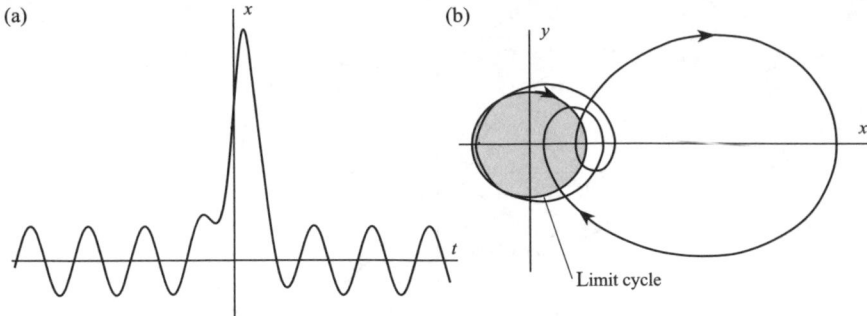

Figure 13.28   Representative time solution (a), and phase diagram (b) for the critical case of tangency between the manifolds.

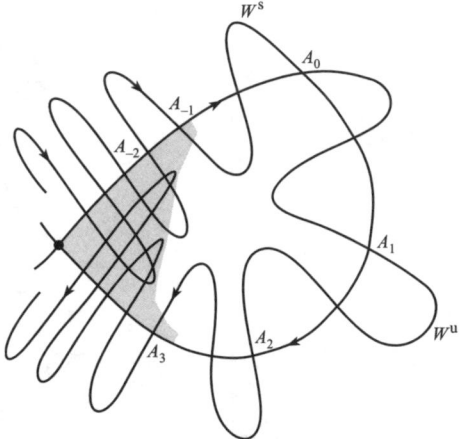

Figure 13.29   Scheme of transverse intersections of stable and unstable manifolds.

of the stable manifold with an interior loop of the unstable manifold which will also generate return points.

The transverse intersections of the manifolds and their returns are difficult to see for the Duffing oscillator because of the crowding of intersections near the fixed point. Figure 13.29 shows the homoclinic intersections in an expanded representation with the stable and unstable manifolds denoted by W^s and W^u. Consider an intersection at A_0. Let the first return forward in time be A_1. By the uniqueness of solutions the point A_1 must lie on both W^u and W^s. If this were not the case, there would exist two distinct solutions starting from A_0 at time $t = 0$. The intersections are also 'orientation preserving' in the sense that W^u crosses W^s in the same direction at A_0 and A_1, implying, by continuity, that there must be at least one further intersection between W^u and W^s between A_0 and A_1. As $t \rightarrow \infty$ there must be an infinite sequence of crossings $A_1, A_2, \ldots$ into the fixed point. Similarly, there must be an infinite sequence of crossings $A_{-1}, A_{-2}, \ldots$ in the reverse direction as shown in Fig. 13.29.

In the shaded region in Fig. 13.29 near the fixed point, the manifolds intersect each other in further points. To understand the implications of these further transverse intesections, consider

Figure 13.30   First returns of a strip S along the stable manifold.

a set of initial values S, shaded in Fig. 13.30, and track their successive returns. The region S contains two transverse intersections of W^{u} and W^{s} which will remain in successive returns of S. As S approaches the fixed point it will be compressed along W^{u} but stretched in the direction of W^{s}. Consider an initial point P_0 in S and on W^{u}, but not on W^{s} (see Fig. 13.30). Suppose that this particular point returns at P_1 and P_2, and also at P_3 where it lies on *both* W^{u} and W^{s} as shown. Since P_3 is also on W^{s} it must reappear again on W^{s} at P_4, and so on. Hence the 'amplitude' of the oscillations of W^{u} must increase as the fixed point is approached. The shaded images of S will be stretched thinly along W^{u}, and reappear *within* S eventually. The process will be repeated for a subset of points within S. Hence initial points in S become scattered along a strip close to the unstable manifold. The same thing happens in reverse close to the stable manifold. This illustrates the chaotic churning effect of homoclinic bifurcation.

The development of the strip in Fig. 13.31 which is a magnification of the strip S in Fig. 13.30. If the strip is further compressed and stretched in Fig. 13.30, then the point P_7 in the sequence must lie on both manifolds as shown in Fig. 13.31(a). The further returns of P_0 will then progress along the stable manifold in a progressively thinner strip as shown in Fig. 13.31(b). In a similar manner a loop of the stable manifold from prior returns must pass through P_0. Points initially close to P_0 but not on a manifold will lie on a strip which is repeatedly subject to compression and stretching. The points in the heavily shaded common region rejoin S, and thus a portion of S goes through the same process again. A subset of this domain will again cover *part* of the heavily shaded regions. It is possible for a later return of a point Q to coincide with Q. If this occurs at the r-th return then a period-r solution exists. In fact in the attractor generated by this process, period-r solutions exist for all r.

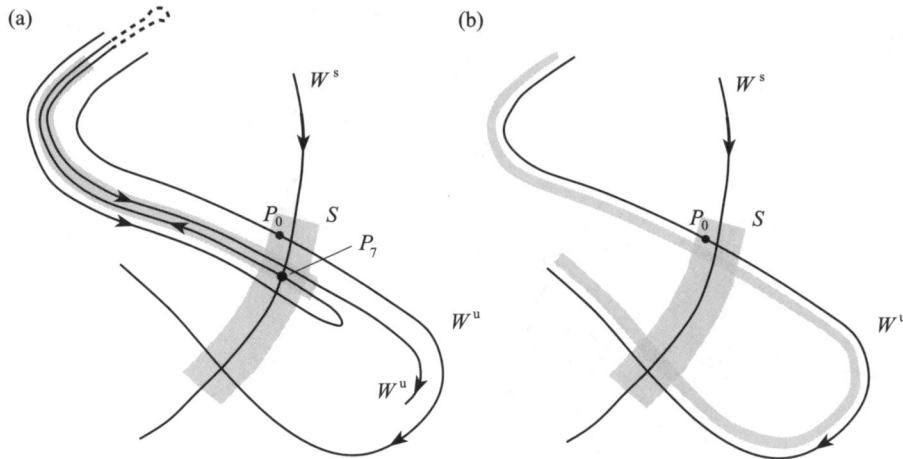

Figure 13.31   Further detail of Fig. 13.30 near P_0: (a) shows the appearance of P_7 and (b) shows further development of the strip close to W^u.

13.9   The horseshoe map

The precise details of the mechanism which creates the attracting set of the strange attractor are difficult to put together for Duffing's equation, but the phenomenon can be accounted for in a general way by analysing the iterated effect of the **Smale horseshoe map**. By using assumptions similar to those involved in discussing the convolutions of the unstable manifold it can be shown that there does exist a set of points having the attracting property, as follows.

Assume for simplicity that homoclinic bifurcation causes a square $ABCD$ (Fig. 13.32) to be mapped in a particular manner, which will be specified, into a horseshoe $A'B'C'D'$. The mapping is assumed to be carried out in this way: the square is stretched in the direction AD, compressed in the direction AB, bent through $180°$, and placed back over the square (Fig. 13.32). Suppose the mapping is repeated for such points as still lie in the square, and that this process is iterated. Figure 13.33 shows the first two iterations. The horizontal shaded strips in the square are *chosen* so as to map onto the vertical parts of the horseshoe. These two vertical strips now map onto the pair of thinner horseshoes. We are interested, at each stage, in the parts of the mapped regions which are shared by the regions from which they are mapped—for the first two stages, shown in Fig. 13.33; these are shown as the hatched regions. After two iterations there are points remaining in 16 'squares'; after the third iteration there will be 64 'squares', and so on. The limit set as the number of iterations tends to infinity turns out to be very much like a **Cantor set**. In one-dimensional terms a Cantor set can be formed as follows. Consider the interval $[0, 1]$. Delete the open interval $\left(\frac{1}{3}, \frac{2}{3}\right)$. Now delete the middle thirds $\left(\frac{1}{9}, \frac{2}{9}\right)$ and $\left(\frac{7}{9}, \frac{8}{9}\right)$ of the remaining intervals, and carry on deleting middle thirds. The limit set of this process is a Cantor set, and the set is uncountable. The limit set of the horseshoe map has a similar, but two-dimensional, structure. The implication is that there exists an uncountable number of points in the initial square which, when treated as initial states at $t = 0$ for iterated first returns, lead ultimately to endlessly repeated scans of a certain set of points—the limit set—which constitutes the

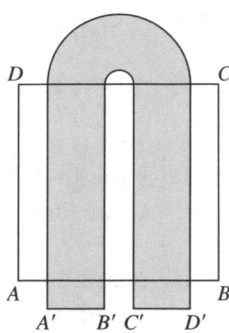

Figure 13.32 The horseshoe map.

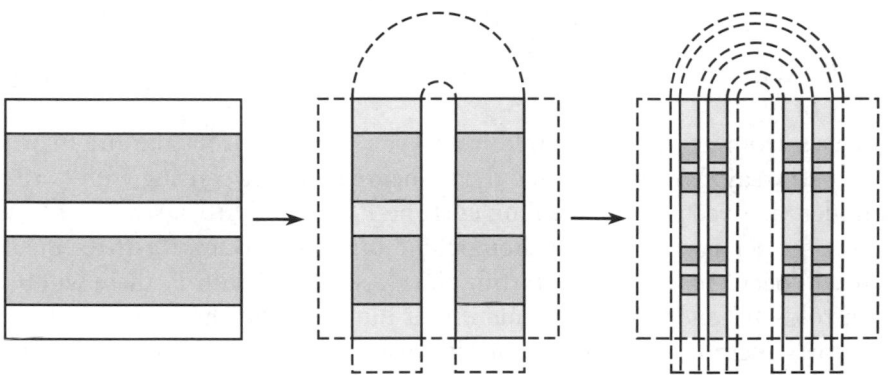

Figure 13.33 Successive horseshoe maps showing generation of the Cantor set.

strange attractor. The associated oscillations will include periodic motions and bounded nonperiodic motions. The elements of this set are distributed on the unstable manifold since horseshoes can be constructed for each loop across the stable manifold. An extensive account of homoclinic bifurcations and horseshoe maps can be found in Guckenheimer and Holmes (1983).

13.10 Melnikov's method for detecting homoclinic bifurcation

We have seen in Section 13.7 how homoclinic bifurcation for the Duffing oscillator leads to homoclinic tangles and chaos. The aim of this section is to explain Melnikov's method for detecting homoclinic bifurcation. This is a global perturbation method applicable to systems which have a *known* homoclinic path in an underlying autonomous system. This system is then perturbed usually by damping and forcing terms, and conditions for which homoclinic manifolds intersect are determined to leading order.

There are various versions of the theory of increasing generality, but here we shall consider systems of the form

$$\dot{x} = y, \qquad \dot{y} + f(x) = \varepsilon h(x, y, t), \tag{13.50}$$

where $h(x, y, t)$ is T-periodic in t, and $|\varepsilon|$ is a small parameter. The unperturbed system is

$$\dot{x} = y, \qquad \dot{y} + f(x) = 0.$$

It is assumed that $f(0) = 0$, and that the origin is a simple saddle with a known homoclinic path $x = x_0(t - t_0), y = y_0(t - t_0)$. Consider the loop from the origin lying in the half-plane $x \geq 0$. Since the system is autonomous we can include an arbitrary time translation t_0, which will be significant in Melnikov's method.

It is assumed that $f(x)$ and $h(x, y, t)$ have continous partial derivatives in each argument up to any required order, with a Taylor series in x and y in a neighbourhood of the origin, namely,

$$f(x) = f'(0)x + \cdots$$

$$h(x, y, t) = h(0, 0, t) + [h_x(0, 0, t)x + h_y(0, 0, t)y] + \cdots,$$

with $f'(0) < 0$ (to guarantee a saddle) and $h_x(0, 0, t) \neq 0, h_y(0, 0, t) \neq 0$, except possibly for isolated values of t.

As ε increases from zero, an unstable limit cycle emerges from the origin with solution $x = x_\varepsilon(t)$, $y = y_\varepsilon(t)$, say. We take as usual the Poincaré sequence starting with $t = 0$ and having period T, and let the fixed point of the limit cycle be P_ε: $(x_\varepsilon(0), y_\varepsilon(0))$ (see Fig. 13.34). Remember that any sequence can be used: if homoclinic tangency occurs for the sequence starting with $t = 0$, it will occur for any other starting time. Associated with P_ε there will be stable and unstable manifolds W_ε^s and W_ε^u: if these manifolds intersect, then homoclinic bifurcation takes place. Melnikov's method investigates the **distance** between the manifolds at a point on the unperturbed homoclinic path where $t = 0$, that is, at $P_0(x_0(-t_0), y_0(-t_0))$.

To approximate to the stable manifold W_ε^s we apply the regular perturbation

$$x_\varepsilon^s(t, t_0) = x_0(t - t_0) + \varepsilon x_1^s(t, t_0) + O(\varepsilon^2),$$

$$y_\varepsilon^s(t, t_0) = y_0(t - t_0) + \varepsilon y_1^s(t, t_0) + O(\varepsilon^2)$$

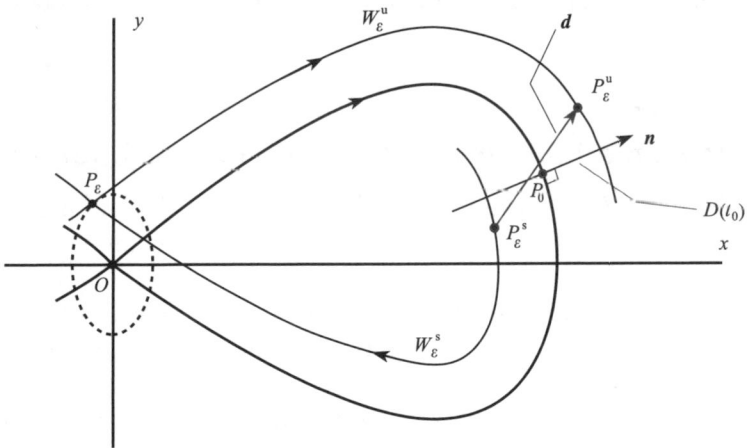

Figure 13.34 Distance function $D(t_0)$ between the manifold W_ε^u and W_ε^s at the point P_0 on the unperturbed autonomous homoclinic path.

to (13.50) for $t \geq 0$. To ensure that W_ε^s approaches P_ε we require

$$x_\varepsilon^s(t, t_0) - x_\varepsilon(t) \to 0, \quad y_\varepsilon^s(t, t_0) - y_\varepsilon(t) \to 0$$

as $t \to \infty$.

Substitute the expansions of $x_\varepsilon^s(t, t_0)$ and $y_\varepsilon^s(t, t_0)$ into (13.50). Then equate the coefficients of ε to zero, using also

$$f[x_\varepsilon^s(t, t_0)] = f[x_0(t - t_0)] + \varepsilon f'[x_0(t - t_0)] x_1^s(t, t_0) + O(\varepsilon^2).$$

It follows that $x_1^s(t, t_0)$ and $y_1^s(t, t_0)$ satisfy.

$$\dot{y}_1^s(t, t_0) + f'[x_0(t - t_0)] x_1^s(t, t_0) = h[x_0(t - t_0), y_0(t - t_0), t],$$
$$y_1^s(t, t_0) = \dot{x}_1^s(t, t_0). \tag{13.51}$$

In a similar manner, for the unstable manifold W_ε^u the perturbation series for x_ε^u is

$$x_\varepsilon^u(t, t_0) = x_0(t - t_0) + \varepsilon x_1^u(t, t_0) + O(\varepsilon^2),$$
$$y_\varepsilon^u(t, t_0) = \dot{y}_0(t - t_0) + \varepsilon y_1^u(t, t_0) + O(\varepsilon^2)$$

for $t \leq 0$, where

$$x_\varepsilon^u(t, t_0) - x_\varepsilon(t) \to 0, \quad y_\varepsilon^u(t, t_0) - y_\varepsilon(t) \to 0$$

as $t \to -\infty$. Thus $x_1^u(t, t_0)$ and $y_1^u(t, t_0)$ satisfy

$$\dot{y}_1^u(t, t_0) + f'[x_0(t - t_0)] x_1^u(t, t_0) = h[x_0(t - t_0), y_0(t - t_0), t],$$
$$y_1^u(t, t_0) = \dot{x}_1^u(t, t_0). \tag{13.52}$$

Note that the stable manifold W_ε^s is defined by the set of points $(x_\varepsilon^s(0, t_0), y_\varepsilon^s(0, t_0))$ for *all* t_0, whilst the unstable manifold W_ε^u is defined by $(x_\varepsilon^u(0, t_0), y_\varepsilon^u(0, t_0))$.

If

$$x_\varepsilon^{u,s}(t, t_0) = [x_\varepsilon^{u,s}(t, t_0), y_\varepsilon^{u,s}(t, t_0)]^T,$$

then the displacement vector at $t = 0$ can be defined as

$$d(t_0) = x_\varepsilon^u(0, t_0) - x_\varepsilon^s(0, t_0)$$
$$= \varepsilon[x_1^u(0, t_0) - x_1^s(0, t_0)] + O(\varepsilon^2),$$

where

$$x_1^{u,s} = [x_1^{u,s}, y_1^{u,s}]^T.$$

If $n(0, t_0)$ is the unit outward normal vector at P_0 (see Fig. 13.34), then the points $P_\varepsilon^{u,s}$: $(x_\varepsilon^{u,s}(0, t_0), y_\varepsilon^{u,s}(0, t_0))$ will not lie exactly on the normal but will be displaced as indicated in Fig. 13.34. We use a distance function $D(t_0)$ which is obtained by projecting the displacement vector $d(t_0)$ onto the unit normal n. Thus $D(t_0) = d(0, t_0) \cdot n(0, t_0) + O(\varepsilon^2)$.

The tangent vector to the unperturbed homoclinic path at $t = 0$ is

$$\{\dot{x}_0(-t_0), \dot{y}_0(-t_0)\} = \{\dot{y}_0(-t_0), f[x_0(-t_0)]\}.$$

Therefore the unit outward normal vector is

$$n(0, t_0) = \frac{\{f[x_0(-t_0)], y_0(-t_0)\}}{\sqrt{\{f[x_0(-t_0)]^2 + y_0(-t_0)^2\}}}.$$

Hence

$$D(t_0) = \mathbf{d}(t_0) \cdot \mathbf{n}(t_0)$$

$$= \frac{\varepsilon[\{x_1^{u}(-t_0) - x_1^{s}(-t_0)\} f[x_0(-t_0)] + \{y_1^{u}(-t_0) - y_1^{s}(-t_0)\} y_0(-t_0)]}{\sqrt{[f[x_0(-t_0)]^2 + y_0(-t_0)^2]}} + O(\varepsilon^2).$$

When $D(t_0) = 0$ homoclinic bifurcation must occur since the distance between the manifolds vanishes to $O(\varepsilon^2)$. However, $D(t_0)$, as it stands, requires x_1^{u} and x_1^{s} but as we shall show, $D(t_0)$ surprisingly does not need these solutions of (13.51) and (13.52). Let

$$\Delta^{u,s}(t, t_0) = x_1^{u,s}(t, t_0) f[x_0(t - t_0)] + y_1^{u,s}(t, t_0) y_0(t - t_0).$$

Note that

$$D(t_0) = \frac{\varepsilon[(\Delta^{u}(0, t_0) - \Delta^{s}(0, t_0)]}{\sqrt{\{f[x_0(0, t_0)]^2 + [y_0(0, t_0)]^2\}}} + O(\varepsilon^2). \tag{13.53}$$

We now show that $\Delta^{s}(t, t_0)$ and $\Delta^{u}(t, t_0)$ do not require $x_1^{s}(t, t_0)$ and $x_1^{u}(t, t_0)$. Differentiate $\Delta^{s}(t, t_0)$ with respect to t:

$$\frac{\mathrm{d}\Delta^{s}(t, t_0)}{\mathrm{d}t} = \dot{x}_1^{s}(t, t_0) f[x_0(t - t_0)] + x_1^{s}(t, t_0) f'[x_0(t - t_0)]\dot{x}_0(t - t_0)$$

$$+ \dot{y}_1^{s}(t, t_0) y_0(t - t_0) + y_1^{s}(t, t_0)\dot{y}_0(t - t_0),$$

$$= y_1^{s}(t, t_0) f[x_0(t - t_0)] + x_1^{s}(t, t_0) f'[x_0(t - t_0)]\dot{x}_0(t - t_0)$$

$$+ y_0(t - t_0)[-f'[x_0(t - t_0)]x_1^{s}(t - t_0)$$

$$+ h[x_0(t - t_0), y_0(t - t_0), t]] - y_1^{s}(t, t_0) f[x_0(t - t_0)],$$

$$= y_0(t - t_0)h[x_0(t - t_0), y_0(t - t_0), t], \tag{13.54}$$

using (13.51) and (13.52). Now integrate (13.54) between 0 and ∞ with respect to t, noting that, since $y_0(t - t_0) \to 0$ and $f[x_0(t - t_0)] \to 0$ as $t \to \infty$, then $\Delta^{s}(t, t_0) \to 0$ also. Hence

$$\Delta^{s}(0, t_0) = -\int_0^\infty y_0(t - t_0)h[x_0(t - t_0), y_0(t - t_0), t]\,\mathrm{d}t.$$

In a similar manner it can be shown that

$$\Delta^{u}(0, t_0) = \int_{-\infty}^0 y_0(t - t_0)h[x_0(t - t_0), y_0(t - t_0), t]\,\mathrm{d}t.$$

The numerator of the coefficients of ε in (13.53) controls homoclinic bifurcation. Let

$$M(t_0) = \Delta^u(0, t_0) - \Delta^s(0, t_0)$$

$$= \int_{-\infty}^{\infty} y_0(t - t_0) h[x_0(t - t_0), y_0(t - t_0), t] \, dt. \tag{13.55}$$

The function $M(t_0)$ is known as the **Melnikov function** associated with the origin of this system. If $M(t_0) = 0$ has *simple* zeros for t_0, then there must exist, to order $O(\varepsilon^2)$, transverse intersections of the manifolds for these particular values of t_0. The points (there will generally be two transverse intersections) P_0 on the *unperturbed* homoclinic path where these intersections occur have the approximate coordinates $(x_0(-t_0), y_0(-t_0))$.

More general versions of Melnikov's method applicable to periodically perturbed systems of the form

$$\dot{x} = F(x) + \varepsilon H(x, t)$$

are given by Guckenheimer and Holmes (1983) and Drazin (1992).

Example 13.6   *Find the Melnikov function for the perturbed Duffing equation*

$$\ddot{x} + \varepsilon \kappa \dot{x} - x + x^3 = \varepsilon \gamma \cos \omega t.$$

Find the relation between the parameters κ, γ and ω for tangency to occur between the manifolds to order $O(\varepsilon^2)$.
In eqn (13.50) we put

$$f(x) = -x + x^3, \quad h(x, y, t) = -\kappa y + \gamma \cos \omega t,$$

so that $T = 2\pi/\omega$. We require the homoclinic solution for the origin of the underlying autonomous system

$$\ddot{x} - x + x^3 = 0.$$

In $x > 0$ this is $x_0(t) = \sqrt{2} \operatorname{sech} t$ (see Example 3.9). Hence

$$M(t_0) = \int_{-\infty}^{\infty} \dot{x}_0(t - t_0)[-\kappa \dot{x}_0(t - t_0) + \gamma \cos \omega t] \, dt$$

$$= -2\kappa \int_{-\infty}^{\infty} \operatorname{sech}^2(t - t_0) \tanh^2(t - t_0) \, dt$$

$$\quad - \gamma \sqrt{2} \int_{-\infty}^{\infty} \operatorname{sech}(t - t_0) \tanh(t - t_0) \cos \omega t \, dt,$$

$$= -2\kappa \int_{-\infty}^{\infty} \operatorname{sech}^2 u \tanh^2 u \, du$$

$$\quad + \gamma \sqrt{2} \sin \omega t_0 \int_{-\infty}^{\infty} \operatorname{sech} u \tanh u \sin \omega u \, du, \tag{13.56}$$

putting $u = t - t_0$ and deleting the odd integrand in the second integral. The integrals can be evaluated using residue theory (an example is given in Appendix D). Gradshteyn and Ryzhik (1994) is also a useful source for integrals which arise in Melnikov's methods. We quote the integrals

$$\int_{-\infty}^{\infty} \mathrm{sech}^2 u \tanh^2 u \, du = \frac{2}{3},$$

$$\int_{-\infty}^{\infty} \mathrm{sech} u \tanh u \sin \omega u \, du = \pi \omega \, \mathrm{sech}\left(\frac{1}{2}\omega\pi\right).$$

Hence, from (13.54)

$$M(t_0) = -\frac{4}{3}\kappa + \sqrt{2}\omega\pi\gamma \, \mathrm{sech}(\frac{1}{2}\omega\pi) \sin \omega t_0.$$

For given κ, γ, and ω, $M(t_0)$ will vanish if a real solution can be found for t_0. Assuming that κ, $\gamma > 0$, this condition will be met if $|\sin \omega t_0| \leq 1$, that is, if

$$\frac{2\sqrt{2}\kappa}{3\pi\gamma\omega} \cosh\left(\frac{1}{2}\omega\pi\right) \leq 1.$$

Put another way, if κ and ω are fixed, then homoclinic tangency will first occur for increasing forcing amplitude when

$$\gamma = \gamma_c = \frac{2\sqrt{2}\kappa}{3\pi\omega} \cosh\left(\frac{1}{2}\omega\pi\right), \qquad (13.57)$$

and homoclinic tangles will be present if γ exceeds this value. For our representative values of $\omega = 1.2$ and $\varepsilon\kappa = 0.3$ (see Section 13.5), the critical forcing amplitude given by (13.57) is $\Gamma_c = \varepsilon\gamma_c = 0.253$ which compares well with the numerically computed value of $\Gamma = 0.256$ (see also Fig. 13.27(b)). Melnikov's method is a perturbation procedure so that it will retain validity if the right-hand side of (13.57) is $O(1)$ as $\varepsilon \to 0$, that is, never too large nor too small.

At the critical value $\gamma = \gamma_c$, $\sin \omega t_0 = 1$, and we can choose the solution $t_0 = \pi/(2\omega)$. Hence the tangency between the manifolds will occur approximately at the corresponding point on the autonomous unperturbed homoclinic path at

$$\left(\sqrt{2} \, \mathrm{sech}\left(-\frac{\pi}{2\omega}\right), -\sqrt{2} \, \mathrm{sech}^2\left(-\frac{\pi}{2\omega}\right) \sinh\left(-\frac{\pi}{2\omega}\right)\right) = (0.712, 0.615)$$

in the section $t = 0$ still for $\omega = 1.2$. This may be compared with the tangency point shown in Fig. 13.27(b). ●

Exercise 13.5
Find the Melnikov function for the equation

$$\ddot{x} + \varepsilon\kappa \, \mathrm{sgn}(\dot{x}) - x + x^3 = \varepsilon\gamma \cos \omega t, \quad \kappa > 0, \ \gamma > 0.$$

For what conditions on the parameters do homoclinic paths exist?

13.11 Liapunov exponents and differential equations

In Section 13.5 the Liapunov exponent was introduced for first-order difference equations as a method of explaining chaos through the exponential divergence of neighbouring solutions. In this section we extend this method to nonlinear differential equations, in particular to the forced Duffing equation of Section 13.3, which has been a theme of this and earlier chapters. The equation is

$$\ddot{x} + k\dot{x} - x + x^3 = \Gamma \cos \omega t. \tag{13.58}$$

It was shown numerically (in Section 13.5) that this equation exhibits period doubling leading to chaotic behaviour for certain values of the parameters k, Γ and ω. The hypothesis behind Liapunov exponents is that there are parameter domains in which the distance between neighbouring solutions exhibits exponential growth in certain directions and exponential decay in other directions. This is the notion of *sensitivity to initial conditions*.

The first step is to express (13.58) as a third-order autonomous system in the form

$$\dot{x} = y, \quad \dot{y} = -ky + x - x^3 + \Gamma \cos z, \quad \dot{z} = \omega. \tag{13.59}$$

but with the restriction $z = \omega t$. An autonomous system must be of order three or more for chaos to appear: chaos is not possible in first order or plane autonomous systems. We will return to this conclusion later. Let $x^*(t) = [x^*(t), \ y^*(t), \ z^*(t)]^{\mathrm{T}}$, which satisfies the initial condition $x^*(t_0) = [x_0^*, \ y_0^*, \ z_0^*]^T$, be the particular solution to be investigated. Consider now a solution $x(t) = x^*(t) + \eta(t)$ which starts near to the initial value of $x^*(t)$: thus $|\eta(t_0)|$ is assumed to be small.

Before we apply the method to Duffing's equation, we consider the general *autonomous* system

$$\dot{x} = f(x). \tag{13.60}$$

of any order. Substitute $x = x^* + \eta$ into (13.60) so that, using a Taylor expansion,

$$\dot{x}^* + \dot{\eta} = f(x^* + \eta) \approx f(x^*) + A(x^*)\eta \tag{13.61}$$

for as long as $|\eta(t)|$ is small, where

$$A(x^*) = [\nabla f] = \begin{bmatrix} f_x & f_y & f_z \\ g_x & g_y & g_z \\ h_x & h_y & h_z \end{bmatrix}$$

is the matrix of partial derivatives of $f = [f, g, h]^{\mathrm{T}}$. It follows that, approximately, $\eta = [\eta_1, \eta_2, \eta_3]^{\mathrm{T}}$ satisfies the linear vector equation

$$\dot{\eta} = A(x^*)\eta \tag{13.62}$$

The evolution of η will depend on its initial value: both exponential growth and exponential decay can occur. Liapunov exponents measure this growth and decay. The computation of exponential growth over long time scales is generally not an option for numerical reasons such as output overflow. Also, there is little hope that any equations can be solved analytically (at least for third order and above), so the numerical procedure which averages growth over shorter time intervals as follows is adopted.

Computing procedure

- (i) choose parameters and initial condition $x^*(t_0)$ (after any transience has been eliminated) in a domain (usually bounded) where a chaotic solution is suspected, and solve (13.60) numerically to find $x^*(t)$ over a time interval $t_0 \leq t \leq t_1$;
- (ii) compute $A(x^*)$ over $t_0 \leq t \leq t_1$ at suitable time steps;
- (iii) since (13.62) is linear and homogeneous we can normalize the initial values so that $|\eta(t_0)| = 1$; choose n initial orthonormal vectors, say

$$\eta_1^{(0)}(t_0) = [1, 0, 0, , 0], \quad \eta_2^{(0)}(t_0) = [0, 1, 0, \ldots, 0], \ldots, \eta_n^{(0)}(t_0) = [0, 0, 0, \ldots, 1],$$

  (or some other orthonormal set such as the eigenvectors of $A(x^*(t_0))$); solve (13.62) numerically for each $\eta_i^{(0)}(t)$ and compute $\eta_i^{(0)}(t_1)$ (the time t_1 is chosen to avoid $|\eta_i^{(0)}(t_1)|$ becoming computationally too large);
- (iv) the set of vectors $\{\eta_i^{(0)}(t_1)\}$ will generally not be orthonormal, but a set of orthonormal vectors $\{\eta_i^{(1)}(t_1)\}$ can be constructed at t_1 using the Gram-Schmidt procedure (see Nayfeh and Balachandran (1995, Section 7.3) for more details);
- (v) repeat the procedure (i) to (iv) at times $t = t_1, t_2, \ldots, t_N$ to generate the vectors $\{\eta_i^{(j-1)}(t_j)\}$ $(j = 2, 3, \ldots, N)$: renormalization takes place at each time;
- (vi) the first approximation to the ith Liapunov exponent is

$$\Lambda_i^{(1)} = \frac{1}{t_1 - t_0} \ln |\eta_i^{(0)}(t_1)|;$$

the second approximation is the *average*

$$\Lambda_i^{(2)} = \frac{1}{t_2 - t_0} \{\ln |\eta_i^{(0)}(t_1)| + \ln |\eta_i^{(1)}(t_2)|\},$$

and so on; the Nth approximation is

$$\Lambda_i^{(N)} = \frac{1}{t_N - t_0} \sum_{k=1}^{N} \ln |\eta_i^{(k-1)}(t_k)|. \tag{13.63}$$

If the limits exist, the **Liapunov exponents** $\{\Lambda_i\}$ are defined by

$$\Lambda_i = \lim_{N \to \infty} \Lambda_i^{(N)} = \lim_{t_N \to \infty} \frac{1}{t_N - t_0} \sum_{k=1}^{N} \ln[|\eta_i^{(k-1)}(t_k)|].$$

The length $d_{ik} = |\eta_i^{(k-1)}(t_k)|$ is the distance between the vectors $x^*(t_k) + \eta_i^{(k-1)}(t_k)$ and $x^*(t_k)$. The steps in the scheme for one component of η are indicated in Fig. 13.35. A **positive Liapunov exponent** Λ indicates that chaos *might* be present: it is a necessary condition but not sufficient.

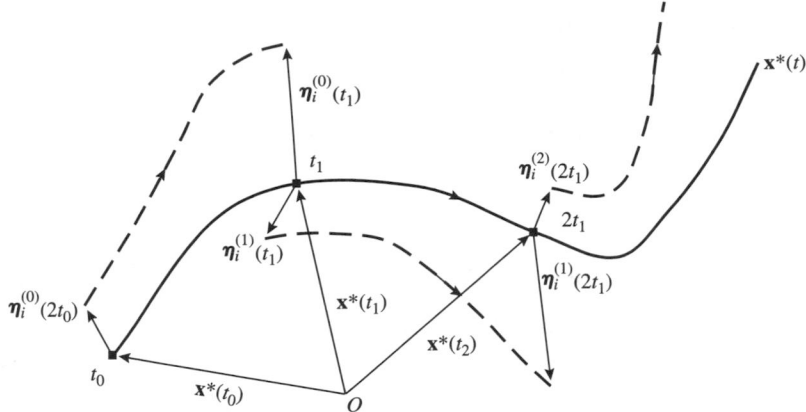

Figure 13.35   The figures shows a solution $x = x^*(t)$. A typical perturbation components $\boldsymbol{\eta}_i^{(0)}(t_0), \boldsymbol{\eta}_i^{(1)}(t_1), \dots$ of the orthonormal sets at each time step. There will be n such components at each step. The dashed curves represent the perturbed solutions.

There are a number of imponderables in the iterative process which can affect its accuracy. These include the choice of parameters, the time steps $t_1, t_2, \dots$, and the number N of iterations. Generally trial and error are required to produce confidence in results.

It is possible to see the thinking behind this definition of the Liapunov exponent if we assume, in a simplistic manner by way of an explanation, that $|\boldsymbol{\eta}_i(t_k)| \sim e^{\mu_i(t_k - t_{k-1})}$. Then, from (13.63),

$$\Lambda_i^{(N)} \sim \frac{1}{t_N - t_0} \sum_{k=1}^{N} \ln[e^{\mu_i(t_k - t_{k-1})}] = \frac{1}{t_N - t_0}(t_N - t_0)\mu_i = \mu_i.$$

Hence a positive μ (exponential growth) implies a positive Liapunov exponential with a corresponding result for $\mu < 0$ (exponential decay).

At first sight it seems to be a contradiction that solutions can diverge exponentially but still be bounded. However, in plane autonomous systems it is possible to have a saddle point inside a closed path (see, e.g., Fig. 2.13). Exponential divergence can be associated with the saddle point but the time solutions will be bounded.

If A is a constant matrix, then the analysis is straightforward and can be illustrated by an example which provides an interpretation of Liapunov exponents.

Example 13.7   *Find the Liapunov exponents of the linear system*

$$\dot{x} = \begin{bmatrix} \dot{x} \\ \dot{y} \\ \dot{z} \end{bmatrix} = \begin{bmatrix} 1 & 2 & 1 \\ 2 & 1 & 1 \\ 1 & 1 & 2 \end{bmatrix} \begin{bmatrix} x \\ y \\ z \end{bmatrix} = Ax.$$

This differential equation was discussed in Example 12.6, where it was shown that its eigenvalues are $4, 1, -1$ so that its general solution is

$$x = \alpha r_1 e^{4t} + \beta r_2 e^t + \gamma r_3 e^{-t},$$

where *orthonormal eigenvectors* are given by

$$r_1 = \frac{1}{\sqrt{3}}[1,\ 1,\ 1]^T, \quad r_2 = \frac{1}{6}[-1,\ -1,\ 2]^T, \quad r_3 = \frac{1}{\sqrt{2}}[1,\ -1,\ 0]^T.$$

Since A is constant, the perturbation η satisfies the same equation as x, that is,

$$\dot{\eta} = A\eta.$$

Hence

$$\eta = \alpha_1 r_1 e^{4t} + \beta_1 r_2 e^t + \gamma_1 r_3 e^{-t}.$$

Let $t_0 = 0$. Since the solution is exact there is no need to iterate: we can choose $0 \le t < \infty$ as the first interval. An orthonormal set can be constructed if $\eta_i(0) = r_i$. Thus for η_1 we put $\alpha_1 = 1$, $\beta_1 = \gamma_1 = 1$ so that $\eta_1 = r_1 e^{4t}$. It follows that

$$\Lambda_1 = \lim_{t \to \infty} \frac{1}{t} \ln|\eta_1(t)| = \lim_{t \to \infty} \frac{1}{t} \ln(e^{4t}) = 4.$$

Similarly the other two Liapunov exponents are $\Lambda_2 = 1$, $\Lambda_3 = -1$. For a constant matrix A, the Liapunov exponents are simply the eigenvalues of A, which confirms the view that Liapunov exponents are generalizations of eigenvalues. This system has two positive exponents but linear systems can never be chaotic. ●

For nonlinear systems, in which the average is taken over N time intervals, the computation is complicated by the need to create a new orthonormal initial vector at each time step. Since the *signs* of the *largest* and *smallest* Liapunov exponents are the main indicators of chaos, it is possible to use a simplified procedure to determine these signs. The discussion will be restricted to *three dimensions*. In the computing procedure above replace (iii), (iv), (v), and (vi) by

- (iii)* on a unit sphere centred at $x^*(t_0)$ select initial values, say, $\eta(t_0) = [\cos i \sin j, \sin i \sin j, \cos j]^T$, where i and j, respectively are chosen as suitable step values over the intervals $(0, 2\pi)$ and $(0, \pi)$ so that the sphere is covered by grid of points; note the *maximum* and *minimum* vales of $|\eta(t_1)|$ over the values of i and j; denote the lengths by $d_1^{(max)}$ and $d_1^{(min)}$. Accuracy will be improved by increasing the number of grid points;
- (iv)* repeat (iii)* with a unit spheres centred at $x^*(t_1), x^*(t_2), \ldots$, and note the lengths $d_k^{(max)}$ and $d_k^{(min)}$ $(k = 2, 3, \ldots)$.
- (v)* compute the averages

$$\overline{\Lambda}_{max}^{(N)} = \frac{1}{t_N - t_0} \sum_{k=1}^N \ln d_k^{(max)}, \quad \overline{\Lambda}_{min}^{(N)} = \frac{1}{t_N - t_0} \sum_{k=1}^N \ln d_k^{(min)}.$$

The numbers $\overline{\Lambda}_{max}^{(N)}$ and $\overline{\Lambda}_{min}^{(N)}$ are the largest and smallest Liapunov exponents and their signs can indicate exponential growth or decay.

Before we apply this procedure to Duffing's equation we can often deduce one Liapunov exponent if the chaotic regime is bounded as seems to be the case for Duffing equation and the Lorenz equations (Problem 13.21). Returning to the general system $\dot{x} = f(x)$, consider the two neighbouring points $x^*(t_0)$ and $x^*(t_0 + h)$ on a solution. For all t_0 these points will follow the same phase path (since the system is autonomous). It follows that

$$\eta = x^*(t + h) - x^*(t) \approx \dot{x}^*(t)h = f(x^*(t))h:$$

this is the tangent approximation to the phase path. In this case there is no need to iterate over time steps since the corresponding Liapunov exponent is given by

$$\Lambda = \lim_{t\to\infty} \frac{1}{t} \ln|\boldsymbol{\eta}(t)| = \lim_{t\to\infty} \frac{1}{t} \ln|\boldsymbol{f}(\boldsymbol{x}^*(t))h|.$$

If the path defined by $\boldsymbol{x} = \boldsymbol{x}^*$ is bounded, then $\boldsymbol{f}(\boldsymbol{x})$ will also be bounded so that $\lambda = 0$ in the limit. Hence, for such bounded chaotic regions one Liapunov exponent will be zero.

Chaos and the Duffing oscillator

We return now to Duffing's equation in the version (13.59). We assume that $t_0 = 0$ and that $t_k = kT$ $(k = 1, 2, \ldots, N)$ (that is, we choose equal time steps). It follows from (13.61) that

$$A(\boldsymbol{x}^*) = \begin{bmatrix} 0 & 1 & 0 \\ 1 - 3x^{*2} & -k & -\Gamma \sin \omega t \\ 0 & 0 & 0 \end{bmatrix}. \tag{13.64}$$

The chosen parameter values are $k = 0.3$, $\omega = 1.2$ and $\Gamma = 0.5$, which are the same as those in the example shown in Fig. 13.15, which is useful for comparisons. Periodically forced systems have a natural time step given by the forced period. Whilst this is not required, we use these steps in which $t_i - t_{i-1} = 2\pi/\omega = T$ for all i, say (sometimes fast growth will demand smaller time steps).

The components of $\boldsymbol{\eta} = [\eta_1, \eta_2, \eta_3]^{\mathrm{T}}$ satisfy

$$\dot{\eta}_1 = \eta_2, \quad \dot{\eta}_2 = (1 - 3x^{*2})\eta_1 - k\eta_2 - \Gamma\eta_3 \sin z^*, \quad \dot{\eta}_3 = 0. \tag{13.65}$$

but since $z = z^* = \omega t$, it follows that $\eta_3 = 0$. Effectively, we need only solve the plane system

$$\begin{bmatrix} \dot{\eta}_1 \\ \dot{\eta}_2 \end{bmatrix} = \begin{bmatrix} 0 & 1 \\ 1 - 3x^{*2} & -k \end{bmatrix} \begin{bmatrix} \eta_1 \\ \eta_2 \end{bmatrix} \tag{13.66}$$

for η_1 and η_2.

These equations are solved using the procedure outlined above for the maximum and minimum values of $|\boldsymbol{\eta}|$, noting that one Liapunov exponent is zero for the reasons explained above. An interpolated solution $\boldsymbol{x}^*$ is obtained numerically, and then eqns (13.65) are solved over each time interval $t_k < t < t_{k+1}(k = 0, 1, 2, \ldots)$, subject to initial conditions for $\boldsymbol{\eta} = [\eta_1, \eta_2]$ (since $\eta_3 = 0$) over 40 sample points on the circumference of a unit circle centred successively at $\boldsymbol{x}^*(t_k)$. At each step the distances $d_k^{(\max)}$ are $d_k^{(\min)}$ over the sample points are computed. Finally

$$\overline{\Lambda}_{\max}^{(N)} = \frac{1}{NT} \sum_{k=1}^{N} \ln d_k^{(\max)}, \quad \overline{\Lambda}_{\min}^{(N)} = \frac{1}{NT} \sum_{k=1}^{N} \ln d_k^{(\min)}$$

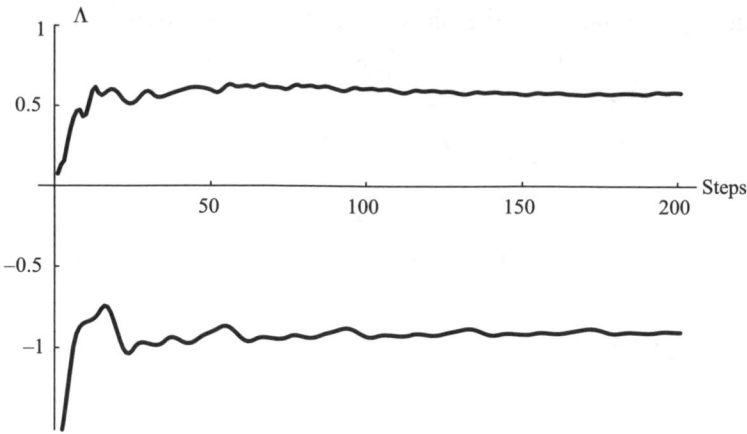

Figure 13.36 The sign of one non-zero Liapunov exponent is positive and the other is negative for Duffing's equation given by (13.58) with $k = 0.3$, $\omega = 1.2$ and $\Gamma = 0.5$. The computation was started with initial conditions $x(0) = -1.30039$, $y(0) = -0.08222$, and computed over 200 steps with $T = 0.2$ which corresponds to a time of $200T = 40$. After 200 steps the maximum and minimum values of $\overline{\Lambda}$ are approximately 0.61 to -0.90.

are computed. The evolutions of the limits are shown in Fig. 13.36 over 200 steps of length $T = 0.2$ or, equivalently, for a time $t = 40$. The limits are approximately 0.61 and -0.90, but the significance is that one is negative and one positive, which indicates the possibility of chaos.

Figure 13.37 (not to scale) shows how a typical circle $|\eta_{t_k}| = 1$ centred at $(x^*(t_k), y^*(t_k))$ is distorted over a time step T into the closed curve shown centred at $(x^*(t_{k+1}), y^*(t_{k+1}))$. The curve is approximately elliptic which has been flattened in one direction, corresponding to exponential decay, and extended in another direction indicating exponential growth. These are very similar to the *principal directions* of an ellipse. These exponential changes are shown by the averaging method in Fig. 2.

If the Duffing's equation has a stable limit cycle then the non-zero Liapunov exponents will be both be negative. Consider the case $k = 0.3$, $\omega = 1.2$, $\Gamma = 0.2$ (see Fig 13.14(a)), for which the equation has a limit cycle in $x > 0$. This case illustrates how the step length can vary T. Consider a typical value $x^* = 1$ on the path in Fig. 13.14(a). Then the matrix in (13.66) takes the value

$$\begin{bmatrix} 0 & 1 \\ 1 - 3x^{*2} & -k \end{bmatrix} = \begin{bmatrix} 0 & 1 \\ -2 & -0.3 \end{bmatrix},$$

and its eigenvalues are $-0.150 \pm 1.406i$ approximately. The real part indicates slow exponential decay so that we must choose a larger T: in this case $T = 4\pi/\omega$ (twice the forcing period) could be chosen, or another value of this order. Note that the eigenvalues have a positive real part over parts of the limit cycle (near to $x^* = 0.4$), but the negative real part dominates on average. The evolution of the two Liapunov exponents is shown in Fig. 13.38. Generally, the Liapunov exponents are not of much interest in stable cases since we can compute η_1 and η_2 over any reasonable time step.

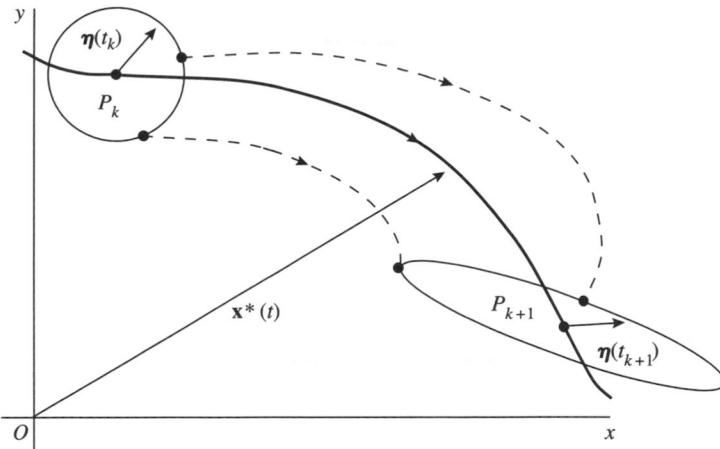

Figure 13.37 This shows the distortion a unit circle centred at $P_k : x^*(t_k)$ when it reaches $P_{k+1} : x^*(t_{k+1})$ (not to scale). It becomes flattened with some paths becoming closer to $x^* t_{k+1}$ and some diverging from $x^* t_{k+1}$.

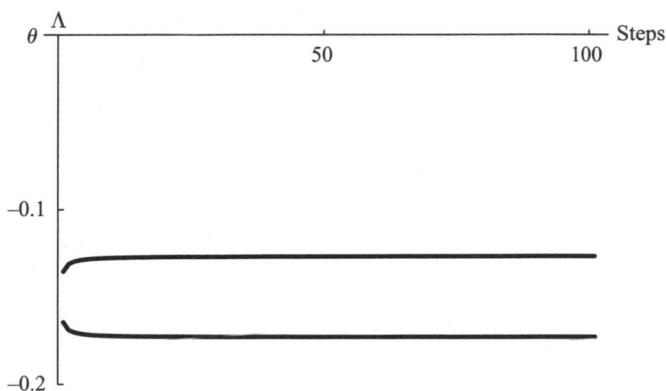

Figure 13.38 Evolution of the Liapunov exponents for Duffing's equation with $k = 0.3$, $\omega = 1.2$, $\Gamma = 0.2$ for which it is known that the equation has a stable limit cycle in $x > 0$. Note the smoother evolution for the stable case. The exponents are approximately -0.127 and -0.173.

More details of this approach can be found in Moon (1987, Section 5.4). Further details of the computation of Liapunov exponents are given also by Rasband (1990, Section 9.2).

The Lorenz equations

The Lorenz equations are given by

$$\dot{x} = a(y - x), \quad \dot{y} = bx - y - xz, \quad \dot{z} = xy - cz$$

(see Problems 8.26 and 13.21). In the latter problem it was shown numerically that chaotic motion is extremely likely. Here we apply the numerical procedure to find the largest and

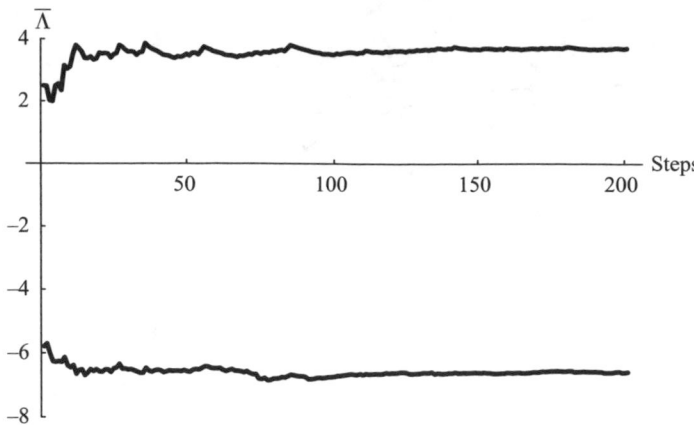

Figure 13.39 Maximum and minimum Liapunov exponents for the Lorenz equations with parameters $a = 10$, $b = 27$, $c = 2.65$ and step length $T = 0.5$, and initial values $x(0) = 0.966944$, $y(0) = -5.25624$, $z(0) = 28.3037$.

smallest Liapunov exponents using (iii)*, (iv)* and (v)* above. Unlike the forced Duffing equation, this is a genuine autonomous system so that the initial values at each step have to be reset on to a unit sphere; also the time t_k can be reset to zero. The parameters are the same as those chosen in Fig. 13.42 (Problem 13.21), namely $a = 10$, $b = 27$, $c = 2.65$. The initial values $x(0) = 0.966944$, $y(0) = -5.25624$, $z(0) = 28.3037$ and a time step of $T = 0.5$ were chosen for $x^*(t)$. In (iii)* in the procedure 200 grid points over the sphere were chosen. The evolution of the maximum and minimum values of approximately 3.80 and -6.61 of $\overline{\Lambda}$ are shown in Fig 13.39. The significant feature is that the exponents are of *opposite signs* which point strongly to chaotic motion. The Liapunov exponents are independent of the initial conditions provided that they are within the bounded chaotic region.

The Lorenz equations display one form of chaos since they exhibit *all* the following features:

- phase paths starting from a certain region do not approach fixed points, periodic orbits, or quasi-periodic orbits (as illustrated in Fig. 13.9(d)) but remain bounded (the exception can be initial points which start on unstable equilibrium points, or unstable periodic orbits): this is known as **aperiodic behaviour;**
- the equations are autonomous and deterministic and contain no random inputs;
- some nearby paths diverge exponentially and others converge exponentially for almost all paths which start in the region as indicated by the Liapunov exponents: as remarked earlier the paths exhibit **sensitive dependence to initial conditions.**

More detailed discussion and explanation can be found in Strogatz (1994, Chapter 9)

An alternative approach uses the notion of **exponential dichotomies** explained by Coppel (1978). The theory of exponential dichotomies gives conditions on the matrix $A(t)$ in the linear equation $\dot{\eta} = A\eta$ in order that solutions can be divided into ones with exponential growth and ones with exponential decay, which accounts for the term dichotomy.

Exercise 13.6
Find the general solution of $\dot{x} = A(x)x$, where

$$A(x) = \begin{bmatrix} 0 & 1 & 0 \\ 1 & 0 & 0 \\ y & x & -3 \end{bmatrix}.$$

Hence determine the Liapunov exponents of the system. (Whilst this *nonlinear* system has positive and negative Liapunov exponents, the system is not chaotic since paths can escape to infinity: they do not have the bounded property of the Lorenz equations.)

13.12 Power spectra

The numerical solution of an ordinary differential equation

$$\ddot{x} = f(x, \dot{x}, t)$$

generates a sequence of values of x at discrete values of t. The 'solution' x is then plotted against t by interpolation through the data using a computer package. The original series of values is known as a **time series,** and could equally well arise from experimental observations. The output from a system might appear *random* or *noisy* but might contain significant harmonics at certain frequencies. For example, we know in Fig. 13.14(b) that the period-2 output will have two frequencies, ω and $\frac{1}{2}\omega$, but the frequency structure of the output in Fig. 13.15 is not clear.

Dominant frequencies in time series can be investigated by looking at the **Fourier transform** of the data. Suppose that we take a sample of N equally spaced values from a time series, say $x = x_0, x_1, \ldots, x_N$. The **discrete Fourier transform** of this sample is defined by

$$X_k = \frac{1}{\sqrt{N}} \sum_{m=0}^{N} x_m e^{-2\pi i k m / N} \quad (k = 0, 1, 2, \ldots, N-1)$$

(there are other definitions of the Fourier transform: care should be taken in comparing results using different computer codes). The original data can be recovered from $\{X_k\}$ by the **inverse Fourier transform**

$$x_j = \frac{1}{\sqrt{N}} \sum_{k=0}^{N} X_k e^{2\pi i k j / N} \quad (j = 0, 1, 2, \ldots, N-1).$$

Generally, X_k is a complex number. We identify the frequency structure of the output by looking at its **power spectrum** $P(\omega_k)$ defined by

$$P(\omega_k) = X_k \bar{X}_k = |X_k|^2.$$

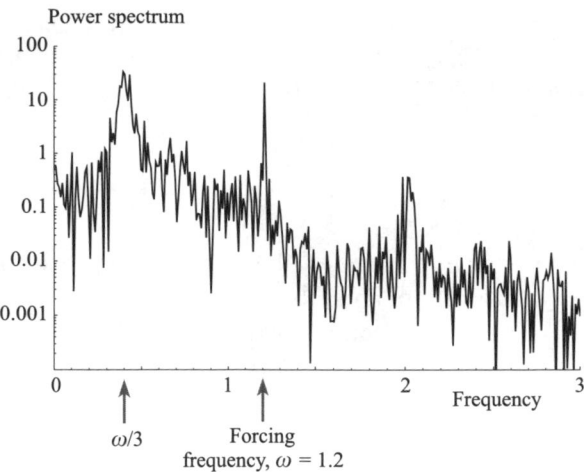

Figure 13.40 Power spectrum of Duffing's equation with $k = 0.3$, $\omega = 1.2$, $\Gamma = 0.5$. The graph is a loglinear plot of $P(\omega_j)$ versus ω_j.

Any dominant frequencies in $\{\omega_k\}$ will be revealed by sharp peaks in the graph of $P(\omega_k)$ against ω_k. If, for example, the harmonic function $x = \cos 3t$ is sampled, say, by a 1000 points over an interval of length 100, then the power spectrum will have a graph with a single sharp peak at the angular frequency 3.

We can look at the numerical output from numerical solutions of Duffing's equation, and take its power spectrum to search for any 'hidden' frequencies. This is done by constructing an interpolation function through the numerical output, and then by sampling this function. Figure 13.40 shows the power spectrum for a solution of Duffing's equation

$$\ddot{x} + k\dot{x} - x + x^3 = \Gamma \cos \omega t,$$

for the chaotic case $k = 0.3$, $\omega = 1.2$, $\Gamma = 0.5$ shown in Fig. 13.15. The figure shows that a large number of frequencies are present, being particularly strong at low values. Also note that some structure is identifiable. The forcing frequency ω is prominent as is a subharmonic of frequency $\omega/3$ and one of its **superharmonics** at $5\omega/3$.

More information on the analysis of time series can be found in the book by Nayfeh and Balachandran (1995).

13.13   Some further features of chaotic oscillations

There have been various attempts to define chaos and what characterizes a chaotic solution of a differential equation. In Section 13.11 it has been explained how exponential divergence of solutions using Liapunov exponents can be an indicator of chaotic output. However, there seems to be, at present, no universally acceptable definition which covers all instances of what we might recognize as a chaotic response.

It has been shown by Brown and Chua (1996, 1998) using examples and counterexamples that there are features which chaos might be expected to have, but which are not all present in any one case. We conclude with a list of some characteristics of chaos described informally. Not all are included in this book, and the list is by no means exhaustive.

Period doubling

As a parameter changes, the periodic oscillation goes through a infinite sequence of pitchfork bifurcations in which the period of the oscillation doubles at each bifurcation. We have already discussed this in relation to the logistic difference equation (Section 13.4) and the Duffing oscillator (Section 13.3).

Smale horseshoes

The horseshoe map is described in Section 13.9, and it is an indicator of the presence of chaos. It arises in homoclinic bifurcation in the Duffing oscillator, and Melnikov's method (Section 13.10) can be used to detect homoclinic bifurcation for periodically forced systems.

Sensitive dependence on initial conditions

In this test of chaos, 'small' changes in initial conditions lead to rapid divergence of solutions. This type of sensitivity is associated with loss of information in systems, and the breaking of links between the past and the future. This dependence has implications for the numerical solutions of equations and the growth of errors. Such solutions which arise under these sensitive conditions should be treated with caution. However, there is a phenomenon called **shadowing** which enhances the usefulness of computed 'orbits' of chaotic systems. Roughly, the shadowing lemma (Guckenheimer and Holmes (1983)) states, perhaps surprisingly, that, under suitable conditions, any computed orbit is an approximation to *some* true orbit of the system.

Autocorrelation function

In this approach the long time average over t of the product of $x(t)$ and $x(t + \tau)$ in a time series is computed. For a periodic response, the autocorrelation function will detect this periodicity. On the other hand for a chaotic response, in which we expect no correlation between $x(t)$ and $x(t + \tau)$, the autocorrelation function will tend to zero as $\tau \to \infty$.

Intermittency

For autonomous systems in three or more dimensions intermittency usually arises from a bifurcation in the neighbourhood of a limit cycle. For a given value of a parameter the system can have a stable limit cycle. For a neighbouring value of a parameter intermittency occurs if the system appears to remain periodic for long stretches but is subject to seemingly random outbursts at irregular intervals. In between these outbursts the system resumes its periodic behaviour. Even though the system drifts away from the periodic orbit, it is quickly re-injected into it, in a similar manner to the near-homoclinic paths of the Rössler attractor (Section 13.4). Some data for intermittency in the Lorenz attractor is given in Problem 13.37.

Problems

13.1 Obtain the solutions for the usual polar coordinates r and θ in terms of t, for the system

$$\dot{x} = x + y - x(x^2 + y^2), \quad \dot{y} = -x + y - y(x^2 + y^2).$$

Let Σ be the section $\theta = 0$, $r > 0$. Find the difference equation for the Poincaré sequence in this section.

13.2 Find the map of 2π first returns on the section $\Sigma: t = 0$ for

$$\ddot{x} + 2\dot{x} + 2x = 2 \sin t$$

in the usual phase plane. Find also the coordinates of the fixed point of the map and discuss its stability. Where is the fixed point of the map if the section is $t = \frac{1}{2}\pi$?

13.3 Let x_1 satisfy

$$\ddot{x}_1 + \tfrac{1}{4}\omega^2 x_1 = \Gamma \cos \omega t.$$

Obtain the solutions for x_1 and $x_2 = \dot{x}_1$ given that $x_1(0) = x_{10}$ and $x_2(0) = x_{20}$. Let Σ be the section $t = 0$ and find the first returns of period $2\pi/\omega$. Show that the mapping is

$$P_{\Sigma}(x_{10}, x_{20}) = \left(-x_{10} - \frac{8\Gamma}{3\omega^2}, -x_{20} \right),$$

and that

$$P_{\Sigma}^2(x_{10}, x_{20}) = (x_{10}, x_{20}).$$

Deduce that the system exhibits period doubling for all initial values except one. Find the coordinates of this fixed point.

13.4 (a) Let

$$\dot{x} = y, \quad \dot{y} = -3y - 2x + 10 \cos t,$$

and assume the initial conditions $x(0) = 4$, $y(0) = -1$. Consider the associated three-dimensional system with $\dot{z} = 1$. Assuming that $z(0) = 0$, plot the solution in the (x, y, z) space, and indicate the 2π-periodic returns which occur at $t = 0, t = 2\pi, t = 4\pi, \ldots$.

(b) Sketch some typical period 1 Poincaré maps in the (x, y, z) space

$$\dot{x} = \lambda x, \quad \dot{y} = \lambda y, \quad \dot{z} = 1$$

for each of the cases $\lambda < 0, \lambda = 0, \lambda > 0$. Discuss the nature of any fixed points in each case. Assume that $x(0) = x_0, y(0) = y_0, z(0) = 0$ and show that

$$x_{n+1} = e^{\lambda} x_n, \quad y_{n+1} = e^{\lambda} y_n, \quad n = 0, 1, 2, \ldots.$$

13.5 Two rings can slide on two fixed horizontal wires which lie in the same vertical plane with separation a. The two rings are connected by a spring of unstretched length l and stiffness μ. The upper ring is forced to move with displacement $\phi(t)$ from a fixed point O as shown in Fig. 13.41. The resistance on the lower ring which has mass m is assumed to be $mk \times (speed)$. Let y be the relative displacement between the rings. Show that the equation of motion of the lower ring is given by

$$\ddot{y} + k\dot{y} - \frac{\mu}{ma}(l - a)y + \frac{\mu l}{2ma^3}y^3 = -\ddot{\phi} - k\dot{\phi}$$

to order y^3 for small $|y|$. (If $l > a$ then a Duffing equation of the type discussed in Section 12.6 can be obtained. This could be a model for a strange attractor for the appropriate parameter values. Cosine forcing can be reproduced by putting $-\ddot{\phi} - k\dot{\phi} = \Gamma \cos \omega t$.)

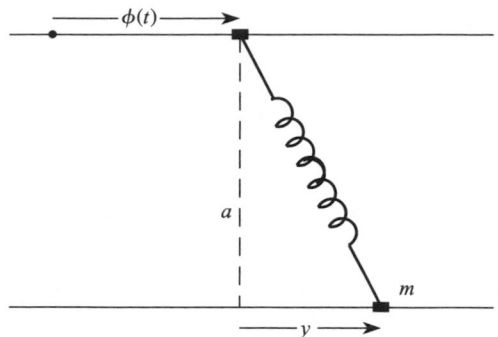

Figure 13.41

13.6 Search for period doubling in the undamped Duffing equation $\ddot{x} - x + x^3 = \Gamma \cos \omega t$ using the form $x = c + a_1 \cos \omega t + a_2 \cos \frac{1}{2} \omega t$, where c, a_1, and a_2 are constants. If frequencies $3\omega/2$ and above are neglected, show that the shift and amplitudes satisfy

$$c[-1 + c^2 + \tfrac{3}{2}(a_1^2 + a_2^2)] + \tfrac{3}{4} a_2^2 a_1 = 0,$$

$$a_1(-\omega^2 - 1 + 3c^2 + \tfrac{3}{4} a_1^2 + \tfrac{3}{2} a_2^2) + \tfrac{3}{2} a_2^2 c = \Gamma,$$

$$a_2(-\tfrac{1}{4}\omega^2 - 1 + 3c^2 + \tfrac{3}{2} a_1^2 + 3 c a_1 + \tfrac{3}{4} a_2^2) = 0.$$

Deduce that for harmonic solutions ($a_2 = 0$), c and a_1 are given by solutions of
(i) $c = 0$, $a_1(-\omega^2 - 1 + \tfrac{3}{4} a_1^2) = \Gamma$,

or

(ii) $c^2 = 1 - \tfrac{3}{2} a_1^2$, $a_1(-\omega^2 + 2 - \tfrac{15}{4} a_1^2) = \Gamma$.
Sketch the amplitude $|a_1|$–amplitude $|\Gamma|$ curves corresponding to Fig. 13.13 for $\omega = 1.2$.

13.7 Design a computer program to plot $2\pi/\omega$ first returns for the system

$$\dot{x} = X(x, y, t), \quad \dot{y} = Y(x, y, t)$$

where $X(x, y, t)$ and $Y(x, y, t)$ are $2\pi/\omega$-periodic functions of t. Apply the program to the following system:

$$X(x, y, t) = y, \quad Y(x, y, t) = -ky + x - x^3 + \Gamma \cos \omega t$$

for $k = 0.3$, $\omega = 1.2$, and Γ taking a selection of values between 0 and 0.8. Let the initial section be $t = 0$.

13.8 Find the equations of the stable and unstable manifolds in the (x, y)-plane of

$$\ddot{x} + \dot{x} - 2x = 10 \cos t, \quad \dot{x} = y$$

for Poincaré maps of period 2π and initial time $t = 0$.

13.9 Apply Melnikov's method to

$$\ddot{x} + \varepsilon \kappa \dot{x} - x + x^3 = \varepsilon \gamma (1 - x^2) \cos \omega t, \quad \kappa > 0, \quad \varepsilon > 0, \gamma > 0$$

and show that homoclinic bifurcation occurs if, for $\omega^2 \ll 2$,

$$|\gamma| \geq \frac{2\sqrt{2}\kappa}{\pi \omega (2 - \omega^2)} \cosh\left(\tfrac{1}{2}\omega\pi\right).$$

13.10 The Duffing oscillator with equation

$$\ddot{x} + \varepsilon \kappa \dot{x} - x + x^3 = \varepsilon f(t),$$

is driven by an even T-periodic function $f(t)$ with mean value zero. Assuming that $f(t)$ can be represented by the Fourier series

$$\sum_{n=1}^{\infty} a_n \cos n\omega t, \quad \omega = 2\pi/T,$$

find the Melnikov function for the oscillator.

Let

$$f(t) = \begin{cases} \gamma, & -\tfrac{1}{2} < t < \tfrac{1}{2}, \\ -\gamma, & \tfrac{1}{2} < t < \tfrac{3}{2}, \end{cases}$$

where $f(t)$ is a function of period 2. Show that the Melnikov function vanishes if

$$\frac{\kappa}{\gamma} = -\frac{3\pi}{2\sqrt{2}} \sum_{r=1}^{\infty} (-1)^r \operatorname{sech} \left[\frac{1}{2} \pi^2 (2r-1) \right] \sin[(2r-1)\pi t_0].$$

Plot the Fourier series as a function of t_0 for $0 \le t_0 \le 2$, and estimate the value of κ/γ at which homoclinic tangency occurs.

13.11 Melnikov's method can be applied also to autonomous systems. The manifolds become the separatrices of the saddle. Let

$$\ddot{x} + \varepsilon \kappa \dot{x} - \varepsilon \alpha x^2 \dot{x} - x + x^3 = 0.$$

Show that the homoclinic path exists to order $O(\varepsilon^2)$ if $\kappa = 4\alpha/5$. [The following integrals are required:

$$\int_{-\infty}^{\infty} \operatorname{sech}^4 s \, ds = \frac{4}{3}; \quad \int_{-\infty}^{\infty} \operatorname{sech}^6 s \, ds = \frac{16}{15}. \, \bigg]$$

13.12 Show that $x = 3^{\frac{1}{4}} \sqrt{(\operatorname{sech} 2t)}$ is a homoclinic solution of

$$\ddot{x} + \varepsilon (\kappa - \alpha x^2) \dot{x} - x + x^5 = 0,$$

when $\varepsilon = 0$. Use Melnikov's method to show that homoclinic bifurcation occurs when $\kappa = 4\sqrt{3}\alpha/(3\pi)$.

13.13 Apply Melnikov's method to the perturbed system

$$\ddot{x} + \varepsilon \kappa \dot{x} - x + x^3 = \varepsilon \gamma x \cos \omega t,$$

which has an equilibrium point at $x = 0$ for all t. Show that the manifolds of the origin intersect if

$$\gamma \ge \frac{4\kappa}{3\omega^2 \pi} \sinh \left(\frac{1}{2} \omega \pi \right).$$

[*Hint*:

$$\int_{-\infty}^{\infty} \operatorname{sech}^2 u \cos \omega u \, du = \frac{\pi \omega}{\sinh(\frac{1}{2}\omega\pi)}. \, \bigg]$$

13.14 Show that the logistic difference equation

$$u_{n+1} = \lambda u_n (1 - u_n)$$

has the general solution $u_n = \sin^2(2^n C\pi)$ if $\lambda = 4$, where C is an arbitrary constant (without loss C can be restricted to $0 \leq C \leq 1$). Show that the solution is 2^q-periodic (q any positive integer) if $C = 1/(2^q - 1)$. The presence of *all* these period doubling solutions indicates chaos. (See the article by Brown and Chua (1996) for further exact solutions of nonlinear difference equations relevant to this and some succeeding problems.)

13.15 Show that the difference equation

$$u_{n+1} = 2u_n^2 - 1$$

has the exact solution $u_n = \cos(2^n C\pi)$ where C is any constant satisfying $0 \leq C \leq 1$. For what values of C do q-periodic solutions exist?

13.16 Using a trigonometric identity for $\cos 3t$, find a first-order difference equation satisfied by $u_n = \cos(3^n C\pi)$.

13.17 A large number of phase diagrams have been computed and analyzed for the two-parameter Duffing equation

$$\ddot{x} + k\dot{x} + x^3 = \Gamma \cos t, \quad \dot{x} = y,$$

revealing a complex pattern of periodic, subharmonic and chaotic oscillations (see Ueda (1980), and also Problem 7.32). Using a suitable computer package plot phase diagrams and time solutions in each of the following cases for the initial data given, and discuss the type of solutions generated:

(a) $k = 0.08$, $\Gamma = 0.2$; $x(0) = -0.205$, $y(0) = 0.0171$; $x(0) = 1.050$, $y(0) = 0.780$.

(b) $k = 0.2$, $\Gamma = 5.5$; $x(0) = 2.958$, $y(0) = 2.958$; $x(0) = 2.029$, $y(0) = -0.632$.

(c) $k = 0.2$, $\Gamma = 10$; $x(0) = 3.064$, $y(0) = 4.936$.

(d) $k = 0.1$, $\Gamma = 12$; $x(0) = 0.892$, $y(0) = -1.292$.

(e) $k = 0.1$, $\Gamma = 12$; $x(0) = 3$, $y(0) = 1.2$.

13.18 Consider the Hamiltonian system

$$\dot{p}_i = -\frac{\partial H}{\partial q_i}, \quad \dot{q}_i = \frac{\partial H}{\partial p_i} \quad (i = 1, 2),$$

where $H = \frac{1}{2}\omega_1(p_1^2 + q_1^2) + \frac{1}{2}\omega_2(p_2^2 + q_2^2)$. Show that q_1, q_2 satisfy the uncoupled system

$$\ddot{q}_i + \omega_i^2 q_i = 0 \quad (i = 1, 2).$$

Explain why the ellipsoids $\frac{1}{2}\omega_1(p_1^2 + q_1^2) + \frac{1}{2}\omega_2(p_2^2 + q_2^2) = $ constant are invariant manifolds in the four-dimensional space (p_1, p_2, q_1, q_2). What condition on ω_1/ω_2 guarantees that all solutions are periodic? Consider the phase path which satisfies $p_1 = 0$, $q_1 = 0$, $p_2 = 1$, $q_2 = 0$. Describe the Poincaré section $p_1 = 0$ of the phase path projected on to the (q_1, p_2, q_2) subspace.

13.19 Consider the system $\dot{x} = -ryz$, $\dot{y} = rxz$, $\dot{z} = -z + \cos t - \sin t$, where $r = \sqrt{(x^2 + y^2)}$. Show that, projected on to the (x, y) plane, the phase paths have the same phase diagram as a plane centre. Show also that the general solution is given by

$$x = x_0 \cos \omega(t) - y_0 \sin \omega(t),$$
$$y = y_0 \cos \omega(t) - x_0 \sin \omega(t),$$
$$z = z_0 e^{-t} + \sin t,$$

where

$$\omega(t) = r_0[1 - \cos t + z_0(1 - e^{-t})],$$

and $x_0 = x(0)$, $y_0 = y(0)$, $z_0 = z(0)$ and $r_0 = \sqrt{(x_0^2 + y_0^2)}$. Confirm that, as $t \to \infty$, all solutions become periodic.

13.20 A common characteristic feature of chaotic oscillators is *sensitive dependence on initial conditions*, in which bounded solutions which start very close together ultimately diverge: such solutions locally diverge exponentially. Investigate time-solutions of Duffing's equation

$$\ddot{x} + k\dot{x} - x + x^3 = \Gamma \cos \omega t$$

for $k = 0.3$, $\Gamma = 0.5$, $\omega = 1.2$, which is in the chaotic parameter region (see Fig. 13.15), for the initial values (a) $x(0) = 0.9$, $y(0) = 0.42$; (b) $y(0) = 0.42$ but with a very small increase in $x(0)$ to, say 0.90000001. Divergence between the solutions occurs at about 40 cycles. Care must be execised in computing solutions in chatotic domains where sensitive dependence on initial values and computation errors can be comparable in effect.]

13.21 The Lorenz equations are given by (see Problem 8.26 and Section 13.2)

$$\dot{x} = a(y - x), \quad \dot{y} = bx - y - xz, \quad \dot{z} = xy - cz.$$

Compute solutions of these equation in (x, y, z) phase space. Chaotic solutions appear near parameter values $a = 10$, $b = 27$, $c = 2.65$: a possible initial state is $x(0) = -11.720$, $y(0) = -17.249$, $z(0) = 22.870$. A sample solution for these parameter and initial values is shown in Fig. 13.42.

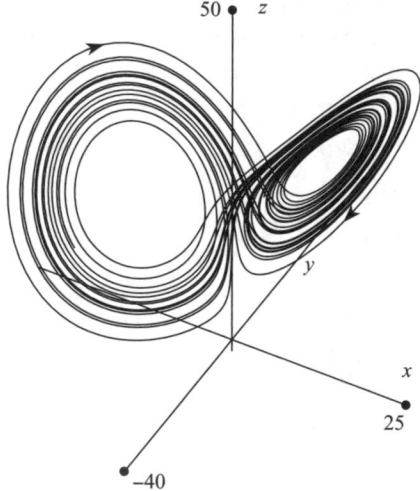

Figure 13.42 A chaotic phase path for the Lorenz attractor.

13.22 Show that the system

$$\dot{x} = -y + \Gamma \sin t, \quad \dot{y} = -x + 2x^3 + \Gamma \cos t, \quad \Gamma > 0,$$

has a limit cycle $x = 0$, $y = \Gamma \sin t$. Find also the time-solutions for x and y of the paths which are homoclinic to this limit cycle. Sketch the phase paths of the limit cycle and the homoclinic paths for $\Gamma = 1$.

13.23 For each of the following functions and solutions plot the Poincaré sequence in the $x, y = \dot{x}$ plane, starting with the given initial time t_0 and given period T.

(a) $x = 2 \cos t$; $t_0 = 0$, $T = 2\pi$.

(b) $x = 3 \cos \frac{1}{2} t$; $t_0 = \frac{1}{2}\pi$, $T = 2\pi$.

(c) $x = \sin t + \sin \pi t$; $t_0 = \frac{1}{2}\pi$, $T = 2\pi$.

(d) The periodic solution of $\ddot{x} - (x^2 + \dot{x}^2)\dot{x} + x = \cos t$, where $t_0 = 0$, $T = 2\pi$.

13.24 Show that

$$\dot{x} + k(1 - x^2 - \dot{x}^2)^2\dot{x} - x = -2\cos t$$

has a limit cycle whose solution is $x_0 = \cos t$. By looking at perturbations $x = x_0 + x'$ where $|x'|$ is small show that the limit cycle has Poincaré fixed points which are saddles.

13.25 Consider the system

$$\dot{x} = y, \quad \dot{y} = [e^{-2x} - e^{-2}] + \varepsilon \cos t.$$

For $\varepsilon = 0$, show that the equation of its phase paths is given by

$$y^2 = 2e^{-x} - e^{-2x} + C.$$

Show that the system has closed paths about the origin if $-1 < C < 0$ with a bounding separatrix given by $y^2 = 2e^{-x} - e^{-2x}$. What happens to paths for $C > 0$? Sketch the phase diagram.

Suppose that the system is moving along the separatrix path, and at some instant the forcing is introduced. Describe what you expect the behaviour of the system to be after the introduction of the forcing. Compute a Poincaré sequence and a time-solution for $\varepsilon = 0.5$ and for the initial conditions, $x(0) = -\ln 2$, $y(0) = 0$.

13.26 Apply the change of variable $z = u + a + b$ to the Lorenz system

$$\dot{x} = a(y - x), \quad \dot{y} = bx - y - xz, \quad \dot{z} = xy - cz,$$

where $a, b, c > 0$. If $s = \sqrt{[x^2 + y^2 + z^2]}$, show that

$$\tfrac{1}{2}s\frac{ds}{dt} = -ax^2 - y^2 - c\left[u + \tfrac{1}{2}(a + b)\right]^2 + \tfrac{1}{4}c(a + b)^2.$$

What is the sign of ds/dt on the ellipsoid

$$ax^2 + y^2 + c[u + \tfrac{1}{2}(a + b)]^2 = \rho, \qquad (*)$$

where $\rho > \tfrac{1}{4}c(a + b)^2$?

Show that all equilibrium points are unstable in the case $a = 4$, $b = 34$, $c = \tfrac{1}{2}$. If this condition is satisfied, what can you say about the attracting set inside the ellipsoid $(*)$ if ρ is sufficiently large?

13.27 A plane autonomous system is governed by the equations $\dot{x} = X(x, y)$, $\dot{y} = Y(x, y)$. Consider a set of solutions $x(t, x_0, y_0)$, $y(t, x_0, y_0)$ which start at time $t = t_0$ at (x_0, y_0), where (x_0, y_0) is any point in a region $\mathcal{D}(t_0)$ bounded by a smooth simple closed curve C. At time t, $\mathcal{D}(t_0)$ becomes $\mathcal{D}(t)$. The area of $\mathcal{D}(t)$ is

$$A(t) = \iint_{\mathcal{D}(t)} dx\,dy = \iint_{\mathcal{D}(t_0)} J(t)dx_0\,dy_0$$

when expressed in terms of the original region. In this integral, the Jacobian

$$J(t) = \det(\Phi(t)),$$

where

$$\Phi(t) = \begin{bmatrix} \dfrac{\partial x}{\partial x_0} & \dfrac{\partial x}{\partial y_0} \\[2mm] \dfrac{\partial y}{\partial x_0} & \dfrac{\partial y}{\partial y_0} \end{bmatrix}.$$

Show that $\boldsymbol{\Phi}(t)$ satisfies the linear equation

$$\dot{\boldsymbol{\Phi}}(t) = \boldsymbol{B}(t)\boldsymbol{\Phi}(t),$$

(note that $\boldsymbol{\Phi}(t)$ is a fundamental matrix of this equation) where

$$\boldsymbol{B}(t) = \begin{bmatrix} \dfrac{\partial X}{\partial x} & \dfrac{\partial X}{\partial y} \\ \dfrac{\partial Y}{\partial x} & \dfrac{\partial Y}{\partial y} \end{bmatrix}.$$

Using Theorem 9.4 (on a property of the Wronskian), show that

$$J(t) = J(t_0) \exp\left[\int_{t_0}^{t} \left(\frac{\partial X}{\partial x} + \frac{\partial Y}{\partial y} \right) ds \right].$$

If the system is Hamiltonian deduce that $J(t) = J(t_0)$. What can you say about the area of $\mathcal{D}(t)$? ($A(t)$ is an example of an **integral invariant** and the result is known as **Liouville's theorem**).

For an autonomous system in n variables $\dot{x} = X(x)$, what would you expect the corresponding condition for a **volume-preserving** phase diagram to be?

13.28 For the more general version of Liouville's theorem (see Problem 13.27) applied to the case $n = 3$ with $\dot{x} = X(x, y, z)$, $\dot{y} = Y(x, y, z)$, $\dot{z} = Z(x, y, z)$, the volume of a region $\mathcal{D}(t)$ which follows the phase paths is given by

$$W(t) = \iiint_{\mathcal{D}(t)} dx\, dy\, dz = \iiint_{\mathcal{D}(t_0)} J(t)\, dx_0\, dy_0\, dz_0,$$

where the Jacobian $J(t) = \det[\boldsymbol{\Phi}(t)]$. As in the previous problem

$$J(t) = J(t_0) \exp\left[\int_{t_0}^{t} \left(\frac{\partial X}{\partial x} + \frac{\partial Y}{\partial y} + \frac{\partial Z}{\partial z} \right) ds \right].$$

Show that $dJ(t)/dt \to 0$ as $t \to \infty$ for the Lorenz system

$$\dot{x} = a(y - x), \quad \dot{y} = bx - y - xz, \quad \dot{z} = xy - cz,$$

where $a, b, c > 0$. What can be said about the volume of any region following phase paths of the Lorenz attractor as time progresses?

13.29 Show that $\ddot{x}(1 + \dot{x}) - x\dot{x} - x = -2\gamma(\dot{x} + 1)\cos t$, $\gamma > 0$, has the exact solution $x = Ae^{t} + Be^{-t} + \gamma \cos t$. What can you say about the stability of the limit cycle? Find the Poincaré sequences of the stable and unstable manifolds associated with $t = 0$ and period 2π. Write down their equations and sketch the limit cycle, its fixed Poincaré point and the stable and unstable manifolds.

13.30 Search for 2π-periodic solutions of

$$\ddot{x} + k\dot{x} - x + (x^2 + \dot{x}^2)x = \Gamma \cos t,$$

using $x = c + a \cos t + b \sin t$, and retaining only first harmonics. Show that c, γ satisfy

$$c(c^2 - 1 + 2r^2) = 0,$$
$$(r^2 + 3c^2 - 2)^2 + k^2 r^2 = \Gamma^2,$$

and that the formula is exact for the limit cycle about the origin. Plot a response amplitude (r) against forcing amplitude (Γ) figure as in Fig. 13.13 for $k = 0.25$.

13.31 A nonlinear oscillator has the equation

$$\ddot{x} + \varepsilon(\dot{x}^2 - x^2 + \tfrac{1}{2}x^4)\dot{x} - x + x^3 = 0, \quad 0 < \varepsilon \ll 1.$$

Show that the system has one saddle and two unstable spiral equilibrium points. Confirm that the saddle point has two associated homoclinic paths given by $x = \pm\sqrt{2}\,\mathrm{sech}\,t$. If $u = \dot{x}^2 - x^2 + \tfrac{1}{2}x^4$ show that u satisfies the equation

$$\dot{u} + 2\varepsilon\dot{x}^2 u = 0.$$

What can you say about the stability of the homoclinic paths from the sign of $\dot{u}$? Plot a phase diagram showing the homoclinic and neighbouring paths.

The system is subject to small forcing $\varepsilon\gamma\cos\omega t$ on the right-hand side of the differential equation. Explain, in general terms, how you expect the system to behave if it is started initially from $x(0) = 0$, $\dot{x}(0) = 0$. Plot the phase diagram over a long time interval, say $t \sim 150$, for $\varepsilon = 0.25$, $\omega = 1$, $\gamma = 0.2$.

13.32 Show, for $\alpha > 3$, that the logistic difference equation $u_{n+1} = \alpha u_n(1 - u_n)$ has a period two solution which alternates between the two values

$$\tfrac{1}{2\alpha}[1 + \alpha - \sqrt{(\alpha^2 - 2\alpha - 3)}] \quad \text{and} \quad \tfrac{1}{2\alpha}[1 + \alpha + \sqrt{(\alpha^2 - 2\alpha - 3)}].$$

Show that it is stable for $3 < \alpha < 1 + \sqrt{6}$.

13.33 The **Shimizu–Morioka equations** are given by the two-parameter system

$$\dot{x} = y, \quad \dot{y} = x(1 - z) - ay, \quad \dot{z} = -bz + x^2, \quad a > 0.$$

Show that there are three equilibrium points for $b > 0$, and one for $b \leq 0$. Show that the origin is a saddle point for all a and $b \neq 0$. Obtain the linear approximation for the other equilibrium points assuming that $b = 1$. Find the eigenvalues of the linear approximation at $a = 1.2$, $a = 1$ and at $a = 0.844$. What occurs at $a = 1$? For $a = 1.2$ and $a = 0.844$ compute the unstable manifolds of the origin by using initial values close to the origin in the direction of its eigenvector, and plot their projections on to the (x, z) plane (see Fig. 13.43). Confirm that two homoclinic paths occur for $a \approx 0.844$. What happens to the stability of the equilibrium points away from the origin as a decreases through 1? What type of bifurcation occurs at $a = 1$? Justify any conjecture by plotting phase diagrams for $0.844 < a < 1$.

13.34 Compute some Poincaré sections given by the plane Σ: $z =$ constant of the Rössler system

$$\dot{x} = -y - z, \quad \dot{y} = x + ay, \quad \dot{z} = bx - cz + xz \quad (a, b, c > 0)$$

where $a = 0.4$, $b = 0.3$ and c takes various values. The choice of the constant for z in Σ is important: if it is too large then the section might not intersect phase paths at all. Remember that the Poincaré sequence

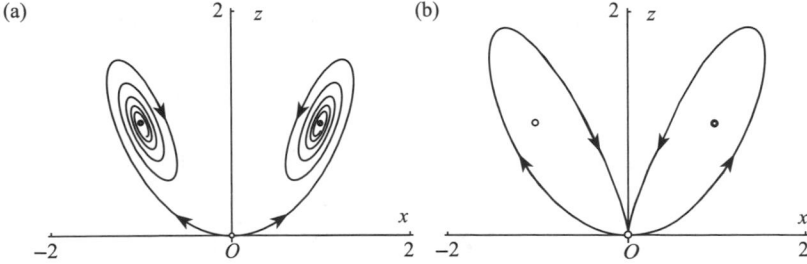

Figure 13.43 The Shimizu–Morioka attractor: (a) unstable manifolds of the origin projected on to the (x, z) plane for $a = 1.2$, $b = 1$ (b) homoclinic paths of the origin for $a \approx 0.844$, $b = 1$. For obvious shape and symmetry reasons, these phase projections are often referred to as **butterflies**.

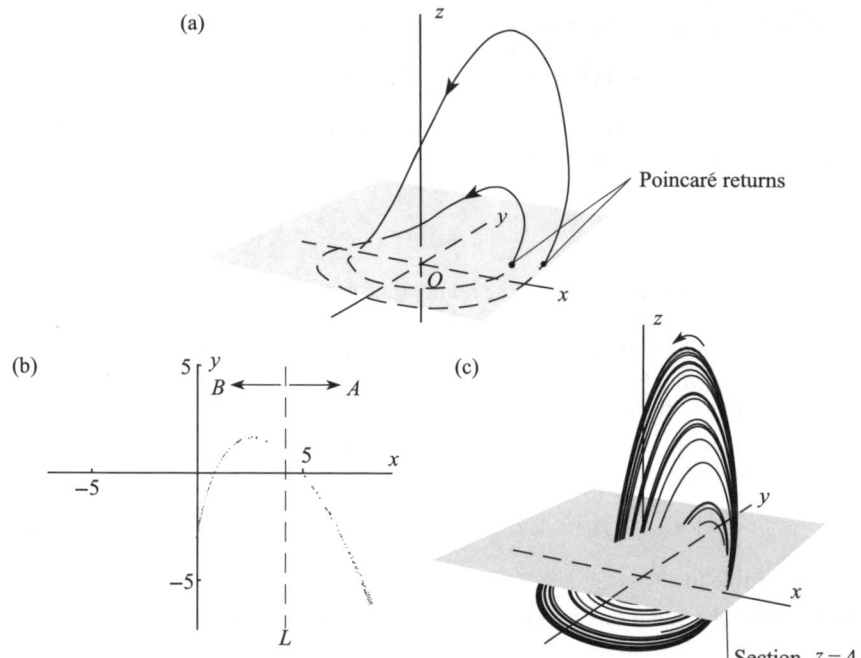

Figure 13.44 (a) The Poincaré section $z=2$ for the period-2 Rössler attractor with $a=0.4$, $b=0.3$, $c=2$. The Poincaré returns must cross the section in the same sense. (b) The section $z=4$ for the Rössler attractor with $a=0.4$, $b=0.3$, $c=4.449$. The returns in the half-plane A on the right of the line L (corresponding to $\dot{z}=0$) occur where $\dot{z}$ is increasing on those on the left in half plane B occur where $\dot{z}$ is decreasing. Each could be a strange attractor of the system. (c) A typical phase path in (b) showing its intersection with the section $z=4$ (see also Figs 13.11 and 13.12).

arises from intersections which occur as the phase paths cut Σ in the same sense. The period-2 solution (Fig. 13.12(b)), with Poincaré section $z=2$ should appear as two dots as shown in Fig. 13.44(a) after transient behaviour as died down. Figures 13.44(a), (b) show a section of chaotic behaviour at $c=4.449$ at $z=4$.

13.35 For the Duffing oscillator

$$\ddot{x} + k\dot{x} - x + x^3 = \Gamma \cos \omega t$$

it was shown in Section 13.3, that the displacement c and the response amplitude r were related to the other parameters by

$$c^2 = 1 - \tfrac{3}{2}r^2, \quad r^2[(2 - \omega^2 - \tfrac{15}{4}r^2)^2 + k^2\omega^2] = \Gamma^2$$

for Type II oscillations (eqn (13.25)). By investigating the roots of $d(\Gamma^2)/d(r^2) = 0$, show that a fold develops in this equation for

$$\omega < \tfrac{1}{2}[4 + 3k^2 - k\sqrt{(24 + 9k^2)}].$$

Hence there are three response amplitudes for these forcing frequencies. Design a computer program to plot the amplitude (Γ) -amplitude curves (r); C_1 and C_2 as in Fig. 13.13. Figure 13.45 shows the two folds in C_1 and C_2 for $k = 0.3$ and $\omega = 0.9$.

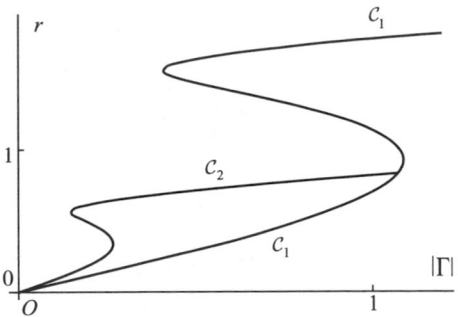

Figure 13.45

13.36 It was shown in Section 13.3 for the Duffing equation

$$\ddot{x} + k\dot{x} - x + x^3 = \Gamma \cos \omega t,$$

that the perturbation $a' = [a', b', c', d']^T$ from the translation $c_0 = \sqrt{[1 - \frac{3}{2}(a_0^2 + b_0^2)]}$ and the amplitudes a_0 and b_0 of the harmonic approximation

$$x = c_0 + a_0 \cos \omega t + b_0 \sin \omega t,$$

satisfy $\dot{a}' = Aa'$ where

$$A = \begin{bmatrix} R(P - \frac{3}{2}ka_0^2 + 3a_0b_0\omega) & -R(Q - \frac{3}{2}ka_0b_0 + 3b_0^2\omega) & R(-6a_0c_0k + 12b_0c_0\omega) & 0 \\ R(Q - 3a_0^2\omega - \frac{3}{2}ka_0b_0) & R(P - 3a_0b_0\omega - \frac{3}{2}b_0^2k) & R(-12a_0c_0k) & 0 \\ 0 & 0 & 0 & 1 \\ -3a_0c_0 & -3b_0c_0 & -(2 - 3r_0^2) & -k \end{bmatrix}$$

where

$$R = 1/(k^2 + 4\omega^2), \quad P = -k(2 + \omega^2 - \tfrac{15}{4}r_0^2),$$
$$Q = \omega(4 - 2\omega^2 - k^2 - \tfrac{15}{4}r_0^2),$$

(see eqn (13.37)). The constants a_0 and b_0 are obtained by solving eqns (13.21) and (13.22). Devise a computer program to find the eigenvalues of the matrix A for $k = 0.3$ and $\omega = 1.2$ as in the main text. By tuning the forcing amplitude Γ, find, approximately, the value of Γ for which one of the eigenvalues changes sign so that the linear system $\dot{a}' = Aa'$ becomes unstable. Investigate numerically how this critical value of Γ varies with the parameters k and ω.

13.37 Compute solutions for the Lorenz system

$$\dot{x} = a(y - x), \quad \dot{y} = bx - y - xz, \quad \dot{z} = xy - cz,$$

for the parameter section $a = 10$, $c = 8/3$ and various values of b: this is the section frequently chosen to illustrate oscillatory features of the Lorenz attractor. In particular try $b = 100.5$ and show numerically that there is a periodic attractor as shown in Fig. 13.46(a). Why will this limit cycle be one of a pair?

Show also that at $b = 166$, the system has a periodic solution as shown in Fig. 13.46(b), but at 166.1 (Fig. 13.46(c)) the periodic solution is regular for long periods but is then subject to irregular bursts at irregular intervals before resuming its oscillation again. This type of chaos is known as **intermittency**. (For discussion of intermittent chaos and references see Nayfeh and Balachandran (1995); for a detailed discussion of the Lorenz system see Sparrow (1982).)

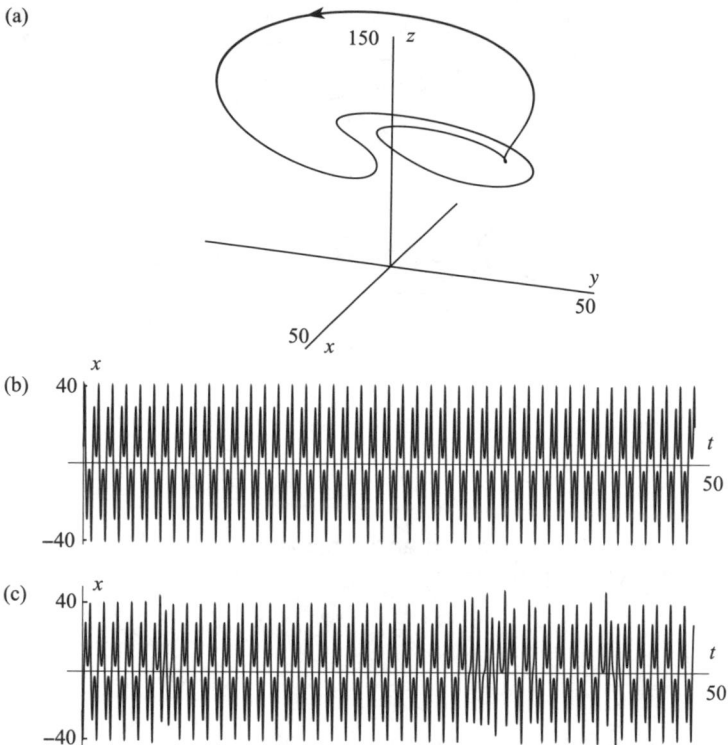

(b)

(c)

Figure 13.46   (a) Phase diagram for the Lorenz system showing limit cycle at $a = 10$, $b = 100.5$, $c = 8/3$; (b) Periodic time solution for x at $a = 10$, $b = 166$, $c = 8/3$: (c) Time solution in (b) being disturbed intermittently as b is increased from 166 to 166.1.

13.38 The damped pendulum with periodic forcing of the pivot leads to the equation (Bogoliubov and Mitropolski 1961)

$$\ddot{x} + \sin x = \varepsilon(\gamma \sin t \sin x - \kappa \dot{x}),$$

where $0 < \varepsilon \ll 1$. Apply Melnikov's method and show that heteroclinic bifurcation occurs if $\gamma \geq 4\kappa \sinh \frac{1}{2}\pi$.

[You will need the integral

$$\int_{-\infty}^{\infty} \sin s \, \text{sech}^2 s \tanh s \, ds = \frac{1}{2}\pi / \sinh(\frac{1}{2}\pi)\bigg].$$

13.39 An alternative method of visualizing the structure of solutions of difference and differential equations is to plot **return maps** of u_{n-1} versus u_n. For example, a sequence of solutions of the logistic difference equation

$$u_{n+1} = \alpha u_n(1 - u_n),$$

the ordinate would be u_{n-1} and the abscissa u_n. The return map should be plotted after any initial transient returns have died out. If $\alpha = 2.8$ (see Section 13.4), how will the long-term return map appear? Find the return map for $\alpha = 3.4$ also. An exact (chaotic) solution of the logistic equation is

$$u_n = \sin^2(2^n)$$

(see Problem 13.14). Plot the points (u_n, u_{n-1}) for $n = 1, 2, \ldots, 200$, say. What structure is revealed?

Using a computer program generate a time-series (numerical solution) for the Duffing equation

$$\ddot{x} + k\dot{x} - x + x^3 = \Gamma \cos \omega t$$

for $k = 0.3$, $\omega = 1.2$ and selected values of Γ, say $\Gamma = 0.2, 0.28, 0.29, 0.37, 0.5$ (see Figs 13.14, 13.15, 13.16). Plot transient-free return maps for the interpolated pairs $(x(2\pi n/\omega), x(2\pi(n-1)/\omega))$. For the *chaotic* case $\Gamma = 0.5$, take the time series over an interval $0 \leq t \leq 5000$, say. These return diagrams show that structure is recognizable in chaotic outputs: the returns are not uniformly distributed.

Answers to the exercises

Chapter 1

1.1 Separatrices are given by $y^2 = 2(1 - \sin x)$.

1.2 Elapsed time is $2\tanh^{-1}(1/\sqrt{3})$.

1.3 $\mathcal{V}(x) = \frac{1}{2}x^2 - \frac{1}{3}x^3$.

1.4 Unstable spiral.

1.5 Every circuit occurs in the time $\pi[(1 - \varepsilon)^{-\frac{1}{2}} + (1 + \varepsilon)^{-\frac{1}{2}}]$.

1.6 The limit cycle is the ellipse $x^2 + \frac{1}{4}y^2 = 1$; period is π.

1.7 Period is $4\pi\sqrt{(a/g)}$.

1.8 No equilibrium points if $\lambda < -1$; one unstable equilibrium point if $\lambda = 0$ or $\lambda = -1$; one stable equilibrium point if $x = \lambda\sqrt{(1 + \lambda)}$, $(\lambda > -1, \ \lambda \neq 0)$; one unstable equilibrium point if $x = -\lambda\sqrt{(1 + \lambda)}$, $(\lambda > -1, \ \lambda \neq 0)$.

1.9 (i) stable spiral; (ii) unstable spiral.

Chapter 2

2.1 Phase paths are given by $\cos x + \sin y = \text{constant}$.

2.2 There are four equilibrium points: $(0, 3)$ and $(3, 0)$ are stable nodes.

2.3 Linear approximation is $\dot{x} = x + 2y$, $\dot{y} = x + 3y$.

2.4 Eigenvalues are -2 and -3.

2.5 (a) saddle point; (c) unstable node.

2.6 There are three equilibrium points $(0, 0)$, $(1, 1)$ and $(-1, -1)$; $(0, 0)$ is a saddle and $(1, 1)$, $(-1, -1)$ are unstable nodes.

2.7 (a) $(1, 1)$ is an unstable spiral; $(-1, 1)$ is a saddle. (b) $(\sqrt{8}, 3)$ and $(-\sqrt{3}, -\sqrt{2})$ are saddles; $(-\sqrt{8}, 3)$ is a stable spiral; $(\sqrt{3}, -\sqrt{2})$ is an unstable spiral.

2.8 The phase diagram has six saddle points and two centres.

Chapter 3

3.2 Index is -2.

3.3 $I_\infty = 4$.

3.4 Equilibrium points at infinity are at the six points $(0, 1)$, $(0, -1)$, $(\pm\frac{\sqrt{3}}{2}, \pm\frac{1}{2})$.

3.5 In polar coordinates the equations are $\dot{r} = r(r^2 - 1)$, $\dot{\theta} = -1$.

3.6 (b) Use Bendixson's negative criterion.

Chapter 4

4.1 Limit cycle with amplitude $a = 2\sqrt{2}$ is stable; the one with amplitude $a = \sqrt{6}$ is unstable.

4.2 Approximate amplitude is $a_0 = 2$.

4.3 The limit cycle has amplitude $a = 32/(9\pi)$.

4.4 The limit cycle has amplitude $a_0 = 2/\sqrt{7}$, and the polar equation of the spiral paths is given by $a^2 = 4/[7 - e^{\varepsilon(\theta-\theta_1)}]$ where θ_1 is a constant.

4.5 Approximately the frequency $\omega = 1$ and the amplitude $a = \frac{1}{2}(15\pi)^{\frac{1}{3}}$.

4.6 The frequency-amplitude approximation derived from the equivalent linear equation is given by $\omega^2 = 3a^2/(4-a^2)$. The exact equation for the phase paths is $y^2 = 1-x^2+Ce^{-x^2}$. One path is $x^2+y^2 = 1\,(C = 0)$ which corresponds to $\omega = a = 1$ in the approximation.

Chapter 5

5.1 The equations for x_0 and x_3 are

$$x_0'' + \Omega^2 x_0 = \Gamma \cos \tau, \quad x_3'' + \Omega^2 x_3 = x_2 f'(x_0) + \tfrac{1}{2}x_1^2 f''(x_0),$$

5.2 $\max |\Gamma| = 0.294$.

5.3 Substitute x_0 and x_1 into (5.32c), and show (using computer software) that the coefficients of $\cos \tau$ and $\sin \tau$ in the right-hand side of (5.32c) are linear in a_1 and b_1.

5.4 For the leading approximation $x = a_0 \cos \tau$, a_0 satisfies $3a_0^3 - 8a_0 + 5 = 0$ (see Appendix E on cubic equations).

5.5 Leading approximation is $x = a_0 \sin \tau$, where $3a_0^3 - 4\beta a_0 + 4\gamma = 0$.

5.6 $a_1 = 5\gamma^3/(6\beta^4)$; $b_1 = 0$.

5.7 $\omega = 1 - \frac{1}{16}(6 + 5a_0^2)a_0^2\varepsilon + O(\varepsilon^2)$.

5.8 One solution is given by $r_0^2 = 4$, in which case $a_0 = -\gamma/(2\omega_1)$, $b_0 = 0$.

Chapter 6

6.1 $x = \cos \tau + \frac{1}{32}\varepsilon[-\cos \tau + \cos 3\tau] + O(\varepsilon^2)$ where $t = \tau + \frac{1}{8}\varepsilon\tau + O(\varepsilon^2)$.

6.2 $x = e^{-t}/t^2$.

6.3 $x = \dfrac{a}{\sqrt{[4a^2-(4a^2-1)e^{-\frac{1}{2}\varepsilon t}]}}\cos \tau$.

6.4 $y_O = 1$, $y_I = 1 - e^{-x/\varepsilon}$.

6.5 $x_O = 2\tan^{-1}[e^{t/k}\tan(2\alpha/k)]$, $x_I = \alpha(1 - e^{-kt/\varepsilon})/k$.

Chapter 7

7.2 $\gamma = \frac{2}{9}\sqrt{[2(1 - \omega^2)^3 + 18k^2\omega^2(1 - \omega^2)]}$.

Chapter 8

8.2 Solutions are given by $x = Ae^{Ct}$, where A and C are constants.

8.3 General solution is

$$\begin{bmatrix} x \\ y \end{bmatrix} = A\begin{bmatrix} 0 \\ 1 \end{bmatrix}e^{-t} + B\begin{bmatrix} 1 \\ 1 \end{bmatrix}e^{-2t} + \begin{bmatrix} \frac{1}{6} \\ \frac{1}{3} \end{bmatrix}e^{t}.$$

8.4 $\Phi(t) = \begin{bmatrix} t^2 & t^3 \\ 2t - t^2 & 3t^2 - t^3 \end{bmatrix}$.

8.5 $x(t) = \dfrac{1}{4} \begin{bmatrix} 2e^t - 2e - t \\ -(1 + 2t) + 9e^{2t} \\ 2e^t + 2e^{-t} \end{bmatrix}$

8.7 $\Phi(t) = \begin{bmatrix} 2e^{-t} & ie^{(1+i)t} & -ie^{(1-i)t} \\ -e^{-t} & 2ie^{(1+i)t} & -2ie^{(1-i)t} \\ e^{-t} & e^{(1+i)t} & e^{(1-i)t} \end{bmatrix}$.

8.9 Straight line paths occur on $x_1 = x_2 = x_3$; $x_1 = x_2$, $x_3 = 0$; $x_1 = -x_2 = -x_3$.

Chapter 9

9.2 $E = \begin{bmatrix} e^{2\pi} & 0 \\ 0 & 1 \end{bmatrix}$, $\mu_1 = e^{2\pi}$, $\mu_2 = 1$.

9.3 $G_1(\alpha, \beta) = 0$ implies $\beta^2 = 4(\alpha - \frac{1}{4})^2$.
$G_2(\alpha, \beta) = 0$ implies

$$4\beta^2 + (8\alpha - 18)\beta - 16\alpha^2 + 40\alpha - 9 = 0, \text{ or,}$$
$$4\beta^2 - (8\alpha - 18)\beta - 16\alpha^2 + 40\alpha - 9 = 0.$$

Chapter 10

10.1 Stable spiral.

10.2 (b) $r = (1 - \frac{3}{4}e^{-2t})^{-1}$, $\theta = -t$.

10.3 Possibility is $V(x, y) = x^4 + 2y^2$.

10.6 Use the function $U(x, y) = x^2 - y^2$.

10.7 $K = \dfrac{1}{2} \begin{bmatrix} 6 & 7 \\ 7 & 1 \end{bmatrix}$.

10.9 $K = \dfrac{1}{6} \begin{bmatrix} 2 & 1 \\ 1 & 2 \end{bmatrix}$.

Chapter 11

11.2 Phase paths are given by $y - \ln|1 + y| = \frac{1}{2}e^{-x^2} + C$.

Chapter 12

12.1 Bifurcations occur at $\lambda = 1$ and $\lambda = 9$.

12.4 There no limit cycles for $\mu > \frac{9}{4}$ or $\mu < -\sqrt{2}$; two for $-\sqrt{2} < \mu < \sqrt{2}$: four for $\sqrt{2} < \mu < \frac{9}{4}$.

12.6 No limit cycles if $\mu > 1$ or $\mu < 0$, and two if $0 < \mu < 1$. If $0 < \mu < 1$, let

$$r_1 = \sqrt{1 + \sqrt{(1 - \mu)}} \qquad r_2 = \sqrt{1 - \sqrt{(1 - \mu)}}.$$

Then r_1 is stable and r_2 is unstable.

Chapter 13

13.2 The limit is $(1, 3)$.

13.4 Instability occurs for $\alpha > \frac{4}{3}$.

13.5 $\kappa \cosh(\frac{1}{2}\omega\pi) \le \omega\gamma\pi$.

13.7 The Liapunov exponents are $4, 1, -1$.

Appendices

A   Existence and uniqueness theorems

We state, without proof, certain theorems of frequent application in the text.

Theorem A1   *For the nth-order system*

$$\dot{x} = f(x, t) \tag{A1}$$

suppose that f is continuous and that $\partial f_j / \partial x_i$, $i, j = 1, 2, \ldots, n$ are continuous for $x \in \mathcal{D}$, $t \in I$, where $\mathcal{D}$ is a domain and I is an open interval. Then if $x_0 \in \mathcal{D}$ and $t_0 \in I$, there exists a solution $x^*(t)$, defined uniquely in some neighbourhood of (x_0, t_0), which satisfies $x^*(t_0) = x_0$.   ∎

We call such a system **regular** on $\mathcal{D} \times I$ (the set (x, t) where $x \in \mathcal{D}$ and $t \in I$). For brevity, we refer in the text to a system regular on $-\infty < x_i < \infty$, $i = 1, 2, \ldots, n$, $-\infty < t < \infty$ as being a **regular system**. Points at which the conditions of the theorem apply are called **ordinary points**.

The theorem states only that the initial-value problem has a unique solution on some sufficiently small interval about t_0, a limitation arising from the method of proof. A question of importance is how far the solution actually extends. It need not extend through the whole of I, and some possibilities are shown in the following example.

Example A1 $\dot{x}_1 = x_2$, $\dot{x}_2 = -x_1$, $x_1(0) = 0$, $x_2(0) = 1$.

$\mathcal{D}$ is the domain $-\infty < x_1 < \infty$, $-\infty < x_2 < \infty$, and I is $-\infty < t < \infty$. The solution is $x_1^*(t) = \sin t$, $x_2^*(t) = \cos t$. This solution is defined and is unique on $-\infty < t < \infty$.   ●

Example A2   The one-dimensional equation $\dot{x} = 3x^2 t^2$, $x(0) = 1$.

$\mathcal{D}$ is $-\infty < x < \infty$, and I is $-\infty < t < \infty$. But

$$x^*(t) = (1 - t^3)^{-1},$$

so the solution is only valid in $-\infty < t < 1$, since $x^*(t)$ approaches infinity as $t \to 1-$.   ●

Example A3   The one-dimensional equation $\dot{x} = 2|x|^{1/2}$, $x(0) = 1$.

A family of solutions is $x(t) = (t + c)^2$, $t > -c$ (so that $\dot{x} > 0$); there also exists the solution $x(t) \equiv 0$. The required solution is shown in Fig. A1, made up from the appropriate parabolic arc together with part of the zero solution (the derivative is continuous at A). Here, $\mathcal{D}$ is the domain $x > 0$, I is $-\infty < t < \infty$. The solution in fact leaves the open region $\mathcal{D} \times I$ at the point A.   ●

Example A4   The one-dimensional equation $\dot{x} = x/t$, $x(1) = 1$.

$\mathcal{D}$ is $-\infty < x < \infty$, and I is $t > 0$. The general solution is $x = ct$, and the required solution is $x = t$. The solution leaves $\mathcal{D} \times I$ at $(0, 0)$. Despite the singularity at $t = 0$, the solution continues uniquely in this case into $-\infty < t < \infty$.   ●

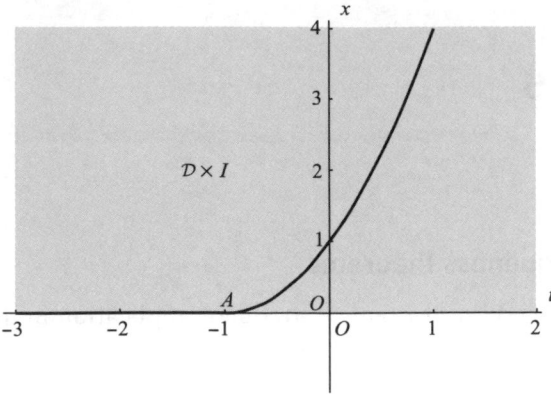

Figure A1

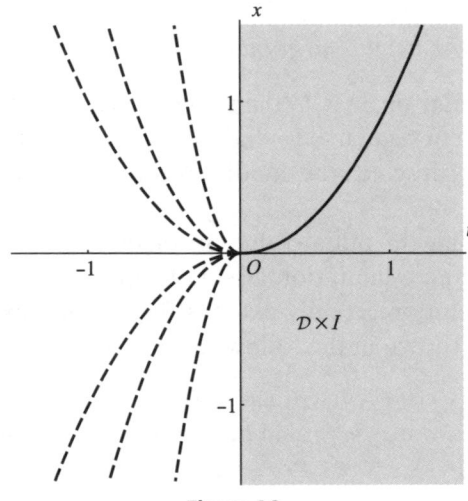

Figure A2

Example A5   The one-dimensional equation $\dot{x} = 2x/t$, $x(1) = 1$.

$\mathcal{D}$ is $-\infty < x < \infty$, and I is $t > 0$. The general solution is $x = ct^2$, and the required solution is $x = t^2$, $t > 0$ (t remaining within I). However, as Fig. A2 shows, there are many continuations of the solution (having a continuous derivative), but uniqueness breaks down at $t = 0$.   ●

For a regular system we shall define $I_{\max}$ (**the maximal interval of existence** of the solution in Theorem A1) as the largest interval of existence of a unique solution x^*. $I_{\max}$ can be shown to be an open interval.

Theorem A2   *Under the conditions of Theorem A1, either $(x^*(t), t)$ approaches the boundary of $\mathcal{D} \times I$, or $x^*(t)$ becomes unbounded as t varies within I. (For example, the possibility that as $t \to \infty$, $x^*(t)$ approaches a value in the interior of $\mathcal{D}$ is excluded.)*   ■

In Example A2, $I_{\max}$ is $-\infty < t < 1$, and as $t \to 1-$, $x^*(t)$ becomes unbounded. In Example A3, $x^*(t) \in \mathcal{D}$ only for $-1 < t < \infty$, and the solution approaches the boundary of $\mathcal{D} \times I$ as $t \to -1+$, at the point $(-1, 0)$. In Example A5, the boundary at $(0,0)$ is approached.

The following theorem shows that *linear systems* have particularly predictable behaviour.

Theorem A3   *Let* $\dot{x} = A(t)x + h(t)$, $x(t_0) = x_0$, *where* A *and* h *are continuous on* $I: t_1 < t < t_2$ (t_1 *may be* $-\infty$, *and* t_2 *may be* ∞). *Then* $I \subseteq I_{\max}$ *for all* $t_0 \in I$, *and for all* x_0. ∎

In this case the unique continuation of the solutions is guaranteed at least throughout the greatest interval in which A, h are continuous.

B   Topographic systems

Theorem B1   *The topographic system of Definition 10.1 has the following properties:*

(i) $V(x, y) < \alpha$ *in the interior of the topographic curve* $\mathcal{T}_\alpha$ *when* $0 < \alpha < \mu$.

(ii)   *There is a topographic curve through every point in the interior of* $\mathcal{T}_\alpha$ *when* $0 < \alpha < \mu$ *(i.e.,* $\mathcal{N}_\mu$ *is a connected neighbourhood of the origin).*

(iii)   *If* $0 < \alpha_1 < \alpha_2 < \mu$, *then* $\mathcal{T}_{\alpha_1}$ *is interior to* $\mathcal{T}_{\alpha_2}$, *and conversely.*

(iv)   *As* $\alpha \to 0$ *monotonically the topographic curves* $\mathcal{T}_\alpha$ *close upon the origin.*

Proof   Choose any α, $0 < \alpha < \mu$. Let P be any point in $\mathcal{I}_\alpha$, the interior of $\mathcal{T}_\alpha$, and A be any point on $\mathcal{T}_\alpha$. Join OPA by a smooth curve lying in $\mathcal{I}_\alpha$. Along OPA, $V(x, y)$ is continuous, and OPA is a closed set. Suppose that $V_P \geq \alpha$. Since $V_O = 0$ there is a point B on OP such that $V_B = \alpha$. But, since B does not lie on $\mathcal{T}_\alpha$, this contradicts the uniqueness requirement (iii) of Definition 10.1. Therefore $V < \alpha < \mu$ at every point interior to $\mathcal{T}_\alpha$.

The properties (ii) and (iii) follow easily from the property just proved.

To prove (iv): choose any $\varepsilon > 0$ and let $\mathcal{C}_\varepsilon$ be the circle $x^2 + y^2 = \varepsilon^2$. V is continuous on $\mathcal{C}_\varepsilon$, so V attains its minimum value $m > 0$ at some point on $\mathcal{C}_\varepsilon$; therefore $V(x, y) \geq m > 0$ on $\mathcal{C}_\varepsilon$. Choose any $\alpha < m$, and construct $\mathcal{T}_\alpha$. No part of $\mathcal{T}_\alpha$ can be exterior to $\mathcal{C}_\varepsilon$ since, by (i), $V(x, y) \leq \alpha < m$ on $\mathcal{T}_\alpha$ and in its interior. Therefore, there exists $\mathcal{T}_\alpha$ interior to $\mathcal{C}_\varepsilon$ for arbitrarily small ε. By (iii), the approach to the origin is monotonic. ∎

Theorem B2   *In some neighbourhood* $\mathcal{N}$ *of the origin let* $V(x, y)$ *be continuous, and* $\partial V/\partial x$, $\partial V/\partial y$ *be continuous except possibly at the origin. Suppose that, in* $\mathcal{N}$, $V(x, y)$ *takes the form, in polar coordinates,*

$$V(x, y) = r^q f(\theta) + E(r, \theta),$$

where

(i) $V(0,0) = 0$;

(ii) $q > 0$;

(iii) $f(\theta)$ *and* $f'(\theta)$ *are continuous for all values of* θ;

(iv) $f(\theta) > 0$, *and has period* 2π;

(v) $\displaystyle\lim_{r\to 0} r^{-q+1}\partial E/\partial r = 0$ *for all* θ.

Then there exists $\mu > 0$ *such that*

$$V(x, y) = \alpha, \quad 0 < \alpha < \mu$$

defines a topographic system covering a neighbourhood of the origin $\mathcal{N}_\mu$, *where* $\mathcal{N}_\mu$ *lies in* $\mathcal{N}$.

Proof Note that the conditions imply that $E(0,0) = 0$, and that E is continuously differentiable except possibly at the origin. Condition (v) together with (i) can be used to show that E is of lower order than $r^q f(\theta)$ (the dominant term, as $r \to 0$). (v) states that the same is true for the derivatives with respect to r.

We have

$$\frac{\partial V}{\partial r} = qr^{q-1}f(\theta) + \frac{\partial E}{\partial r}. \tag{B1}$$

From conditions (iii) and (iv), f attains its minimum value $m > 0$ on the closed set $0 \le \theta \le 2\pi$:

$$m = \min_{[0,2\pi]} f(\theta) > 0. \tag{B2}$$

Condition (iv) implies that given any $\varepsilon > 0$ these exists $\delta > 0$ such that in a region $\mathcal{R}_\delta$ lying in $\mathcal{N}$ and defined by $0 < r \le \delta,\ 0 \le \theta \le 2\pi$,

$$\left|\frac{\partial E}{\partial r}\right| < \varepsilon r^{q-1}.$$

Put $\varepsilon = \frac{1}{2}qm$, and remove the modulus sign to obtain

$$\frac{\partial E}{\partial r} > -\varepsilon r^{q-1} = -\frac{1}{2}qmr^{q-1} \quad \text{on } \mathcal{R}_\delta. \tag{B3}$$

From (B1) and (B3): on $\mathcal{R}_\delta$

$$\frac{\partial V}{\partial r} > qr^{q-1}m - \frac{1}{2}qmr^{q-1} = \frac{1}{2}mqr^{q-1},$$

so

$$\frac{\partial V}{\partial r} > 0 \quad \text{on } 0 < r \le \delta,\ 0 \le \theta \le 2\pi. \tag{B4}$$

Therefore, from (B4) and the continuity of V at the origin where $V = 0$, it follows that $V > 0$ on $\mathcal{R}_\delta$.

In Fig. B1, denote by $\mathcal{C}_\delta$ the circle $r = \delta$. $\mathcal{C}_\delta$ is a closed set and V is continuous on $\mathcal{C}_\delta$, so V attains its minimum value μ at some point on $\mathcal{C}_\delta$, and it is positive at this point:

$$\mu = \min_{\mathcal{C}_\delta} V(x, y) > 0. \tag{B5}$$

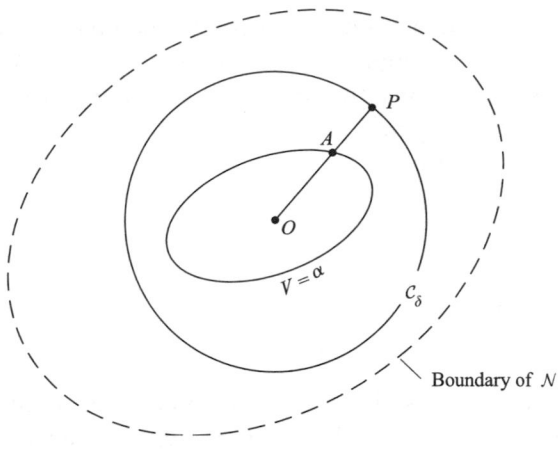

Figure B1

Let P be any point on $\mathcal{C}_\delta$ and OP a radial straight line. OP is a closed set and V is continuous on OP. By (B4), V is strictly increasing from its zero value at the origin to its value $V(x_P, y_P)$ at P, where, by (B5),

$$V(x_P, y_P) \geq \mu > 0. \tag{B6}$$

Choose any value of α in the range

$$0 < \alpha < \mu.$$

By the Intermediate Value theorem there is a unique point (since V is strictly increasing) $A:(x_A, y_A)$, on OP such that

$$V(x_A, y_A) = \alpha.$$

The continuity of V ensures that as P moves around $\mathcal{C}_\delta$, the corresponding points A lie on a continuous closed curve $\mathcal{T}_\alpha$, and the Inverse Function theorem ensures that the curve $\mathcal{T}_\alpha$ is smooth.

Clearly, as $\alpha \to 0$ monotonically, the curves $\mathcal{T}_\alpha$ close steadily on to the origin. They therefore form a topographic system in a neighbourhood $\mathcal{N}_\mu$ of the origin defined as the set of points $V(x, y) < \mu$. ∎

C Norms for vectors and matrices

The **norm** of a vector or matrix is a positive number which serves as a measure of its size in certain contexts. It is indicated by double vertical lines, $\|x\|$ when x is a vector and $\|A\|$ for a matrix A. In this book we use the **Euclidean** norms, defined as follows.

For a real or complex vector x of dimension n with components $x_1, x_2, \ldots, x_n$ we define

$$\|x\| = \left(\sum_{i=1}^{n} |x_i|^2 \right)^{1/2}, \tag{C1}$$

where $|x_i|$ signifies the modulus of x_i in cases where it is complex. If x is a real vector, $\|x\|$ is the usual measure of its length or magnitude: the modulus signs in (C1) may be omitted. For real vectors in two or three dimensions, the equation $\|x\| < c$, where $c > 0$, defines a circular or spherical neighbourhood of the origin.

The vector norm $\|x\|$ has the following properties ((i) to (iv) constitute the **axioms** for a norm):

(i) $\|x\| \geq 0$ for all x.

(ii) $\|x\| = 0$ if, and only if, $x = 0$.

(iii) $\|\alpha x\| = |\alpha| \|x\|$, α a real or complex number.

(iv) $\|x + y\| \leq \|x\| + \|y\|$. This is the **triangle inequality**. If x and y are real and three-dimensional, then it states that one side of any triangle is of length less than or equal to the sum of the lengths of the other two sides. The inequality can be extended, step by step; for example

$$\|x + y + z\| \leq \|x\| + \|y\| + \|z\|.$$

(v) If x is a function of a real variable t, and $t_2 \geq t_1$, then

$$\left\| \int_{t_1}^{t_2} x(t) \mathrm{d}t \right\| \leq \int_{t_1}^{t_2} \|x(t)\| \mathrm{d}t$$

(proved by interpreting the integral as a sum and using (iv)).

For an $n \times n$ matrix A with elements a_{ij} we define the **matrix norm**

$$\|A\| = \left(\sum_{i=1}^{n} \sum_{j=1}^{n} |a_{ij}|^2 \right)^{1/2}, \tag{C2}$$

in which the elements may be real or complex. $\|A\|$ has the same properties, (i) to (v) above, as does $\|x\|$.

The vector and matrix norms are compatible with the frequently needed inequality:

(vi) $\|Ax\| \leq \|A\| \|x\|$.

The reader should be aware that other norms are in general use, devised in order to simplify the analysis for which they are used; for example (Cesari 1971) the following norms:

$$\|x\| = \sum_{i=1}^{n} |x_i|, \quad \|A\| = \sum_{i=1}^{n} \sum_{j=1}^{n} |a_{ij}|,$$

have all the properties (i) to (iv) above. In two dimensions the equation $\|x\| < c$ $(c > 0)$ defines a diamond-shaped neighbourhood of the origin.

D   A contour integral

The following integral arose in Melnikov's method (Section 13.7), and is a typical example of infinite integrals in this context:

$$I = \int_{-\infty}^{\infty} f(u)\, du,$$

where $f(u) = \operatorname{sech} u \tanh u \sin \omega u$. Generally these integrals can be evaluated using contour integration and residue theory. Considered as a function of the complex variable $z = u + iv$, the integrand $f(z) = \operatorname{sech} z \tanh z \sin \omega z$ has poles of order 2 where $\cosh z = 0$, that is, where

$$e^{2z} = -1 = e^{(2n+1)i\pi}.$$

These poles are located at $z = (n + \frac{1}{2})i\pi$, $(n = 0, \pm 1, \pm 2, \ldots)$. We use residue theory to evaluate I by choosing the rectangular contour C shown in Fig. D1.   The rectangle has corners at $z = \pm R$ and $z = \pm R + i\pi$. One pole at $P : z = \frac{1}{2}i\pi$ lies inside the contour.

By Cauchy's residue theorem (Osborne 1998), the contour integral around C taken counterclockwise is given by

$$\int_C f(z)\, dz = 2\pi i \quad (\text{residue at } z = z_0),$$

where $z_0 = \frac{1}{2}i\pi$. The residue at $z_0 = \frac{1}{2}i\pi$ is the coefficient of $1/(z - z_0)$ in the Laurent expansion of $f(z)$ about $z = z_0$. In this case

$$f(z) = \frac{\sinh(\frac{1}{2}\pi\omega)}{(z - \frac{1}{2}i\pi)^2} - \frac{i\omega \cosh(\frac{1}{2}\pi\omega)}{(z - \frac{1}{2}i\pi)} + O(1)$$

as $z \to \frac{1}{2}i\pi$. Hence

$$\int_C \operatorname{sech} z \tanh z \sin \omega z\, dz = -2i\pi[i\omega \cosh(\frac{1}{2}\pi\omega)] = 2\pi\omega \cosh(\frac{1}{2}\pi\omega).$$

We now separate the integral around C into integrals along each edge, and then let $R \to \infty$. It can be shown that the integrals along BC and DA tend to zero as $R \to \infty$, whilst the integral

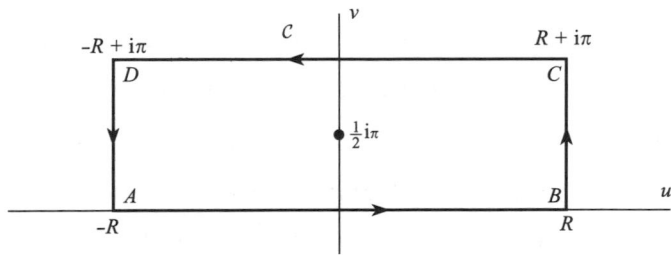

Figure D1

along AB approaches the required infinite integral I. Along CD, $z = u + i\pi$ so that

$$\int_C f(z)\, dz = \int_R^{-R} \mathrm{sech}(u + i\pi)\tanh(u + i\pi)\sin(u + i\pi)\, du$$

$$= -\int_{-R}^{R} (-\mathrm{sech}\, u)(\tanh u)[\sin(\omega u)\cos(i\omega\pi)$$

$$+ \cos(\omega u)\sin(i\omega\pi)]\, du$$

$$= \int_{-R}^{R} \mathrm{sech}\, u \tanh u \sin \omega u\, du \cdot \cosh \omega\pi$$

$$\to I \cosh \omega\pi,$$

as $R \to \infty$. The exponential behaviour of the integrand as $u \to \pm\infty$ guarantees covergence. Hence

$$(1 + \cosh \omega\pi)I = 2\pi\omega \cosh(\tfrac{1}{2}\pi\omega)$$

from which it follows that

$$I = \int_{-\infty}^{\infty} \mathrm{sech}\, u \tanh u \sin \omega u\, du = \frac{2\pi\omega \cosh(\tfrac{1}{2}\pi\omega)}{1 + \cosh \omega\pi} = \omega\pi\, \mathrm{sech}\left(\tfrac{1}{2}\omega\pi\right).$$

E   Useful results

E1   *Trigonometric identities*

$$\cos^2 \omega t = \tfrac{1}{2}(1 + \cos 2\omega t); \quad \sin^2 \omega t = \tfrac{1}{2}(1 - \cos 2\omega t)$$

$$\cos^3 \omega t = \tfrac{3}{4}\cos \omega t + \tfrac{1}{4}\cos 3\omega t; \quad \sin^3 \omega t = \tfrac{3}{4}\sin \omega t - \tfrac{1}{4}\sin 3\omega t$$

$$\cos^5 \omega t = \tfrac{5}{8}\cos \omega t + \tfrac{5}{16}\cos 3\omega t + \tfrac{1}{16}\cos 5\omega t$$

$$\sin^5 \omega t = \tfrac{5}{8}\sin \omega t - \tfrac{5}{16}\sin 3\omega t + \tfrac{1}{16}\sin 5\omega t$$

$$(a \cos \omega t + b \sin \omega t)^2 = \tfrac{1}{2}(a^2 + b^2) + \tfrac{1}{2}(b^2 - a^2)\cos 2\omega t + ab \sin 2\omega t$$

$$(a \cos \omega t + b \sin \omega t)^3 = \tfrac{3}{4}a(a^2 + b^2)\cos \omega t + \tfrac{3}{4}b(a^2 + b^2)\sin \omega t$$

$$+ \tfrac{1}{4}a(a^2 - 3b^2)\cos 3\omega t + \tfrac{1}{4}b(3a^2 - b^2)\sin 3\omega t$$

$$(c + a \cos \omega t + b \sin \omega t)^2 = \tfrac{1}{2}(a^2 + b^2 + 2c^2) + 2ac \cos \omega t + 2bc \sin \omega t$$

$$+ \tfrac{1}{2}(a^2 - b^2)\cos 2\omega t + ab \sin 3\omega t$$

$$(c + a \cos \omega t + b \sin \omega t)^3 = \tfrac{1}{2}c[3(a^2 + b^2) + 2c^2] + \tfrac{1}{4}a[3(a^2 + b^2) + 12c^2]$$

$$\times \cos \omega t + \tfrac{1}{4}b[3(a^2 + b^2) + 12c^2]\sin \omega t$$

$$+ \tfrac{3}{2}c(a^2 - b^2)\cos 2\omega t + 3abc \sin 2\omega t$$

$$+ \tfrac{1}{4}a(a^2 - 3b^2)\cos 3\omega t + \tfrac{1}{4}b(3a^2 - b^2)\sin 3\omega t$$

E2 *Taylor series*

The Taylor series of $f(x)$ about $x = a$ is given by

$$f(a+h) = f(a) + f'(a)h + \tfrac{1}{2!}f''(a)h^2 + \cdots + \tfrac{1}{n!}f^{(n)}(a)h^n + \cdots.$$

Particular series are

$$\cos t = 1 - \tfrac{1}{2!}t^2 + \tfrac{1}{4!}t^4 - \cdots \text{ for all } t.$$

$$\sin t = t - \tfrac{1}{3!}t^3 + \tfrac{1}{5!}t^5 - \cdots \text{ for all } t.$$

$$e^t = 1 + t + \tfrac{1}{2!}t^2 + \tfrac{1}{3!}t^3 + \cdots \text{ for all } t.$$

$$(1+t)^\alpha = 1 + \alpha t + \tfrac{\alpha(\alpha-1)}{2!}t^2 + \tfrac{\alpha(\alpha-1)(\alpha-2)}{3!}t^3 + \cdots \text{ for } |t| < 1 \text{ unless } \alpha \text{ is positive integer.}$$

E3 *Fourier series*

Let $f(t)$ be a periodic function of period $2\pi/\omega$ for all t. Then its Fourier series is given by

$$f(t) = \frac{1}{2}a_0 + \sum_{n=1}^{\infty}(a_n \cos n\omega t + b_n \sin \omega t),$$

where

$$a_n = \frac{\omega}{\pi}\int_{-\pi/\omega}^{\pi/\omega} f(t)\cos n\omega t \, \mathrm{d}t, \ (n = 0, 1, 2, \ldots), \quad b_n = \frac{\omega}{\pi}\int_{-\pi/\omega}^{\pi/\omega} f(t)\sin n\omega t \, \mathrm{d}t, \ (n = 1, 2, \ldots).$$

E4 *Integrals (mainly used in Melnikov's method)*

$$\int_{-\infty}^{\infty} \operatorname{sech}^2 u \tanh^2 u \, \mathrm{d}u = \tfrac{2}{3}$$

$$\int_{-\infty}^{\infty} \operatorname{sech}^4 u \, \mathrm{d}u = \tfrac{16}{15}$$

$$\int_{-\infty}^{\infty} \operatorname{sech} u \tanh u \sin \omega u \, \mathrm{d}u = \pi\omega \operatorname{sech}\left(\tfrac{1}{2}\omega\pi\right) \text{ (see Appendix D)}.$$

$$\int_{-\infty}^{\infty} \operatorname{sech} u \cos \omega u \, \mathrm{d}u = \pi \operatorname{sech}\left(\tfrac{1}{2}\omega\pi\right)$$

$$\int_{-\infty}^{\infty} \operatorname{sech}^2 u \cos \omega u \, \mathrm{d}u = \frac{\pi\omega}{\sinh(\tfrac{1}{2}\omega\pi)}$$

$$\int_{-\infty}^{\infty} \operatorname{sech}^3 u \cos \omega u \, \mathrm{d}u = \tfrac{1}{2}\pi(1+\omega^2)\operatorname{sech}\left(\tfrac{1}{2}\omega\pi\right)$$

E5 *Cubic equations*

In the cubic equation

$$a_0 x^3 + 3a_1 x^2 + 3a_2 x + a_3 = 0,$$

let $a_0 x = z - a_1$. The equation reduces to the form

$$z^3 + 3Hz + G = 0.$$

Then, if the equation has real coefficients,

(a) $G^2 + 4H^3 > 0$, one root is real and two are complex;
(b) $G^2 + 4H^3 = 0$, the roots are real and two are equal;
(c) $G^2 + 4H^3 < 0$, the roots are all real and different.

References and further reading

The qualitative theory of differential equations was founded H. Poincaré and I.O. Bendixson towards the end of the nineteenth century in the context of celestial mechanics, particularly in the context of the n-body problem in Newtonian dynamics. Henri Poincaré (1854–1912) has had enormous impact on the development of the subject, and the many methods and theorems carrying his name acknowledge his contributions. At about the same time A.M. Liapunov (1857–1918) produced his formal definition of stability. Subsequent work proceeded along two complementary lines: the abstract ideas of topological dynamics developed by G.D. Birkhoff (1884–1944) and others in the 1920s, and the more practical qualitative approach exploited particularly by Russian mathematicians in the 1930s, notably by N.N. Bogoliubov, N. Krylov, and Y.A. Mitropolsky. Much of the earlier work is recorded in the first real expository text, by A.A. Andronovand C.E. Chaikin published in 1937 with an English translation first published in 1949. An account of the development of the subject from celestial mechanics can be found in the book by F. Diacu and P. Holmes (1996).

Since this time the subject has been generalized and diversified to the extent that specialized literature exists in all its areas. The advent of computers which could be easily programmed in the 1960s and 1970s caused an explosion of interest in the subject, and a huge literature. Many hypotheses could be readily confirmed by computation, which encouraged further confidence in qualitative methods. The further development of symbolic computing in the 1980s through software such as *Mathematica, Maple* and dedicated packages, has increased further the ability to represent phase diagrams, solutions, Poincaré sequences, etc. graphically using simple programs. The basis of the subject has been broadened and now often goes under the name of **dynamical systems**, which generally includes initial-value problems arising from ordinary and partial differential equations and difference equations. The other major development has been an appreciation that, whilst systems arising from ordinary differential equations are *deterministic*, that is, future states of the system can be completely predicted from the initial state, nonetheless the divergence of neighbouring solutions can be so large that over a period of time the link with past states of the system can be effectively lost computationally. This loss of past information is *deterministic chaos*. The landmark paper in the appreciation of this phenomena is generally accepted to be the work by E.N. Lorenz (1963) on a meteorological problem. Since the 1960s there has been an enormous growth in the interest in systems which exhibit chaotic outputs, which is reflected in the titles of more recent texts.

The bibliography below contains those works referred to in the text, and others which provide further introductory or specialist reading, often with extensive lists of references.

Abarbanel DI, Rabinovich MI, and Sushchik MM (1993) *Introduction to nonlinear dynamics for physicists.* World Scientific, Singapore.

Abramowitz M and Stegun IA (1965) *Handbook of mathematical functions.* Dover, London.

Acheson D (1997) *From calculus to chaos.* Oxford University Press, Oxford.

Addison PS (1997) *Fractals and chaos: an illustrated course.* Institute of Physics, Bristol.

Ames WF (1968) *Nonlinear differential equations in transport processes.* Academic Press, New York.

Andronov AA and Chaikin CE (1949) *Theory of oscillations.* Princeton University Press.

Andronov AA, Leontovich EA, Gordon II, and Maier AG (1973a) *Qualitative theory of second-order dynamic systems.* Wiley, New York.

Andronov AA, Leontovich EA, Gordon II, and Maier AG (1973b) *Theory of bifurcations of dynamical systems on a plane.* Halstead Press, New York.

Arnold VI (1983) *Geometrical methods in the theory of ordinary differential equations.* Springer-Verlag, Berlin.

Arrowsmith DK and Place CM (1990) *An introduction to dynamical systems.* Cambridge University Press.

Ayres F (1962) *Matrices.* Schaum, New York.

Baker GL and Blackburn JA (2005) *The Pendulum.* Oxford University Press.

Barbashin EA (1970) *Introduction to the theory of stability.* Wolters-Nordhoff, The Netherlands.

Bogoliubov NN and Mitroposky YA (1961) *Asymptotic methods in the theory of oscillations.* Hindustan Publishing Company, Delhi.

Boyce WE and DiPrima RC (1996) *Elementary differential equations and boundary value problems.* Wiley, New York.

Brown R and Chua LO (1996) Clarifying chaos: examples and counter examples. *Int. J. Bifurcation Chaos,* **6**, 219–249.

Brown R and Chua LO (1998) Clarifying chaos II: Bernoulli chaos, zero Lyaponov exponents and strange attractors. *Int. J. Bifurcation Chaos,* **8**, 1–32.

Carr J (1981) *Applications of center manifold theory.* Springer-Verlag, New York.

Cesari L (1971) *Asymptotic behaviour and stability problems in ordinary differential equations.* Springer, Berlin.

Coddington EA and Levinson L (1955) *Theory of ordinary differential equations.* McGraw-Hill, New York.

Cohen AM (1973) *Numerical analysis.* McGraw-Hill, London.

Copson ET (1965) *Asymptotic expansions.* Cambridge University Press.

Coppel WA (1978) *Dichotomies in stability theory,* Springer, Berlin.

Crocco L (1972) Coordinate perturbations and multiple scales in gas dynamics. *Phil. Trans. Roy. Soc.,* **A272**, 275–301.

Diacu F and Holmes P (1996) *Celestial encounters.* Princeton University Press.

Drazin PG (1992) *Nonlinear systems.* Cambridge University Press.

Ermentrout B (2002) *Simulating, analyzing, and animating dynamical systems: a guide to XPPAUT,* Siam Publications, Philadelphia.

Ferrar WL (1950) *Higher algebra.* Clarendon Press, Oxford.

Ferrar WL (1951) *Finite matrices.* Clarendon Press, Oxford.

Gradshteyn IS and Ryzhik I (1994) *Table of integrals, series, and products.* Academic Press, London.

Grimshaw R (1990) *Nonlinear ordinary differential equations.* Blackwell Scientific Publications, Oxford.

Guckenheimer J and Holmes P (1983) *Nonlinear oscillations, dynamical systems, and bifurcations of vector fields.* Springer-Verlag, New York.

Hale J (1969) *Ordinary differential equations.* Wiley-Interscience, London.

Hale J and Kocak H (1991) *Dynamics and bifurcations.* Springer-Verlag, New York.

Hayashi C (1964) *Nonlinear oscillations in physical systems.* McGraw-Hill, New York.

Hilborn RC (1994) *Chaos and nonlinear dynamics.* Oxford University Press.

Hill R (1964) *Principles of dynamics.* Pergamon Press, Oxford.

Hinch EJ (1991) *Perturbation methods.* Cambridge University Press.

Holmes P (1979) A nonlinear oscillator with a strange attractor. *Phil. Trans. Roy. Soc.,* **A292**, 419–448.

Hubbard JH and West BH (1995) *Differential equations: a dynamical systems approach.* Springer-Verlag, New York.

Jackson EA (1991) *Perspectives in nonlinear dynamics, Vols 1 and 2.* Cambridge University Press.

Jones DS (1963) *Electrical and mechanical oscillations*. Routledge and Kegan Paul, London.

Jordan DW and Smith P (2002) *Mathematical techniques*, 3rd edn. Oxford University Press, Oxford.

Kevorkian J and Cole JD (1996) *Multiple scale and singular perturbation methods*. Springer-Verlag, New York.

Krylov N and Bogoliubov N (1949) *Introduction to nonlinear mechanics*. Princeton University Press.

La Salle J and Lefshetz S (1961) *Stability of Liapunov's direct method*. Academic Press, New York.

Leipholz H (1970) *Stability theory*. Academic Press, New York.

Logan JD (1994) *Nonlinear partial differential equations*. Wiley-Interscience, New York.

Lorenz EN (1963) Deterministic nonperiodic flow. *J. Atmospheric Sci.*, 20, 130–141.

Magnus K (1965) *Vibrations*. Blackie, London.

Mattheij RMM and Molenaar J (1996) *Ordinary differential equations in theory and practice*. Wiley, Chichester.

McLachlan NW (1956) *Ordinary nonlinear differential equations in engineering and physical sciences*. Clarendon Press, Oxford.

Minorsky N (1962) *Nonlinear oscillations*. Van Nostrand, New York.

Moon FC (1987) *Chaotic vibrations*. Wiley, New York.

Nayfeh AH (1973) *Perturbation methods*. Wiley, New York.

Nayfeh AH and Balachandran B (1995) *Applied nonlinear dynamics*. Wiley, New York.

Nayfeh AH and Mook DT (1979) *Nonlinear oscillations*. Wiley, New York.

Nemytskii VV and Stepanov VV (1960) *Qualitative theory of differential equations*. Princeton University Press.

Nicolis G (1995) *Introduction to nonlinear science*. Cambridge University Press.

O'Malley RE (1974) *Introduction to singular perturbations*. Academic Press, New York.

Osborne AD (1998) *Complex variables and their applications*. Addison Wesley Longman.

Pavlidis T (1973) *Biological oscillators: their mathematical analysis*. Academic Press, New York.

Pielou EC (1969) *An introduction to mathematical ecology*. Cambridge University Press.

Poston T and Stewart I (1978) *Catastrophe theory and its applications*. Pitman, London.

Rade L and Westergren B (1995) *Mathematics handbook for science and engineering*. Studentlitteratur, Lund, Sweden.

Rasband SN (1990) *Chaotic dynamics of nonlinear systems*. Wiley, New York.

Reissig R, Sansone G, and Conti R (1974) *Nonlinear differential equations of higher order*. Nordhoff, Leiden.

Rosen R (ed.) (1973) *Foundations of mathematical systems, Volume III, Supercellular systems*. Academic Press, New York.

Sanchez DA (1968) *Ordinary differential equations and stability theory*. Freeman, San Francisco.

Simmonds JG (1986) *A first look at perturbation theory*. Krieger Publishing, Florida.

Small RD (1989) Population growth in a closed system. In *Mathematical modelling: classroom notes in applied mathematics*, edited by MS Klamkin. SIAM Publications, Philadelphia.

Sparrow C (1982) *The Lorenz equations: bifurcations, chaos, and strange attractors*. Springer-Verlag, New York.

Stoker JJ (1950) *Nonlinear vibrations*. Interscience, New York.

Strogatz SH (1994) *Nonlinear dynamics and chaos*. Perseus, Massachusetts.

Struble RA (1962) *Nonlinear differential equations*. McGraw-Hill, New York.

Thompson JMT and Stewart HB (1986) *Nonlinear dynamics and chaos*. Wiley, Chichester.

Ueda Y (1980) Steady motions exhibited by Duffing's equation: a picture book of regular and chaotic motions. In *New approaches to nonlinear problems in dynamics*, edited by PJ Holmes. SIAM Publications, Philadelphia.

Urabe M (1967) *Nonlinear autonomous oscillations*. Academic Press, New York.

Van Dyke M (1964) *Perturbation methods in fluid mechanics*. Academic Press, New York.

Verhulst F (1996) *Nonlinear differential equations and dynamical systems*, Second edition. Springer, Berlin.

Virgin LN (2000) *Introduction to experimental nonlinear dynamics*. Cambridge University Press.

Watson GN (1966) *A treatise on the theory of Bessel functions*. Cambridge University Press.

Whittaker ET and Watson GN (1962) *A course of modern analysis*. Cambridge University Press.

Wiggins S (1990) *Introduction to applied nonlinear dynamical systems and chaos*. Springer-Verlag, New York.

Wiggins S (1992) *Chaotic transport in dynamical systems*. Springer-Verlag, New York.

Willems JL (1970) *Stability theory of dynamical systems*. Nelson, London.

Wilson HK (1971) *Ordinary differential equations*. Addison–Wesley, Reading, MA.

Wolfram S (1996) *The Mathematica book*. Cambridge University Press.

Index